2/96

ALTERED FATES

ALSO BY JEFF LYON

*Playing God
in the Nursery*

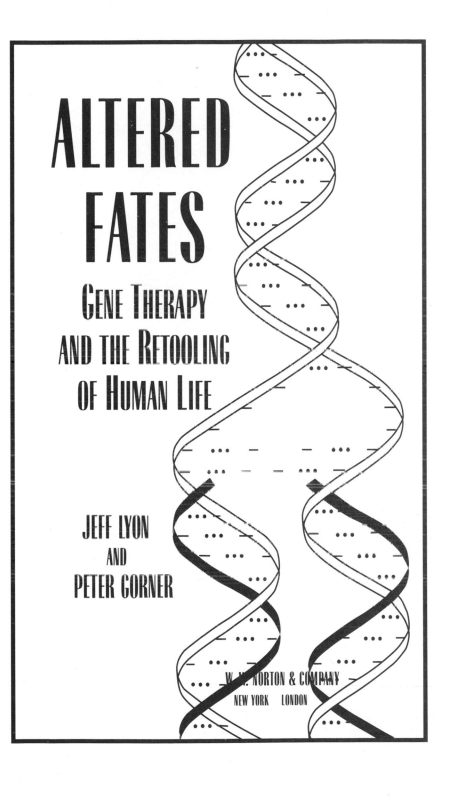

ALTERED FATES

FATES

GENE THERAPY AND THE RETOOLING OF HUMAN LIFE

JEFF LYON
AND
PETER GORNER

W. W. NORTON & COMPANY
NEW YORK   LONDON

*To our wives,
Bonita and Jackie,
and to Lindsay, Derek,
Peter, and Jeremy
for more reasons
than we can say.*

First Edition

The text of this book is composed in Trump Medieval
with the display set in Radiant Bold Extra Condensed
Composition and manufacturing by The Haddon Craftsmen, Inc.
Book design by Charlotte Staub

ISBN 0-393-03596-4

W. W. Norton & Company, Inc., 500 Fifth Avenue, New York, N.Y. 10110
W. W. Norton & Company Ltd., 10 Coptic Street, London WC1A 1PU

1 2 3 4 5 6 7 8 9 0

CONTENTS

Authors' Note 7

Introduction 11

PART I

1. A Gathering of Geese 17

2. The Long and Winding Road 38

3. A Simple Little System 59

4. The Horror 86

5. The Understudy Steps In 105

6. The Road Forks 120

7. The Troika Forms 138

8. Gene Transfer Is Born 162

9. Gene Therapy Meets the Public 176

10. Victory 202

11. September 14, 1990 227

12. One Errant Cell 241

13. Belt and Suspenders 259

14. Aftermath 277

PART II

15. A Matter of the Heart   299

16. Jekyll and Hyde   322

17. Triumph or Tragedy?   349

18. The Thief of Breath   383

19. It Takes a Worried Man   407

20. Delivering the Goods   428

21. Something in the Blood   457

22. Gypsy Fortune-tellers of Technology   480

23. The Ultimate Frontier   507

24. Prophecy   531

Acknowledgments   569

Notes   571

Index   613

Authors' Note

As this book went to press, the race to find BRCA1, the first gene known to cause breast cancer—which since 1990 had a dozen teams of researchers worldwide frantically tearing apart the long arm of chromosome 17 in search of mutant DNA—ended in a triumph, of sorts, on September 14, 1994. The journal *Science* announced it had accepted a paper for publication in its October 7 edition that reported the cloning of BRCA1 by a huge team led by veteran gene hunter Mark Skolnick, of the University of Utah Medical Center in Salt Lake City. Skolnick's costars in the drama were many and varied, and included more than 40 colleagues from the medical center; the National Institute of Environmental Health Sciences; Myriad Genetics, Inc., a Salt Lake City biotech company with which Skolnick is affiliated; the pharmaceutical giant Eli Lilly and Company of Indianapolis, presumably interested in marketing a screening test for BRCA1; and McGill University in Montreal.

After the customary victory celebrations that found the winners toasting themselves and the losers diplomatically gushing plaudits like "beautiful" and "revolutionary," it quickly became apparent that the newly captured gene was going to raise more questions than it answered. That BRCA1 confers susceptibility to breast and ovarian cancer in certain beleaguered families is beyond dispute—a woman who carries a defective copy stands an 85 percent chance of developing cancer before the age of 65—but the gene mutation was found in only about half the hereditary cancer families studied. That fact, which had been predicted beforehand, led researchers to hunt for a second major gene, BRCA2, that would account for most of the rest of the hereditary families. Sure enough, a second paper, in the September 30, 1994, issue of *Science*, announced that the companion gene had been mapped to the long arm of chromosome 13 by David Goldgar of Utah and Michael Stratton and

Doug Easton at the Institute of Cancer Research in Sutton, Surrey, England.

Together, BRCA1 and BRCA2 may be responsible for between 5 and 10 percent of breast and ovarian cancers that are believed to be hereditary, but the gene appears to play no role in the much more common "familial" and "sporadic" cases that comprise the rest. Like the muscular dystrophy gene, the long-sought BRCA1 is huge—about 100,000 base pairs long, or ten times the size of a normal gene—and thus vulnerable to a wide variety of possible mishaps. The Utah researchers spotted it, in fact, when they discovered a frame-shift mutation carried by three members of one Utah family who had breast cancer but not by two healthy members of the same family. Frame-shifts are dead giveaways for mutations in genes because they cause the misreading of codons, the three-nucleotide code words that specify amino acids leading to proteins, resulting in a garbled message that starts in the wrong place and ends up ordering up a defective product.

The researchers, however, became even more curious about the gene when they kept finding more mutations—five in toto—in afflicted families. After the first frame-shift, they found another; then something called a "stop codon," which disrupts a third of the gene's protein from being manufactured; then a "missense substitution," which replaced a weakly effective amino acid with a supercharged one; and, finally, a "regulatory mutation," that prevented transcription. Such a panoply of problems connected to one gene convinced the Utah gene hunters they had found BRCA1.

Earlier, the gene had been presumed to be a tumor suppressor because its home-length of chromosome 17 was found to be missing in about half of all sporadic and familial cancers studied. But it really may be serving some other function. Within weeks of being discovered, BRCA1 was starting to be viewed as an important gene, conserved throughout mammalian evolution, that serves as a traffic cop to control chemicals called transcription factors that communicate with DNA to switch on and off other genes needed to regulate proteins in the cell.

Still, it was puzzling that Skolnick could find no indication of mutated BRCA1 genes in sporadic tumors. Perhaps the mutations are there, but merely elusive so far. Perhaps another as-yet-unidentified tumor suppressor gene on chromosome 17 is critical. Moreover, a few women in the study who carried mutated BRCA1 genes were still cancer-free in their eighties, meaning that through diet or some other environmental factors they had somehow dodged the bullet.

All these considerations raised questions. Was this really BRCA1?

And if a genetic test is imminent, what would be the value to society of yet another test merely telling women—and presumably their insurance companies—that they had an elevated risk of getting breast cancer at some time in the future no matter what they did.

Introduction

This book is really two books. On the one hand, it is the tale of what was arguably the most audacious medical experiment in history, the first U.S. government–sponsored attempt to reprogram the genetic code of a living human being. This epic story, which we have re-created with as much attention to nuance and texture as possible, is a chronicle of reach and overreach, persistence and folly, brilliance and bluster. Ultimately, it tracks the halting but irreversible journey of medicine through a once forbidden door.

Our attempt to define what lies on the other side of that door occupies the second half of the book. As scientists begin to apply the principles of gene therapy—an umbrella term we use here to signify the fruits of molecular medicine in their many forms—to the understanding, diagnosis, and treatment of various human diseases, the outlines of the golden age of medicine to come are slowly taking shape. We have tried to trace the contours of this fast-approaching millennium and—because it cannot be ignored—to consider how the technology will most likely be applied, for better or worse, to a panoply of traits outside those that relate to illness.

The art of manipulating human genes, of deliberately altering the fates of individuals and—inevitably, it seems to us—of the species as a whole, is the incarnation of utopian longings that have captivated humanity for eons. The quest for eternal youth, the Faustian myth of abrupt reversal of fortune, and our own century's flirtation with the perfectibility of man are themes that find realization in the modern alchemy of gene therapy. Yet, as always when fairy tales come to life, there is a dark side. Even in these first, hesitant days of human gene therapy, a period sure to be looked upon one day with the same amused nostalgia with which we view the initial flights of the Wright brothers at Kitty Hawk, many thoughtful people are frightened by the power of the tools that science has placed in our hands. Will we comport ourselves with grace and wisdom as we put them to use? Or will we create the same hash with them that we have made with other Promethean technologies?

As journalists who have been covering the possibilities of gene therapy for the *Chicago Tribune* since the mid-1980s, we have been both stunned by the shining promise the new technology affords and troubled by its implications. As we continue to ponder these issues, we find ourselves marveling at how smoothly such an incredible idea has stitched itself into the social fabric. How is it that the general public has come to embrace something so cosmic as gene replacement so easily? Is it that the daily diet of the fantastic and the bizarre to which our society is exposed has left us incapable of awe? Or is it something else?

A decade ago, we and a very few other science journalists had the beat pretty much to ourselves. As we struggled to master the principles of molecular biology and recombinant DNA technology in order to translate their complexities to the public in our 1986 *Tribune* series, "Altered Fates: The Promise of Gene Therapy," we were never allowed to forget that readers knew little, if anything, about genes. Our editors granted us great latitude, but a little nervously; we were constantly reminded, for instance, to couple our use of the term "DNA" with a parenthetical "deoxyribonucleic acid", as if that addition somehow made the arcane clearer to an audience whose only acquaintance with the subject remained a distant biology course back in school.

It is difficult to appreciate how far the science has come in a strikingly brief time. When we started this project in the wake of our series, few disease-causing genes had been discovered and the mechanisms under consideration for replacing or overriding them therapeutically were primitive and only starting to undergo testing in animal experiments. Any clinical use of gene therapy was considered by most experts to be unlikely, if not impossible, until well into the next century.

The picture, of course, has changed dramatically in less than a decade. Genetic discoveries dominate the headlines today, with newfound disease genes announced to the world on what seems to be a weekly basis. Second-, third-, and fourth-generation techniques for conveying healthy versions of critical genes into ailing bodies are under development even as this book approaches publication. The number of human gene therapy experiments now under way in the United States exceeds sixty, and other nations are well into their own research programs. Throughout the world, newly ordained gene therapists are gingerly shooting genes into human brains, livers, lungs, bones, biceps, quadriceps, skin, arteries, and lymph nodes, working out kinks in the treatment, refining procedures. Universities and medical centers race one another to establish gene therapy departments, while virtually all the industrialized nations have joined hands in an unprecedented effort to tease out and decipher the thousands upon thousands of genes that make up the human animal.

The public, often characterized as intrinsically conservative and fearful of the unknown, has, in our experience, reacted positively to the prospect of gene therapy. Rather than shrink from it, most people tend to become excited about the potentialities, often putting a personal spin on it, asking such questions as "Will cancer really be cured within my lifetime?" It would be overstating things to say that society is becoming gene literate, but no longer do people need the crutch of "(deoxyribonucleic acid)" when confronted with the term "DNA," and years of media coverage have given them at least a superficial familiarity with the fundamentals of molecular genetics. When the NFL Hall of Famer O. J. Simpson was charged with the grisly murders of his wife and a friend in the summer of 1994, the common reaction was to ask not what DNA fingerprinting was but rather when it would begin to be applied in the Simpson case. There is a growing sense that with molecular biology in general and gene therapy in particular all things are possible.

This climate of expectation is not looked upon as healthy by everyone. There are those who believe that gene therapy's promise has far outpaced its reality. One of the leading and more thoughtful overseers of the field, Nelson Wivel, head of the federal government's Office of Recombinant DNA Activities, went so far as to suggest to us that the revolution had been "oversold," that it was long on testimonials and short on substance. Despite major advances in diagnosis and a plethora of marvelous laboratory techniques inconceivable a decade ago, measurable treatment advances were not coming fast enough, contended Wivel: the field needed time, perhaps another decade or more, to get over the many fundamental obstacles facing it and make an impact on the lives of ordinary people.

Wivel is a smart man and protective of the dream. He is right to warn that excitement must be tempered with the recognition that the perfection of any new medical technology, especially one so complex as gene therapy, with its exquisitely orchestrated machinations conducted on a scale unimaginably small, is going to take time. The history of medical progress, whether in the development of antibiotics, in the perfecting of bone marrow transplants, or in the psychopharmaceutical conquest of depression, provides abundant evidence that the pace of advance is measured in decades, even generations, not in mere months or years. To demand otherwise of gene therapy is unreasonable and ultimately destructive. Let there be no doubt that, in time, the technology will fulfill most, if not all, of the extravagant claims made on its behalf.

It is imperative to point out, however, that the real danger may lie not in overanticipation but in undersurveillance. Disturbing signs already exist that the biowizardry of gene therapy is vulnerable to expropriation and misuse, perhaps on a catastrophic scale, by corporate, political, and institutional interests, whose interests may or may not coincide with those of society at large. A movement is afoot to weaken, perhaps emasculate, the agencies charged with controlling the development of gene therapy, and we argue in this book that to allow such an effort to prevail would be a grave mistake, in both the short and the long run.

We live in a time of epochal events. The Soviet empire crumbles, apartheid melts away, the Palestinians and Israelis work toward peace, transmitting an inextinguishable message of hope to a world long since cynically resigned to frightful outcomes. Nowhere else is the potential for extraordinary, transfiguring change so great as in the scientific arena, and of the myriad pathways down which science is at present hurtling—whether in the exploration of space, in the dissection of matter, in the cultivation of artificial intelligence—none will have so powerful and immediate an effect on the human population as those being followed in molecular biology and gene therapy. We have sought to render intelligible these fascinating, if sometimes esoteric, explorations so as to help prepare whoever happens to read this book for the adventure that lies just ahead.

August 1994

PART I

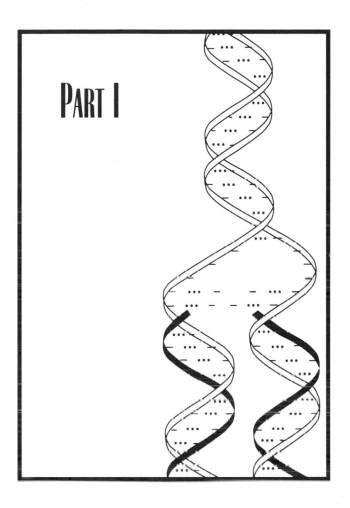

1

A GATHERING
OF GEESE

High in the leaden autumn skies over Long Island's picturesque Cold Spring Harbor Laboratory, geese from the surrounding bird sanctuary dipped and veered in noisy formation. Their primal honking signaled their readiness for the annual trek that would take them to the marshes of Chesapeake Bay and points south. Far below them, another pilgrimage was under way, that of the tight, sometimes owlish society of human scientists, a species only slightly less given than its winged cousins to flocking behavior, rude honking, and a disturbing tendency to discriminate by plumage.

More than three hundred strong they came, these scientists, the cream of the world's biologists and geneticists. Many were shockingly young, outfitted indifferently in wrinkled sport coats and unconscionable ties worn over Dockers and deck shoes, their determined grunge contrasting sharply with the more staid tailoring of the elder statesmen among them. Clutching coffee cups and blinking sleep from their eyes—the hour was not yet nine—the aggregation filed into Cold Spring Harbor's main auditorium until no more seats remained and then spilled, as if obeying the laws of fluid dynamics, into an overflow hall, filling each of its rows as well.

Travel and conferences are regular features of the scientist's life, but the symposium that had drawn distinguished researchers from as far away as Italy, the United Kingdom, China, and Germany on this sodden morning in late 1992 was far from routine. It had been organized to pay homage to an extraordinary new field of science called gene therapy, which, despite its deceptively prosaic name, promises to rewrite not merely the practice of medicine but, in the fullness of time, the evolutionary course of the human race itself. Most of those in attendance appreciated the delicious irony of the occasion. For as recently as three years earlier, the field had been widely regarded as, at best, a will-o'-the-wisp and, at worst, an outright joke, its practitioners little better than bumblers and flakes.

It was significant that Cold Spring Harbor was playing host to the symposium because the people at the prestigious, century-old research compound on Long Island's north shore—the "University of DNA," where scientists had been unlocking the secrets of the molecule of heredity for better than four decades—had been notably skeptical about gene therapy throughout the 1980s. For the lab to have opened its doors to the two-day conference was the ultimate validation not only of gene therapy but of its chief pioneer, impresario, and champion—William French Anderson.

And thus it was that, as wild geese formed great swooping V's overhead, the silvery-haired Anderson assumed the podium before his once contemptuous peers. Here, in this temple of DNA, where a sinuous design suggestive of the molecule's double strands girdles the auditorium's ceiling, he called up his first color slide and began a triumphant overview of the burgeoning field that had consumed him for nearly half a century. Few who had not shared his tortuous struggle could know how sweet the moment was, or what a long, strange trip it had been.

As a bookish young man growing up in Tulsa, Oklahoma, French Anderson used to daydream about becoming one of the great names in science. It is a common enough fantasy among boys, the bulk of whom inevitably wipe out on the reefs of calculus, physics, or chemistry. But Anderson was no ordinary boy. There were reasons to think he might someday deliver the goods.

For one thing, intellectually he tested off the charts. Very soon after Anderson's birth, on the last day of the dust bowl year of 1936, his father, Daniel French Anderson, a government engineer with the Southwest Power Administration, and his mother, LaVere, a journalist with the *Tulsa World* and an author of juvenile books, came to the realization that they had a child prodigy on their hands. With their eager assistance, the little fellow mastered reading and arithmetic well before entering kindergarten.

Along with the gift of brains, young Anderson had the self-discipline of a Trappist monk. Unlike his two older sisters, he shunned contact with playmates, preferring to lose himself in college-level texts. He kept meticulous diaries and displayed a precocious, even compulsive drive to agendize his life before he had begun to live it. "I was a weird little kid," he recalls, not without a trace of pride.

In high school, while other young men dreamed of driving fast cars and dating cheerleaders, Anderson remained a devoted grind. Although he developed a passion for cross-country running, setting several state records at shorter distances, he never fell in with the locker-room crowd. On the team bus, he would sit in the back, reading the likes of Immanuel Kant, oblivious to the horseplay around him. The very archetype of a straight arrow, he never drank, smoked, cursed, or flirted. Instead, he spent his early adolescence auditioning the various branches of science for a field worthy of his soaring aspirations. Astronomy was fascinating, but inherently limiting. "I figured there was no sense in it," he says retrospectively. "We are stuck down here on Earth, and what you can see is always going to be distorted by the atmosphere." Physics, too, beckoned for a time, but he found it too abstract. More lasting proved to be an infatuation with archaeology. While still in high school, Anderson began correlating Roman numerals with their Egyptian and Babylonian antecedents, and studying how the ancients used them to do arithmetic, a project that was to earn him publication in several archaeological journals. But immersion in the past could never satisfy one with such a hunger to sculpt the future. As he prepared to set out for Harvard at the age of seventeen, the would-be Einstein of the plains was still casting about for his true vocation.

It was the summer of 1954, a golden time for youthful dreamers. In the preceding months, two events had taken place half a world away in England that were to have immense influence on young French Anderson. At Cambridge University, a pair of scientific upstarts named Francis Crick and James Watson had electrified the scientific community by announcing their discovery of what a gene would look like if you could actually see one under a microscope. And at Oxford, the British physician Roger Bannister had just run the first four-minute mile.

The sheer brilliance of Bannister's feat stoked the kiln of Anderson's ambitions. To him, the notion of the physician-athlete, the Renaissance scholar, was irresistible. He resolved that he would not merely excel at science. He would try out for the Harvard track team and become a world-class runner.

He found himself even more profoundly affected by what he read of Crick and Watson's work. By means of a quasi-photographic technique called x-ray diffraction, which creates a kind of silhouette image of molecules by shooting concentrated x-ray beams at them, the two young scientists were able to infer that DNA, the stuff that genes are made of, takes the molecular form of a double helix. In other words, the backbone of a DNA molecule winds around and around in three dimensions like a spiral staircase, enabling it to perform prodigious biochemical feats, such as the magical conjuring of more of itself and the synthesis of all the body's tissues and fluids. The finding promised to explain how life perpetuates and maintains itself through the timeless governance of the genes.

Here was grist for a young prodigy out to make his mark. There are really just two central questions in all of science: "How does the universe work?" and "What accounts for life?" Everything else is so much spin-off. Galileo, Newton, and later scientists had made great progress in answering the former question. Now it was clear that the biological sciences were on the verge of answering the second, and Anderson wanted very much to be a part of that. (Neither he nor anyone else, of course, could have foreseen that, by the late twentieth century, scientists would be chipping away at two still-greater mysteries, previously consigned to metaphysics on the assumption they were inaccessible to science: "Where did the universe come from?" and "How did life begin?")

When Anderson began thumbing through the Harvard catalog and spied a course entitled Biophysical Chemistry 184, taught by the eminent biochemist John T. Edsall, his seduction was all but complete. The course description, dry and terse as it was, nevertheless shook him. It warranted that the student would learn how the human body functions at the molecular level. *At the molecular level!* Where hormones and enzymes dart like bullets and arteries flow with tiny cellular barges laden with oxygen; where antibodies and macrophages patrol the body's portals, devouring trespassers whole like Hollywood monsters; where muscle cells contracting by the millions are able to move enormously disproportionate weights, embodying Archimedes' wondrous lever; where sperm and ova take midnight swims; where the junctions between neurons crackle with noble current. The quintessential arena of life where, as the old tire commercials used to have it, the rubber meets the road.

At night, Anderson contemplated this incredibly complex and beautiful symphony, the thousands of simultaneous physical processes going on at nearly the speed of light—imperceptibly, automatically, eternally—all orchestrated faultlessly by the genes. And then his mind would wander to the darker side—the occasions when the enterprise does not execute perfectly, when one or more of these physical processes keep breaking down because the gene in command is somehow flawed. Anderson had read about such things, how the consequences could be devastating, leading to frightening, incurable diseases, transmittable to any descendant unfortunate enough to inherit the errant gene. Since the beginning of time, hapless souls had suffered horribly and died young because of genetic mistakes passed along to them by their parents, mistakes that could be due to sudden changes—mutations—in the parent's genes or to the legacy of nameless ancestors from time out of mind. One's genetic fate, as Anderson well knew from his biology texts, hinges on a lottery whose outcome is determined at the instant of conception— to wit, by which of a male's 300 million or more sperm happens to fertilize which of a female's 100,000 eggs. If, by the luck of the draw, a sperm and/or egg containing bad genes from the parental hopper should come together at conception, the human being resulting from this union will pay a terrible price. Where in nature could be found a more telling example of existential indifference than in this random damnation of innocents without possibility of redress?

And now a thought began to drift across the panoramic screen of Anderson's fancy. What if you could somehow descend into the heart of a cell and repair or replace a malfunctioning gene? Would it cleanly and permanently cure genetic disease? The more he thought about it, the more intrigued he became. As his last boyhood summer in Oklahoma drew to a close, Anderson experienced a revelation. He knew at last what he was born to do. He would major in biochemistry and go on to medical school. He would journey to England and implore Watson and Crick to let him apprentice at their feet. He would, he resolved in his romantic fervor, help kindle a new field of medicine that would work at the level of the genes. And then he would go forth to heal the world.

"Those were the two things I was going to do with my life," Anderson recalls wryly. "I was going to be in the Olympics, and I was going to cure defective molecules."

It is a cool, wet afternoon in early 1992. The rain has been falling since dawn on the suburban Washington community of Bethesda, Maryland, a one-industry town where the industry happens to be human health. Bethesda is the site of the nation's renowned naval hospital. More important, it is home to the National Institutes of Health, arguably the world's leading biomedical research establishment, a sprawling complex of government buildings dedicated to the understanding and the treatment of disease. One of these buildings, a cavernous red structure called the Warren G. Magnuson Clinical Center or, as locals prefer, Building No. 10, said to be the largest brick edifice in the world, has, for nearly a generation, housed the laboratory of Dr. William French Anderson, pediatrician, molecular biologist, blood specialist, director of the molecular hematology division of the National Heart, Lung, and Blood Institute, and mastermind of history's first successful attempt to transfer corrective genes into the body of a human being.

The day finds Anderson in an upbeat mood. Little more than a year has passed since he and his colleagues accomplished their Promethean feat. Many associates expected Anderson to spend the remainder of his career resting on his laurels. The annals of science are replete with tales of researchers who, having pulled off the professional coup of their lives, slip thereafter into a complacent torpor and are never heard from scientifically again. But that is not Anderson's style. Aptly, if somewhat fatuously, referred to in the press as the father of gene therapy, Anderson is hell-bent on inseminating the world with his zeal for the fledgling field. He has made it his mission to become a gene therapy circuit rider, traveling from research institution to research institution mentoring budding gene therapists and breathing life into their experiments. He has also made it clear that this will be his last year at the National Institutes of Health. He is soon to take a position as director of the University of Southern California's new gene therapy institute.

Still, NIH has been Anderson's home virtually all his working life. Anderson's lab at NIH, with its twenty-five postdoctoral fellows, graduate students, and technicians, takes up an entire corridor on the maze-like seventh floor. It is not of imposing size as genetic-engineering research programs go, but Anderson likes it that way. "He values intimacy," says his former senior fellow and top assistant, Martin Eglitis. The air in the corridor reeks of test animals and formaldehyde, and a row of obstructions, in the form of cardboard boxes, stainless steel tanks, and refrigerators, makes passage difficult. The hallway gives access to a series of small suites where the researchers, almost all of them young, irreverent, and highly motivated, often work far into the night, seven days a week. Walk into one of these tiny laboratories, and you might think there is nothing going on. Much of what molecular biologists do is uneventful and often undetectable by the layman. Molecular research is unobtrusive, offering little in the way of pyrotechnics. Its practitioners work in a world small enough to be housed in a pillbox, a domain so Lilliputian that it is to the visible world as a twenty-five-cent piece is to the North American continent. But appearances deceive. In this unprepossessing laboratory and many others scattered around the planet, experiments are under way that will change the face of medicine, and society, forever.

On this day, Anderson has been caught in the downpour without an umbrella. Soaked to the skin, he begins toweling off his thatch of nearly white hair in the doorway of his narrow office. The walls of the office are covered, Brown Derby–style, with framed photographs of every lab assistant he has had in more than twenty years as a laboratory chief. There also are portraits of children who have died from genetic blood diseases, a haunting reminder of the limitations of basic research.

Anderson is a restless, high-strung man with a face that is blandly, boyishly handsome, with a trim, sedulously trained body that masks his nearly sixty years, and with a tendency to stammer when he gets excited. (In high school, he tried to cure himself of the affliction by speaking with stones in his mouth, like Demosthenes, the ancient Greek orator.) He never did become an Olympic athlete, though he was an accomplished collegiate runner and captain of the Harvard track team. "My hamstring muscles let me down," he says ruefully of the failure, for which he has compensated, in his typically overachieving way, by turning himself into a fourth-degree black belt in the Korean martial art of tae kwon do. But if one youthful obsession has eluded him, Anderson has fought hard to consummate the other. After a lonely, often bitter quest, the boyhood dreamer from the Tulsa oil flats has achieved a scientific milestone of incalculable magnitude.

With his coresearchers Steven Rosenberg, the powerful and driven chief of surgery at the National Cancer Institute, R. Michael Blaese, the NCI's canny, self-effacing chief of cellular immunology, and the earnest, resourceful senior clinical investigator Kenneth Culver, Anderson has succeeded in piercing the once impregnable fortress of the human cell to affect the outcome of the genetic lottery. In one remarkable, twenty-month period ending in early 1991, the sometimes feuding colleagues achieved a series of stunning precedents. First, they implanted a harmless bacterial gene into the living human body as a dry run, thereby overcoming a psychological hurdle every bit as daunting as the sound barrier was to early jet pilots. In the two decades that molecular biologists had been learning how to cut, glue, mix, zap, and otherwise manipulate the genetic material of everything from fruit flies and tobacco to pigs and salmon, the genes of humankind itself had remained virtually inviolate.

That fateful step achieved, the team moved rapidly on to gene therapy itself, tackling first a rare, inherited killer of children and then attacking cancer.

Anderson finishes drying off his hair and tosses the towel neatly onto his desk. He is jubilant and wants the world to know it. "The flood-gates," he declares, "are wide open. We've got the green light.

"We're going to have to proceed a step at a time. But we are at the beginning of what promises to be the most exciting time in the history of medicine. What an incredible time to be alive!"

And indeed it is. Even as Anderson was voicing these sentiments, scenarios for gene therapy experiments were pouring in to the national regulatory bodies, proposals to harness gene therapy to treat everything from heart disease to brain cancer to AIDS. By the end of 1993, only three years after Anderson's first experiment, upwards of fifty gene therapy protocols, as such proposals are called, had been submitted for approval, an explosion of research unprecedented in the history of medicine.

For Anderson, gene therapy is nothing less than a crusade. His protégé Eglitis recalls how, as long ago as the late 1960s, Anderson gave a talk in which he prophesied the coming of a golden age of medicine when human ills would be entirely banished by the clean, almost surgical replacement of malfunctioning genes. "This was before anyone had actually isolated the first human gene," marvels Eglitis. "The next speaker arose and said, 'Gee, it's wonderful that we have such energetic young scientists like French, but this is all pie in the sky. Nothing of this sort is ever going to happen in our lifetimes.' Well, who's laughing now?"

Anderson has never been the sole researcher trying to bring gene therapy to the bedside. Others of imposing stature are in the game. There is Anderson's most persistent and vocal critic, the redoubtable Richard Mulligan, and his Boston-based group working on curing blood diseases and cancer at the Massachusetts Institute of Technology and the Harvard Medical School in conjunction with Johns Hopkins, in Baltimore, home of the celebrated cancer geneticist Bert Vogelstein. There is Anderson's former colleague Ronald Crystal, recently departed from NIH for academe and private industry. Crystal, in addition to attacking cystic fibrosis and an inherited form of emphysema, is masterminding ways to get genes to the brain to fight neurological and psychiatric ailments. In Philadelphia, the University of Pennsylvania's James Wilson pursues gene therapy for cholesterol-ravaged arteries as well as cystic fibrosis, while across the state, at the University of Pittsburgh, Michael Lotze seeks ways to treat cancers of the breast, colon, kidney, and skin. At Boston Children's Hospital and at Massachusetts General Hospital, meanwhile, the geneticist Lou Kunkel and the neurologists James Gusella and Robert Brown are developing gene-based treatments for such inherited horrors as muscular dystrophy, Huntington's disease, amyotrophic lateral sclerosis (ALS), neurofibromatosis, and Alzheimer's disease.

The University of Michigan boasts the gene sleuth extraordinaire Francis Collins, whose wide-ranging research interests include breast cancer, Huntington's disease, and cystic fibrosis. Also in Ann Arbor are the husband-and-wife gene therapists Gary and Elizabeth Nabel, who are zeroing in on melanoma, AIDS and, cardiovascular diseases. At the University of Wisconsin is Jon Wolff, exploring cutting-edge treatments for muscular dystrophy and Parkinson's disease, while at the University of Iowa, Mike Welsh chips away at cystic fibrosis. In Chicago, Northwestern University's Teepu Siddique and the University of Chicago's Jeffrey Leiden and Nicholas Vogelzang are stretching the envelope toward applying gene therapy to ALS, heart disease, and cancer.

To the west, the Utah investigators Ray White and Mark Skolnick utilize the unique genealogical records of the Mormons to make important discoveries about the location of disease genes. On the West Coast, San Diego has emerged as one of the hubs of gene therapy research. At the University of California at San Diego, for example, a team headed by the Austrian-born Theodore Friedmann and the neurospecialist Fred Gage is pressing gene therapy for diseases of the brain and working closely with the pediatrician William Nyhan, renowned as the world's greatest genetic sleuth. Literally across the road from Friedmann and Gage, the virologist Inder Verma explores gene-based cancer therapies at the Salk Institute for Biological Studies. Also in San Diego is a Friedmann protégé, Douglas Jolly of the biotech firm Viagene, who is pursuing a promising gene therapy strategy for AIDS. Up north in the Bay Area, Berkeley's Mary-Claire King has single handedly forced gene hunters to join her in hot pursuit of the breast cancer gene that she alone for the past twenty years has believed to exist; at the University of California at San Francisco, Charles Epstein probes the mysteries of Down's syndrome; and at Stanford, David Cox and Richard Myers are working on Down's and other diseases. In Seattle, at the Hutchinson Cancer Research Center, hemophilia has been targeted by A. Dusty Miller, and Hodgkin's disease and lymphoma are the focus of the research of Miller's colleague Friedrich Schuening.

To the south, in Houston, the chest surgeon Jack Roth leads the gene therapy effort against lung cancer at the famed cancer center M. D. Anderson, while Baylor University's Howard Hughes Institute houses a veteran family of gene researchers in Savio Woo, Fred Ledley, John Belmont, Thomas Caskey, and Arthur Beaudet. In Memphis, Malcolm

Brenner, of St. Jude Children's Research Hospital, has used gene transfer techniques to unlock many of the mysteries of leukemia and other dreaded childhood cancers. And at the University of Miami, the psychiatric researcher Janice Egeland seeks to understand the genetics behind manic-depressive disorder, working with colleagues from Yale University and NIH.

After initial misgivings harking back to the eugenics atrocities of World War II, gene therapy research is starting to move in Europe. Laboratories in Britain, France, and Italy are making important headway against diseases ranging from schizophrenia and sickle-cell anemia to Parkinson's disease, cystic fibrosis, and inborn disorders of the immune system. Canada's gene therapy effort has been humming for years under the aegis of Alan Bernstein, Lap-Chee Tsui, Ron Worton, George Karpati, and others. And the work of Rosenberg, Culver, and Blaese continues unabated. They have branched out into newer gene therapy trials of their own, serving as consultants to nations ranging from China to India.

The discoveries made by some of these individuals have been more crucial to gene therapy's embryonic development than those of W. French Anderson. In truth, Anderson's ideas have never been the most seminal of those seeking the grail of therapeutic gene replacement. But he has unquestionably been the most driven, pushing on with daring— his critics would say foolhardy—intensity, rationalizing his failures, leapfrogging intermediate research steps, paying little heed to accusations that he has forsaken the time-honored rule of science that you understand fully what you're doing before you start experimenting on people. Brilliant, obsessive, seemingly dedicated to making utopia a reality for his patients, Anderson has been maneuvering manically on half a dozen fronts at once. But in being first to bring gene therapy from the lab bench to the bedside, he has seen his tenacity pay off.

Even his enemies would concede that.

It is easy to understand why gene therapy has fired the imaginations of researchers everywhere. The professional stakes are very high. A Nobel Prize and a place in the scientific pantheon no doubt await the practitioner who refines and enlarges the technique. But of vastly greater importance are the implications of gene therapy for the treatment of human disease.

Nearly four thousand known medical conditions can be blamed directly on the influence of defective genes. These range from relatively familiar scourges, such as cystic fibrosis, muscular dystrophy, sickle-cell anemia, and hemophilia, to an array of lesser-known afflictions with challenging names, such as xeroderma pigmentosum, citrullinemia, achondroplasia, and neurofibromatosis. Each is caused by a single deficient gene, a physiological error so minute that it beggars comprehension how something so small could spell the difference between robust health and a life blighted by illness and premature death.

As if single-gene diseases were not enough, genetic conditions may also be polygenic, that is, they may arise from the interaction of two or more improperly functioning genes. Such infirmities may even involve whole communities of genes, as in Down's syndrome and other chromosomal abnormalities. To further complicate the picture, there are ailments caused by genes acting in concert with certain environmental agents—viruses, say, or foods and toxic chemicals. Notable among such "multifactorial" disorders are diabetes, high blood pressure, multiple sclerosis, asthma, and spina bifida.

Until recently, inherited disease remained on the periphery of public consciousness. Genetic illness was something that always happened to someone else. Even if members of one's own family line showed a disconcerting tendency to die young from some damnably recurrent cause, there was little or nothing one could do about it. The only rational response seemed fatalistic acceptance.

But the last decade has brought a major turnaround. Even as medicine closes in on the technology for healing such intractable ailments, the number of illnesses whose underlying genetic cause has been identified has snowballed. Hardly a week passes without an announcement in the scientific press of the discovery of still another gene responsible for human disease. In some cases, these findings forge a link between genes and ailments once thought to have no such connection. Genetic illness, consequently, has become the vital concern of everyone. Most, if not all, of us harbor deleterious genes that can shorten or oppress our lives, or the lives of our children, and we may all ultimately profit from ongoing research into gene therapy.

The list of diseases now thought to be gene-based is startling. Recent research has heightened the likelihood that inherited predisposition underlies mental illness; such conditions as schizophrenia, manic depression, and alcoholism seem to have an important, and possibly decisive, genetic component. Alzheimer's disease almost certainly is rooted in the failure, late in life, of a crucial gene in brain cells. So is the debilitating palsy of Parkinson's disease. Amyotrophic lateral sclerosis, the terrifying, immobilizing disease that took the lives of Lou Gehrig, Senator Jacob Javits, and the actor David Niven, is caused by genes, as are the most prevalent forms of arthritis and blindness.

Of even broader significance to the general population, science is demonstrating a correlation between genes and the primary human killers—heart disease, cancer, high blood pressure, and stroke. One's chances of developing a life-threatening cardiovascular problem are dictated to a large extent by genetic endowment. Having the wrong genes can trigger a complex series of molecular events that can result in the clogging of arteries, with often tragic results. Similarly, our prospect of acquiring cancer appears to hinge on the environmental assaults visited upon certain growth-regulating genes we all harbor within our cells. Though it is not, primarily, an inherited affliction, cancer is in the truest sense a disease of the genes.

Researchers have concluded that virtually all human illness, even infectious disease, has some relationship to genetic endowment. Why during the most virulent plague does some portion of the populace survive? Are some people more susceptible to certain microorganisms and not to others because of inherited characteristics of their immune systems?

"I start with the premise that all human disease is genetic," says the Stanford University Nobel laureate Paul Berg, whose $125 million Beckman Center for Molecular and Genetic Medicine is deeply involved in long-range gene therapy research. "You can sit here for an hour, and you can't get me to conclude that any disease that you can think of is not genetic."

Are aging and death themselves forms of genetic "disease"? Are there longevity genes that determine life span, and mortality genes programmed to switch off at a certain time, resulting in the eventual demise of an organism? A growing number of scientists think so. Clinical trials with drugs, diets, and other regimens are under way in hopes that the aging process can be thwarted and that instead of merely living longer, which is all medicine has heretofore managed to do for us, we may actually fulfill the dreams of the alchemists of yore and postpone aging itself.

In theory, most of our ancient physiological nemeses may one day be treated and cured by gene therapy. Those illnesses that may not, for certain reasons, be amenable to direct gene replacement will almost certainly be brought under control by gene therapy's spin-offs, including new drug technology to replace missing or defective hormones, enzymes, and other proteins, and even more exotic concepts, such as the switching on or turning off of certain genes.

Meanwhile, well before the advent of gene therapy, molecular biology has begun to revolutionize the science of prenatal diagnosis and genetic screening. Parents no longer need dwell in agony and ignorance before discovering whether their baby will suffer from a genetic defect. A sample of fetal DNA now enables geneticists to detect the presence of hundreds of inherited diseases in time for parents to consider terminating a pregnancy. Newer technologies even do away with the need for abortion. In the ultimate extension of genetic screening, problems may now be avoided *prior to* pregnancy by an analysis of female eggs and early-stage pre-embryos consisting of a few cells.

In a similar fashion, adults wishing to know whether they carry certain deleterious genes that they might transmit to offspring increasingly can receive definitive answers from a simple blood test. They may soon also be able to learn, if they wish, what ailments they themselves will be susceptible to in later life, allowing them to alter their lifestyles accordingly. For the 600,000 American women with a predisposition to breast and ovarian cancer, for example, regular clinical breast exams and mammograms are not an issue but a lifesaver. If one is at a high risk of developing heart disease or lung cancer, it would make even more sense than usual to avoid fatty foods and cigarettes.

With all that is going on in molecular biology, it is small wonder that, although formidable technical obstacles remain, a rush of exhilaration has swept through the scientific community. For eons, genetic diseases have remained incurable and, in the main, untreatable. They have been utterly beyond the reach of the dramatic progress made in other areas of modern medicine. Now, all at once, medical science is acquiring the ability to reverse, with one clean stroke, a patient's apparently inflexible genetic destiny. Given the laboratory milestones achieved already, and the gathering momentum of genetic research, it seems but a matter of time before gene therapy takes its rightful place beside the other two miracles of twentieth-century medicine—antibiotics and organ transplantation. When that day comes, no longer will the human race be slave to the spin of the roulette wheel that accompanies the biological act of conception.

The number of potential beneficiaries is awesome. While only a small fraction, perhaps 15 percent, of the public suffers from the rarer forms of genetic disorder, the percentage of those affected swells hugely if one takes into account maladies that appear later in life, such as cancer, cardiovascular disorders, Alzheimer's disease, depression, and diabetes. Heart and other circulatory ailments affect more than two million people in the United States each year, and cancer afflicts a million more. Together, they are responsible for nearly a million and a half deaths a year, or approximately 75 percent of all deaths.

Genetic flaws are responsible for a fifth of all infant mortality, half of all miscarriages, and up to 80 percent of all mental retardation. They account for 40 percent of the pediatric admissions to hospitals and more than half of the adult admissions—the number would be higher but for the ranks of genetically ill individuals who die before reaching adulthood. The expense of caring for the 1.2 million people hospitalized annually for just the more uncommon forms of genetic pathology runs to several billion dollars a year. If one counts the millions more afflicted with genetically influenced heart problems, malignancies, and neuropsychiatric disorders, not to mention the large number of retarded individuals who must be housed in long-term care facilities at a cost of up to $30,000 a year, the price of care becomes a major contributor to a national health care bill that is running out of control. In contrast, gene-based treatments may offer a relatively inexpensive way of attacking illness.

It is against this backdrop that the pioneers are vigorously pursuing the goal of gene therapy. So far have they come, so fast, that the biological sciences today stand at a threshold similar to the one the physical sciences confronted at the dawn of the atomic age. Lamentably, the parallels with nuclear research are more than superficial. While genetic manipulation promises a future that is electrifying in its prospects for good, it carries with it an apocalyptic potential for disaster. Those who propose to tinker with humanity's genetic code are spawning some of the most profound ethical issues that science has ever faced.

For example, should gene alteration be used to cure disease exclusively in the patient at hand, or, if it becomes technologically possible, should the cure be extended to the patient's sperm or egg cells as well, to ensure that his or her descendants will also be free of the disease? This is not so readily answered in the affirmative as it might appear.

Mistakes are often made in medicine, and presumably gene technology will not be exempt from them. But whereas a doctor's error in treating appendicitis will harm only an individual patient, a gene therapist's gaffe, if it finds its way into the gametes, or sex cells, could subtly, over many years, be transmitted to an entire family line; and, should it spread through the exponential process of mating, as have countless mutations in the past, it could eventually contaminate a portion of the human gene pool.

On another level, the same technology that would permit medicine to manipulate disease genes could presumably be used to alter genes unrelated to health. It might become possible someday, for instance, to create smarter people by instilling in them the genes for higher intelligence, or stronger people, taller people, svelter people, and so on. But such an emphasis on perfection is uncomfortably reminiscent of the discredited eugenic ideologies of the earlier part of the century that urged the breeding of human beings like racehorses, a point of view whose nightmarish extreme found expression in the Nazi scheme to create a super race by selective inbreeding and extermination of genetic "inferiors." Even a benign use of eugenics—to screen out defective embryos before making test-tube babies, for example—could have its pitfalls. The technology might be available only to the rich, creating disparities among people that make today's gulf between advantaged and disadvantaged look narrow.

The ethical concerns do not end there. Prenatal testing has aggravated the controversy over abortion, and screening of individuals for their genetic weaknesses could, if the information became public, hamper their ability to obtain insurance and employment. Clearly, much remains to be sorted out before the full potential of gene therapy and related technologies come on-line.

Conceptually, gene therapy is simple, though the techniques for carrying it out are exceedingly complex. "It's basically a transplant," says Stuart Orkin, a Harvard physician and molecular researcher and a major player in the early development of the field. What sets the procedure apart from engrafting a new heart or kidney is that the anatomical parts to be installed—namely, genes—are molecules only a few hundred thousand atoms in size, too small to be seen with the most powerful electron microscopes. Something so diminutive would by all rights seem inconsequential; in fact, genes are the sine qua non of life. They contain the master plan directing development of an entire plant or animal from a single fertilized cell. They are in total command of every heartbeat, breath, muscle contraction, and thought, every act of digestion, circulation, elimination, and immunity that a creature will ever experience in its lifetime. Genes also are a bridge to posterity. They direct the construction of sperm and ova, then hop aboard for the closest thing the cosmos has to offer in the way of time travel, carrying the blueprints of life to the next generation, and the next, in the exquisite process we call heredity.

The genes are present everywhere in the anatomy. Each person is endowed with an estimated 100,000 separate genes, strung side by side, rather like pearls on a string, along twenty-three pairs of sinuous entities, chromosomes. An identical set of these 100,000 genes—the full set is known collectively as the genome—is tucked into the nucleus of all but a small fraction of the 100 trillion cells that make up the human body. (The lone exceptions are the red blood cells, which lose their nuclei to gain room for their oxygen cargo.) If at first glance it seems incomprehensible how so many ships could have gotten into so many bottles, remember that it all starts with one fertilized egg containing a full complement of genes. Subsequent cell division takes care of the rest.

Genes are able to perform their feats of magic because of a chemical code engraved upon each one, a code that differs in sequence from gene to gene. This code tells tiny factories in our cells how to manufacture proteins, the huge and diverse class of substances that make up living tissue and regulate all of the body's functions. If one knows this, one can begin to understand how genetic diseases come about. Occasionally— not that often, when one considers that there are roughly three billion chemical units in the genome and therefore three billion chances for something to go wrong every time a cell divides—a mistake can creep in. The mistake is often slight, perhaps just one unit of the code deleted or misspelled. Yet, if a gene's recipe is wrong in the slightest respect, then its corresponding protein will be similarly improper—and a chain of circumstances is set in motion that can, if the protein is essential enough to survival, lead to serious illness and death.

If a gene should become mutated in someone's body, say, during exposure to x-rays, it typically causes no harm to its owner, for the simple reason that the person has millions upon millions of additional cells with fully operational versions of the same gene. The loss of gene function in a single cell is as indiscernible in the vast firmament of healthy cells as a burned-out bulb is in a Las Vegas marquee. Only when the mutation affects the so-called germ line, that is, when it occurs in a testicle or ovary cell making sperm or ova, or within one of the gametes itself, does the danger of disease loom. Again, however, the brunt is borne not by the individual at hand but by that person's offspring. Should the mutation-bearing sperm or ovum become involved in conception, the resulting embryo will grow up with the mistake in each of its cells. The only exception to this generational buck-passing occurs in cancer, where the corruption of a lone cell is enough to trigger illness, that is, a tumor, in the same individual who experienced the mutation.

What gene therapy seeks to do is introduce normal genes into a patient's cell nuclei to repair, replace, or compensate for the defective ones. Threading the needle of the cell nucleus and its chromosomes is a daunting task, something like trying to make a hole in one in golf. For better than ten years, French Anderson and his peers at other institutions inserted genes into a variety of animals in preparation for the initial human trial. They succeeded in getting a few balls to land on the green. But for a long time they could not get a ball to drop into the cup.

Summer comes to the Boston area with a rare brilliance. On certain days, the sun bathes the old city with rich gold pigment, so that it seems to come alive like the pages of an illuminated text. Nowhere else is the effect so striking as in Cambridge, across the Charles River, where the radiance falls on the sprawling, architecturally diverse campus of the Massachusetts Institute of Technology with special intensity, as if to assist, if only metaphorically, in the great battle against unenlightenment that never ceases there. The order of the day, every day, at MIT is the pursuit of the ineffable. MIT scientists run the cosmos in reverse, sending the tape counter spinning back to the big bang. They muck around inside black holes and attenuate the universe into string. Befitting its importance, the quest to understand genes has been given a place of honor at MIT, the Whitehead Institute for Biomedical Research, a ten-story, privately funded, state-of-the-art facility dedicated exclusively to expanding the frontiers of molecular biology. Since its opening, in 1984, the Whitehead has attracted some of the most revered names in the field, thanks in no small measure to the recruiting skills of its first director, David Baltimore, himself a Nobel laureate and voyageur into the molecular mysteries of malignancy and viruses. The institute's resident superstars include Robert Weinberg, a recognized world leader in the study of cancer genes; Rudolf Jaenisch, whose specialty is creating genetically engineered animal models for human illnesses; and Eric Lander, a key figure in the charting of the world's first working map of the human genome. And in the middle of this all-pro lineup is the wild, protean intellect of Richard Mulligan.

Mulligan is a certified boy wonder who runs his own twenty-one-person laboratory at the Whitehead. At the age of twenty-five, he made gene therapy technologically possible, becoming the first molecular researcher to create a workable system for infiltrating genes into living cells. At thirty, he was the recipient of a MacArthur Foundation Fellowship, the "genius" award, which granted him $150,000 a year for five years, to do with as he pleased. Among his assets is a home, on Martha's Vineyard, where his neighbors included the late Jacqueline Onassis. Now in his late thirties, a full professor, married and a new father, Mulligan is a loose-limbed blond giant with a Civil War general's beard, a man indifferent to fashion but passionate about the quality of his food and drink, a man who says exactly what he thinks, thinks exactly what he pleases, and is devoted almost to the point of self-mortification to the often exacting rigors of the scientific method. It is he who has been most critical of French Anderson's research style.

On one of those brilliant summer days, Rich Mulligan sits in his office in the Whitehead, nursing a cup of coffee. He is in a talkative mood. His animation contrasts sharply with the gloom he displayed in the past when he was extremely wary and restrained in discussing the prospects of gene therapy. His own research had hit a wall, and, finding little to buoy him in work that was proceeding elsewhere, he lapsed into a kind of naysaying funk in which he began to predict that consummation of the first human experiment in gene correction might be a good deal further off than people had hoped. But now the funk has dissipated.

"I'll admit I was cautious," Mulligan says. "So little had been done, so many imponderables remained, it seemed embarrassing to talk about gene therapy as if it were a legitimate field. To do so would have invited ridicule from colleagues.

"But there's a lot going on now," he says. "It's perfectly reasonable to talk of gene therapy as a field, as a serious and imminent form of treatment."

Rich Mulligan is a congenital stickler. "He hates bullshit," says a colleague. "He just hates it." That gene therapy is off the ground is indisputable. But Mulligan is always careful to distinguish between the *kind* of gene therapy he would like to see—clean, chaste gene therapy in which the patient is cured because the science is exacting, by his own definition—and the kind of muddy, first-round gene therapy he finds swirling all around him, a kind that he feels borders on the spurious and that he dismisses as built on an empire of hype.

The draftsman of this second and, by Mulligan's lights, inferior brand of treatment is the same person whom the media have lionized as the progenitor of gene therapy. None other than French Anderson.

Oil and water. Fire and ice. Maggie and Jiggs. For nearly a decade, Anderson and Mulligan rebuked each other in private, in public, and in print, circling and parrying, as they vied to take the helm of the fledgling methodology. A perhaps superficial explanation for their differences lies in their contrasting roles and perspectives. Anderson is the embodiment of the physician-researcher, eager to bring treatments on-line to help sick and dying patients. To him, it could be argued, science is an instrument to be subordinated to clinical goals. Mulligan, on the other hand, is the pure researcher, the lofty, tenured Ph.D., to whom science is a church with sacraments and pieties. Such a man, some claim, is insulated from the anguished cries of the ill. Like so many conceits, the comparison is simplistic and far too neat. It has some truth to it, but more than a little falsehood. It makes Mulligan sound as if he had little humanity, which is patently false.

The bottom line, however, is that Richard Mulligan finds French Anderson scientifically repugnant, even while he is giving him his due as a politician and expediter. The ultimate in left-handed compliments occurred when Mulligan agreed to cosponsor the Cold Spring Harbor gene therapy conference with Anderson—and then failed to appear for all but one session.

"What French has done is gain the acceptance by the public of the idea of gene therapy," Mulligan says. "That's positive. That's good. But in a technical sense, he has not spawned any good experiments. We still don't know if gene therapy is really safe or not. We're still not a step ahead clinically. The science so far has been abysmal.

"The clinical guys have opened the door for experiments that are going to be successful," Mulligan concedes magnanimously, but quickly returns to his Prince of Darkness theme. He snaps, "I think, though, it's becoming clear to people that this scientific field should be based on science, not on hype."

Just who is right and wrong in these affairs has never been clear. One of Mulligan's occasional research partners, Harvard's Stuart Orkin, predicted a number of years ago that the initial gene therapy experiment would be "something of a stunt" and that society should expect no encores of significance for some time to come. And yet the Anderson team's opening experiments have been followed by a rash of applicants seeking to apply gene therapy to a litany of dread diseases. Many applications have come from people considered eminent in both medicine and biology.

Moreover, whereas several years ago, few would dare go on the record with the prediction that gene therapy was in anything but the most embryonic stage, all at once the technology is taken seriously. Leroy Hood, former chairman of the California Institute of Technology biology department and winner of the Albert Lasker Prize—among the most prestigious awards in biomedical science—is willing to make some bold promises. "Over the next twenty to forty years, we will have the potential for eradicating the major diseases that plague the American population," he says. "Our goal is not to help people live forever. But we will let them stay healthy and productive through their entire natural lifetimes."

But throughout all the meanderings of scientific opinion, only one voice has never wavered in its true belief. It belongs to William French

Anderson. His unceasing and at times quixotic sprint after the rabbit of gene therapy has gotten him accused of everything from showboating to bad science, not just by Mulligan but by many others. Undeniably, though, it is he who now stands at center stage.

It is a complex man who has arrived at the pinnacle of his craft. "Uncritical enthusiast" is the pejorative epithet one patronizing Nobel laureate has applied to him. The pressure has taken its toll on Anderson over the years. On occasion, he has displayed a siege mentality reminiscent of the Nixon White House, peppering his conversation with references to his "enemies." But if at times Anderson sounds like Richard Nixon, his true doppelgänger seems to be Jerry Brown, the ex-governor and presidential candidate with the New Age bent.

Anderson likes to pad around his laboratory in stocking feet. When he was at NIH, before leaving for USC, he had the window shades in his office blacked out and frequently sat for long periods in total darkness, explaining that the resulting sensory deprivation encouraged deeper meditation. At the home he shared in suburban Maryland with his wife, the noted pediatric surgeon Kathryn Anderson, he often spent evenings floating in their backyard pool doing what he calls "subconscious thinking." The yard there was screened by a thirty-foot thicket of bamboo that Anderson planted to shield himself from prying eyes.

And then there are the daily performance charts. Each day for the past twenty years, Anderson has evaluated his productivity and assigned himself points on a color-coded calendar taped to one of his filing cabinets. Good research earns two points. Making speeches and reading scientific journals merit one point. And administrative and other nonscientific pursuits get no points. At the end of the month, the points are totaled, and months with thirty or more points are tinted green. Those with between twenty and thirty points become yellow, and months with a score below twenty are colored red. "I do these charts to keep myself on track," Anderson explains. "Too many reds, and it means I'm doing something else other than the things I truly value. I enjoy so many different pursuits that I am easily distracted. If I don't keep myself doing the primary things, the charts make me aware of it and I can get back to business." He has agonized through long periods of reds, as he followed other pursuits to expel energy, including formula-one auto racing, archaeology, sports medicine, and, of course, tae kwon do. No longer an active competitor, he has become an instructor as well as chief physician to the American tae kwon do team, filling that role at the Seoul Olympics, in 1988.

A reporter once accompanied Anderson to the martial arts class he taught for years on Friday nights in an NIH fitness center down the road from Building No. 10. On the way, Anderson told the writer that he likes tae kwon do because it is something he can do in a near-transcendental state, without really concentrating on it. That evening, Anderson gave his class a demonstration of tae kwon do performed on autopilot. He stood stock still while facing a wall, making his mind an apparent blank. Suddenly his arm shot out and his knuckles smote the wall. He turned back to the class, oblivious to the pain. "Try not thinking of what to do," he said. "Just do it."

The next morning Anderson confided that he does research the way he does tae kwon do. Science, he said, "is something best done without thinking, something transcendent, and intuitive." Statements like that drive more punctilious rivals like Mulligan positively crazy.

Anderson's spontancity somehow coexists with the unquenchable passion for organization that has been a part of him all of his life. He recalls the day he first walked into his adviser's office as a Harvard freshman, clad in cowboy boots and a western hat, and speaking in his native dust bowl twang. "I had my four college years all planned, my entire life diagrammed out, really, down to the last detail. My adviser thought it was the funniest thing he had ever seen."

But Anderson had the last laugh. He has followed almost to the letter the plan he worked out as a teenager. As he vowed to himself, he did attend medical school. Likewise, he went on to take subsequent training in molecular biology, even wangling, improbably, a two-year graduate fellowship at Cambridge University's Trinity College with his idol Sir Francis Crick. Now he has fulfilled the ultimate fantasy of his youth, to lead a direct assault on genetic disease. A few years ago, Anderson ran into his former adviser again. The adviser blurted, "Hey, I remember you. You're French Anderson, the guy who had his whole life planned out. By God, you went ahead and did exactly what you said you were going to do, didn't you?"

Anderson relates this story with relish, recognizing he is an unusual breed and making no apologies for it. "What can I tell you?" he says. "I was weird back then, and I'm weird now."

2

The Long and Winding Road

As long as human beings have had the power to reason, they have suspected that there must be some sort of machinery of heredity. Our prehistoric ancestors could hardly have been blind to the way children resemble their parents and grandparents, right down to the family nose or chin. In time, it became clear that certain diseases could be inherited as well. Hemophilia was well known to the Old Testament Hebrews who, in the Talmud, excused from circumcision newborns whose mothers had previously lost sons or brothers to uncontrollable bleeding. The authors of the Talmud correctly recognized that the ailment is passed on from mothers to sons, though they could not have known the cause is a fault in the gene for a substance known as factor VIII, which is vital to blood clotting.

Experience with agriculture bolstered the case for a hereditary mechanism. The archaeological record shows that for at least six thousand years farmers have known how to breed plants and animals with desirable characteristics to enrich the quality of their crops and livestock. Yet, despite this early familiarity with breeding, millennia were to pass before the precise way in which biological traits are handed down ceased to be an impenetrable mystery. Surely the traits themselves were not passed along in literal fashion. One does not inherit as physical objects a parent's bushy eyebrows, hemorrhoids, or gout. But how is the tendency to manifest such characteristics transmitted from generation to generation? What sort of invisible messages are involved?

It was this instrument for transferring organic title that a young

Swiss biochemist named Friedrich Miescher stumbled upon in 1869 while working in his University of Basel laboratory, high above the Rhine River. Miescher, who studied white blood cells, seems to have made a fetish of examining objectionable fluids. First, he isolated white cells in the pus wrung from surgical bandages; then he turned to the semen of Rhine salmon. After numerous experiments, he discovered something strange. When he broke the white cells down with the enzyme pepsin, their nuclei released a substance he had never seen before, a gummy, whitish sap he called nuclein. Miescher hadn't a clue what the stuff was good for, and he lived another twenty-six years without ever finding out. But he rightly sensed that it was significant. The gooey strands that he coaxed from the heart of his cells were what we today call deoxyribonucleic acid, or DNA—the long-sought vehicle of heredity and the supreme regulator of life.

Even as young Miescher was puzzling over his white cells, another investigator knelt in a monastery garden just 360 miles away, laboring to tease out the timeless laws governing inheritance. This solitary individual, the monk Gregor Mendel, used pea plants to infer the existence of invisible units of patrimony, which he called factors (they were not given the name *genes* until 1909, by the biologist Wilhelm Johannsen).

It was Mendel's intuition that plants and animals have a pair of genes for each individual trait, with one of every pair contributed by the organism's mother and the other by its father. He also noted that some genes are dominant, that is, they *express*, or manifest, their trait if inherited in a single copy from either parent. Other genes are recessive, meaning they must be present in two copies, one from each parent, before they make their presence known. Mendel went on to calculate the universal mathematical ratios governing the two kinds of inheritance, a mammoth task that consumed years of patient observation. As he discovered, children have a 50 percent chance of inheriting a dominant trait from an affected parent, whereas recessive traits are expressed in offspring only 25 percent of the time.

Mendel's findings were published in 1866 and sent to 133 scientific associations. Even Charles Darwin got a copy, although there is no evidence that he ever read it. If he had, he would have realized that here was the mechanism by which the theory of evolution worked. But the response to Mendel from the research establishment was dead silence. The monk's work was to gather dust until 1900, when three other European investigators unwittingly duplicated it and discovered, to their chagrin, that Mendel, dead for sixteen years, had trod the same terrain a generation before.

It is one of the more curious and regrettable twists of history that two men, Mendel and Miescher, should, within a span of only three years and a distance of just a few hundred miles, have made the key discoveries upon which modern biology is based and thereafter been utterly neglected by their peers. Nonetheless, Mendel and Miescher set the tone for the two parallel lines of research, genetics and biochemistry, that were to proceed independently of one another for many decades before finally converging in our era as the new field of molecular biology.

"Biochemists weren't interested in genetics," the Nobel laureate George Beadle once noted, "and geneticists weren't interested in biochemistry." The two specialties "were like miners tunneling toward each other in the dark."

It was a geneticist, Thomas Hunt Morgan, who first demonstrated that genes for specific traits could be "mapped" to their location on the chromosomes. Beginning in 1906, he and his high-powered coterie at Columbia University studied mutations in the lowly fruit fly and found that these mutations, which ranged from unusual eye coloring to bizarre wing deformities, were trackable by the pattern in which they were inherited. For example, if in the normally red-eyed fruit fly population an aberrant white-eyed fly occasionally appeared, and the trait was confined exclusively to males, then it could be safely assumed that the gene for eye color must be on the X, or female, sex chromosome. (Females have two X chromosomes, and thus can offset the mistake, but males with only one X and one Y, have no such compensating backup copy.) Morgan's team was also struck by how often the same two traits from one parent were inherited together by offspring. This linkage of traits, the team decided, must reflect close proximity of the genes on the chromosomes. Adjacent genes on the same chromosome would be inherited together except for occasional lapses when, during the shuffling of genes that occurs during the creation of sperm or ovum (a system nature builds in to ensure that siblings differ to a degree from each other), the genes were separated. Moreover, the linkage could be used to chart the relative position of genes along a chromosome. In fact, you could reduce the phenomenon to a mathematical formula: the closer the genes, the rarer this separation. With such an empire of inferences, Morgan's team was able to assemble a crude map of all four pairs of fruit fly chromosomes, an operation that calls to mind the familiar parable of the blind men feeling the contours of an elephant.

Meanwhile, across the Atlantic, a British physician, Sir Archibald Garrod, was applying the notion of genes to illness, specifically to a disease of the joints called alkaptonuria, in which the urine turns black because of an excess of a substance called homogentisic acid. Garrod noted that the parents of patients were almost always first cousins. Since the odds of a child's inheriting two recessive genes go up markedly if the parents are related, for the simple reason that members of a family share many of the same genes, Garrod concluded that alkaptonuria is caused by a defect in a gene. Moreover, he correctly assumed in a 1909 paper that the gene must be for an enzyme that in normal people neutralizes homogentisic acid but that when present in mutant form allows the substance to build up dangerously.

More than thirty years later, in 1941, Garrod's hypothesis was verified by two Stanford scientists, the geneticist Beadle and biochemist Edward Tatum. Beadle and Tatum conducted an ingenious series of experiments on bread molds in which they zapped the molds with x-rays, hoping to cause mutations in their genes. When some of the mutants thereafter began starving, Beadle and Tatum showed that it was because they had lost the ability to make specific digestive enzymes, demonstrating conclusively that the function of genes is to direct the production of enzymes and other proteins.

While a great deal was being learned about genes, scientists had only the sketchiest notion of what substance genes were made of and no understanding at all of how they worked their magic. These longstanding mysteries began to unravel in 1944 when an elfin American scientist named Oswald T. Avery identified DNA as the key component of the genes. Avery, as unlikely a world-class scientist as ever lived, grew up playing trumpet on the steps of a Bowery church to attract souls for his evangelist father to save. A wee man, weighing no more than one hundred pounds, he never married, never traveled, in fact, scarcely ever left his laboratory at the Rockefeller Institute, in upper Manhattan. It was there that he and his colleagues used two strains of *Pneumococcus* the variety that causes pneumonia and a harmless kind—to make the case for DNA.

It had been observed that, when mixed together, the dangerous pneumococci could transform their benign cousins into lethal fellows. This would occur even after the dangerous germs had been killed in a bath of boiling water. Avery employed a basic scientific technique, the process of elimination, to coax nucleic acid out of the closet. He systematically removed parts of the virulent bacteria until all that was left to cause the transformation was the DNA. Conversely, when he took away the DNA, the transformation stopped short. What was happening, we know in retrospect, was that the genes, that is, the DNA, in the virulent bacteria, were being absorbed by the chromosomes of the harmless strain, enabling the latter to start manufacturing weaponry that would make it deadly—namely, a smooth, sticky outer coat that allows pneumococci to adhere to lung cells while shielding them from antibodies.

Now that science knew the cloth from which genes were cut, it was only a matter of time until someone identified the signature design, and from there it was but a step to figure out how these ephemeral objects worked. In 1952–53, a gawky former quiz kid from Chicago named Jim Watson teamed up with a brash physicist from Cambridge University named Francis Crick to unmask the DNA molecule in its deceptive simplicity. Their strategy was to build Tinkertoy-like models of what they thought the molecule might look like, on the basis of fuzzy footprints it left when subjected to an x-ray imaging technique. Of the various structures suggested by this will-o'-the-wisp portraiture, the only one that met all the known physical, behavioral, and biochemical conditions relevant to genes was a double helix.

In what was surely one of the most reassuring moments in scientific memory, when researchers were finally able to visualize the DNA molecule, in 1988, with the aid of the newly developed scanning tunneling electron microscope, it looked precisely the way Watson and Crick had predicted it would. Etched on a grainy photograph was a double helix, looking for all the world like interwoven strands of yarn.

DNA, as we now know, is an awesomely versatile material, so flexible that each cell holds six feet of it wound infinitely tighter than the core of a golf ball. If uncoiled and knit together, all the DNA in your body would stretch from Earth to the planet Pluto some 1,600 times. Yet, despite its pliancy, DNA is resilient enough to withstand assault from dioxin, aflatoxin, cosmic rays, oxygen free radicals, pesticides, microwaves, and the thousands of other injurious environmental agents that bombard us from minute to minute.

DNA is virtually immortal. Genes extracted from fossils millions of years old can be made to work if plugged into bacteria and given the right stimulus. It is possible that the actual appearance of dinosaurs and other prehistoric beasts may one day be revealed beyond dispute by an analysis of their long-preserved genes. In a fascinating application of DNA's persistence, a project is now being proposed to make a working library of Abraham Lincoln's genes taken from hair and bone samples preserved after his assassination, in 1865. The chief goal is to see whether the Great Emancipator suffered, as has been suspected, from Marfan's syndrome, an inherited disease that is marked by a tall, lank appearance and defects of the eyes and cardiac vessels. The genetic mutation was identified in the summer of 1991.

So eternal is DNA that the actual primordial molecule that gave rise to life eons ago may still be in circulation. Some of us may be walking around with bits of it in our cells. By the same token, DNA is a living museum of the story of life. Among the chromosomes of all higher animals are relics of genes from lower or extinct species, genes that are no longer in use. Very likely, for example, we have genes for gills tucked away in our cells, but they became gradually inoperative when our ancestors began to develop lungs. The body keeps such genes around much the way a railroad hangs on to abandoned sidings.

DNA is interchangeable among species. You cannot chemically distinguish between mushroom DNA, horse DNA, and human DNA. The difference between a paramecium and a prime minister lies in the way that the four key chemicals in DNA are lined up in a row. This quartet of chemicals, or nucleotide bases, as they are properly called, have scientific names—adenine, thymine, guanine, and cytosine—but even molecular biologists refer to them only as A, T, G, and C. The entire code of life is written in these same four letters, the sequence varying from gene to gene. The order the chemicals take in a given gene denotes the recipe for one of the estimated 100,000 proteins required by our bodies. For example, the formula for human lung surfactant, the substance that enables us to breathe in and out, starts with GAA GTG AAG GAC GTT TGT and continues on for hundreds of permutations of the same four letters. Just the same four chemicals, but the billions of possible combinations easily allow for the fantastic diversity of life. All the human beings who have ever lived represent fewer than 1 percent of the potential nucleotide arrangements.

Think of DNA as a twisted rope ladder with As, Gs, Ts, and Cs spaced regularly along two side rails made of sugars and phosphates. Every rung of the ladder consists of a nucleotide glued end to end to its complement on the other side. Why do we say complement? Because each base will bond only with one other kind. Hence an A joins only with T and a G only with C. The rungs thus formed are known as base pairs. This affinity is of paramount importance. (When Watson and Crick first determined that DNA takes the form of a double helix, they assumed that A hooked up with G and T with C, because A and G were of the same family of chemicals, called purines, and T and C were of another chemical class, called pyrimidines. Because purines and pyrimidines are of different molecular size, however, the rope ladder would be wider at some points and narrower at others, resulting in a lumpy helix. But this could not be, because the x-ray diffraction research showed that the side rails were the same distance apart at any point, approximately twenty angstroms. Ultimately, Watson deduced that if A mated with T and G with C, the helix would be smooth and equidistant, and it would jibe with a previous observation by the Columbia University chemist Erwin Chargaff that the amount of adenine in DNA seemed always to equal the sum of cytosine, as did the amount of guanine match that of thymine.)

Around and around wind the two strands of the rope ladder, with millions of As, Ts, Gs, and Cs inscribed (figuratively, of course) on either side. They occur serially, rather like the notes in a very lengthy symphonic score. This two-stranded coil of DNA, a sinuous molecule much longer than it is wide—twenty angstroms is only 50 trillionths of an inch across—is known, in its entirety, as a chromosome. There are forty-six, or twenty-three pairs, of these chromosomal rope ladders stashed in the nucleus of each human cell, and every one contains enough information in the form of nucleotides that if the entire sequence were written out in linear form, it would fill a thousand books the size of the Manhattan telephone directory!

Each chromosome is partitioned into several thousand genes, and both chromosomes in a pair are homologous; that is, they house genes for the same protein at corresponding sites along their length. A chromosome's genes are physically separated from one another by chemical punctuation and regulatory sequences that convey to the cell important information, such as how much to make of each protein, when to make it, and when to cease production. These regulatory snippets are crucial to a gene's harmonious function, it seems. They may even dictate idiosyncratic differences between people. For example, Kareem Abdul-Jabbar's height or Albert Einstein's intelligence may not have been due to their having genes different from everyone else's. Rather, it probably reflects a difference in the way the same genes were regulated, so that, in effect, they got a bigger dose.

Of the myriad attributes of DNA, one sets it off from all other materials. Alone among substances, DNA can reproduce itself. The process by which this happens depends on DNA's double helical structure and the mutual attraction of As, Ts, Gs, and Cs. It was Crick who first saw how the helical shape could facilitate self-copying. He suggested that during cell division the helices unzip, separating each chromosome into two discrete strands, leaving the nucleotide bases along the strands bereft of their complements. Floating loose in the nuclei of cells at any given time are millions of unattached molecules of A, C, G, and T. When one of these free nucleotides spies its natural partner along a DNA strand, it bellies up and bonds to it. It is a mating dance of elemental and prodigious proportions, for the millions of trolling bases manage to locate companions in the span of a thousandth of a second. On completion of this complex ritual, each strand has once again become two, with the freshly matched base pairs providing the rungs and adhesive for a new rope ladder. Forty-six of these micromiracles unfold with every cell division.

But reproduction is only half the story, for DNA exists not only to duplicate itself but to store the blueprints for making protein. It is essential that these blueprints not leave the nucleus of the cell, lest they get lost. The nucleus is, metaphorically speaking, a reference library whose volumes may never leave the premises. How, then, do the instructions for manufacturing protein get outside the nucleus, to the cell's cytoplasm, where the actual job of protein assembly takes place in tiny factories called ribosomes? The answer again lies in the double helical structure of DNA.

Sometimes the helix unzips only partway in order to call up a protein. The two strands separate to expose a limited, gene-sized segment of DNA. As before, free-floating nucleotides come flocking to bond with the exposed nucleotides; however, this time the nucleotides belong to a chemical cousin of DNA, ribonucleic acid, or RNA. At length, a complementary strand of RNA, corresponding to the nucleotide pattern embedded on the DNA template, is forged—or *transcribed.* This strand of RNA carries the message to the ribosomes in the form of a code.

The cracking of the genetic code was carried out in the 1960s at the National Institutes of Health by a group led by a shy, secretive genius named Marshall Nirenberg, who later won a Nobel Prize for the feat. It was a magnificent team effort, a quantum jump in the history of molecular biology that would enable humans to tamper at last with their own genetic clockworks, and point the way to medical treatments based on the transfer of genes. But the groundwork for the development of gene therapy had already been laid nearly twenty years earlier.

Even as William French Anderson was completing his high school career in Tulsa, notching such a phenomenal record that his teachers were forced to take his test scores off the grading curve to prevent a classroom mutiny, another precocious young man, just a few years Anderson's senior, was making seminal discoveries that would one day open the door to gene transfer. Joshua Lederberg had graduated from Columbia University in 1944, at the age of nineteen, and earned a doctorate in microbial genetics from Yale just three years later—a feat all the more remarkable for his having spent two of those years not working on his Ph.D. but attending medical school. Lederberg had fully intended to become a physician, but a three-month spell spent in 1946 assisting the brilliant Edward Tatum in a series of biochemical experiments on bacteria turned unexpectedly into his life's work.

Tatum was interested in extending to higher organisms the pioneering research that he and George Beadle had begun on bread mold. It may seem strange to think of bacteria as a "higher" form of life, but compared with bread mold, they are complex creatures indeed. To understand the working of genes, one does not want to aim too high, by tackling, say, a woodchuck. It clouds the issue to sift through too many genes at once. Hence, the living systems that twentieth-century geneticists have studied have necessarily been rather simple ones, organisms like viruses, bacteria, yeast, worms, and fruit flies.

Tatum and Lederberg made the breakthrough discovery that bacteria sometimes have sexual encounters. Not sex in any customary sense of the word, nothing that might cause the earth to move, but rather a primitive type of union called conjugation, in which a connection is formed allowing DNA to be piped from one bacterial cell to another. After all, take away foreplay, the cigarette, and the like, and that is the bottom line in conjugal sex. Specifically, the two scientists found that certain strains of the *Escherichia coli* bacteria, whose customary address is the human gut, fertilize one another to produce crossbreeds possessing the genetic qualities of both sires.

A powerfully built man with a bulletlike cranium, Lederberg went on to accept a position at the University of Wisconsin, where he joined forces with his wife, Esther, herself a distinguished geneticist, and a twenty-three-year-old doctoral candidate named Norton Zinder. The trio did not know it, but their emphasis was about to shift from bacterial sex to bacterial violence.

The investigators had another conjugation experiment in mind. They wanted to mix two separate strains of *Salmonella* bacteria, the kind that cause typhoid fever, in hopes that they would mate. Each strain they chose lacked an ability, possessed by the other, to make a specific amino acid required for survival. Neither could survive in a laboratory culture that did not provide the two amino acids as food supplements. However, reasoned the Lederberg team, if conjugation took place, at least some of the progeny of the parent salmonella should be able to grow in laboratory cultures that did not contain the amino acids.

To the scientists' delight, a number of offspring proceeded to thrive in the barren environment.

Yet, when a routine control experiment was carried out, the researchers got a shock. The two strains of bacteria were placed in a laboratory device called a U-tube, so that each specimen was isolated from the other by a porous glass filter that permitted fluid from the cultures to pass back and forth but no cells. The researchers found that the survival trait was transmitted from one strain to the other even when there was no sexual contact between the bacteria. How could this be? wondered the Lederbergs and Zinder.

They labored to explain the incongruity. Perhaps one strain had spontaneously mutated. But that was clearly not the case. Whatever was altering the DNA was present in the liquid in the U-tube, a filtered liquid, remember, that could contain nothing so large as a bacterial cell. Well, then, they reasoned, possibly loose fragments of DNA containing isolated genes had somehow floated through the liquid to the second strain and become incorporated in the bacterium's chromosome. But this possibility, too, was ruled out by pouring into the cell cultures an enzyme called DNase, which destroys loose DNA. Transmission of the survival trait continued unabated.

Intuition now came into play. Perhaps a virus was at work, for viruses are small enough to pass through the tiniest filter. And indeed, on examination, the donor salmonella proved to be carrying a virus called P22, which belongs to a predacious race of viruses called bacteriophages—eaters of bacteria. Phages make their home inside bacteria, where they use the host's cell machinery to create hundreds of progeny before *lysing*, or destroying, the host in a molecular explosion that shoots offspring out to infect new bacteria. Phages are patient as well as cunning. Sometimes they bide their time before taking over the cell, secreting their own genes among those of the bacteria until the moment to seize power. When that moment comes, when the host is weakened by malnutrition or some other factor, they reemerge from the DNA like a fifth column, cutting their genes free from the bacterial chromosome. In doing so, they are slipshod surgeons, often stripping away their bacterial host's genes while carving out their own. These pirated genes then become part and parcel of the phage's offspring. When the young viruses leave the cell to infect neighboring bacteria, they pass the expropriated genes on along with their own as they settle into the new DNA.

And that is exactly what was happening to the salmonella in Lederberg's laboratory. P22 was acting like a honeybee, conveying genes from one germ to another as if carrying pollen. Actually, it would be more apt to compare the tadpole-shaped P22 to a sperm cell. As the renowned Scottish molecular geneticist William Hayes has observed, "It is surely one of the more bizarre manifestations of evolutionary adaptation, that a potentially lethal virus should acquire the redeeming function of a gamete, rescuing some of its victims' genes for posterity."

What the researchers had discovered by accident, they now set out to duplicate on purpose. They knew that E. coli has genes for special enzymes that enable it to process milk sugar. What, they asked, would happen if a phage that had recently infected E. coli were put together with a strain of bacteria that lacked the capacity to break down this sugar, known as galactose? Would the second strain acquire the ability to digest galactose?

The idea seemed rather like voodoo, bringing to mind South Sea cannibals who believe they can acquire their adversary's courage by eating the unfortunate fellow's heart. Yet, sure enough, when the experiment was carried out, the galactose-intolerant bacteria suddenly began dining on milk sugar as if they were born for the job. The phage, of a variety labeled lambda, had transplanted or, as Joshua Lederberg called it, transduced E. coli's genes into the new bacteria in a way that permitted the migrant genes to continue to function in their new abode. In so doing, a genetically deficient organism had acquired a fresh gene in a way that was adaptable to laboratory manipulation.

Lederberg recognized the implications almost immediately. If viruses could transport bacterial genes, why couldn't they do the same for human genes? What if they could be induced to soak up genes that would be therapeutic to people and deposit them in the chromosomes of genetically deficient individuals? In a piece that he wrote for the magazine Participatory Evolution, Lederberg spoke of the future concocting of "genetic vaccines . . . viruses especially developed to carry correct genetic information to the body cells of patients with certain specific defects."

Other farsighted individuals began speculating, among them Lederberg's Rockefeller colleague Rollin Hotchkiss, who in 1965 wrote of a future when cellular "farms" would mass-produce genes, "delivering them up to viruses that can then profusely replicate them and infectiously transfer them to . . . sick patients."

At the forefront of the soothsaying was a newcomer to the National Institutes of Health who made no secret of his preoccupation with the idea of gene therapy. The young man, thirtyish and newly married, had arrived at NIH with a gold-plated résumé. Four years of biochemistry at Harvard, where his instructors had included James Watson. Two years at

Cambridge University's famed Cavendish Laboratory, working with Francis Crick, investigating the mechanism of the genetic code. Harvard Medical School, where he trained in pediatrics. Two years spent working with Luigi Garini, again on the genetic code, using bacterial models. When, in July of 1965, he joined the high-powered team that, under the guidance of the quietly determined Marshall Nirenberg, was deciphering the code, he was immediately penciled in to take over the bench of another youthful genius, Philip Leder, who would go on to become one of America's leading molecular cancer researchers. In the next nine months, the young man would so distinguish himself that Nirenberg selected him in April of 1966 to read the team's paper presenting their findings before two thousand peers. The young man's name was William French Anderson.

A visionary to the bone, Anderson has always prided himself on his ability to imagine the future. It is easy to dismiss him as the senator from Shangri-la, as many have done. But he has a considerable track record as a prophet to back him up. There was a day in Chicago early in 1968 when he electrified an international symposium on mental retardation by limning a picture of an impending medical Age of Aquarius. The agent of this epiphany, he asserted, would be—what else?—gene therapy. Delegates to the symposium, sponsored by the Joseph P. Kennedy Foundation, heard Anderson predict that the first attempt to repair genetic defects was merely a few years away. To an audience concerned with brain abnormalities, many of them genetically induced, the prospect was more than heartwarming. It was downright seductive.

As his listeners hung on every word, Anderson described how investigators had recently developed many of the techniques that would be necessary for genetic repair, and he noted that some manipulations were possible even then. This was an exaggeration. Researchers had yet to master any of the complex techniques for transferring genes from one organism to another. Still, Anderson presciently suggested that disarmed viruses might be enlisted to do the job. And, flying in the face of a fear that continues to haunt molecular geneticists today, he raised the possibility that gene therapy might be extended to all the descendants of an afflicted person by a kind of treatment known as germ line modification. Cells from the patient's testes or ovaries might be grown in tissue culture, he said, and receive the same implant of therapeutic new genes that the patient's other cells were undergoing. If the altered sex cells were thereupon returned to their owner, any sperm or eggs they secreted in the future would carry the same correction and liberate offspring from whatever family curse had driven the patient to seek treatment in the first place.

Anderson, with the bit in his teeth, hungered for a larger audience. In the summer of 1968, he submitted a paper to the prestigious *New England Journal of Medicine*. Entitled "Current Potential for Modification of Genetic Defects," the paper amplified on his prediction of the early advent of gene therapy and ventured that blood diseases such as sickle-cell anemia and beta thalassemia might be among the first human ailments to be so treated. But the *Journal*'s editor at the time, Franz Ingelfinger, considered the article too speculative and rejected it.

As sometimes happens, the popular press was quicker than the scientific media to sense the winds of change. In a 1967 piece entitled "Good Genes for Bad," *Newsweek* predicted that genetic diseases might eventually be treated "much the same way a mechanic fixes a car by replacing a burned-out spark plug." The article based its forecast on the work of an Oak Ridge National Laboratory biochemist and physician named Stanfield Rogers.

For years, Rogers had studied an organism called the Shope papilloma virus, which dwells in the cells of rabbits, in whom it induces warts and cancers. The Shope virus, which belongs to the viral family that also produces genital warts in human beings, has a singular peculiarity, as Rogers discovered. Through one of its genes, it can boost an animal's production of an enzyme called arginase. Arginase has a police function in the body. Its job is to hold in check blood levels of an important amino acid known as arginine. But the process had gone out of control in Rogers's rabbits. Spurred by the Shope virus, the rabbits were overproducing arginase and neutralizing arginine with wild abandon. Their blood levels of arginine were unnaturally low.

Rogers wondered, Could the same thing happen in humans? To find out, he drew blood from lab workers who regularly handled Shope viruses. He was stunned to learn that half of them had significantly lower arginine levels than people who had never been exposed to the virus. Rogers had stumbled on evidence that viruses could transform the DNA of not only germs but people. Interestingly, though, none of the lab workers was in ill health. Apparently, a high arginase–low arginine reading was not injurious to the body, nor did the virus itself seem to cause any harm.

It occurred to Rogers that he had potential medical dynamite in his hands. Here was a virus capable of delivering a functioning gene to man, a gene that would lower levels of a key enzyme and do it safely. Unfortunately, it was a gene that nobody seemed to need. There were no cases in the medical annals of anyone with dangerously elevated levels of arginine. Rogers later wrote, "It was clear that we had uncovered a therapeutic agent in search of a disease."

The irony did not prevent Rogers from telling *Newsweek* that gene therapy using viruses would eventually become reality. "I'm pathologically optimistic," said the Tennessee researcher, "so I'll estimate that we should be able to apply these techniques in the foreseeable future. I'd like to see it done in five years, but it may take fifty."

Not even Rogers, the optimist, suspected that the first attempt at human gene transfer was only three years away. Or that he would be the one to try it. In 1969, Rogers learned through Joshua Lederberg about two young sisters with grossly elevated blood levels of arginine who had been found living in the German city of Cologne. It soon became clear why nature has dictated that arginase hold the line against arginine. Arginine is critical to the process by which the body gets rid of ammonia. Normally, ammonia is converted to arginine and finally urea, which the body can then excrete. But if there is already too much arginine in the bloodstream, a paradoxical event occurs. Our systems sense it and refuse to make any more. The result is that we are unable to convert ammonia. It is like a thermostat placed too close to a heat vent. The thermostat keeps shutting off the furnace, making the rest of the house too cold. What happens to someone with too much arginine isn't pretty. Unconverted ammonia builds up to toxic levels and causes mental retardation, epilepsy, and spasticity.

At the age of five, the older of the two German girls was already suffering severe symptoms. The sisters plainly had two copies of the mutant gene. They were making no appreciable amounts of arginase whatsoever.

To Rogers, the karma must have seemed perfect. Just when he had found a cure without a disease, the disease had serendipitously come along. He got in touch with the girls' doctors, and, he has said, "with much trepidation but great hope [we] decided to try to stop the progression of the disease with the Shope virus." As a cautious prelude, they first added some of the virus to a culture made from one of the girls' cells. The results were encouraging. The arginase content of the cells shot up to something approximating normal levels. Rogers and company steeled themselves. It was time to step out on the high wire and try the procedure in the girls.

The experiment was crude. Lacking the technology to emasculate the virus genetically, they were forced to inject raw Shope into the girls' bloodstreams. Whether it was because the physicians were nervous about trying an unproven and potentially dangerous therapy upon children and therefore used too little virus to make any difference—the dose was one-twentieth the amount found to be harmless to mice—or because the therapy could not have succeeded under any circumstances, owing to basic conceptual errors, the treatment did not take. The medical team noted no effect whatsoever on the children's health. A year later, after a second, larger dose, the older child showed some tentative signs of improvement, but Rogers himself admitted these could have been due to the girl's special diet designed to control ammonia.

In the meantime, the experiment became public in 1971 and had a profound impact on Rogers's career. His peers rose to censure him, denouncing the procedure as premature. As a result, his funding soon dried up, and a chastened Rogers gave up his dream of gene therapy and turned instead to working with plant viruses. (As it turned out, Rogers's audacity was echoed a decade later by a UCLA researcher named Martin Cline, who would also be rebuked severely.) Nevertheless, some of the same authorities heaping criticism on Rogers went on to predict that gene transfer would one day have a place in medicine.

"In our view," wrote Theodore Friedmann and Richard Roblin in 1972, after condemning Rogers for being impetuous, "gene therapy may ameliorate some human genetic diseases in the future. For this reason, we believe that research directed at the development of techniques for gene therapy should continue." Friedmann, by the way, followed his own counsel avidly. A decade later, he was in the thick of the gene therapy race.

Even as the Rogers debacle unfolded, a paper underscoring the potential of gene therapy was published in 1971 by Carl Merril and colleagues at the National Institutes of Health. What the Merril group claimed to have done was "cure" human cells taken from a patient with galactosemia, a rare genetic disorder in which the victim lacks the enzyme for processing milk sugar. Newborns with this fault can suffer brain damage and even death if they are not quickly taken off milk. The cultured cells were exposed to lambda phages containing the E. coli genes for converting milk sugar—the same phages that Lederberg used to transfer the genes between bacteria. But now the virus was carrying the gene into human cells and, according to Merril, was causing the cells to make the missing milk sugar enzyme. The experiment seemed to reinforce the case for gene manipulation.

Was gene therapy just around the corner? Appearances to the contrary, it was not. A number of obstacles were standing in the way.

First, the avenues heretofore explored had severe limitations. The

Shope papilloma virus was a fluke. It defied probability that other viruses would be found that just happened to be carrying the right genes to correct even a tiny fraction of the four thousand human hereditary diseases. As for phages, they picked up only bacterial genes, and did so randomly at that. Real gene therapy would depend on the genetic engineering of viruses to take up the desired healthy human genes and ensuring that these viruses transferred the genes into living people safely and effectively. By any measure, the means for accomplishing this seemed far off.

Beyond these daunting hurdles was an even more fundamental stumbling block. There were not yet any genes to implant. In fact, in the early 1970s, no one had yet isolated the first human gene. And without genes, obviously, there can be no gene therapy. Boosters of genetic intervention were in the same position as Jules Verne, who in the 1870s was able to conceive of space travel in his book *From the Earth to the Moon* but could offer no credible description of how a space vehicle might work.

Development of ways to put new genes into genetically ill persons proceeded in fits and starts over the next two decades. But the real explosion in molecular biology over the same period came in the second realm—the realm of the discovery of genes. As of this writing, some 2,700 human genes have been mapped to their approximate location on the chromosomes, and slightly more than 1,000 have been isolated outright, or "cloned." The count contrasts sharply with the sparse number of genes that had been mapped in 1973, when the hunt was just beginning. At that time, scientists believed they knew the chromosomal address of only 219 genes, and had actually cloned none. Seventy percent of all genes then mapped were on the X chromosome, simply because it is the easier chromosome to map. If a trait is limited to males, as it is in red-green color blindness or hemophilia, then, as we have seen, the gene for that trait can be assumed to be on the X chromosome. Today, however, each chromosome has been extensively mapped, and several, such as the sixteenth and the twenty-first, have been almost totally explored.

Molecular biology has become as addicted to numbers and statistics as professional baseball, thanks primarily to one man—Victor McKusick, of Johns Hopkins University. Geneticist, epidemiologist, and cardiologist, McKusick is a courtly figure with a fringe of thinning white hair, a pair of trademark rounded spectacles, and a highly polished East Coast accent. In his long career, he has contributed to more than one thousand research papers and trained hundreds of younger medical geneticists. But he is by far best known for his labor of love, *Mendelian Inheritance in Man*—a catalog of all the genetic traits that have been detected up to the present time. The catalog includes not only genes that have been definitively mapped but human traits thought, but not proven, to be genetically based, for example, the tendency toward migraine. First published in a hardcover edition in 1966, the catalog is updated weekly by McKusick on a computer database. When the discovery of a new gene is announced, McKusick moves swiftly to confirm the finding and add it to his list.

As of June 1993, McKusick had inventoried more than 6,200 actual and putative genes, relying on reports in scientific journals. The number, which had grown by 500 over the preceding year, sounds impressive, but it amounts to only slightly more than 5 percent of mankind's total. By comparison, 99 percent of *E. coli*'s genome has been mapped, as has 90 percent of the genome of the lowly nematode worm. Definitive maps exist for many food plants, ranging from beans and papaya to lettuce and tomatoes ("Salad is virtually completely mapped," jokes Eric Lander of the Whitehead Institute). Although exploration of our genome has lagged behind that of other species, it is accelerating tremendously under the Human Genome Project, a government-funded plan to map and sequence the entire genetic endowment of man over the next decade and a half, at a projected cost of $3 billion.

As a physician, McKusick is interested in cataloging genes primarily for what they can tell us about disease. Nearly all the genes listed in *Mendelian Inheritance in Man* can be assumed to have both normal and abnormal counterparts. Any protein—be it an enzyme, hormone, or structural protein—can be faulty or missing entirely. If it is a critical enough protein, the result will be illness. Yet the effort to find disease genes is in its infancy. Of the 2,700 genes that have been mapped, barely 1,500 are known to be involved in disease. Among the more recent additions to the list, however, are the genes responsible for colon cancer, Huntington's disease, hypospadias, spinal muscular atrophy, Charcot-Marie-Tooth syndrome, hyperactivity, Duchenne's muscular dystrophy, myotonic dystrophy, amyotrophic lateral sclerosis (better known as Lou Gehrig's disease), cystic fibrosis, neurofibromatosis, and fragile X syndrome.

As laboratories raced to isolate and manipulate genes, the technology they urgently required was being invented almost overnight, just a step ahead of the need. It called to mind old cliff-hanger movies in which the hero, chased by natives, comes upon an abandoned biplane in a field and takes off in the nick of time. In 1968, molecular biologists first learned that bacteria are not utterly helpless in the face of bacteriophages, that they possess a diabolic set of weapons called restriction enzymes, which literally shred invading phages to pieces by slicing apart their DNA. It did not take long for scientists to realize they could exploit these enzymes, which will obligingly cut the DNA of any species. Each of the more than four hundred restriction enzymes recognizes a particular nucleotide sequence and makes its incision whenever and wherever it encounters that sequence. By carefully harnessing specific restriction enzymes, then, one can sever DNA at any desired point, allowing the insertion of alien segments of DNA, even DNA of other species. The first attempt to perform such gene "splicing" was not long in coming. In 1972, Stanford's Paul Berg, David Jackson, and Robert Symons used a restriction enzyme called Eco RI to put genes from the lambda phage and E. coli into a virus normally found in monkeys (called SV40, for simian virus). For molecular biology, this was the shot heard round the world. It was the first successful laboratory application of genetic engineering, and a decade later it brought Berg the Nobel Prize.

In 1973 came an even more astounding demonstration of this nascent ability to recombine DNA. This time, Herbert Boyer and Stanley Cohen, of Stanford, took DNA from a higher creature, the African clawed toad, and placed it inside the genetic material of an E. coli bacterium. They soon observed that the toad DNA was being copied and passed along as the bacterial cells divided. Each succeeding generation of E. coli received the same piece of toad DNA, creating more and more copies, like an assembly line gone mad. The science of copying genes, or cloning, had been born.

By 1975, another development had appeared, in the form of a chemical "glue" to go along with the "scissors" supplied by restriction enzymes. Biochemists had for some time known about a class of enzymes called ligases, the opposite of restriction enzymes in that they tie molecules together. They now learned of a particular variety that would mend a rip in DNA. Overnight, scientists no longer had to make overlapping cuts in their pieces of DNA to get them to recombine, as they had hitherto been forced to do. They could now fit segments together end to end, like sections of pipe.

With the rapid-fire advent of cutting, pasting, and cloning, all the tools were in hand for the isolation of genes. When one considers the eons that have been required for human technology to develop, it is extraordinary that so many innovations should have become available to molecular biologists in the span of only five or six years. It is as if fire, the telescope, and the car had all been invented in the same decade.

Remarkable or not, it is a fact that, by 1975, molecular biologists had essentially all the implements they needed to go about capturing that most elusive quarry—the gene.

Not surprisingly, 1975 was the watershed year when the first mammalian gene was isolated and cloned. It proved to be a gene for rabbit globin, and was identified by a quartet of Harvard researchers consisting of Argiris Efstratiadis, Fotis Kafatos, Allan Maxam, and Tom Maniatis.

As Harvard's team reported its deed, a brilliant achievement involving the "capture" of a snippet of RNA on its way to the ribosomes, William French Anderson was already chafing to move molecular biology forward into the treatment of disease. By that time, Anderson had written or coauthored nearly two hundred scientific papers, most of them dealing with his medical specialty, the blood. Specifically, they concerned the synthesis of hemoglobin, the molecule in red cells that transports oxygen through the body. There was the occasional whimsical side trip, wherein the restless Anderson indulged his passion for archaeology, as he did in his 1971 paper "Arithmetic in Maya Numerals," published in *American Antiquity*. But most Anderson contributions bore such arcane titles as "Translation of Rabbit Haemoglobin Messenger RNA by Thalassaemic and Non-thalassaemic Ribosomes" and "Hemoglobin Switching in Sheep and Goats: Change in Functional Globin Messenger RNA in Reticulocytes in Bone Marrow Cells."

Globin, the key protein in hemoglobin, was Anderson's forte, and he worked to unveil the secrets of its initiation, expression, and regulation at the molecular level. With others at NIH, he conducted lengthy investigations of the globin molecules and the complex families of genes that produce them. The effort culminated in 1977, when the team he was part of identified the human alpha globin gene, which was localized to chromosome 16. Landmark achievement though its identification was, it was not the first human gene to be isolated. Earlier that same year, Maniatis had bagged the gene for beta globin, the protein chain that is amiss in sickle-cell anemia and beta thalassemia (also known as Cooley's anemia).

Anderson was unusual among molecular researchers in that he had distinguished himself in a parallel career as a clinical researcher. Soon after leaving Nirenberg's group in 1967, he persuaded NIH to give him his own laboratory. In 1968, he created the hematology section and began the concerted study of hemoglobin diseases. Along the way, he was instrumental in developing drugs called iron chelators, for children suffering from beta thalassemia, a disease in which the red blood cells die at an accelerated pace, producing severe anemia. In order to give such youngsters the hemoglobin molecules their bodies cannot make properly, doctors traditionally have prescribed blood transfusions. However, new blood can overload their system with iron, endangering their liver and heart. Most thalassemic children now sleep each night while hooked to an intravenous pump that infuses them with a drug called Desferal (desferrioxamine), which is an iron chelator. A chelating agent is a molecule with high affinity for another chemical, in this case iron. When given, it combines with iron, leaching it out of the body so it may be excreted in the urine. Anderson pioneered the development of Desferol—work that earned him the 1977 Thomas B. Cooley Award for Scientific Achievement, given by the Cooley's Anemia Blood and Research Foundation for Children—and he enthusiastically champions the drug today.

But the lifesaving IV pump, besides being a pain to patients and their families, is a dangerously mixed blessing—serious side effects can lead to blindness and deafness, and demand the constant monitoring of patients' eyes and ears. Moreover, medical bills are colossal: more than $15,000 a year and rising.

Painfully aware that, in any case, such treatments were only buying his young patients time, Anderson the pediatrician yearned all the more to become Anderson the gene therapist. For all his extravagant talk of a golden age to come, the fact was that the science of gene replacement had scarcely progressed beyond square one. Yet and still, by 1975, he could stand the wait no longer. Abruptly, he dropped the bulk of his nonmolecular research and switched to the pursuit of gene therapy, handing three-quarters of his lab space over to NIH colleagues and declaring, to their amazement, that he intended henceforth to specialize in a field that, by the world's standards, did not exist.

Anderson's early gene therapy efforts proceeded slowly. The technology was too new and his training, long on theory but short on practice, left him ill prepared to spearhead the nascent discipline. To compensate for his deficiencies, he traveled to Cold Spring Harbor Laboratory, on Long Island, where Jim Watson was now director, for a crash course in genetic engineering. Then, in 1976, an opportunity presented itself that Anderson, a believer in such things, took for a sign. He encountered a Rockefeller University researcher named Elaine Diacumakos who had recently perfected the ability to inject soupçons of molecular material into cells by means of glass needles finer than human hair.

"When she showed me the technique the first time, it was fantastic," he recalls. "So I said, 'Have you ever tried it with DNA?' And she said she hadn't, but that she felt the stress on the cell would be such that it wouldn't work. I said, 'All we have to do is work out a better way to do it.' "

With that, Anderson began laboriously teaching himself how to make even tinier needles in his lab, needles small enough to pierce the walls of the cell nucleus yet large enough to let pass a loop of DNA containing a gene without tearing the DNA apart. The technique brainstormed by Anderson became known as microinjection. At a time when other methods of getting genes into cells of higher animals, such as the use of viral vectors, were still primitive, it seemed to offer a reliable way to study the results of gene transfer. There commenced a backbreaking, eyestraining period in which Anderson spent hours hunched like a jeweler over his microscope, set on a rickety table in his office, trying to improve his technique. By 1979, he had progressed to the point where he was able to inject the gene for an enzyme called thymidine kinase into a cell that lacked it. The cell began expressing the enzyme, thereby "curing" the deficient cell. For good measure, he also introduced the gene for globin, causing the cell to begin formulating the blood protein, albeit at a very weak level. These feats quickly earned him newspaper headlines—it marked the first time a gene had been microinjected into a cell so that it would express—and gave him his initial taste of fame.

Still, the recognition could not blind Anderson to the fact that the field of gene therapy seemed to be at a dead end. Microinjection was an impossibly inefficient way to deliver genes, and hence it held little, if any, clinical promise. More than half an hour was needed to alter a single cell, whereas treatment of an actual patient would require injection of healing genes into diseased cells by the billions. Moreover, the needle had a tendency to destroy the cell. Only about seven cells out of a hundred seemed to survive the procedure intact. Finally, Anderson's hope that the infant science might lead to the treatment of beta thalassemia and other blood diseases was dashed by the poor results he was getting. The beta-globin genes were expressing, but at a rate far below a therapeutic threshold.

By the end of 1979, a disillusioned French Anderson had all but hung up his gene therapy spikes. "I was just plain discouraged," he says in retrospect. In another of the quixotic shifts that have characterized his career, he turned his energies with a vengeance toward sports medicine and his favorite pastime, tae kwan do, almost abandoning gene therapy research altogether. But a series of events at Stanford University and elsewhere were eventually to bring him back into the fold.

3
A SIMPLE
LITTLE SYSTEM

If one measures a man by the stature of his enemies, French Anderson's chief nemesis, the MIT biologist Richard Mulligan, does Anderson proud. A molecular Sikorsky, Mulligan would finally open the door to gene therapy by designing the fleet of tiny vehicles, or vectors, used to airlift genes into the nuclei of diseased cells.

For Mulligan, the decisive moment occurred in 1979, when he was a twenty-five-year-old doctoral candidate in biology at Stanford. Tall, quirky, and something of a party animal, Mulligan had been regarded as an underage genius as an MIT undergraduate, a genuine accomplishment in a place where great gifts and eccentricities are taken for granted. Heading to Palo Alto for his graduate studies, Mulligan was determined to crash into the big leagues of biochemistry. That meant the stellar, tightly knit team devoted to DNA that had been assembled over thirty years by the master enzyme chemist Arthur Kornberg, a Nobel laureate and one of the founding fathers of the science. "Recombinant DNA was the most exciting discipline I could think of," Mulligan recalls. "It wasn't formally taught anywhere, but you could learn it by joining a lab and following the research interest of the group. So that's what I did at MIT and planned on doing at Stanford."

Mulligan already knew the mentor he wanted—Paul Berg, who himself had been Kornberg's postdoc, in the elite tradition that finds one Nobel Prize–winning biologist begetting another. Mulligan recalls how once he set his sights on Berg, he simply wouldn't stop pestering him. Hungry and determined, Mulligan worked his way into Berg's lab by the

time-honored expedient of constantly cornering Berg in his lair and begging, even camping outside and politely refusing to move until the gentle Berg finally let him in.

Mulligan wasted no time making his reputation. While French Anderson was back in his Bethesda lab muttering to himself and tediously
injecting genes by hand into cells one at a time, Richard Mulligan was
learning, in effect, how to paint walls with a roller instead of a tiny
brush. His 1979 experiment was so remarkable that it swept across biochemistry—and into Anderson's passionate awareness—with the speed
of juicy gossip. Mulligan's chimeric feat was to splice a rabbit gene for
globin into a monkey virus and then to unleash the virus on the kidney
cell of a monkey. When Mulligan did that, the kidney cell did something
that nature never intended any kidney cell to do: instead of urine, it
started making gobs of globin molecules, meekly following the orders of
the transplanted rabbit gene. The experiment marked the first time anyone had effectively moved genes from one species of mammal into the
cells of another, and by itself was enough to make the scientific world
take notice of young Richard Mulligan.

But in truth he had done much more. Mulligan had, in effect, pirated
the monkey virus and made it his slave. He had "got cooking," to use his
favorite phrase, and chemically carved out the virus's harmful replication genes from its midsection, as if scooping the seeds from the center
of a cantaloupe. In their place, he had inserted the rabbit hemoglobin
gene. And it worked!

While such a freakishly mutant cell couldn't live very long, it nevertheless represented something almost cosmic. This was no experimental union of irrelevant microbial genes. It was the forced handoff of a
medically important gene from one higher organism to another. Nor was
the transfer a lucky accident. It was achieved by the intentional use of a
carefully contrived miniature tank car.

Here was the ability to change life at its most basic, a mastery over
nature that for the first time could give scientists the means to discern
and manipulate the very molecular roots of disease. Mulligan, the archetype of the hot young molecular biologist, had boldly embarked on the
path that could, conceivably, lead anywhere and cure anything. Here
was power. Here was hope. And here, as well, was the kind of research
that scared the hell out of a lot of people.

Five years earlier, in 1974, the gene splicers themselves
had been jittery enough to call for a self-imposed moratorium on even
the simplest recombining of DNA. The warning, in fact, had focused on

the very lab Mulligan was working in, Berg's, which had been the first actually to recombine the genes of two different microbial species. Berg's creation, or "construct," as it's known in the trade, involved the marriage of two viruses—the bacteria eater, lambda phage, and SV40, the monkey cancer virus—that had opened up a new era in biology. But when word got out that Berg next planned to splice his hybrid into *E. coli*, the fast-breeding laboratory workhorse (which he called "the best understood living organism on our planet"), the notion of mixing a cancer-causing virus with a microbe whose home was the human gut frightened many colleagues. They implored Berg to hold off and petitioned the National Academy of Sciences. Things were moving just too fast, they said, demanding that a committee investigate the risks of genetic engineering.

Stung, Berg canceled the experiment. But true to his nature, he did more than that. He took the moral reins. First, he volunteered to chair the National Academy's committee, which imposed a worldwide prohibition on certain types of experiments until the dangers could be weighed. Next, the committee proposed that a permanent government committee oversee recombinant genetic research. The result on October 7, 1974, was the NIH's Recombinant DNA Advisory Committee, a unique body of scientists and nonscientists that became so renowned for toughness and fussiness that chagrined scientists have dubbed it "The RAC."

None of this happened in a vacuum. The public got wind of what the gene splicers were up to after a widely publicized meeting of 150 concerned scientists, sixteen reporters, and four attorneys amid the rustic splendor of the Asilomar Conference Center, near Monterey, California, in February 1975. The scientists came away from Asilomar recommending that experiments with recombinant molecules should continue, providing that biological and physical containments were followed. More meetings and headlines soon ensued, showing that by opening up their science to public scrutiny the gene researchers were determined not to suffer the guilt and self-hatred experienced by nuclear physicists after the top-secret building of the atom and hydrogen bombs. The biologists' self-restraint marked the first time that any major branch of science had halted its researches out of fear. Yet, noble or not, the actions were to elicit a rude response. Within a year or two, when the scientists completed the requisite safety studies and reassured themselves and federal officials that they had molecular biology under control, they were shocked to find that much of society didn't believe them. By going public with their fears, the splicers had inadvertently precipitated an "Andromeda Strain" public relations disaster, replete with doomsday scenarios of mutant strains of savage microbes escaping from labs and eating Earth or inducing unknown plagues in an unsuspecting populace.

The worries have never really gone away, though the furor quieted after federal safety guidelines were written in 1976 and the labs built special containment facilities at great cost. The biologists helped out by designing experimental microbes that were too puny to survive outdoors, but their efforts were not enough. Fears persist even though not a single environmental or public health accident, big or small, has been blamed either on molecular biology or on biotechnology in the thirty years since the field got cooking.

As a young bench Turk at Stanford, Mulligan watched the post-Asilomar wrangling of his elders with bemusement and didn't miss a beat. He just tended to his test tubes doing what he could legally do, cutting out viral genes and pasting them back together in new ways, determining which signals the viruses used to transcribe their messages, and how he could meddle more productively to get viruses to transduce, or transfer, genes into various types of cells. In fact, his 1980 doctoral thesis, "Transduction of Genes into Animal Cells," laid out the scheme that Mulligan has since pursued with remarkable single-mindedness.

Today, high dramas at the submicrocellular level are always being staged in Mulligan's lab at MIT's Whitehead Institute, though the performances must be imagined, and the critical reviews may take months or years to appear. But unlike clinician-researchers such as Anderson with sick patients waiting impatiently in the wings, pure researchers like Mulligan represent modern-day Magellans who are charting the vast molecular unknown of the human body. They are the polar and African explorers of our time, and their field is wide open (many of the "grand old men" of molecular biology are only in their forties) and ripe for magnificent findings. Biology, as the distinguished physician-essayist Lewis Thomas once declared, is one stupefaction after another.

The stupefactions of discovery all hinge on the same tools—gleaming glassware, plastic petri dishes, tissue culture flasks, electrophoresis gel beds, microcapillary suction pipettors, and endless racks of tiny, arrowhead-shaped polypropylene containers called Eppendorf microcentrifuge tubes that hold scant microliters of enzymes, viruses, and other solutions. Texts and reference books compete for valuable shelf space with Christmas-colored bottles of reagents and scarlet fetal calf serum. Kitchen microwave ovens and bartender trays of crushed ice are perched next to Vortex mixers, variable rotators, and machines costing upwards of $60,000 that can analyze several thousand nucleotide sequences a day, a labor that formerly took months. Other machines can mete out the four chemicals of DNA in any sequence required, constructing portions of genes to order.

As batteries of experiments are performed, boxlike machines softly hum, centrifuges whirl, and shiny incubators make no sound whatsoever, though their red lights gleam optimistically. Inside, countless billions of microbial lab assistants, *E. coli* bacteria, grow up cozily and happily with new genes spliced in, forming colonies, making foreign proteins, and dividing like mad in gamy little dishes of agar jelly that smell like a high school boy's gym locker. Other odors—of chemicals, disinfectant, coffee—hang in the air, as, against the pressures to make fundamental finds and beat their rivals into print, grad students and postdocs live in tune with their experiments, jiggling temperatures, juggling reactions, downing junk food and instant coffee, nodding and dozing over hardbound lab notebooks. As music plays on the radio, chemicals are silently catalyzing, absorbing, adsorbing, dissolving. Encouraging cell colonies are spotted, singled out, grown up in cultures, finagled with, tossed out in disgust. Regrown. Strange manual skills are in evidence as scientists sit interminably before big steel chimneys (sterile air hoods), their fingers darting in front of glass windows designed to protect the microbes from the scientist, not vice versa—"the latest in salad bar technology," one researcher dryly calls it.

The role of Mulligan or any other senior scientist in a virus vector lab is to conceptualize molecules, motivate students, evaluate the struggle, produce big discoveries, and hustle for more funds. He tries to avoid direct contact with his main experimental animal, the mouse—"We, um, don't relate," he murmurs. In fact, except in key situations when he gets his hands dirty, Mulligan practices what he calls "think science." He holds bull sessions to dream up concepts, checks data, regulates chemical recipes as students come to him for advice, prepares papers, and, especially, draws pictures of *circles*—infinitely complex circles with all sorts of chemical molecules spliced into them. Each ring represents a construct of a model gene delivery vector, an individual viral particle that has been so weakened and genetically engineered that it is not really a virus at all any more, but merely a cargo hold for the crucial DNA sequences that may make sick people healthy for the first time.

Viruses have been auditioning for the role of gene delivery systems ever since Joshua Lederberg proposed them in the middle 1950s. It is their singular talent to invade the cells of host organisms and set up shop. In fact, the very survival of a virus depends on how well it can infiltrate cells. As long there has been life on Earth, parasitic viruses have moved on wind and water, plants and people, spreading pestilence. Like Blanche DuBois, they have always depended on the kindness of strangers.

But when he started designing pathogens to deliver new genes, Mulligan discovered that ordinary viruses did not work all that well. They performed adequately enough in tissue culture but left much to be desired when he tried to dispatch genes to the cells of living mice. The reason is fundamental. Ordinary viruses, whose genetic core is composed of DNA, do not operate by worming their way into the nucleus of the host cell. Instead, they are content to merely hijack the cell's protein-making machinery and turn it into a virus factory, eventually killing the cell in the process.

Enter a peculiar and mysterious family of viruses—retroviruses—that don't work this way. Among nature's oddest entities, they insinuate themselves into the cell's nucleus, integrating with the DNA itself, thus becoming part of the host animal forever. Curiously, the host cell survives and even thrives, replicating normally while apparently unaware that it has been taken over by a foreign power. It is a superb evolutionary strategy for the virus, which no longer has the burden of finding new hosts to infect, and when the day comes that the retrovirus has multiplied by normal cell division to the point that it finally loudly proclaims its presence to the host immune system, it is too late. The virus has spread too far.

Not surprisingly, retroviruses are responsible for some dreadful diseases. Long known to cause cancerous tumors in animals and in the last decade discovered to be responsible for at least two rare human forms of leukemia and lymphoma, retroviruses have played a major role in the development of genetic engineering and in studies of the genetic origins of some cancers. But they were virtually unknown to the public until the early 1980s, when the AIDS retrovirus, HIV, was identified. In fact, that identification came about only because of decades of work with animal retroviruses by scientists of whom the public has rarely heard.

How do retroviruses perform their black magic? They can do it because nature has wired them up in reverse; hence their name. Their genetic information is in the form of RNA, not DNA, which means that in expressing their genes, they make DNA strands of themselves rather than the usual RNA strand. Using cellular and viral proteins, the DNA strand easily splices itself randomly into the host cell's own DNA, enabling it to hide out undetected and increase its numbers exponentially as the cell routinely divides.

Although these insidious traits have made retroviruses the odds-on favorite delivery vehicles for gene therapy, they still present problems, and tinkering with them demands the utmost in sophistication from specialists like Mulligan. Retroviruses possess only a limited space into which foreign genes may be spliced. This means that only the key DNA sequences—the protein-coding sequences and a few chemical signals to regulate them—may be inserted. Mulligan has spent years searching for the smallest "bite-sized chunks," as he calls them, that "are necessary and sufficient" to do the job.

Another major drawback is that retroviruses will infect only dividing cells, making them unsuitable for the treatment of diseases of organs such as the brain and central nervous system, where the cell populations generally do not divide.

However, the biggest weakness of retroviruses as gene delivery vehicles is their innate sloppiness. They insert genes willy-nilly into chromosomes. Genes end up everywhere without rhyme or reason, occupying environments that may be so alien to them that they refuse to work, or landing unceremoniously in some supersensitive spots along the genome. If they come to rest next to a gene that predisposes toward cancer, called a proto-oncogene, they may trigger the formation of a tumor. If they crash-land on top of a working gene, they may render it inoperative, and if this happens enough times it could halt some particularly important function of the body. These are remote possibilities, but possibilities nonetheless.

Despite these disadvantages, modified retroviruses have become the primary vectors for gene therapy.

Many retroviruses are responsible for cancer (in animals, for the most part), and in fact the retroviruses Mulligan works with are primarily tumor viruses; yet he can make the vicious things sound almost benign, in the manner of a gun owner talking about guns. "Retroviruses don't kill target cells like DNA viruses do," he says, bragging about his pets. "They often don't even make cells sick, and they'll get genes into lots of different kinds of cells. That's the trick—to leave the genes that allow the virus to act like a virus, but to remove the genes that let it spread. Those are dangerous genes."

Mulligan came to this arcane art circuitously, but has to admit that it probably was in his own genes. Born in New Jersey, Mulligan and his brother, Jim, older by three years, grew up in a scientific home. Their father was a professor of electrical engineering at New York University who moved his family to Rockville, Maryland, when Richard was in the seventh grade, in order to become a top official with the National Academy of Engineering. As a sixties student at Woodward High, Rich Mulligan was something of a terror.

"I was a yippie," he recalls. "Not a hippie, but a yippie. You remember Abbie Hoffman? I was a cynical, politicized, smart-ass type. Rather than being sympathetic to all the anti-Vietnam War causes, I was more into pointing out the absurdity of it all. We'd go down to D.C. and visit the North Vietnamese mission and rent atrocity films of American soldiers doing terrible things to North Vietnamese. Then we'd show the films in school. We also set up a big locker that had all the Communist books and propaganda. We were a bunch of cynical guys who spent a lot of time discussing matters with the principal. We weren't disciplinary problems, just real cocky and above it all.

"I loved science, though. I had a chemistry teacher who had a Ph.D., and I thought that was real cool. But biology was terrible. Absolutely. The gym teacher taught biology. I had no interest in that. Mathematics was my thing. My father really was a genius in mathematics, and he started teaching us early. Integrals, calculus—I just aced everything and always was the best in the class. They were fun puzzles."

In point of fact, Mulligan was almost lost to biology. He entered MIT to become a mathematician and found that the math courses there were, for him, just as easy. Too easy. "After a while, the wonderful puzzles began to get boring," he says. "I started looking for something else."

Biology suddenly started to seem germane. "It involved living things. Helping people. I never wanted to go to medical school, but I liked the idea of human biology. My ideas weren't well formed at the time, but I was tired of mathematics, and biology appealed to my curiosity. I was curious about how living things grow and develop, how the molecules work, and what goes wrong," says Mulligan, echoing the thoughts of a young William French Anderson almost two decades before.

Recalls Mulligan, "So I went around to MIT professors and said I wanted to do practical research in the field of gene translation—how the cell makes proteins from the messenger RNA. I probably thought of that because it was the first thing I learned. There's a system for doing that."

Mulligan approached many different scientists who weren't impressed—"It's funny now, because they're colleagues," he muses. Eventually, though, Mulligan wandered into the lab of Alexander Rich. There everything changed. Mulligan found Rich "unbelievable." He was a biophysicist, an x-ray crystallographer, a Renaissance man loose in an age of narrow specialties. "He dabbles in all kinds of things," says Mulligan of Rich. "But he's the reason that I got turned on to science. I still talk to him weekly and see him socially. He's in his sixties, but if I'm having a big party on Martha's Vineyard, he's the first guy I invite."

Mulligan was only eighteen at the time, still just a pup, but Alexander Rich threw open the portals. Mulligan could learn molecular biology in the lab, if he really wanted to. Rich assigned him to a brilliant post-doctoral fellow from Wales, Brian Roberts, who happened to be an expert in gene translation. The two hit it off immediately and enjoyed raucous times, which for Mulligan usually meant impromptu basketball games and beer.

"It was remarkable. I was given total respect in that lab, as if I really knew what I was doing," he remembers. "So I responded by working really hard all the time." He didn't take a lot of courses. "I didn't have to. I was working with these guys. They trained me totally. They were unique, and remain so today."

During this period, Mulligan and other fledgling genetic engineers were doing basic scut work—physically mapping the genes of viruses. "It was before DNA sequencing was possible, and very few restriction enzymes were around," recalls Mulligan. "We'd take a piece of DNA, and it would be the only piece we had access to, because we couldn't clone genes yet. The DNA would be SV40, because we were able to purify the virus particles. We'd cut up the piece and ask, What part of the genome encodes for what viral protein? Real basic stuff. But I began to get the idea that maybe we could put things into the viral DNA and use it as some sort of vector. That led people to tell me, 'Berg is talking about this kind of stuff. Why don't you go and see Berg?' "

Mulligan's persistence (and, he admits, a phone call from Alexander Rich) finally persuaded Berg to accept him as a doctoral candidate.

"He's my hero," Mulligan says simply. "Just fabulous. Exciting as hell, stimulating as hell, yet kind of laissez-faire. Everything I now do is basically from him: the notion of making things happen; doing the perfect experiment, not only the one that tells you something but the one that is also so pictorially beautiful that everyone will understand it; not just getting the answer, but making it obvious. Dramatic!"

In 1980, however, the same year Mulligan was completing his work at Stanford, and French Anderson was losing heart, an affair unfolded that threatened the future of the entire field of gene therapy. Down the California coast, at UCLA, a distinguished cancer researcher and blood disease specialist named Martin Cline was about to become the first researcher to insert new genes into living human beings since Stanfield Rogers's abortive efforts with the two German sisters a decade before.

Cline was a brilliant, Harvard-educated physician who could do it all, as the sportscasters say. His widely admired bedside skills, honed by years of caring for mortally ill patients, were matched, to an unusual degree, by a flair for dreaming up pioneering research. As chief of the hematology and oncology unit at UCLA through much of the 1970s, he led many important experiments, helping to trailblaze the techniques of bone marrow transplantation and refine chemotherapy for lymphoma and the leukemias. His résumé was as thick as a marble slab, listing more than two hundred scientific publications, and his grantsmanship was enviable, attracting to his laboratory more than $1 million a year in funding from NIH and other sources—testimony to the esteem in which he was held by fellow scientists who reviewed the grants. Now, in the mid-1970s, he astutely perceived that the future of medicine belonged to the manipulation of genes. "The next phase of research in my field was quite obviously going to be in molecular biology," he later recalled, "so I decided to learn something about it."

For a mentor, he chose a UCLA colleague, Winston Salser, a tall, bewhiskered biologist whose specialty was working with recombinant DNA. Salser's tutelage of Martin Cline began in the summer of 1976, and soon the restless Cline obtained a sabbatical from his department to study in Salser's lab. Cline needed little time to master the basic techniques of genetic engineering. He jumped quickly into harnessing them to his research interests. Initially, his goal was to identify the genes that go awry in certain forms of leukemia. His strategy was straightforward: whenever he isolated a new gene, he put it into normal cells to see whether the cells became leukemic. In the process, he gained an easy familiarity with the methods then in use for introducing genes into cells: microinjection, the technique that had so occupied French Anderson; electroporation, which involved shooting a small current into the cell membrane to make it accept incoming DNA; and calcium precipitation, a technique in which calcium phosphate is employed to induce the membrane to open windows through which the DNA can slip.

But the leukemia work was tedious. One afternoon, Salser put a pointed question to Cline. "What," he asked, "would be another important thing for us to do with these techniques?" Cline reflected for a moment or two and replied, "To put normal genes into the bone marrow of patients with sickle-cell anemia or beta thalassemia." Cline recalls the moment vividly, for the thought he had just uttered to Salser was soon to father the deed. "That's when it all began," Cline says of the string of events that over the next several years ended his useful life in molecular biology and badly tarnished his medical career.

Gene therapy for inherited blood disorders had been a private dream of Cline's for some time. But now the idea began to assume concrete form. Little more than a year had passed since Tom Maniatis, recently transplanted to Caltech from Harvard, reported isolating the beta globin gene, the offender in both sickle-cell disease and beta thalassemia. With the gene now in hand to work with, and the advent of workable, though inefficient, methods to transfer genes into blood cells, gene therapy for the two blood diseases suddenly seemed, at least by Cline's lights, to have entered the realm of possibility.

Aware that the same thought had probably occurred to others, Cline set to work getting into the game. Other activities were placed on the back burner as he and Salser began to insert genes into the bone marrow cells of mice. They utilized marker DNA to determine whether the genes had been taken up by the cells. These early forays into cultured cells merely duplicated work that had been accomplished at Columbia University and elsewhere. But soon Cline and Salser took the daring and innovative step of moving into live mice. Extracting bone marrow from the rodents, the two scientists treated the marrow cells with calcium phosphate, then exposed the cells to genes for the enzyme thymidine kinase (TK), taken from the herpes simplex virus and chosen for their ability to act as markers. Next the cells were returned to the mice, whose remaining marrow had been destroyed by radiation to make room for the altered cells. Now the trick was to tell how many marrow cells actually took up the viral TK gene. Here is where the TK gene served so well. As Cline knew from his cancer research, the enzyme blocks the action of the chemotherapy drug Methotrexate. Ordinarily, Methotrexate is deadly to bone marrow cells, but the viral TK enzyme functions like a bulletproof vest, making the cells resistant to the drug. To learn how efficiently the genes were absorbed by the mouse cells, the UCLA researchers simply gave the mice Methotrexate and then noted how many cells survived the bath. Cells that had not taken up the TK gene would die.

In Cline's view, the results seemed promising. A third to a half of the test mice gave evidence that some of their cells had been altered. But the change could be detected only in small numbers of cells, reflecting the low rate at which the calcium phosphate technique gets genes into cells. Moreover, the effect did not last very long. Under 15 percent of the animals showed expression of the gene for as long as several months. The transience of the results reminded the researchers that they had not hit many stem cells—the progenitor cells that give rise to all the red and white blood cells in the body and constantly replenish the blood-making system. Failure to get the gene into sufficient stem cells dooms the experiment, because any mature blood cells that take up a new gene eventually die, and expression of the gene dies with them. Convert a handful of stem cells, however, and nature becomes an ally. Not only are these unique cells immortal, but any new genes they acquire will be passed along to all the blood cells they go on to make. Or so most scientists believe.

In spite of the experiment's lackluster showing with regard to stem cells, Cline felt the time was ripe to move the research into human beings. In May of 1979, well before the results of the mouse experiments were published in *Nature* and *Science,* Cline and Salser applied for permission from UCLA to try gene therapy in people. Cline was notorious among colleagues for his bluntness and self-assurance. He was also known to be a very ambitious, competitive man, and it surprised no one that he should make a move to become the first authorized investigator to try gene therapy. Cline insists, however, that he felt justified on scientific grounds in seeking approval for a human trial. Says he, "The mouse experiments had gone on for two years, and we had done several hundred animals, with no observed toxic side effects. When I thought about doing people, I was relying on my background in chemotherapy, and my knowledge of how you introduce new therapeutic drugs and techniques. I envisioned that it would probably take at least a decade to perfect gene therapy techniques in humans, but I thought it appropriate to try the equivalent of a phase one study."

A phase one study, in the parlance of medical research, examines the safety of a new drug or procedure. Such studies precede tests of a new treatment's effectiveness, which are known as phase two trials. Cline says of his proposed phase one study, "We wanted to see if there was any toxicity connected with the procedure and to ascertain the tolerable dose level. We also wished to determine whether, in fact, the genes got in. Those were the objectives of the first experiment."

The Cline protocol proposed to give patients suffering from hemoglobin diseases healthy beta globin genes tandemed with viral TK genes. As in the mice, this would be accomplished by extracting marrow cells from the patient. But the cells would be subjected to a density separation procedure that would "enrich" for stem cells—in other words, assure a greater than normal number of stem cells in the brew. The cells would then be treated with calcium phosphate and exposed to the beta globin and TK genes for several hours before being returned to the patient intravenously. Later it would be possible to evaluate the rate of gene transfer by removing some marrow cells and treating them with Methotrexate.

It was a bold proposal that Cline and Salser put before UCLA's institutional review board (IRB), especially in light of the controversy surrounding recombinant DNA research that still raged five years after Asilomar. At a time when many critics remained unconvinced of the safety of placing recombinant genes into bacteria, here were investigators seeking consent to put such genes into *people*.

Devised expressly to deal with such issues, the IRB process has a moral authority that transcends the concerns raised by Asilomar. The impulse behind it stretches back to Nuremberg and the international soul-searching that followed the Nazi research outrages, and it was nourished by later infamies, such as the U.S. Public Health Service's notorious Tuskegee study in which black men afflicted with syphilis were allowed to go untreated, even after the introduction of penicillin, so that the natural course of the disease could be studied. The IRB at UCLA took its job of protecting human subjects very seriously, so much so that Cline's protocol sat dead in the water for months as committee members wrangled over it. Contributing to the delay was many committee members' unfamiliarity with the science of genetic engineering, a circumstance that even today rankles Cline. Never known for his tact, Cline says, "Most of the people on that committee did not have the expertise to review that kind of proposal. I'd have to say I didn't think they were terribly competent."

Whatever his opinion of his peers on the IRB, in September 1979, at their behest, Cline obediently altered his protocol to satisfy anxieties about the use of recombinant DNA. His experiment, he assured the panel, would not involve recombinant molecules. To be sure, the therapy required huge numbers of genes, making it necessary, prior to the actual procedure, to knit together unrelated DNA—to wit, the human globin gene, the herpes TK gene, and a loop of bacterial DNA called a plasmid, which would allow the construct to be copied millions of times inside bacteria. But Cline vowed that just before going into the patient he would snip apart the whole construct with restriction enzymes. The loose genes would still constitute a mosaic of human, viral, and bacterial DNA. But in the strictest sense they could no longer be called recombinant. Cline had given in. So far, he was staying within the lines.

The target disease for the experiment was originally to be sickle-cell anemia, a recessive blood disorder endemic to black people. One in 500 African-Americans suffers from the crippling disease, in which red blood cells assume an uncharacteristic crescent shape and, because angles now exist where there used to be curves, begin hooking onto one another instead of sliding inoffensively off. The ensuing gridlock in smaller blood vessels leads to swelling, severe pain, and oxygen starvation in important tissues. About half of all sickle-cell victims do not live to be forty.

But problems plagued sickle cell as a model for a first experiment. The Tuskegee study had made many African-Americans wary of research that was carried out exclusively on blacks. Moreover, the federal government had recently embarked on a screening program for sickle-cell anemia in which carriers of the gene were counseled not to have children, because of their one-in-four chance of conceiving a child with the disease. Some African-American activists saw the effort as a plot to hold down black reproduction in the name of medicine. In this highly charged atmosphere, Martin Cline had to consider whether he really wanted to single out African-Americans as guinea pigs for something as experimental as gene therapy. Eventually, Cline bowed to these sensitivities. "I thought it best not to start in sickle-cell disease," he says, "because I thought there might be complaints that I was trying an unproven therapy in black people. I thought it more reasonable to turn to thalassemia."

But there was trouble here, too. The Los Angeles area lacks enough beta thalassemia patients to sustain a large-scale trial of gene therapy. In pursuit of an adequate supply, Cline turned eastward, to the nations surrounding the Mediterranean. The disease afflicts Mediterranean ethnic groups: Italians, Greeks, and peoples of the Levantine coast. Its name, in fact, derives from Thalassa, the ancient Greek term for the Mediterranean. In early 1980, with his protocol still languishing before the UCLA IRB, despite his having given way on the crucial recombinant issue, Cline took a fateful step. In a series of letters, he initiated contacts with doctors he knew in Israel, Italy, and Greece. Would they be interested in collaborating with him on a gene therapy experiment that summer?

Two doctors responded favorably. They were Israel's Eliezer Rachmilewitz, chief hematologist at Mount Scopus Hospital in Jerusalem, and Italy's Cesare Peschle, hematology director at the University of Naples. Both men had access to an extensive number of thalassemia patients, and both were receptive to the idea of gene therapy, although the Cline proposals still had to go through the proper channels.

By performing the experiments overseas, Cline would be putting himself beyond the reach of American overseers. He denies, however, that such an end run was his motivation. "I wasn't escaping oversight," he insists. "It was suggested that I was, but I wasn't. I went to both the head of my department at UCLA and the vice chancellor for science and told them what I was going to do and why. It would have been very expensive and impractical to bring the patients back to the U.S. to do the experiment, so I told them I would do the patients there. The response was 'That's fine as long as you are not doing it to avoid university regulations.' "

That June, while his proposals hung fire in Italy and Israel, Cline flew to Naples to confer with Peschle and his colleagues. His baggage included a Styrofoam cooler full of test tubes packed in dry ice. Some of the test tubes housed beta globin genes. Others contained herpes TK genes. Still others held recombinant hybrids of the two. Cline was ready for anything. If approval came through from the Israelis or the Italians, he was equipped to proceed with gene therapy. If not, he would be able to perform lab experiments with his new collaborators to cement their fledgling relationship.

The Italians wished to chew on Cline's proposal some more, so the American researcher journeyed on to Jerusalem, where his wife and son joined him. There he was vigorously grilled by the Israeli equivalent of an IRB, the hospital's Helsinki committee, named for the city where an international accord was first signed forbidding unethical human research. The questioning continued for days, during which time Cline assured the committee that he would follow the same protocol he had put before UCLA authorities; in other words, he would use no recombinant DNA. This proved to be a crucial point in his winning approval.

At 7 A.M. on July 10, 1980, Cline received formal consent to perform gene therapy on Israeli soil. He and Rachmilewitz roared into action. Within moments the patient had been summoned to Mount Scopus Hospital, and by 11 A.M. she lay on an operating table.

Her name was Ora Morduch, and she was desperately ill with beta thalassemia. As a result of a fault in the way her genes coded for beta globin, her red blood cells were incapable of normal survival. Thalassemic people's red cells die at an alarming rate, causing the body to compensate frantically by stepping up the production of bone marrow. Excess marrow then deforms the bones and inflates the face into a grotesque, outsized, gargoyle's mask. Transfusions can allay some of the anemia, but they contribute to runaway iron overload, the detritus of billions of blood cells that have perished and overwhelmed the system's waste removal capabilities. The iron builds up in the heart, weakening it and causing wildly irregular heartbeats. In 1980, when iron-clearing chelation treatments were not in widespread use, beta thalassemia frequently killed its victims by their early twenties. Ora was twenty-one.

Cline says he was very up-front with the woman, informing her of the procedure's risks and, above all, its limitations. He would not be able to get the beta globin gene into many of her marrow cells, he said, and few, if any, of the cells he did hit would be stem cells. Most likely, the procedure would make no difference in her clinical picture. Nevertheless, Ora gave the go-ahead, figuring she had little to lose.

First, bone marrow was extracted from her hip. Then she was wheeled off to radiology, where radiation would kill some of the marrow in her leg to make space for the new cells. Cline, in the meantime, retired to the hospital laboratory to mix the patient's marrow cells with the beta globin genes he had brought with him. As he stood over the lab table preparing to treat the cell's membranes with calcium phosphate, he reviewed a decision he had made the night before in his hotel room after a lengthy internal struggle. Available to him were both the recombinant versions of the globin and herpes genes and the same genes in their chopped-up, nonrecombinant form. Why use the nonrecombinant DNA, he had asked himself, when his mouse data suggested he would get better expression of the globin gene if he used the recombinant material? Furthermore, when the genes were cut by restriction enzymes, it left them with sticky ends. This meant that as soon as they were placed into cells, the loose genes would latch onto one other again like Velcro. Since the genes were destined to recombine inside the patient's body no matter what he did, Cline considered the choice clear. He would go ahead and use his recombinant genes.

But he had promised everyone—the California IRB, the Israelis, the patient herself—that he would use nonrecombinant genes. To do otherwise was breaking his pledge. It was a decision he regrets. "I'm embarrassed about this part of the story," he admits. "I guess 'hubris' is the word that would come to mind. I thought the restrictions on recombinant molecules really didn't make much sense. I regret it now because it is one of the factors that has kept me out of the gene therapy field ever since. Still, intellectually the situation has not changed. We now know the recombinant material would have made no difference and was harmless."

Five hours later the cells were ready for reimplantation in Ora's body. As they dripped into her arm from a vinyl bag, Cline and Rachmilewitz watched nervously. But there were no complications. From start to finish, the entire process of gene therapy on Ora Morduch had taken eight hours. A day later, the procedure was repeated on the young woman.

For several more days, Cline remained in Jerusalem. He arranged for blood and marrow samples from Ora to be sent to him at UCLA in ensuing months for further study. Then he caught a plane for Naples, where he hoped to treat a second patient.

Italy in 1980 had no institutional review boards. Approval of human experimentation at that time was the province of departmental chairmen. Within days, Cline had won the necessary permission of Peschle's boss and undertook to treat a sixteen-year-old girl from Turin named Maria Addolorata at the University Polyclinic in Naples. As before, he used recombinant DNA but informed no one. And as before, the patient tolerated the treatment well.

Cline viewed the two experiments as only the first in a series of gene transplants that he would conduct as time went on. "I thought the experiments would go on rather quietly for a number of years," he says. "I would come back to the United States and treat more people there." As fate would have it, however, on July 16, the day after Cline conducted the Italian procedure, UCLA reviewers rejected his proposed gene therapy experiment, believing that far more animal work was needed. Cline learned of the rebuff upon his return from abroad.

Soon his woes multiplied. As word got out of his activities, he was greeted by an outpouring of fury from the scientific community, which viewed his actions as rash and unethical. Clinical researchers, in the prevailing view, must not be the sole arbiters of when history-making clinical research ought to begin in human subjects. As the furor raged, Cline was censured by NIH, which stripped him of much of his federal research funding and for several years afterward attached a report of its finding to his applications for financial support. Subsequently, Cline resigned as chief of hematology-oncology at UCLA, and he was forced by circumstances to abandon the field of gene therapy.

Leaving aside, for the moment, the ethical issues, what about the two young women whom Cline treated? Were they helped at all? Were they harmed? In the main, they were unaffected by the procedure. The genes seem to have passed into oblivion without any consequences, positive or negative.

Ora Morduch survived for twelve more years, the beneficiary of chelation treatments finally introduced in Israel in the early 1980s that rid her body of iron and took the burden off her heart. She died in 1992 of food poisoning that she contracted on a trip to Eastern Europe. Maria Addolorata, also a beneficiary of chelation, made a dramatic recovery and reportedly is in good health today. She continues to live in Italy.

Although his experiments produced no therapeutic results, Cline insists they yielded valuable data that were drowned out by negative publicity. "We continued to find the TK marker genes expressed in blood and bone marrow for at least four months," he says. "The experiments fulfilled my initial objectives, which were to show you could put genes into people and they would be retained for a while without toxicity. But when we went to write up the results, our paper was rejected by all the journals, some without review. So I've never been able to publish the results."

In his own defense, Cline contends that he did not know, as he prepared for the experiments that summer, that he needed permission not only from the host countries but also from UCLA to conduct research abroad. "I thought as long as I went through an appropriate human protection committee and followed the guidelines, that would be adequate. I thought I had sufficient approval. I would not have flagrantly violated the law." He describes himself as "naive at the time."

Whether he was an ignorant victim is open to question, however, since he did not observe the guidelines with respect to recombinant DNA. But he certainly has paid a price, losing his status as a marquee player in medical research. At the age of sixty, he remains a professor at UCLA and is active at the lab bench. But his focus has necessarily narrowed. He studies the genes that control the transition of leukemia from the chronic to the acute phase, a pursuit that is hampered, he says, by his lack of adequate funding. "There are others doing a lot better work," he says.

He talks sadly of his forced exile from gene therapy. "I couldn't work effectively in the field any more," he says. "And though I still do interesting things, that was terribly exciting to me. I was on a high intellectually back then, thinking of the possibilities of gene therapy and feeling myself well ahead of anybody at the time. It is very frustrating to see your career go up in flames."

To the molecular biology community, Cline remains a symbol of the dangers of scientific rashness. "The experiment was destined to failure," observed a highly respected biologist, Tom Maniatis, who gave Cline the beta globin genes for research, not suspecting that he wanted to put them in human patients. Other experts echoed the views of Columbia's Richard Axel, whose team developed the gene transfer technique Cline used. In published accounts, Axel said there was no scientific basis for expecting the experiment to work in people. On the recommendation of the President's Commission for the Study of Ethical Problems in Medicine, NIH moved to tighten federal control over the future development of gene therapy. It created the unique Working Group on Human Gene Therapy, under the aegis of the RAC, to provide a national oversight panel of scientists and nonscientists to screen the first human gene therapy proposals as they came along. Its name was later changed to the Human Gene Therapy Subcommittee, and it was slated to play a fateful role.

Gene researchers were tired, as the 1980s dawned, of reacting defensively to foes outside their community. But now, it seemed, they had to worry about insiders as well. Observes Paul Berg, "All we learned from Martin Cline's experiment was that it didn't work. But what didn't work? Experiments are supposed to tell us how to do the next experiment. His experiment didn't, and therefore it got creamed."

It was the mid-1970s all over again. Once more, the Martin Cline debacle held molecular biology's feet to the public fire for many months. Richard Mulligan just kept on playing with his viruses, unperturbed. He might have been Mendel working in his monastery garden.

It was time for Mulligan to start thinking about his postdoctoral studies. The usual practice was for young scientists to diversify, to try their hand at some different kind of research. "But I didn't want to do it," Mulligan asserts. "I was convinced that gene transfer was what I wanted to do. I lived and breathed gene transfer."

Still, he was torn. Very conscious of protocol and turf, he had a sense of guilt about continuing to center his efforts on SV40. "It was Paul's [Berg's] virus," he says. "I didn't want to step on my boss's toes. The average cocky graduate student thing is to say, 'Hey, I did the work here. That means I can go off and compete with these guys.' But I didn't feel that way. I had so much respect for Paul that I didn't feel that would be appropriate."

Apart from that, Mulligan had come to believe privately that the SV40 system was too limiting, that it would be incapable of the strategies that would be required by gene therapy. SV40 is a prodigious infector of cells. But after entering a cell, it does things the gene therapist would prefer it not do, such as make a million copies of itself before blowing up the cell and dashing off to invade again. Mulligan forced himself to start thinking about other systems. Was there a viral type that did not destroy the cell, that was not so zealous in seeking to replicate itself? It turned out there was. It was called the retrovirus, and as dray horses for gene therapy, retroviruses made sense to Rich Mulligan.

His gifts as a vector designer for Berg soon brought overtures from

MIT. Although his field is by no means crowded, Mulligan's major intellectual competition—at least in his view—came from Inder Verma, the quietly amiable East Indian virologist at La Jolla's Salk Institute. While devoting half his efforts to deciphering and understanding oncogenes, Verma, too, was doing pioneering work with the transfer of genes into retroviruses. He collaborated with Ted Friedmann's group at UCSD and trained such future players as his onetime postdoc Dusty Miller, now of Seattle's Fred Hutchinson Cancer Research Center, who later teamed up with French Anderson.

All during the 1980s, Mulligan and Verma played Alphonse and Gaston from coast to coast, respectfully topping each other with discoveries about which viral genes to leave in and which to take out in the race to make bigger, better vectors. Verma worried greatly about their safety, though, and expressed fears that they might be used prematurely. "It's a matter of personal feelings," he says. "A clinician, such as Ted Friedmann, will say, 'Look, this patient is fifteen. He's gonna die. He's got six months at best. What's the harm in trying it on him?'

"Well—and I'm afraid this may sound terrible—I think one should be able to die in grace, just as one lives in grace. It's the job of the clinician to save his patient. I'm not a clinician. I don't have the same feeling for patients. I'm very worried about all the things that might go wrong if we start experimenting on people too soon."

After Mulligan earned his doctorate, in 1980, he was corralled by MIT and returned east intrigued by the chance to work on the tempting family of retroviruses with the celebrated tumor cell biologist David Baltimore. "We were glad to get him back," Baltimore says of his young protégé. "He is a rare intellect in biology."

As a hot property, Mulligan suffered no shortage of sponsors. He soon landed a unique arrangement. As David Baltimore's boy, he got lab space by making SV40 vectors for Phil Sharp at MIT's elite Center for Cancer Research. In the meantime, thanks to John Potts, chief of medicine at Massachusetts General Hospital, Mulligan began earning the paycheck of a fellow in medicine merely for sitting with endocrine specialists and telling them about genes. "Before long, I was making more as a postdoc than anyone else, because I had this position as a medical fellow at Harvard. So I was working in Phil Sharp's lab for David Baltimore and being paid by John Potts."

The payoff for science came in 1983 when Mulligan, Baltimore, and Richard Mann, Baltimore's graduate student, presented the field with what would become the state-of-the-art way to deliver new

genes into cells. By then, Mulligan was well along in his work with retroviruses. To his delight, he had found they made a much better cantaloupe than SV40 did. Improving on his earlier concept, he extracted
most of the genetic material from one of his retroviruses and spliced in a
so-called marker gene—the gene for GPT, an *E. coli* enzyme. Mulligan's
goal was a vector that not only would take up the inserted gene sequence
and express it but would also allow him to tell which cells had taken up
the new sequence. If the marker could tell him that, he would end up
with a gene delivery system much superior to Martin Cline's fusion
method.

Any virus particle consists of a dense inner core of genes, with a piece
or two of protein wrapped around, a sort of protective jacket that also
lets the virus attach itself to the receptors on specific kinds of host cells.
"A bit of bad news in a protein coat," is how the late British Nobelist Sir
Peter Medawar aptly described them. By itself, a virus is essentially
inert, mostly a bundle of molecules that floats docilely in some bodily
fluid until it happens on an inviting cell. Then chemical signals flash,
switch on the viral genes, and command the particle to invade and
spread the infection like molecular wildfire.

A retrovirus's basic genetic material is packaged in a protective envelope of knobby proteins, known as glycoproteins. Like other viruses, a
retroviral particle drifts asleep in the body until it bumps against the
outer membrane of an appropriate host cell. If the cell does not possess a
receptor molecule that recognizes the glycoproteins on the retrovirus,
the cell will be immune to infection, the virus will just keep on drifting,
and the encounter will be like two ships passing in the night. But if the
envelope of the particle binds to the cell receptor, much as a key fits a
lock, the virus rouses. The RNA molecule enters the cytoplasm of the
infected cell and is reverse transcribed, or chemically flip flopped, into a
DNA form called a provirus, which squirms into a host chromosome at
random, then stitches the chromosome up again, neat as a pin. Stop-
start sequences are generated at the beginning and end of the provirus
invader—long terminal repeats (LTRs), these are called—which contain
powerful signals for expressing the retroviral genes. The LTR sequence
that governs transcription is called the promoter. Another LTR sequence, the enhancer, acts like a little supercharger and boosts the frequency at which transcription occurs. Next to this is another sequence,
known as psi, that provides the crucial chemical signals necessary for
the future packaging of the retroviral genes into an infectious particle.

Then come the three main genes that code for viral proteins and
whose names were whimsically donated to science by David Baltimore.

For no particular reason, he says, he tagged them *gag, pol,* and *env.* Gag codes for the RNA core of the virus; pol, for the reverse transcriptase enzyme; and env, not surprisingly, for the protective envelope. The simplest way to think about this is that all the genes are in the middle, and the chemicals that control everything—the "controlling regions" with their LTRs, promoters and enhancers—are at either end. Crucially, if one carves out the gag, pol, and env genes, the controllers do not seem to care much about what new genes have been plunked in between them. They will mechanistically go through their acts, reacting chemically and making things happen, as if the genes were bona fide viral ones.

When vector specialists set about making constructs for gene therapy, they start virtually from scratch; it's rather like taking a Swiss watch apart and rebuilding it so that it ticks but doesn't tell time. Scientists want to retain the outer shell of the virus, the envelope, so that it will be recognized by receptors on the target cell they're hoping to infect. They also want to keep the regulatory signals and transcriptase enzyme that let the virus copy its RNA into DNA and squirrel itself away in the host chromosome. But in the middle the new DNA provirus bears no resemblance to its former self. Its "dangerous" gag genes are gone, replaced by the gene that is to manufacture healing proteins to cure the defective cell. The very *last* thing scientists want is for their vector to make regular viral proteins and continue the life cycle of the virus. Or, even worse, to raise the possibility that it could link up with some other viral fragments that happen to be stored in the genome from previous infections or inherited from parents and reconstitute itself in infectious form.

Engineering starts at the DNA stage—the provirus itself—because the retroviral genetic material is very hard to work with in its RNA form. A virus of choice for Mulligan is the Moloney murine (mouse) leukemia virus, a cancer virus that kills mice and is a tough and powerful little beastie. Mulligan begins by gutting it chemically to remove the gag, pol, and env genes and splices in the key "bite-sized" chunks of the desired gene. He also may add a selectable marker gene, such as the neomycin resistance gene from *E. coli*, because it is easy to test for it later to see how many target cells have indeed been infected.

However, by gutting the provirus and inserting new genes into what essentially is becoming an extrachromosomal piece of DNA, Mulligan must now provide the vector with the psi sequence it needs to build a new package around itself to give it a body. He also must include the sequences necessary for integration and expression. Why is this so? Because without a shell able to dock with cell receptors, and the requisite enzymes and proteins to enter a target cell and be incorporated into the cell's genome, the virion (a single viral particle) would be unable to infect anything and the healing process would never get off the ground. But how to give a virus a shell—in other words, make it infectious—without allowing it to replicate dangerously?

Obviously, the gene products that have been deleted must be provided in some other way. Mulligan had been pondering a way to do this since his days at Stanford. In 1984, he and his colleagues at MIT revolutionized gene therapy when they finally figured it out. Mulligan's system has been adopted by all vector designers because it makes the vector safe to use.

What the MIT team did was create a kind of body stocking, or special "packaging" or "helper" cell, composed of a mouse cell that has been hollowed out and contains the proteins (not the genes) needed by any viral particle to build its own body shell, yet lacks the psi sequence that would allow the helper itself to package its own RNA and reproduce.

Packaging cell lines thus produce what are known as empty virions that contain no viral RNA. The helper is thus like a crippled prisoner who shows his buddies how to escape their cells but can't go along.

The MIT team members conjured up a two-part system for gene therapy: they stripped a retrovirus of its own genes, and into this empty vessel they tucked the genes they wanted to transfer. Then they adapted mouse cells to be reservoirs for key viral proteins that the cells would dole out as needed by the vector. The vector has at its side all the ingredients it needs to refashion itself into a fully clad virus and to continue its normal lifestyle.

Obvious, pictorially beautiful, dramatic—the system met all of Mulligan's criteria for great science. A living entity that was entirely artificial; a vector that could infect target cells mercilessly and deliver new genes to the nucleus; a virus that would then dutifully stop dead, like a truck that shuts itself off. It could go no further and cause no damage, because it lacked the genes to do so. The only thing left intact in the host cell's DNA was the corrected gene that the biologists hoped to transfer.

Instantly, the system was seized upon by every laboratory in the world that was interested in gene therapy. But Mulligan, only thirty years old, was not about to fold his tent and let others take over. Having facilitated the transfer of genes, he was determined to remain in the game as a player.

The problem he shared with all would-be gene therapists in those early days was that the roster of diseases to which the new therapy might be applied was disconcertingly short. The choice called to mind

Henry Ford's famous offer to purchasers of a Model T: they could have any color they wanted, so long as it was black. The only afflictions accessible to the new art were those involving the blood system and its manufacturing hub, the bone marrow. The primary reason was that gene hunters had barely begun to isolate genes. Thus far, the list was confined to those for globin and a few other blood products. Of course, marrow had ancillary advantages that made it uniquely suited to the limited state of the art. It could be easily removed for infection by the new retroviral vectors. Afterwards, it could be given back to the patient relatively simply with the reasonable hope that it would make its way unaided to its destination, the interior of bones, where it could be expected to form a new colony of healthy founder cells. This is in stark contrast to diseases affecting large expanses of fixed tissue—the muscles, say, or organs such as the liver, lungs, or brain—where cells could not be removed en masse for viral infection in the laboratory and where, even if therapists had a way to deliver genes to such remote sites in the body, the number of cellular targets was too vast to offer any chance of affecting sufficient quantities to make a difference. The final virtue of blood diseases as a model system for gene therapy lay in the fact that blood-making cells in the marrow divide with wanton abandon, and Mulligan's vectors, reflecting the preference of retroviruses for dividing cells, would naturally infect them avidly.

In the early 1980s, it appeared to Mulligan and others that, although the playing field was severely circumscribed, they still had room to maneuver. There are some three hundred incurable hemoglobin disorders, each of them theoretically curable by the insertion of corrective DNA. Gene therapy pioneers looked forward to tackling such killers as sickle-cell anemia, beta thalassemia, and alpha thalassemia. It seemed just a matter of getting the genes in for globin and standing back while nature took its course. But the researchers were about to encounter unexpected difficulties.

Protein molecules are not the long, flat, slinky affairs that some people might imagine. They are architecturally diverse, three-dimensional structures whose form tends to follow their function. No exception is the hemoglobin molecule, which assumes the shape of a rough cube. This cube consists of four linked chains of amino acids called globins, which fold in on each other to produce the geometry and, in so doing, offer points at which oxygen molecules can bond and hitch rides to the tissues where they are needed. Two of these identical chains, each 141 amino acids long, are called alpha globins, while the other two, 146 amino acids long, are known as beta globins.

An example of what can go wrong is presented by sickle-cell anemia. In the normal hemoglobin molecule, the sixth amino acid in the beta chain is supposed to be glutamic acid. But in a person with sickle cell, because of a mutation in the beta globin gene—the code reads GTG instead of GTC—the amino acid valine has been erroneously substituted at the sixth position. One amino acid wrong out of 574 does not sound like much. But the oxygen-carrying ability of hemoglobin is so critically dependent on its architecture that the subtlest change can wreak havoc. The amino acid change affects the bonding attraction of the cube's sides, and when the cube is not carrying oxygen, when its parent red blood cell is on its way back to the heart for more oxygen cargo, it tends to collapse in on itself into the sickle shape. Thus one tiny alteration in a gene is enough to ruin a human life.

The problem in beta thalassemia is more complex. There are more than fifty possible mutations or nucleotide deletions that can produce the same result, the premature death of red corpuscles. But for the gene therapist, deciding which of this bewildering array of errors applies to an individual patient is irrelevant. Getting a corrected version of the gene into the patient's red blood cells would, with one stroke, nullify the error, whichever one it was. Or so it would seem in theory.

But it is not that simple. Nature has played a trick on those who would dare usurp its power. In the hemoglobin molecule, as we have seen, the cube is composed of alpha and beta chains, which must be made in precisely the same amounts, lest the sides become of different lengths, causing the molecule to become unstable and incapable of carrying out its oxygen-binding functions. But these chains, the amount of whose production is so crucial, are actually assembled by genes on two different chromosomes, the alpha chain by genes on chromosome 16 and the beta chain by genes on chromosome 11. It is as if one pant leg were made by Hart Schaffner & Marx, and the other by Levi Strauss. Somehow, the body seems perfectly able to coordinate the production of these geographically detached genes. But the novice gene therapists of the early 1980s were dismayed to find that they were not so adept. Try as they might, working in the test tube and in mice, they could not duplicate the precise regulation required of newly inserted beta globin genes necessary to precisely equal the amount of alpha globin protein being made by the cell at any given time. It was, in short, a mess.

There was a further obstacle beyond regulation. Despite repeated attempts to target their gene payloads to the stem cells in bone marrow that replace and replenish the blood system, gene therapists had been thwarted by their basic inability even to distinguish stem cells from any of the other myriad blood cell types floating around in blood and marrow. And if they couldn't see them, they couldn't target them. Mulligan had hoped he could get around this impediment by making his viruses so effective, so aggressive, that they would infect all the cells in a bone marrow sample, thus ensuring that if any stem cells were present, he had to hit them. It was the shotgun approach. But such proficiency eluded him at first. Later, when he had mastered the art of getting genes into 100 percent of cells, he still encountered problems. Stem cells were not the compliant factories he and others had hoped. They were downright recalcitrant. "We ran into a lot of problems getting the genes to work after they arrived," says Mulligan. "We had tremendous difficulties in getting long-term consistent expression of the transferred genes into new cells. By the mid-80s, it didn't look like we were going to be able to cure very much."

It is characteristic of Richard Mulligan that, instead of leaping ahead two squares, as French Anderson would later do, or dropping the blood diseases altogether and moving, with fickleness, on to something else, he persisted, like the inveterate laboratory drudge he is, in trying to understand what was eluding him. This is the way of the basic researcher, the only way that Mulligan knows.

Observes his admirer David Baltimore, "I think if you deeply examine the divisions in the gene therapy field, you will sense the different sensitivities among physicians and biologists. Only the physicians must deal directly with desperate patients."

Yet in an interview a few years ago, Mulligan admitted that he once came very close to trying a human experiment on his own. It happened when he was at Stanford and ran into a buddy, a former MIT classmate and fellow biologist, Richard Parker, who was suffering from blood cancer. "He was a good guy," Mulligan says. "He was one of us—he went to Caltech and worked with Tom Maniatis, followed by a stint in Mike Bishop's lab at UCSF as a postdoc. Richard came down with a T-cell tumor so aggressive that he was given only six months to live. A bunch of scientists got him into a treatment program at Stanford that was experimental, but looked promising."

When Mulligan bumped into Parker, "he was terribly sick." He says, "We rekindled our friendship and kept in touch all the rest of his life, which ended in 1988. Parker was super-courageous and very interested in his own cancer chemotherapy.

"He mentioned that he was going to have an autologous bone marrow transplant and that the difficulty with those things is that you never know why they don't work. When you take cells out of the patient, you must purge them to knock off any residual cancer cells, then blast the patient with radiation and reimplant the cells. Then you hope they engraft."

Parker and Mulligan came up with the same idea. "If we were to mark his marrow cells with bacterial genes when they were in the lab, and then reintroduce them, we might be able to track them. Richard volunteered to be the patient."

Torn by his friend's plight, Mulligan called his scientist acquaintances for advice, and Parker called his friends, and both asked whether such a procedure was possible. The general opinion was that it was. But the ethical climate had not reached the state where such a gene transfer would be permissible.

"I listened to George Santos, who's chief of the transplant unit at Johns Hopkins and a very wise man," Mulligan recounts. "He said, 'Richard, we just can't do it. Not yet. It's not ethical, even though it would tell us something very important for future treatments.' I deferred to him.

"But," said Mulligan at the time, "I wouldn't be surprised to see a bacterial gene tagged to an autologous bone marrow transplant soon."

On September 9, 1991, a youngster at St. Jude's Children's Research Hospital, in Memphis, suffering from an abdominal cancer, neuroblastoma, was given a marker gene in just such a transplant. Society, as Mulligan had predicted, had changed its mind.

But his anecdote illustrates something that even he didn't seem to realize. When Mulligan was personally touched by a dying friend, he was willing to try anything to help him. Physicians such as Anderson and Rosenberg routinely find themselves on the horns of that dilemma, and the horns are too sharp to be ignored.

Although he did go on to experiment with many other things, Mulligan has never stopped working out the basic biology of gene regulation in the hemoglobin disorders. Quick fixes and being first to do gene therapy, he insists, were never part of his plan. "I was staking my professional career on the development of this new type of therapy," he says. "I wanted it to be done right—and that means basing my findings on solid, repeatable, predictable science."

Nonetheless, there is no denying that back in the middle of 1984, even though he then had the gene delivery system that Martin Cline had lacked, Richard Mulligan still could do nothing about blood diseases, or about any other ailments, for that matter. Like W. French Anderson before him, he had hit a stone wall.

4

THE HORROR

If the East Coast seemed to be littered with thwarted gene therapists, a team out west was rehearsing resolutely for the big event. It had a horrible disease, had cloned the responsible gene, and, heaven knows, had no shortage of desperate families willing to try anything.

There are few afflictions worse than the Lesch-Nyhan syndrome. Imagine the subtle urge that sometimes makes you chew your fingernails. Now amplify that feeling a millionfold, and you have a sense of the bizarre compulsion for self-destruction that bedevils two thousand American families that must live day to day with sons suffering from an enzyme deficiency that crushes their bodies and minds.

Such boys must remain immobilized all the time—their hands and feet strapped in restraints—lest they mutilate themselves or those who love them. Totally dependent, often mentally retarded, they are nonetheless keenly aware of their plight. "All these children have bright, understanding eyes," wrote William Nyhan, the California physician who discovered the disease in 1962. "They relate unusually well to people and are felt by those closest to them to understand everything that is said."

Lesch-Nyhan syndrome probably has been around for millennia and may account for folktales about boys whose ranting, vomiting, thrashing about, and uncontrollable biting behavior was attributed to demonic possession. How incongruous it seems that some of the best research into such an ancient and violent sickness occurs in a setting as lush and lovely as the oceanside campus of the University of California at San

Diego. But that is largely due to Nyhan. A Johns Hopkins transplant who followed the sun in 1969 to become chairman of pediatrics at UCSD, he has helped a lesser-known branch of the University of California gain a worldwide reputation for genetics research and treatment. As perhaps medical science's premier genetics detective, Nyhan routinely diagnoses and manages the most complex biochemical disorders of childhood—enigmatic maladies so rare that they may afflict only 1 child out of 120,000. Obviously, most physicians will spend their entire careers without ever seeing such a mystifying youngster, let alone have the foggiest idea how to care for him. That is why parents who lose in nature's lottery often move to California to be near Bill Nyhan; why children are sent tens of thousands of miles to see him; and why, despite his renown, he remains the sort of old-fashioned pediatrician whom frantic mothers rouse from sleep at three in the morning.

"The kids I treat are the kind that a lot of doctors don't like to see, because they get so sick," Nyhan says. "It's not unusual for a child with a disease like propionic acidemia to go to bed perfectly well, and wake up at death's door. That's why you can be in the business as long as I have and people are still waking you up in the middle of the night."

After Nyhan and one of his medical students at Hopkins, Michael Lesch, identified the syndrome that bears their name, the enzyme deficiency that causes it was pinpointed by Nyhan's colleague J. Edwin Seegmiller in 1967. But it wasn't until 1982 that new genetic-engineering techniques allowed the scientists (and another group at Baylor University) to isolate the gene responsible for the defective enzyme from the surrounding material on the female X chromosome. This is a scourge that mothers unwittingly pass on to their sons.

With the gene in hand, and with a waiting list of eager patients scattered around San Diego, a UCSD research team led by Theodore Friedmann, assisted by the virologist Inder Verma at the nearby Salk Institute, spent the next half decade trying to perfect gene therapy for one of medicine's most complicated syndromes.

Born in Vienna in 1935, Ted Friedmann attracts honors like a magnet. Elected to Phi Beta Kappa at the University of Pennsylvania, awarded a Guggenheim Fellowship, plus teaching and research fellowships at Harvard, where he received his medical training, followed by fellowships to Cambridge (to study biology) and Oxford (pathology), Friedmann comes across as a no-nonsense type. Thoughtful, serious, something of a scold, the slightly built pediatrician-geneticist radiates a sternness that seems to border on hauteur. With interviewers, Friedmann often plays the bad cop to the amiable Nyhan, but there is no denying his status as a virtuoso researcher and leading gene therapy theorist—albeit one who nearly drowned while dog-paddling upstream against an abominable disease. During the mid-1980s, Lesch-Nyhan was widely regarded as the first disorder likely to be attacked with gene therapy, and Friedmann doubtless felt many times that he was carrying the entire weight of West Coast genetics on his narrow shoulders as he swam. But while French Anderson, Richard Mulligan, and other eastern competitors were stymied by the regulation of globin genes in blood diseases, Friedmann and the Lesch-Nyhan team were confounded by the problem of what organ to target with healing genes.

Nyhan discovered what he calls "the special plague" when a four-year-old boy named Michael was brought by his parents to the emergency room of the Harriet Lane Home, the children's wing of the Johns Hopkins Hospital, in Baltimore. "Young Michael's immediate problem was blood in his urine," Nyhan recalls, "but he had a lot of other problems as well." Nor, as it was to turn out, was he the first child in the family to be so stricken. Like his older brother, Edgar, before him, Michael had been born normal, but started to deteriorate before his first birthday. At about eight months, his mental growth had just stopped. The strength in his sturdy arms and legs faded; they became flaccid. Eerie spasms wracked his body. He screamed in torment from attacks of arthritis. Michael's parents had been through all this before. They took the infant Michael, like Edgar, from doctor to doctor—each of whom was stumped. The doctors performed test after test, shook their heads, shrugged, and told the parents their son had cerebral palsy and they didn't know why. At home, even sleep brought no relief to the tortured family, which had long ago learned to dread nightfall. Evenings filled their baby boys with terror. When the children's mother prepared Michael for bed, he acted as if he were scared of his own naked body—as if somehow it posed a threat.

Then, about the time he started to teethe, strange new feelings appeared to sweep over him. He seemed compelled to bite himself—to chew at his lips, to gnaw at his fingers, and even to lash out at those who tried to comfort him. Michael, like Edgar, got worse and worse, becoming so quaky with palsy that he could neither sit nor stand without assistance. The mystery deepened when bloody urine brought him to the hospital in 1962. A meticulous young intern on duty prepared a urine sample, examined it under a microscope, and was startled to discover that he could spot no germs at all, no bugs that could be causing Michael's acute infection. However, as the intern could plainly see, there was something else—something bizarre. Michael's urine was teeming with strange, sharp-edged microscopic crystals.

Well, reasoned the intern, jagged crystals can tear tissue and cause the bleeding that sometimes appears in urine. Maybe that was it. The intern reported his suspicion to his resident, and the two of them dug out a book that let them identify crystallized body substances on the basis of their appearance. Michael's crystals looked very much like those of cystine, one of the twenty amino acid building blocks that come together in long chains and bend and gracefully fold in particular structures to form the countless proteins the body needs to stay alive. But cystine had no business being in Michael's urine. Its presence indicated that he suffered from a disease called cystinuria, a kidney disorder in which excess cystine is secreted and forms stones and crystals in the urinary tract. Cystinuria is one of the most common causes of crystals and stones in children.

Next morning, to confirm their diagnosis, the intern and his resident toted Michael's urine sample and microscopic slides to Bill Nyhan's laboratory. They knew that Nyhan, then a thirty-seven-year-old assistant professor of pediatrics who had earned a doctorate in biochemistry as well as in medicine, liked puzzles. In fact, he was becoming famous for solving them. Boston-born, Nyhan attended Harvard until joining the Navy during World War II. After the war, he earned his medical degree at Columbia University and then went to the University of Illinois as a graduate student.

"I got into research after I became a pediatrician," explains the lanky man, whose laconic manner reminds one vaguely of the late actor Gary Cooper. "I always had been interested in scientific investigation, but wasn't sure exactly what I should investigate. The answers probably could come through learning biochemistry, so I undertook a fellowship that kept getting me in deeper and deeper. The man I studied with was working on the biochemistry of tumors, so that led me to cancer research. I was seeing patients, too, and the idea that there were people with problems that could be solved in a chemical way was irresistible to me."

After serving his internship and residency at Yale University and the Grace–New Haven Community Hospital, Nyhan got a job at Hopkins. He abruptly immersed himself in what was to become his life's work—genetics—when he tried to save the life of a comatose baby boy whose blood held far too much glycine, another amino acid, the presence of which indicated an intolerance to certain constituents of protein. "But which ones? That was my problem," Nyhan recounts.

To solve it, he painstakingly isolated each of the twenty amino acids that make up all proteins and fed them to the baby one by one, checking the reaction. Within weeks, he diagnosed the child as the world's first known case of what is called propionic acidemia—an incredibly rare inherited metabolic disorder that results from the inability to digest the amino acids threonine, isoleucine, and methionine, which, like most amino acids, are commonly found in food. When these amino acids built up in the baby's body, it was unable to dispose of them. The result was cataclysmic—acidosis, lethargy, mental and physical retardation, coma, and, despite Nyhan's best efforts, death.

This may not happen, Nyhan eventually discovered, if a baby is put on a rock-rigid diet low in protein and given massive doses of vitamin B_{12}—a chemical juggling act that permits digestion and, hence, growth. Such pioneering therapies have since become standard for babies suffering from a host of similar genetic metabolic disorders. The best known is phenylketonuria, or PKU, in which a child's intellect will be destroyed unless the child is put on a diet low in phenylalanine, a common ingredient of dairy products and certain other foods. Feeding such children is tricky. A baby cannot survive without a certain amount of protein, yet these youngsters will die or become mentally disabled if they get too much. One must vigilantly control their diets to determine the minimum amount of protein they need to develop normally.

Nyhan's specialties, perhaps because they are so uncommon, have jawbreaking or vividly descriptive names: argininosuccinic aciduria, citrullinemia, ornithine transcarbamylase deficiency, maple syrup urine disease—the list goes on and on. Citrullinemia and ornithine transcarbamylase deficiency, for instance, are conditions in which waste ammonia builds up in the body. In all such disorders, a malfunctioning gene permits toxins to accumulate in the bloodstream and in the brain, causing mental and physical devastation. Formerly fatal within days, if not hours, such diseases often may be controlled by tightrope-walking regimens of diet, vitamin therapy, and antibiotics. But children with such a disease are, "walking time bombs." Only good new genes would truly cure them.

No disease, however, is worse than the one Nyhan uncovered after Baby Michael was admitted to Hopkins with cystinuria. By that time, Nyhan and his protégé Mike Lesch were deep into the study of disorders of amino acid metabolism. "When they brought the boy's urine to us," Nyhan recalls, "they said, 'Prove to us that this kid has cystinuria,' because that's what the crystals looked like. So we ran the assay, but he didn't have excess cystine at all. Nor did he have too much lysine, ornithine, and arginine, the amino acids that normally accompany cystinuria. His urine showed only normal amino acids."

That was indeed puzzling, even to Nyhan.

"The next step," he says, "was to identify the crystals chemically. They turned out to be crystals of uric acid. And that was a sure sign of gout."

Gout? In children? That didn't make sense. Gout usually affects middle aged men whose bellies bear witness to their prosperity and love of rich food. For inexplicable reasons, gout is often viewed as humorous, characterized by a cartoon victim with a swollen big toe. "But there really is nothing funny about gout," Nyhan observes. "It hurts like hell. Crystals of uric acid settle in the joints and cause excruciating pain. If deposited in the kidneys, the crystals may bring on renal failure and death."

Uric acid is a molecule that removes from the body excess nitrogen bases called purines, which are involved in the synthesis of genetic material, DNA and RNA. Like everything else, uric acid is itself subject to breakdown and excretion in normal metabolism. Yet the concentration of uric acid in the blood of Baby Michael was very high, and that in his urine was four times greater than normal. Ironically, before Nyhan pieced together this pediatric picture of gout, uric acid had already been linked to mental performance, but in a far different way. Many famous men suffered from gout—Alexander the Great, Benjamin Franklin, Isaac Newton, Charles Darwin, Martin Luther, John Calvin, and John Milton. Studies in recent years have even equated high performance levels and success to slightly elevated levels of uric acid.

None of this applied to Michael or his brother, of course. They had gout because they had been born without a gene for an enzyme whose absence causes uric acid to accumulate within the body. Enzymes—the enormous class of protein molecules that operate as catalysts, speeding up and slowing down the countless chemical reactions that occur every instant of our lives—are requisitioned constantly by the DNA code. They regulate the rate of all metabolic processes, doing everything from transmitting nerve impulses to assembling genes. Life cannot exist without enzymes, and the most minuscule defect in one of them can wreak unbelievable havoc, as Lesch-Nyhan syndrome so clearly demonstrates. All of us have the functioning Lesch-Nyhan gene in every cell of our bodies. Indeed, it is so ubiquitously active that it is called a housekeeping gene. The enzyme it makes has an impossibly difficult, but to biochemists precisely descriptive, name—hypoxanthine guanine phosphoribosyl transferase (HGPRT).

Although every one of our 100 trillion cells produces some HGPRT, the genes in the brain produce fifty times more than those anywhere else. So the brain relies heavily on this housekeeping gene. Elsewhere, with too little HGPRT, the buildup of insoluble lipids, or fats, can cause arthritis and kidney damage. Its near-absence in the brain leads to retardation, palsy, and violent behavior.

The normal function of HGPRT is to salvage purines that would otherwise be adrift in the wastes of uric acid, saving them for the production of new DNA and RNA. "In the absence of HGPRT," Nyhan says, "the body machinery gets cranked up and oversynthesizes purine; some twenty times more uric acid than normal is produced." Excess production of uric acid, however, cannot explain in itself the bizarre features of the Lesch-Nyhan syndrome.

Since the 1960s, gout has been effectively treated by drugs, such as allopurinol, that control the formation of uric acid. In fact, the use of modern pharmaceuticals to prevent the accumulation of a metabolic product leading to clinical disease is best exemplified in the treatment of gout, Nyhan says. Symptoms result only after prolonged accumulation of high concentrations of uric acid in the body fluids. Therefore, the long-term goal of gout therapy must be to rid the body of excessive uric acid, not merely to treat gout only when the joints ache. To the biochemist, allopurinol works because its chemical structure was designed in the lab to resemble the enzyme hypoxanthine's. That, in turn, inhibits the action of the enzyme xanthine oxidase, which produces uric acid from hypoxanthine and xanthine.

"Allopurinol is a remarkable drug," notes Nyhan. "It readily lowers the amount of uric acid in both blood and urine. It puts patients on their feet. People who were bedridden can now go dancing."

But as Nyhan pressed on in his investigation, he was to discover that the problems in children like Michael and Edgar were far more complex than common gout, and although allopurinol and other drugs effectively reduce uric acid concentrations and allow such Lesch-Nyhan youngsters to live well into their thirties instead of dying after a few, miserable years as they used to, the drugs have no effect on the neurological symptoms, even when treatment is started shortly after birth.

Michael's inborn lack of the HGPRT enzyme, deduced Nyhan, probably was poisoning him and somehow destroying his brain. Once again, Nyhan realized, he might be looking at a brand-new childhood disease, even though the symptoms seemed unrelated and were scattered all over the map—unusual biochemistry (crystals in the urine), severe retardation, neurological abnormalities (palsy), and bizarre aggressive behavior.

"I asked to see the boy, and Mike Lesch and I examined him," Nyhan says. "What really struck us was that he had a badly mutilated lip and fingers. He had done it to himself, and it was the first self-mutilation I had ever seen in my life."

"This, as it has turned out since we found that first case in 1962, is an absolutely uncontrollable compulsion," declares Nyhan, his voice rising in frustration. "These kids are ferocious and quick the way they do this. You must keep them in physical restraints—hands and feet tied—all the time. A Lesch-Nyhan kid gets loose, and he might amputate a finger—he'll chew his fingers to pieces. Let him loose for just a short time, and there's blood everywhere."

Michael's case exemplifies how a biochemical disorder can affect the brain and manifest itself in retardation and self-destruction. There was much to be learned from this hapless child, Nyhan knew, "especially about the brain and behavior. So we dropped all other projects. Mike Lesch went to work on this full-time." That meant that for the first time since the nightmare had begun for Edgar, Michael, and their parents, doctors actually investigated the household. Lesch did the legwork, going to the child's home, one of those tidy red brick row houses with carefully maintained marble steps near the Baltimore waterfront that give the place its unique colonial-era flavor. Michael's parents, both twenty-five years old, lived in this house, owned by the boy's maternal grandfather. One of Michael's mother's sisters lived there, too, with her children. The family, thrilled that someone finally was trying to help them, quite willingly gave Lesch blood samples for uric acid analysis. He found nothing abnormal, but in conversation learned about Edgar for the first time, who seemed to have the same, weird disease. That electrified Lesch. It meant it was genetic. Edgar was institutionalized at Rosewood, a state facility for the mentally disabled just outside Baltimore. Nyhan and Lesch brought Edgar to Hopkins for intensive study and medical care.

For more than a year, the two physicians went over both boys with every medical, genetic, and biochemical tool they had. The brothers' metabolisms proved virtually identical. Gout it was and had to be. But why? The boys produced enormous quantities of uric acid—at least ten times more than the amount that had previously been reported in any gout patient. Knowing this much at least, Nyhan and Lesch published their findings in the April 1964 issue of the *American Journal of Medicine*. They concluded that the condition was a new disease.

But then, to their amazement, other previously bewildered doctors with similar patients immediately responded from all over the world. The pattern of inheritance rapidly became plain—only males get it, probably from their mothers. In one extended family alone, Nyhan found fifteen males with the disease. No female victims have ever been discovered.

"An utter nightmare," Nyhan calls the Lesch-Nyhan syndrome. "Absolutely the worst disease I treat.

"Over the years, I've tried to bring families together for mutual support, to try to help them cope with this plague, but they decline. I think it's because each of these kids has awful behavior, but they're all different. And the mothers figure that if their son got together with another Lesch-Nyhan kid, he'd pick up something even *more* awful."

A particularly chilling aspect is that, unlike other mentally disabled youngsters who may injure themselves and seem oblivious to pain, children with the Lesch-Nyhan syndrome are granted no such peaceful oblivion. "These kids *hurt!* They scream in pain as they bite themselves. It's just terrible. They really are happy only when protected from themselves by restraints."

As babies, these children scream all night until their parents are taught how to restrain them securely in bed. But when they are tied down, a poignant behavioral change occurs. The change is somewhat bizarre. "They are unusually engaging children while they are restrained," Nyhan says. "They have a good sense of humor and smile and laugh easily." They remind him of the principal character in Stravinsky's ballet *Petrushka*, with "their tragicomic air, exaggerated posturing, and mittenlike protective coverings on their hands."

Nyhan is tormented by the fact that his patients are not blessed with the balm of unawareness. "The nightmare is that some of them have normal intelligence. Lots of others are *not that* retarded. They know their plight! And it obviously puts a complete limitation on their lives. I had one kid who learned to use crayons and pencils. And then, all of a sudden, he learned he could put his pencil in his eye. He missed, the first time. But it scared the hell out of him, and out of me. So now he's deleted all that handwork from his life."

Nyhan marvels at the ingenuity these children display in hurting themselves when freed from their bonds. Placed in a wheelchair, some Lesch-Nyhan youngsters find they can jam their fingers into the spokes of wheels. Braces prescribed for cerebral palsy get turned into weapons. Children may find hot water faucets and scald themselves. They vomit compulsively—and sometimes voluntarily, Nyhan believes—and are particularly prone to wrecking family gatherings, such as birthday parties.

"They will suddenly strike anyone who comes near them," Nyhan says, frowning. "They kick, hit, and break the eyeglasses of nurses and doctors who care for them. They develop such disconcerting habits as pinching the breasts of their mothers or nurses or grabbing for genitals.

"I remember we had a boy here in the hospital for several months, and his mother came from out of state to visit him at Christmas. Somebody had given him a toy fire truck as a gift. She walked into the room, and he hit her with it.

"On the other hand, they are just as often remorseful about having produced injury. But then, as they get older and learn to speak, they become verbally aggressive as well. They love to swear, to scream, to shock."

The publication of Nyhan's landmark paper in 1964 describing the appalling syndrome was to bring some solace to Felice Weiner, a New Jersey woman whose son was relentlessly turning into a grotesque parody of the normal youngster he had once been.

Weiner knew something was terribly wrong with her son when he was only three months old. His eldest brother, Brock, and the twins, Brett and Scott, had not developed like this. "Craig wasn't progressing, but was going backwards," she explains. "He didn't want to sit up any more, but wanted to lie down flat. He started to clench his fists and pull his neck to one side. He didn't want to eat."

As if this weren't bad enough, the boy developed an annoying habit. "He'd put his fingers in mouth, and I couldn't get them out. He'd bite down and leave the marks of his gums on his fingers." As the infant teethed, the urge grew worse.

"He didn't want to be in the nude," Weiner recalls, shuddering. "He was afraid he'd hurt himself. At night, he'd start to scream when he felt the feelings coming on. He'd hold on to the sides of the crib to keep himself from biting his hands. He could hold on until the moment I got there. Then he'd go for his hands."

She learned to react fast. "I'd grab them."

"When the weather was nice, I would take Craig out-doors," Weiner related during an interview at her tidy San Diego ranch-style home a few years ago. "People in the neighborhood would grab their children and run away. I couldn't understand it. They were very nice people."

Weiner's confusion led her to seek psychotherapy. Why should her son be considered a pariah? "My therapist finally explained that people were afraid it might happen to them," she relates, still hurt by the callousness. "Craig had been an absolutely normal, beautiful, chubby little boy. Nobody knew what the Lesch-Nyhan syndrome was. How much do people know about it today?"

Living with this every day would drive most mothers to distraction, but Weiner was one tough New Jersey housewife. She never stopped searching for an answer. For three years, like the parents of Michael and Edgar, she took Craig to doctor after doctor. . . . Nothing. But as Nyhan's bizarre scientific paper worked its way through the health care establishment, pediatricians could finally give Weiner a diagnosis, albeit one that offered no hope. "The doctor in New Jersey told me Craig would be dead by the time he was five," she says, matter-of-factly. "Before he died, his head would swell to three times normal size."

And that supposedly was that. "They told me to institutionalize my son," the mother goes on. "I said no. Absolutely not. Of course, I was young and strong then, and I had made a commitment. It's one I'll stick to until the day I die."

When William Nyhan left Hopkins, the Weiner family fol-lowed him to San Diego to be near the world's leading specialist in their son's disease. Craig, a sweet and lonely boy, grew into an amiable man who especially enjoys receiving visitors. He is thirty now. Tormented by savage demons since infancy, he sits hunched in his wheelchair all day long with his hands and feet tied and watches Padres and Chargers games on TV. At night, he lies spread-eagled in his bed, strapped in re-straints. He sleeps that way because he feels threatened when untied. He doesn't want to hurt himself.

"Can't control it. Can't help it," Craig muttered by way of explana-tion during an interview, his speech garbled by cerebral palsy and slurred because his front teeth have been extracted to protect him. Craig has many doctors, and sometimes he has time to warn them to tie him up during his frequent examinations. Otherwise, when the urge comes, if they're not fast enough, he may land a haymaker.

"I love you, Mom," he used to tell Felice Weiner several times a day as he was growing up. And he would ask her, "How long will I be like this? Will it go on forever?"

Stress and pressure took their obvious toll on the Weiner family. In 1970, Felice suffered a stroke that left her paralyzed on her left side and confined her to a wheelchair. Although she clearly needed home health aides to help care for Craig, she acceded to letting in outsiders only because she was left-handed, "which really made things tough." In the early 1980s, Felice's husband, Herman, was forced to have both legs amputated, because of failing circulation, and himself became confined to a wheelchair. He died in 1987.

Felice, though, refused to bemoan her fate. What mattered was Craig and the family's commitment to keep him at home, where he belonged, rather than locked in some institution. His daily care she had down to a precise science. "He has tubes in his stomach, so I can flush his kidneys with water every day," she said, ticking off the list. "He eats normally, but I have to be careful he doesn't choke when I feed him. He vomits a lot.

"He needs total care. He has to be lifted, dressed, diapered. Despite all the medicines, he gets sick—bladder infections are common. Kidney stones, too. Very painful. We have to be careful.

"But I'm proud of my son," she says summing up, her face a beaming smile. "He has come a long way. Does Craig look so horrible to you?" she asked, the challenge implicit in her eyes. "Is he disgusting? He's not like the children I've seen in institutions—alone, sitting in the corner, playing with their feces. Craig's clean. He's neat. He's loved."

She paused. This hurt her. "The worst thing is that he knows too much," she confided softly. "He cannot talk as well as you or I, but he can make himself understood. He knows exactly how he's supposed to be taken care of. He knows that he has these compulsions and can't stop them."

Her eyes dropped. "If he didn't know, and if he didn't love, and if he didn't *feel*, it would hard enough. But this way, it's impossible."

The battle to free people like Craig Weiner from the shackles of a dreadful inherited disease began in the late 1960s when scientists made the novel discovery that they could sometimes locate genes on individual chromosomes by uniting human cells with mouse cells in a test tube. The process, called cell fusion, involved a shotgun marriage in which a little biochemical trickery could create a hybrid cell containing a full set of mouse and human chromosomes. As these bastard cells divided, they tended to kick out the human DNA, until one human chromosome was left. With patience, the researcher could amass a collection of cell colonies, each containing a different human chromosome. The game plan then called for feeding the colonies a diet that would sustain them if they made a particular protein whose gene was being sought. When only a few colonies remained, each containing the same human chromosome, one could safely assume that the gene in question was on that particular chromosome.

The Lesch-Nyhan team at UCSD set out to clone the HGPRT gene by using chemistry that selected it out of DNA. The scientists were very clever. They took special mouse cells, which had been engineered to contain no HGPRT—absolutely none—and fused them with human cells containing the gene and placed the mixture in a petri dish containing a special culture medium called HAT (it is composed of hypoxanthine, aminopterin, and thymidine).

HAT was chosen because it is the mortal enemy of cells that cannot make the HGPRT enzyme. Cells lacking HGPRT are unable to survive in HAT medium. However, a tiny fraction of Ted Friedmann's mutant mouse-human cells had soaked up the HGPRT gene. The point was that Friedmann had a way to select out the cells he wanted.

When millions of mouse-human cells were dumped into HAT, only those cells that produced HGPRT were able to live and prosper. Eventually, by chemically eliminating all the fragments of other genes situated on the relevant bit of chromosome, the Friedmann team ended up with cells that could survive in HAT only because they were making HGPRT—the cells held the HGPRT gene. These Friedmann could spot, cull, and grow up in culture. He soon had millions of carbon copies of thriving colonies of mouse-human cells, each of which held the human HGPRT gene. With plenty of cells to work with, he set about retrieving the human gene from the mouse DNA.

Genes spaced along DNA strands are separated by vast stretches of nucleotide sequences that seem to be gibberish—endless expanses of mixed-up As, Cs, Ts, and Gs that don't code for anything or seem to do any useful work. Nobody knows the purpose of such "junk" DNA, as it is called by biologists; perhaps it represents the remnants of genes once important to evolution, or genes that didn't quite make it, or bits of old viral genes. But it was known that such sequences often are found next to functioning genes. Genes, in fact, may be seen as isolated bits of sense floating in a veritable sea of nonsense. Radioactive probes existed for such sense, and Friedmann hoped that one of them might enable him to fish out at least a piece of the HGPRT gene from the mouse DNA.

Because luck favors the prepared mind, the first gene piece snatched by Friedmann turned out to be part of the ubiquitous HGPRT gene sequence he desired. With this in hand, it was easy for the team to use the piece of the gene as a probe to find the rest of it in a sample of human DNA. Friedmann was in business. He had the gene.

Next, the plan was to mix the gene with white blood cells from a Lesch-Nyhan patient in tissue culture and see whether the cells would accept the good genes and start producing HGPRT. But no matter what the scientists did, no matter what chemical tricks they tried, the genes stubbornly refused to enter the human cells. They just plain refused to fuse.

Enter Inder Verma's retroviruses. In 1984, the Friedmann team was able to ferry healthy HGPRT genes into new cells. And, after taking bone marrow from a mouse, the researchers showed they could infect the marrow with human HGPRT genes. When returned to the mouse, the animal's marrow cells would carry the gene and sometimes make human HGPRT.

This feat by the Friedmann team marked the first time a foreign gene was successfully placed in bone marrow by means of a virus as a vector, and it is viewed as one of the most important experiments in early gene therapy.

By 1985, the Lesch-Nyhan picture was brightening considerably. The team had been able to design a virus that successfully implanted the HGPRT gene into human bone marrow—at least in tissue culture. This was important because, like all other would-be gene therapists, the UCSD researchers were limited to marrow, for it was the only self-renewing body substance they could remove, repair genetically, and replace in hopes of curing a disease. All this work was leading toward an attempt to inject the genes into a sample of bone marrow taken from a Lesch-Nyhan victim, before giving him back his marrow and hoping for the best.

But despite the advances in Lesch-Nyhan—the discovery of the putative cause of the syndrome and the ability of doctors to treat some symptoms pharmacologically—a craggy scientific problem still loomed. Although the disease occurs when there is a total absence of HGPRT in all cells, the Lesch-Nyhan patient's aggression, retardation, and palsy are blamed on the disruption of normal functions of the brain.

Yet, no matter how closely pathologists have looked, on autopsy no brain defect has ever been found.

"Lesch-Nyhan brains look normal," Nyhan says. "We just can't see anything wrong." Although this strange fact would seem to be dismaying if one were looking for a defect, it actually buoyed Nyhan and his fellow researchers. Perhaps because nothing seemed to be awry in the brain, the disease need not be permanent. Perhaps the symptoms could even be reversed. If scientists could just get some fresh new HGPRT genes into the bodies of these youngsters, the new genes might stabilize their metabolism and cure them.

Once again, though, the story would not be that simple. In the Lesch-Nyhan syndrome, the problem is not with uric acid itself. Concentrations of uric acid in the blood of Lesch-Nyhan boys have been lowered with allopurinol, and yet the self-mutilation and other aggressive behavior have continued. As Nyhan has written, "Uric acid is not a purine that is found in the central nervous system. It is the end-product molecule. It may reflect events in the nervous system, but it cannot reveal their exact nature."

Nonetheless, he believed, if some way could be found to give new HGPRT genes to Lesch-Nyhan boys, those nervous system "events," whatever they were, might become normalized. What's pleasing about the theory is that doctors could avoid some of the difficulties inherent in treating blood diseases, in that they wouldn't have to perform even the most minor regulation of HGPRT genes in Lesch-Nyhan patients. If they could get the genes to replicate in hundreds of billions of marrow cells, more and more HGPRT would be produced. No matter. The genes needn't be delicately tuned. They could all stay on all the time. The body loves HGPRT. Gushing genes couldn't do any harm, and might do a lot of good.

Appealing though it was, that notion aroused considerable skepticism among other scientists, particularly rivals on the East Coast. The idea of introducing the genes into bone marrow didn't bother them— they were trying to do that themselves. If all went well, the new HGPRT genes, once in place in billions of marrow cells and transfused back into a Lesch-Nyhan victim, could send the enzyme coursing through his bloodstream. But so what? The trouble lay deeper than that. How did the West Coast guys plan to get new genes into the brain?

In the mid-1980s, when dozens of gene splicers were laboring to get new genes to work in bone marrow and blood, the brain as a target seemed as distant as the planet Mars. In the first place, the brain has evolved a highly efficient mechanism—the blood–brain barrier—to monitor substances in the bloodstream to screen out toxins. In the bargain, genes might get screened out, too. Hence correcting a defect elsewhere in the body might have absolutely no effect on the brain. Moreover, brain cells (which actually are neurons connected by glial cells) do not proliferate after they mature, as bone marrow cells do, so genetic material introduced into one brain cell would not spread to others by ordinary cell division. For that matter, retroviruses will infect only dividing cells, so the leading gene transfer system would be useless if one were trying to do something about a Lesch-Nyhan brain.

The big issue in Lesch-Nyhan, and in hundreds of similar diseases, was a lack of understanding of the biochemical mechanism underlying the disorder, even if, with the isolation of HGPRT, scientists thought they knew the cause.

No one grasped these problems better than Ted Friedmann. "We're painfully aware of the warts in this model. We don't know the nature of the brain defect," he conceded in an interview. "Maybe there's a defect in an organ like the liver that leads to the accumulation of uric acid, or the failure to make something that the brain needs. Many diseases that masquerade as central nervous system diseases are really diseases of the liver."

The model for Friedmann and Nyhan was another inherited childhood disorder, PKU, which is actually a liver disease. "It's not a brain disease at all, yet it shows up as one," Friedmann pointed out. "So with Lesch-Nyhan, there's no assurance the defect is in the brain."

Bill Nyhan and Ted Friedmann continued to look longingly at PKU in their efforts to wipe out the Lesch-Nyhan syndrome. Nyhan just had a gut feeling that a similar treatment might work.

"Lesch-Nyhan is an apt model for gene therapy," Friedmann believed for a long time. "Biochemically, the HGPRT enzyme is very useful. It lets you ask questions and develop techniques for following genes around when you put them into new places."

Nyhan pointed out, "The HGPRT gene has been useful for all kinds of studies in genetics. This enzyme has kept more people in work than any other enzyme I can think of."

Moreover, Nyhan and Friedmann realized, cells in the body do somehow find ways to talk to one another. There is evidence that they help cure one another, too. This line of logic led to the possibility that if the

California doctors could help the bone marrow of a Lesch-Nyhan boy, the brain might get the word.

Farfetched? Nobody knew whether it was possible. But there was a way to find out, and, being good scientists, Friedmann and Nyhan were determined to pursue it, even if the results might dash their hopes. They decided to attempt a bone marrow transplant in a Lesch-Nyhan patient. If they gave him new HGPRT genes—not freshly made in their laboratory but volunteered by a sibling—and the genes helped, well, that would mean it was indeed possible to use bone marrow to repair the brain.

By 1985, bone marrow transplant techniques had improved sufficiently for Nyhan to propose the idea to a Lesch-Nyhan patient, twenty-two years old, who was fortunate enough to have a sibling whose tissues matched his own. Their family readily consented to the experiment—Lesch-Nyhan is so horrible that parents will try anything—and journeyed from the East Coast to California, though not to San Diego but to Los Angeles. Nyhan had prevailed on the best bone marrow specialist he knew, UCLA's Robertson Parkman, to attempt the procedure. Parkman, who later made history as a prominent member of the federal panel that regulated genetic-engineering experiments, was comfortable with the Lesch-Nyhan trial. He removed bone marrow from the brother and, after cleansing and treating it in customary fashion, infused it into the body of the Lesch-Nyhan patient. For the first time, the young man had a blood-making system that contained normal HGPRT genes.

At first, the results were tantalizing.

Friedman and Nyhan watched hopefully. The patient's body accepted the bone marrow from his brother without fuss and soon began producing the missing enzyme. A few days after the transplant, the recipient asked for his restraints to be taken off. He was able to sleep without them. He also began to feed himself. And he stopped vomiting, which had been one of his symptoms.

However, then the roof fell in.

"He seemed improved for about a week," Nyhan remembers. "But then he developed a kidney stone, which often happens in this disease. The urologist tried to handle it without surgery—they instrumented him from below, which, as you can imagine, is painful at best. At worst, it didn't work. So he finally went into surgery and had the stone removed. In the course of that, the vomiting returned, and he had to go back into restraints."

The bottom line was that even though the patient had received new HGPRT genes, they hadn't made a lasting difference. This was devastating news to everyone involved, though many other scientists were not surprised. Until the brain defect was figured out, they didn't see much point in bone marrow transplants.

But the Nyhan-Friedmann team disagreed. Friedmann viewed the result of the experiment as nothing more than a setback. "An experiment doesn't fail," he said, defensively. "An experiment tells you something."

Nyhan had mixed feelings. "Before we did the transplant," he recalls, "I took the hypothetical position that if we put the enzyme into marrow and let it circulate in the blood, it ought to fix the head. I keep thinking about PKU. You and I don't have PKU, because we have the necessary chemical in our liver that keeps our head normal. The kids who don't have it become mentally retarded. And if I came along and transplanted a liver into them at age twenty, they'd still be retarded. The damage has already been done. But if I'd transplanted the liver at two weeks, I'd fix them.

"Also, if you're as intelligent as our patient is, and you spend your life mutilating yourself, and even if now you've got a little enzyme that would normally protect you, it might be that those patterns of behavior are so ingrained that you're never going to get rid of them."

All of which set him to thinking. What would happen if he could get the HGPRT genes into a patient very early on—before the devastating symptoms appeared? Most Lesch-Nyhan youngsters, like the Baltimore brothers and Craig Weiner, seem normal at birth. It takes time for the damage to show up. Why not treat a Lesch-Nyhan baby the way PKU babies are treated in infancy with their special diets?

And if it turned out that the defect itself is based in the brain, perhaps gene therapy techniques would someday allow the delivery of healthy HGPRT genes there. "There's a definite window," Nyhan says. "These youngsters develop normally for six to eight months. What I would dearly love would be able to treat somebody who was in that window."

The disease is so rare that finding such a baby would not be easy. Even by 1993, Nyhan had not found a candidate. "If the syndrome has struck once in a family, we do prenatal diagnosis, and when there's an affected fetus, the family generally chooses an abortion," he observes. "But that certainly is not the case all the time. Some people are so opposed to abortion that they wouldn't consider it under any circumstances."

After having waited for so long, patients like Craig Weiner and their families still have no choice but to keep the faith—only gene therapy might end their nightmare.

"Maybe it wouldn't return him to the way he was, but I'd take anything to improve the quality of his life," Craig's mother says. "I'm doing fine, as long as I have him. but I'd like to see him get well before I go. I'm not getting any younger. Anything that wouldn't kill him, I'd be willing to try."

As she speaks, Craig sits strapped in his wheelchair, his head resting on three pillows. Wracked by fierce spasms, he used to snap his chin forward. He would scream from the pain, so Felice built up the cushion of pillows for a headrest. Thus he has sat, tied down, his head on the pillows, day after day, watching TV for thirty years.

"Who's perfect, anyway?" Felice asks. "I look at him and I see beauty. I see that he speaks, he smiles, he loves. He's a human being. Having him has been a joy."

But if one asks Craig Weiner about the Lesch-Nyhan syndrome when his mother is out of hearing range, he minces no words. "I want to die," he mutters. "I'm sorry I was born."

5

THE UNDERSTUDY STEPS IN

Short of a scientist's worst nightmares—being stripped of all funding, losing years of work in a laboratory fire, getting wrongfully charged with fraud because of the sins of an unscrupulous collaborator—it is difficult to imagine a greater blow to the spirits of gene therapy researchers than the news that the Lesch-Nyhan syndrome was beyond the reach of their embryonic technology.

Reactions ranged from keen disappointment to a sense of having been poleaxed. Hardest hit, obviously, was the Friedmann-Nyhan consortium, once very hopeful that it was closing in on a cure for the disease. In calling the question surgically, the transplant specialist Robertson Parkman had forced the lab partners to face the disheartening conclusion that, when all was said and done, they had spent four years on a wild-goose chase. With little charity, Stuart Orkin, a rival at the Harvard Medical School, declared, "The California people have a problem. They gave their patient a bone marrow transplant and he didn't rise like Lazarus. Lesch-Nyhan is not a good model for gene therapy."

To be sure, the efforts of Friedmann and his colleagues had contributed much to the understanding and methodology of the new field. By "curing" defective cells in the test tube, and then learning how to insert the HGPRT gene into progressively more complicated systems, up to and including live mice, they had shown that the delivery of a foreign gene was both possible, potentially effective, and, so far as anyone could tell, safe. But from a purely clinical point of view, they had spent a great deal of time and money only to dash the hopes of families like the Weiners, whose sons were every bit as cursed at the end as they had been at the beginning.

For those who were keeping score—and there was no shortage of scoffers with pencils at the ready—it was strike two against gene therapy. First hemoglobin diseases and then a severe metabolic disorder had proven to be more than a match for the infant science of gene transfer.

If gene therapy were to survive with its dignity and credibility intact, if it were not to be hooted prematurely into oblivion, a new and more convincing disease model would have to be found. This one would have to be unequivocally treatable by means of bone marrow. It would ideally require little, if any, regulation of the implanted gene. Above all, it would have to come from the still-meager inventory of ailments for which the culprit gene had been isolated. But where could an illness be found that satisfied this list of preconditions so conveniently?

Upon meeting Alison Ashcraft for the first time, one would never guess there was anything wrong with her. She looks like any of the healthy, suntanned, affluent teenagers who dot the winding streets of Laguna Hills, California. But the attractive, diminutive high school senior possesses a world title she would prefer not to have. At the age of twenty, Alison is the oldest-known survivor of the rare genetic disorder adenosine deaminase (ADA) deficiency, whose noxious effects have decimated her body's immune system.

Unable to mount more than the feeblest defense against a microscopic world filled by what the physician-essayist Lewis Thomas calls "little round things"—viruses, bacteria, protozoa, fungi, and other nasties—Alison for much of her life has walked a razor-thin line between health and dire, potentially mortal illness. For her, common afflictions like the flu may represent a catastrophe. A bout of chicken pox can be fatal. She has a ten thousand times greater risk of incurring cancer, notably leukemia, than the average person. Always there has been the haunting fear that a case of sniffles might metamorphose into pneumonia. She has contracted the serious lung disease no fewer than eighteen times.

For years following the diagnosis of her illness, at age seven, Alison was forced to avoid associating with large numbers of people in enclosed spaces, because it heightened her chances of infection. She missed great chunks of the school year as a result of illness; this put her under great pressure to avoid falling behind academically. And with unsparing regularity, there was the *doctoring*. Immune globulin infusions. White cell monitoring. Thyroid tests. Her bedroom, testament to all these trials, resembles nothing so much as a petting zoo run by Madame Tussaud.

Lifeless eyes belonging to hundreds of stuffed animals peer at visitors from every corner of the room. Legions of bears, dogs, pandas, and Cookie Monsters, rabbits, tigers, porkers, and frogs have taken over the premises, dangling from shelves, gawking from closets, and forming huge mounds on the bedspread. The collection might seem the result of girlish excess; in fact, it is a monument to bribery, the benevolent parental kind. Each button-eyed, felt-tongued beast stands for a time in Alison's young life that doctors have had to draw her blood.

Life with ADA deficiency has been no less a high-wire act for Ashanthi DeSilva, of suburban Cleveland. Ashi, as her parents call her, has endured several brushes with death since entering the world in 1986. The most harrowing occurred when, at the age of two, her platelet count dropped precipitously, threatening her with a fatal bout of internal bleeding. Chronic coughs have wracked her chest and uncontrollable vomiting her digestive tract. The facial expression she routinely presents to the world, dour and suspicious, reflects a childhood spent on the examining table.

The word "Dickensian" comes to mind to describe the plight of Ashi's parents, Raja and Van. The couple, who are from Sri Lanka, have two other children, each likewise beset by serious illness. Their oldest, Anoushka, is confined to a wheelchair, the victim of a mysterious, encephalitis-like sickness that struck when she was a year old, leaving her with multiple physical disabilities. The youngest, Dilani, also suffered a severe fever and convulsions as a baby and today can neither talk nor sit up, a condition aggravated by serious learning problems. But it is Ashi, the least incapacitated of the three children, who is paradoxically in the most danger. ADA deficiency can kill at any time.

Van DeSilva's professional training as a nurse—specifically a geriatric nurse—has served her well as she tends her three children in the two-bedroom, two-bath high-rise apartment the family shares in the pleasant, middle-class suburb of North Olmstead. But clinical skills do not immunize her against the emotional impact of the remarks that sometimes pass from Ashi's lips.

"Mommy," the little girl said mournfully one afternoon, "you should have got yourself a life before you had me and the others. You shouldn't have had a child like me."

Despite the daily tribulations that disrupt Raj and Van's domestic life, the two have accepted their lot with a certain equanimity, lavishing attention and affection on each of their children while seldom, if ever, seeming overwhelmed. "Love sustains us," Van says simply.

Like the Lesch-Nyhan syndrome, ADA deficiency is so rare that very few doctors will ever see it in a lifetime of practice. But Ashi DeSilva is only one of two ADA deficiency patients under treatment at Cleveland's Rainbow Babies and Children's Hospital, part of the Case Western Reserve University medical complex.

The other is Cynthia Cutshall, a pixie-faced, personable, blond-haired youngster from the Ohio city of Canton, some fifty miles south of Cleveland. Cindy began displaying symptoms of the ailment in 1984, when she was three years old. By the age of four, she was developing unusual conditions that cried out for explanation. Peculiar strains of pneumonia, for example, and leukopenia, a reduction in white blood cells that leaves the body open to infections. In kindergarten, she typically missed four of every five school days because of illness.

"The doctors originally said she had an inactive immune system that would kick in when she was a little older," recalls Cindy's mother, Susan, who, improbably enough, is a nurse like Van DeSilva, assigned to critical care at Canton's Altman Hospital. "But when she was five years old, they finally diagnosed it as ADA deficiency—a late-onset form, because she didn't show it at birth, like a lot of kids who have it."

ADA deficiency is caused by a flaw so obscure that it strains the mind to contemplate it. Embedded deep inside each of our cells—within the nucleus, a command center no larger than 1/12,000 of an inch—are our twenty-three pairs of chromosomes. Buried still deeper in this infinitesimal universe, on chromosome 20, is a gene containing a complex recipe that tells bone marrow cells how to make adenosine deaminase, the enzyme whose job it is to direct the conversion of a cellular waste product called deoxyadenosine into the harmless substance inosine. Were this bit of chemical defanging not to take place, we would all drown in a toxic accumulation of deoxyadenosine.

At any given moment, the body is streaming with a vast army of enzymes, each controlling a vital biochemical reaction. However, imperfect organisms that we are, we all have a few Achilles' heels concealed somewhere within our genomes. Because of mutations in various genes, we each make inadequate amounts of one or another enzyme, and sometimes specious versions that will not work at all. If we are lucky, the biochemical processes that hang in the balance will be minor enough that we can get by with merely a degree of discomfort—a siege of gas upon drinking milk, say, or a special susceptibility to tooth decay; a touch of hay fever, perhaps, or a tendency toward seasickness. But in Alison, Ashi, and Cindy, the consequence of genetic mutation is far more grave. The ADA enzyme is crucial to life and health, and the girls' recipe for this pivotal substance contains a deadly misprint, requisitioning one wrong amino acid out of a list of several thousand ingredients. Because of this minuscule mistake, magnified trillions of times, their bodies cannot make sufficient quantities of the essential protein, and the result is that deoxyadenosine fails to get neutralized as it should.

Here is what happens. When the compound starts rattling around the anatomy like a loose cannon, it combines with the phosphorous present in the body to form something called deoxyadenosine triphosphate, a substance lethal to immune cells—particularly to the master white blood cells of immunity called T cells. T cells, the clarions that alert the rest of the immune system to dispatch killer cells and antibodies against a foreign entity, are so named because, after being sired in the marrow by stem cells, they migrate for processing to the thymus—the walnut-sized primary gland of the lymphatic system tucked away in the upper chest, just below the thyroid—before ending up at their barracks in the lymph nodes and spleen. In an ADA-deficient child, this system switches off and the thymus gland withers away from disuse. The lack of a single gene ends up wiping out a child's defense against physical invaders. Such children are, for all intents and purposes, as vulnerable to infection as if they had AIDS, the only difference being that AIDS is acquired and transmittable whereas ADA deficiency is inborn and noncontagious.

Part of a larger class of some eighty ailments lumped under the umbrella term "severe combined immunodeficiency" (SCID), ADA deficiency strikes but once in every 150,000 births, and kills so efficiently within the first few months of life that, in all the world, there are fewer than thirty living victims of the disease. By far the most famous SCID child was a Houston boy named David who spent most of his twelve years in a sterile, plastic-enclosed environment, experiencing physical contact with no one, not even his parents. The plight of the so-called bubble boy was played out in media as diverse as *Life* magazine and the *National Lampoon*. He was even portrayed in the movies by an actor incongruous in the role, John Travolta. Onscreen, Travolta's David went to school by closed-circuit television, hoodwinked teachers, earned academic honors, and charmed pubescent girls. In real life, David was a pathetically isolated youth whose hospital confinement severely deprived him of experience of the physical world, distorting his concepts of space, depth, and distance. Though exceedingly bright, he described trees as being brown rectangles with green circles on top and refused to believe that the buildings he could see from his window at Texas Children's Hospital had any backs to them.

David died in 1984, as the result of a failed bone marrow transplant from his sister undertaken in the hopes of liberating him from his plastic prison. It is ironic that as he was dying, scientists elsewhere had already begun turning to immune deficiencies as the model for gene therapy. Soon there would be, as Stuart Orkin wryly put it, "more people working on ADA deficiency than there are patients who have it."

Among them was a rejuvenated William French Anderson. For several years, Anderson had been looking for an opening, imparting very little of substance to the development of gene therapy. His initial flirtation with microinjection was now but a wearisome memory. During this dry spell, Anderson's main contribution to the field was a handful of slightly futuristic articles. With an intuition that was at once informed and yet vaguely fanciful, calling to mind Leonardo da Vinci writing on manned flight, his monographs detailed the mechanics of how gene therapy would most likely be accomplished when the enabling technologies were finally worked out. In one piece, coauthored with the NIH ethicist John Fletcher, Anderson indulged his passion for philosophy and rules of conduct, focusing on the ethical dilemmas raised by gene therapy and suggestions for resolving them.

Ever the cheerleader, the restive Oklahoman undertook in early 1984 to prepare a review article on the field's prospects for *Science* magazine. In the article, he predicted that the first disease treated by the technology he had for so long championed would, in fact, be an immunodeficiency. The leading candidate was adenosine deaminase deficiency, the gene for which had recently been cloned by researchers in three different laboratories on two continents. Among the labs had been Orkin's at Boston Children's Hospital. But the gene had also been isolated by a Dutch graduate student named Dinko Valerio, at the Institute of Applied Radiology and Immunology in the Netherlands, and by a University of Cincinnati team headed by John Hutton, dean of that institution's School of Medicine. Anderson began calling around for their latest data.

By now, Richard Mulligan had announced to the molecular biology world the successful end of his long march toward a workable retroviral vector, and Anderson was itching to get back into the game. As he made his calls, he found himself daydreaming, not about the mundane matter of writing the *Science* piece, but about using the new gene himself in a clinical gene therapy setting. But any such effort would require a copy of the gene, and if Anderson had to come up with one from scratch, it could take months to personally tease it out of the mass of DNA on chromosome 20. Molecular biologists long ago realized that, even among rival labs, a certain amount of sharing and barter was in everyone's interest, at least until such time as there would be biological mail-order houses to which one could simply send away for things like genes and probes. Thus, emulating Indians of the plains, labs would frequently exchange this or that gene for something else, a piece of data, perhaps, or a new method for sequencing DNA.

Anderson was ready to deal. But if his eagerness to acquire a clone of the ADA gene was detectable in his voice, no one seemed to pick up on it. Not one of the labs he spoke with offered to share the ADA gene with him. That is, until he rang up his colleague Hutton in Cincinnati. Hutton had not only cloned the ADA gene; he had gone to the great trouble of sequencing it in its entirety. Yet, amazingly, he had no plans to do anything further with it. "They didn't have the resources to do gene transfer," says Anderson, recalling the conversation. "They were only going to use the gene for diagnostic purposes. So I asked John, 'Can I have it?' And he said, 'Sure.' And then he sent it to me! With that one call, my lab suddenly became an ADA player."

Anderson, mindful to a fault, as always, of the trappings of decorum, felt compelled to telephone a surprised Orkin at Harvard to announce that he was no longer a sideline observer and would not expect Orkin and his partner, Mulligan, to share any more lab results with him. What he didn't have to mention was that the Orkin team had been working on the ADA gene for two years and that Anderson had obtained it in just two days. "That just inflamed Stu and Rich Mulligan," as Anderson remembers it. The Orkin/Mulligan version is quite different, however. Both of them sternly deny having been angered that Anderson had taken a shortcut. What *did* nettle them, the two men say, is Anderson's implication that they would care, one way or another. They insist that at no time have they ever considered themselves to be in competition with French Anderson, that their aim has never been to get involved in a race to be first to try gene therapy. According to them, they were interested only in completely understanding the gene and its properties before trying to put it into a human being.

And indeed, both Mulligan and Orkin have hewn quite faithfully to that ethic, maintaining a stately research pace over the past decade, neither rushing their science nor submitting a single protocol proposing a gene therapy experiment until Mulligan finally did so in 1993. Nevertheless, the relationship between their laboratories and Anderson's has since 1984 been rancorous, to say the least. In time, it became downright hostile.

Race or no, both the Mulligan/Orkin consortium and Anderson's laboratory soon began trying in earnest to get the ADA gene first into ADA-deficient cells and, shortly afterwards, into mice. It turned out that the mice had other ideas.

Even as the researchers geared up for these animal experiments, however, one thing was clear from the start. Unlike the Lesch-Nyhan syndrome, ADA deficiency could unquestionably be approached through the bone marrow, the home office of the immune system; that made it a very attractive disease model for gene therapy. Investigators needed no further proof than the fact that, where a tissue-matched donor existed, the preferred treatment for ADA deficiency was a bone marrow transplant. Such a transplant has the power to create a complete immune system in the recipient, fashioned from the donor's immune system. The reasons are readily understandable. Marrow from an unaffected donor, by its very nature, contains healthy blood-making stem cells, those long-lived, all-powerful progenitor cells that, something like sourdough bread starter, are continuously manufacturing the various white and red blood corpuscles of the body. Donor stem cells travel rather well and upon transplant are capable of repopulating a patient's entire body with healthy blood cells; and inasmuch as each of these stem cells presumably contains two normal genes for the ADA enzyme, all cellular offspring will possess the healthy genes as well. Any T cells subsequently manufactured would thus be armed and ready to neutralize toxic precursors, amounting to an outright cure for the illness.

Yet bone marrow transplantation remains an expensive, agonizing, and often dangerous ordeal. Even though the first person to receive a bone marrow transplant was a SCID child who is still alive a quarter of a century later, many patients succumb to a bizarre complication in which the donor tissue treats the recipient's body like an alien threat and attacks it, often with fatal results. The phenomenon is known as graft-versus-host disease. Moreover, good tissue matches are hard to find. Some 75 percent of all ADA-deficient children, including Alison Ashcraft, Ashi DeSilva, and Cindy Cutshall, lack a compatible tissue donor, and mismatched bone marrow transplants still produce an alarmingly high mortality rate.

Much of what we today know about the bone marrow and stem cells is due to the trailblazing work of the University of Minnesota immunologist Robert Good, who performed that initial bone marrow transplant in July of 1968 on a five-month-old SCID patient named David Camp. Good used tissue from the boy's older sister, although she was not a perfect match. Not only did her tissue type not complement David's as precisely as it should have; her blood type was way off. She belonged to the blood group O, while David fell into the type A group, which certainly did not bode well. Still, young David was failing fast, so Good decided to remove a billion marrow cells from the five-year-old girl and implant them in her baby brother's belly. David responded spectacularly as his sister's cells became established in his bone marrow, reconstituting his immune system and giving him normal white cells. The transplant had some quirky consequences. For example, it switched David's blood type from A to O, and to this day he carries in his white blood cells the characteristic X-X sex chromosome pair of his sister. But the David Camp experiment marked the first complete correction of an inborn error of metabolism by bone marrow transplantation. It showed that both the B- and T-cell populations, the twin pillars of the immune system, could be reconstituted by transplant, proving that the defect in a disease like ADA deficiency could be cured by an influx of new stem cells via gene therapy.

The seemingly insurmountable problem facing gene therapists was science's inability to tell a bona fide human stem cell apart from all the other cells in a bone marrow sample. Hence, researchers could only fire blindly at them as they supposedly bob around in the vast cellular ocean of red marrow that normally lines the body's skeletal crests and long bones. In fact, the very existence of stem cells has been inferred more than proven. Each kind of tissue in the body is said to have its own precursor cell, but as of this writing there has been no way to prove the truth of this hypothesis. No scientist has ever been able to snare a stem cell and tear it apart, experiment with it, and watch it differentiate. Nevertheless, in blood at least, the evidence points to their existence as potent little factories—a single founder cell can produce perhaps a million mature blood cells. T cells, the primary site of ADA deficiency, perform three different jobs: they directly attack foreign invaders, such as viruses and bacteria; they reject transplanted tissue; and they call out the guard in the form of other voracious immune cells when they sense that the body has been violated. The B cells produce antibodies on demand when the T cells tell them to, and hence are virtually helpless when their T-cell triggers are defective, as in ADA deficiency. A different class of stem cell, meanwhile, is believed to divide into oxygen-carrying red blood cells, platelets for clotting, bone-growth cells known as osteoclasts, and various breeds of ravenous scavengers called macrophages, which devour Lewis Thomas's little round things.

Much of what science postulates about stem cells comes from mouse research. In 1985, Beatrice Mintz, of the Fox Chase Cancer Center, in Philadelphia, showed that genetically disabled mouse fetuses, engineered with faulty T cells, could be saved by transplants of healthy cells. And by patiently whittling down the possibilities, she determined that in mice, at least, it takes only *one* good stem cell to make good blood. Great excitement followed in 1988 when a Stanford University team led by the pathologist Irving Weissman reported dramatic success in identifying mouse stem cells, using special vehicles called monoclonal antibodies to separate these unique cells from ordinary ones. A year later, Weissman succeeded in nearly transferring his stem cell identification method from mice to humans, thus giving the entire field of bone marrow gene therapy a shot in the arm, and in 1991 he triumphantly announced that human stem cells had been purified—a feat that as of 1994 awaits reproducibility, the cornerstone of true scientific discovery.

Back in 1984, though, the stem cell was an ephemeral thing, a physiological will-o'-the-wisp. Ensuring that a payload of normal ADA genes was hitting stem cell targets, as would be essential to the success of gene therapy for ADA deficiency, would be like trying to send a postcard to the Flying Dutchman.

But while the gene therapy guild grappled with the stem cell problem, it could rest assured that ADA deficiency more than satisfied the second prerequisite of a good disease model for its burgeoning technology; that is, in stark contrast with beta thalassemia and other hemoglobin disorders, the illness posed little in the way of a regulation problem. Since victims of the immune misfortune were virtually devoid of any ADA enzyme, whatever minuscule production could be squeezed from their cells by implanted genes would be gravy. And the common wisdom among immunologists was that less than 10 percent of the normal production would suffice to cure victims completely.

As it came to be conceived by researchers, gene therapy for ADA deficiency shared some aspects of a marrow transplant. A desperately ill child would be anesthetized on an operating table. A physician would punch a stubby, evil-looking needle into the unconscious youngster's body about twenty times to remove marrow from the cavities of the long bones. Now would come the genetic manipulation, enhanced greatly by Rich Mulligan's newly perfected retroviral vectors. The thick, red marrow would be rushed to a nearby laboratory, where a gene therapist would painstakingly cleanse the cells in a dish and then mix in a solution containing billions of identical particles of a genetically engineered retrovirus. The altered virus particles would, of course, retain their fabled power to infect cells, but their ability to migrate to other cells and spread viral disease would have been removed and each disabled virus would contain a curative gene. As it went about its natural business of parasitizing cells, the virus would randomly insert the healthy genes into the chromosomes of the child's marrow sample, and these cells would be patiently cultured and grown. To complete the original strategy, the child's doctors would then give the marrow back to the youngster in a simple transfusion of his or her own blood. The entire procedure would take only a matter of hours. In days to follow, the child would become sicker before the rebirth of immunity as the new genes struggled to survive in a tumultuous chemical environment. But gradually, if all went well, the various kinds of white cells would stagger to life, wearing figurative bulletproof vests to protect them from toxic biochemicals, and go prowling for rampaging viruses and bacteria to kill. Metabolism over many months would slowly adjust and the child would, for the first time, start to feel truly well.

Even as the Anderson and Mulligan laboratories began squaring off for what would prove a rancorous battle to make gene therapy happen, the prospects for patients like Alison Ashcraft, Ashanthi DeSilva, and Cindy Cutshall were already in the process of brightening, however—and from an entirely different direction.

It had long been conjectured that one might treat certain inherited diseases by simply injecting into the victim's body the protein that happens to be missing, thereby avoiding the more difficult line of attack represented by gene therapy. The notion appeals to the pragmatic intellect—the sound one hears is the clean, economical strop of Occam's razor—and indeed, so-called protein replacement therapy had been a feature of the medical landscape for some time. The use of insulin shots to treat diabetes offers the primary example of how drugs might be mobilized to substitute for the homegrown variety when patients cannot muster enough of their own essential proteins. But attempts to pipe in the raw protein missing in most other genetic diseases had failed because of a basic problem: the proteins tended to degrade swiftly as they passed through the bloodstream, as a result of antibody formation and general wear and tear. The former might not be a factor in ADA deficiency, but the latter definitely was.

In late 1985, a breakthrough was announced at Duke University, where the pediatrician-biochemist Michael Hershfield, in conjunction with scientists at the start-up biotechnology firm Enzon, Inc., of New Jersey, had learned how to coat the ADA enzyme with a waxy protective envelope, made of a polymer called polyethylene glycol, that keeps it intact as it circulates in the blood serum. In a short time befitting a potentially lifesaving drug, safety trials were arranged for what had become known as PEG (for polyethylene glycol) ADA. According to the protocol, five children suffering from ADA deficiency were to receive an experimental regimen of the preparation to see whether they could tolerate it without ill effect. Among them was a recently diagnosed Cindy Cutshall, whose parents were offered the choice of charter membership in the PEG-ADA club. "We were leery," admits Susan Cutshall, "because we had no idea what the side effects would be."

Fortunately for Cindy, the negatives of PEG-ADA proved, at least in her case, to be minimal. Early returns on the drug showed it to be relatively safe—although not without some risk, as later data revealed. Several youngsters placed on PEG-ADA therapy (including Alison Ashcraft) went on to develop antibodies to it, despite their weakened immunity, and although most of them quickly lost this undesirable response when taken temporarily off the drug, at least two youngsters did not and eventually succumbed to their insidious disease.

But safety was not the central issue surrounding PEG-ADA. Efficacy was. Would the new drug boost immunity, even though it stays outside T cells? The answer was not entirely clear, but it seemed to be yes, up to a point.

Deoxyadenosine, the toxic metabolite that destroys T cells, ducks from cell to cell rather slowly. Consequently, high levels of circulating ADA enzyme floating outside are able to dispose of the toxin before it accumulates inside the cells, where it does its damage. And this beneficial effect will occur even though the cell interior itself contains no protective enzyme. "PEG-ADA seems to bathe every T cell and acts like a kidney dialysis machine that cleanses the circulation," says Hershfield. "But it has a half-life of only thirty days, so must be maintained by weekly shots."

For Alison Ashcraft, the initiation of PEG-ADA therapy in August of 1988 marked a reversal of fortune after a dreary odyssey. The second of two children of Corinne and Aaron Ashcraft, Alison was plagued with inexplicable ill health almost from the day she was born.

"She was sick frequently," recalls Corinne. "Earaches. Nose and upper respiratory infections. Pneumonias. And she didn't get over things as quickly as she should. Usually with kids one dose of medicine is enough to take care of it, but with Alison we'd always have to go back for a second, third, even fourth round."

Alison's troubles continued to mount. She was not growing properly. Her stature was exceptionally small. The doctors attributed her size to the debilitating effects of her constant illnesses. "She was iron deficient," adds Aaron. "All she wanted to do was sleep. She would sleep up to twenty-two hours a day."

Only after years of running from physician to physician were the Ashcrafts finally steered to an allergist. The allergist, David Cook, performed no fewer than thirty-six scratch tests on a section of Alison's back. He tested her for reaction to a variety of foods, grasses, pollen, and animal hairs. When the testing was completed, Cook called the Ashcrafts into his office and began shaking his head. "This is very puzzling," he said. "She doesn't show a positive reaction to anything." Aaron and Corinne traded glances. "That's good, isn't it?" asked Aaron cautiously. Cook looked at them solemnly and shook his head. "Everybody gets at least a small reaction to something," he said.

What the allergist had stumbled on was Alison's inability to muster antibodies to anything at all. Providentially, Cook also happened to be an immunologist. He asked for permission to extract a large sample of Alison's blood for study, which Corinne and Aaron readily gave. They did not hear from him again for several days. The Ashcrafts will never forget the moment. It was a Friday in late May of 1981, and they were about to take Alison and her older brother, David, then eleven, to Yosemite Park for the weekend. Cook's phone call caught them as they were loading up the car. "Please come to my office immediately," he demanded, refusing to explain further. When he had them in his office, he sat them down excitedly and said, "I think we've got it. I think I have found the source of Alison's problem. She appears to have an immunodeficiency."

Soon, Alison's condition was stabilized by a monthly regimen of infusions of immune globulin, a blood extract rich in antibodies. These infusions provided her with temporary immunity, although the supply had to be replenished regularly or her defenses would revert to near-zero. It is a tribute to Alison's endurance—and to her ability to produce a slight amount of ADA enzyme on her own (2 percent of normal)—that she lived to see PEG-ADA. The drug has changed her life, and that of most of the handful of children in the world who have been taking it. "I believe I lead a normal life now," Alison declared not long ago, "and that's something I never thought would be possible. I was afraid before. I'm not afraid now."

At the age of twelve, when the authors first met her, she was a sweet child, though somewhat subdued, as befitted someone who had never known what her next infection held in store. But years later she has grown into a lovely, cheerful, outgoing young woman eager to embrace the future. "My friends catch things now that I never catch," she says exuberantly. "I feel so strong! I really do. Before, I felt there was something missing. Now I have all this energy and am able to do all these things. It's almost like a miracle."

The miracle, however, comes with no guarantees. Although each day allows Alison to set a new longevity record for a patient with ADA deficiency, her thymus function surely will decline with maturity—it does so in normal people, and Alison never has had much to begin with. Eventually, her body may produce few T cells for PEG-ADA to save. In that case, only gene therapy would offer her a chance at permanent cure.

Duke's Hershfield, who has monitored more than a dozen ADA children on the drug over the course of eight years, has called PEG-ADA a "milestone." But in spite of the preparation's powers, many in the field concur that the ideal solution is gene therapy. Intervening with new genes at the point at which T cells are freshly stamped out by the bone marrow's stem cells would keep the patient's immune system intact from the outset. As Billie Holiday used to sing, "God bless the child that's got its own." So, too, might fate smile more broadly on a child with rugged, genetically sound T cells, rather than on one with defective cells spared from harm by a gossamer curtain of PEG-ADA.

Problems, furthermore, have arisen with PEG-ADA that throw its long-range effectiveness and value into question. Some of the children on the drug have shown little, if any, immunological improvement, and a few have even died despite receiving the drug. Moreover, the success rate of the drug is based on a relatively brief experience. Whether it will continue to boost immunity over a period of years, and whether unanticipated side effects may show up over time, is currently unknown. Finally, the cost of PEG-ADA has become astronomical, challenging the ability of all but the wealthiest or most heavily insured recipients to pay for it.

The Cutshalls have been less than impressed with PEG. "Cindy hasn't done all that differently on PEG," says Susan Cutshall. "She seems to have had better endurance. On the other hand, she has had more sinus infections, but they have been easily treated with antibiotics." The most telling incident involves Cindy's first confrontation with chicken pox, which happened shortly after she went on the drug. In spite of the PEG-ADA, she had to be hospitalized and treated with the antiviral drug acyclovir.

The other thing that bothers them about PEG-ADA is the painfulness of the shots. "The shots are in the hip, and very painful" says her mother. "It's three ccs and it takes twenty seconds." Ask little Ashi

DeSilva about the PEG-ADA shots, and she says, "They always make me cry."

By the late 1980s, despite their wonder drug, ADA-deficient patients like Alison, Ashi, and Cindy would find themselves inevitably swept into the orbit of the high-flying French Anderson. He was fast becoming the driving force behind gene therapy, and for these children, as well as for their stricken brothers and sisters around the globe, gene therapy seemed to hold out the last, best hope.

Though Alison Ashcraft has often expressed her desire to be an actress or a marine biologist someday, she has long known in her heart that medical science is going to have to buy her that someday. In principle, she accepts the inevitability of gene therapy. "PEG-ADA helped take the urgency away," she has said. "But if they can show they can do gene therapy safely, I'd be all for it. I'd like to be cured."

The road to gene therapy, however, has been a long and tortuous one. On the afternoon that he importuned John Hutton to send him the ADA gene, French Anderson little appreciated just how arduous the journey would be. All he knew, as he awaited the arrival of the gene, which would come by express delivery in a nest of Styrofoam and dry ice, was that in spite of more than fifteen years in pediatric medicine, he knew next to nothing about ADA deficiency. "I'd barely even heard of it until 1984," he concedes.

Endlessly resourceful, he soon found a way out of that box.

6

THE ROAD FORKS

With John Hutton's ADA gene safely stowed in his NIH refrigerator, French Anderson began prepping himself for the ADA race with the youthful intensity of the cross-country runner. There was, however, that discomfiting basic problem to be dealt with: his virtual ignorance about ADA deficiency. This he sought to remedy by scouring NIH's labyrinthine community of physician-scholars for a knowledgeable collaborator. At first, he came up empty. But his surgeon-wife, Kathy, had not spent a lifetime networking in pediatrics for nothing, and her experience proved invaluable. "Talk to Mike Blaese at the Cancer Institute," she advised her husband. Blaese, she assured him, was the guru of childhood immune diseases.

Anderson had heard of this immunologist, but not in the context of ADA deficiency. Rather, Blaese had been touted as the leading expert regarding an extremely rare but disastrous hereditary immune disorder known as Wiskott-Aldrich syndrome. As luck would have it, however, Wiskott-Aldrich proved similar to ADA deficiency.

Anderson and the big, bearlike Blaese hit it off immediately. Within a short time of their meeting, Blaese was teaching Anderson the ins and outs of T-cell biology and cooperating with him in joint lab meetings during which they mapped out the possibilities of gene therapy for ADA deficiency. Remembers Blaese, "I had my lab switch over to working full-time on the problem, and we very quickly established T-cell lines from ADA-deficient kids." Anderson also forged an alliance with Princeton's Eli Gilboa, whom he terms his "mentor in retroviral gene transfer." Within a year, the new team had succeeded in duplicating what the Boston group of Stuart Orkin and Richard Mulligan had already done, that is, transfer a test gene into a strain of mice. The gene was not therapeutic. It was merely a selectable marker that could be identified in the lab later to determine whether the retroviral transfer of the foreign gene had worked in the test animal. But it was a start. And it had leveled the playing field between the NIH group and the Orkin-Mulligan coalition.

The collaboration could not have come soon enough for Blaese, a warm, rock-solid pediatrician with a human touch and cerebral gifts, but a man who does not seem to ignite the passions that Anderson does. In his early fifties, Blaese is a self-described "card-carrying immunologist," who has been steadily lengthening the lives of his young patients—from three years to their teens and more—since he signed on as an NIH researcher, eventually to proceed to taking care of more than half the Wiskott-Aldrich youngsters in the nation. The children, doomed as they generally were, deeply moved him. Like most specialists in rare diseases, Anderson works in a tight little community. "I've spent a lot of time hugging parents and crying in the hallways when their kids die," he admits.

Only males are born with the Wiskott-Aldrich syndrome. The result of a flawed X chromosome, the ailment allows the boys to produce no more than a token array of immune cells and antibodies. Blaese had been treating such boys since 1966 for hemorrhages, eczema, and repeated infections and been watching them like a hawk for their tendency to develop cancer, particularly leukemia and lymphoma. Such kids, Blaese had written, "are a study in paradoxes." They have normal B-cell levels yet poor antibody responses, because the antibodies they make have greatly shortened life spans. They cannot mount responses to such common microbes as candida and those that cause mumps, diphtheria, or tetanus. On the other hand, Wiskott-Aldrich children have normal proportions of T cells in their bloodstream, and their cells seem to respond normally to nonspecific antigens—or generalized threats.

Blaese studied this disease with Robert Good and Thomas Waldmann, both among the fathers of modern immunology. The field has concentrated on exploring the links between immune system breakdowns and the genesis of cancer—of Blaese's first dozen Wiskott-Aldrich patients, five developed cancer, a rate he calls "screamingly high."

A native Minnesotan whose mother was a registered nurse, Blaese entered the University of Minnesota Medical School in 1961, upon his graduation with highest honors from Gustavus Adolphus College, in St.

Peter, one of the fine private schools that grace the Twin Cities area. By no means as committed to science as French Anderson, young Blaese didn't know what he wanted to do and opted for medicine only after falling asleep in engineering classes. That brought him to the giant university. "I was a freshman medical student, very much in love, who needed a job so I could support my wife," he says.

At the university, Blaese started wandering the halls of the medical school looking for work, he says, "Bob Good offered me a job, and I did student research with him for four, wonderful years. He really got my juices flowing. And that led me to pediatrics because the diseases are clean, if you will. They're not cluttered up with fifty years of alcohol and tobacco and other things." Blaese contributed to several papers with Good, as the senior scientist studied bone marrow transplantation and cellular engineering to correct immunodeficiency diseases like ADA deficiency. T cells were to prove to be the lasting link between the two men. Just as T cells a generation earlier had compelled Good to perform history's first successful bone marrow transplant on a SCID child, T cells would lead Blaese to an even more revolutionary therapy for the same sort of youngster. What Good had tried to do with marrow, Blaese would try to do with genes.

After completing his training, Blaese moved to the National Cancer Institute in 1966 to work with Thomas Waldmann. A protein metabolism pioneer with more than four hundred scientific papers to his credit, Waldmann spearheaded a lab of superachievers in the exciting days at NIH when genes that controlled the immune response in white cells were being cloned. The Bethesda immunologists, arguably the strongest group in the world, began by unraveling the sequence of events by which blood cells communicate with one another in response to an invader. In 1973, for instance, Alan Rosenthal and Ethan Shevach discovered that the drama opens when a patrolling macrophage tips off a helper T cell that it has found something odd to eat. This occurs, as Robert Gallo's lab discovered by accident in 1976, because T cells talk to one another by secreting a protein the team named interleukin-2, or IL-2. The finding of IL-2 and subsequent interleukins (or "between-white cells") immediately presented the possibility of someday switching T cells on or off. In an AIDS patient who has lost most of his T cells to HIV, for instance, one would want to switch them on; in such disorders as rheumatoid arthritis or blood cancer, one would want to switch them off.

As he labored in his small lab at the National Cancer Institute, Blaese found himself drawn to the gigantic lab just around the corner. It was a powerhouse run by Steven Rosenberg, NCI's chief of surgery, and Blaese became a fan as he tracked Rosenberg's pioneering and controversial efforts to learn how to supercharge the human immune system and unleash it against cancer. Their friendship would bear fruit soon, but in the meantime Blaese's lab was running full-tilt with French Anderson's in the race to treat ADA deficiency.

In the summer of 1985, a postdoctoral fellow in Blaese's lab, Don Kohn, joined with an Anderson fellow, Phil Kantoff, who had constructed an admirable retroviral vector called SAX, and showed that they could cure ADA deficiency by using SAX to ferry healthy ADA genes into T cells cultured from a boy named Jimmy Fox, who died of the disease. The genes made ADA in the test tube—not for very long, but long enough and in sufficient quantities to be considered a therapeutic dosage if the cells were to be given back to a patient, at least in the estimation of the NIH scientists. The feat involved technical maneuvering by a cancer-causing human retrovirus, HTLV-1, which induced the child's cells to keep dividing in culture, thus solving the key problem of longevity in vitro. Because of the cancer-virus trick, which seemed to turn the game into an exhibition rather than a real advance for patients, rival scientists weren't impressed by the work. One of them called it "a parlor trick."

In any case, when it came to trying to put the genes into living tissue, the field was stalled. No matter what any team tried, the ADA gene refused to go reliably into mice. Occasionally it worked, but most of the time it just refused to. Nobody knew why.

At that point, in the mid-1980s, a great schism occurred, the ramifications of which are still felt today.

Up in Boston, the Mulligan-Orkin team dedicated itself to solving the problem, snipping the ADA gene, toying with the virus, whittling enhancers and promoters, slowly working the gene into the mouse system. That was the only way, Mulligan maintained, to gain the basic scientific understanding that ultimately would lead to safe and efficient gene therapy. Ironically, it later turned out that the version of the gene Orkin had clone had an infinitesimal mistake that coded for a defective protein, thus hampering their research efforts.

But by then, in Bethesda, the Anderson-Blaese team had decided there wasn't much future in mice when one wished to cure human beings. They knew that, by longstanding scientific convention, the well-traveled road leading to human experimentation was to be unremittingly linear. One went from tissue culture to mice to primates. Only then did one move on to people. Each step had to lead to the next. Shortcuts were not acceptable to the scientific community, because failure in one system rarely led to success in another. If one wanted to be considered a real scientist, one trod the beaten path, solving each problem as it came along before proceeding to a new level. The Anderson-Blaese group knew all this, and yet it was about to break with tradition.

Despite its failure with mouse systems, the NIH team threw away the book and struck off in a new direction. At a gene therapy conference at the Medical College of Wisconsin in Milwaukee, Anderson explained the crucial point: "selective advantage," he called it.

"When you do a matched bone marrow transplant on a kid with ADA deficiency and check him a year later, all the marrow cells are his own, except the T cells. Those are from the donor, and they're enough to do the job. The new stem cells have a *specific advantage* and overgrow the patient's own abnormal cells and repopulate the patient and make him healthy.

"Theoretically, if we could get new ADA genes into stem cells of these patients, those cells should have a *selective growth advantage.* So you should only have to be able to get in just a few cells to cure the patient," Anderson argued.

"The difficulty is there's no animal model for ADA deficiency," he said. "The only way to tell is in a patient. So you have a vicious circle. You don't want to give new genes to a patient until you're sure they'll work. And you won't know if they'll work until you give them to a patient."

Because of this, he said, his lab had disbanded its mouse research and gone directly up the ladder to monkeys, which he called "a clinically applicable model." Let other labs work with mice. Anderson and Blaese wanted to save kids.

"Nobody knows what may happen when new genes enter the body of a human being—it's a total black box, despite what anyone tells you," Anderson conceded. "But test-tube and animal research can only tell us so much. Eventually you have to try it in a person. The potential benefits to a patient have to justify the risks. And as of now, the risks are unknown."

Ominous as that sounds, there was nothing new here, Anderson insisted. "The same criteria are required for gene therapy as for any new experimental procedure on humans—that the gene reach the area of pathology, correct it, and be safe. Delivery. Expression. Safety. Those are the basic issues."

By this time, however, Anderson was dealing with an issue even more basic, one that threatened to end his career as a gene therapist before he'd even come up to bat. The skyrocketing national deficit had prompted passage of the 1985 Gramm-Rudman-Hollings deficit reduction act and sent a chill of fear through NIH and other basic medical research laboratories that depended on public funding. In February 1986, Jack Orloff, scientific director of the heart institute, called an emergency meeting of lab chiefs and announced that the NIH budget would be cut by a walloping $236 million that year. The chiefs were free to try to come up with their own funding, and Orloff suggested they look at the Technology Transfer Act of 1986, a little-known law that might nevertheless be helpful. The act, which stemmed from concerns about foreign competition and the feeling that NIH lethargy was keeping great discoveries from leaving the labs, instructed government scientists that it was their patriotic duty to collaborate with private industry to speed publicly funded research to the public. NIH and other agency scientists were allowed a minimum of a 15 percent share of any royalties, not to exceed $100,000 a year, and counseled that under the act such arrangements—known as cooperative research and development agreements (CRADAs)—did not constitute conflicts of interest.

CRADAs did constitute a moral quandary, however, that ultimately could affect all of biomedicine. In the past, academic researchers dreamed about Nobel Prizes; but now when one made an important discovery, one called a venture capitalist and started a company. Even Nobelists seemed to have no shame about lending their names, as the line between basic research and applied technology became thinner. Because of the genetic revolution, for the first time, biologists had a chance to make a buck, and they felt if they didn't protect the fruits of their discoveries, somebody else would steal their ideas and market them. CRADAs formalized this point of view and extended it to the federal government establishment, where it struck at the very heart of the spirit of open research and the lab-to-lab communication that were NIH hallmarks. When Marshall Nirenberg was cracking the genetic code, he reportedly used to lock his work in a safe each night to protect it from rivals, but that was the exception, not the rule. Openness and sharing of data were standards of behavior at NIH. The Technology Transfer Act, no matter how altruistic the intent, placed dollar signs before the eyes of biomedical researchers. Under federal law, government scientists were not only allowed to work out commercial patent rights and exclusive licensing rights to the commercialization of discoveries that the public had paid for; they were *supposed* to do it. Many basic researchers considered the act outrageous. In the past, NIH had tough rules about what outside work a scientist could do, with payments and hours worked rigidly controlled. More important, all NIH research had to remain in the public domain. A company could apply to the government for the rights to the fruits of research the company had sponsored, but there were no guarantees. Now this was all going to change. Typical was the warning issued by Anthony S. Fauci, director of the National Institute of Allergy and Infectious Diseases, who told *Science* magazine that he feared an outbreak of "CRADA fever." It especially worried Fauci that when a federal scientist joined forces with a biotech company, CRADAs actually required that certain proprietary data be kept secret. Fauci said, "For the first time in 21 years at NIH, I detect an inkling of hesitation among scientists about sharing information."

At the lab chiefs' meeting, Anderson quickly calculated that the new NIH funding cuts would cost him 38 percent of his budget for the remainder of the current fiscal year. He was in deep trouble. Desperate, he told Orloff about a call he had received the preceding December from a man named Wallace Steinberg, chairman of Healthcare Ventures, a venture-capital fund in New Jersey. Steinberg wanted to get in on the ground floor of gene therapy and start a company, and he wanted Anderson to head it. Anderson had received many such calls over the years, but as the quintessential basic researcher who virtually had grown up at NIH in an atmosphere of scientific discovery untainted by commercial gain, he viewed such business ventures with disdain. "But I suddenly realized that I was going to lose four of the best people in my lab," he recalls. "I couldn't let that happen." Orloff suggested he check out Steinberg.

Whenever Anderson has a question about finances related to his job—be it about travel vouchers, expenses, or potential conflicts of interest—he puts everything on the table where anyone can see it and writes lots of memos. This time, Anderson dispatched a blizzard of memos to his research colleagues at NIH, seeking their advice about the funding cuts and asking what they were planning to do. Then he happened to receive a call from a woman named Deeda McCormick Blair, who for years had been with the Lasker Foundation working behind the scenes as a lobbyist in behalf of congressional support for science. Blair assured Anderson that Steinberg was trustworthy and that his firm had a philosophy Anderson would be able to live with. After more checking, Anderson met with Steinberg and agreed to cooperate with the start-up company but firmly declared that he would not run it. His sole interest, he maintained, was to help bring gene therapy on-line. Steinberg quickly agreed to Anderson's terms, and in July 1987 the company called Genetic Therapy Inc. was born with a $2.5 million investment by Healthcare Ventures. Soon GTI, based in nearby Gaithersburg, Maryland, raised another $5 million financing and started shopping for a president who would meet with Anderson's satisfaction. Anderson was unhappy with several candidates, until he met Jim Barrett.

Barrett, who was just turning forty, blended unusual talents—he had a doctorate in biochemistry and an M.B.A. and had spent thirteen years working his way up to the presidency of the clinical laboratory division of SmithKline Beckman, the pharmaceutical and medical-equipment giant. Barrett left SmithKline in 1982 to take over a troubled little biotech company, Bethesda Research Labs. At the time, the firm had dozens of research projects in the works. Barrett whittled them down, arranged a merger with a profitable partner, and turned around the company, which was renamed Life Technologies. Sales soon soared to more than $120 million, the company went public in 1986, and with its success Barrett acquired a reputation as a miracle manager. The vibes were good when he and Anderson met, and GTI finally had its first boss.

NIH scientists have today signed more than two hundred CRADA agreements with private industry, but the first was Anderson's deal with GTI, worked out in the spring of 1988. It seemed open and above board, with GTI taking all the risks and potentially making money down the line. Essentially, the company got exclusive commercial rights to any gene transfer technology developed by French Anderson. NIH, in return, would receive payments and royalties once the company started to sell or license products. Anderson, who was not allowed to take a salary or hold any stock in GTI, would be eligible to earn royalties up to $100,000 a year—which he quickly was to defend as "a pittance," compared with what he could earn if he actually could bring the revolutionary therapy into existence. Of course, all bets were off if he left NIH. Then he could become filthy rich. But what mattered at the time was that he had a commercial collaborator that would offer him the same laboratory resources he had lost as a result of NIH funding cuts. In fact, three senior staff fellows from Anderson's lab, who would have been forced out, quickly signed on at GTI, starting a trend that continues today.

Although it would be several years before the company made any money, GTI's existence freed Anderson to make big plans. He viewed the company as a welcome adjunct, a functioning lab capable of growing the large quantities of gene-carrying viruses he would need to perform experiments. But nipping at his heels was the appearance of conflict of interest, the recurring charge that he had funneled government-sponsored genetic engineering to a single company. In late 1990, the nipping got nastier when GTI hired none other than the chairman of the august committee that governed human genetic engineering, Gerard J. McGarrity. Low-key, charming, and a capable cell biologist who looks like Phil Donahue, McGarrity had made no secret of his enthusiasm for gene therapy, or that he was a buddy and admirer of French Anderson. In fact, McGarrity quietly used his considerable clout in the days ahead to show his support, in the process helping to make his friend's dreams come true. Nor did the mobilizing of powerful forces by little GTI go unnoticed in the volatile cosmos of biotechnology, where start-up R & D companies are a nickel a dozen. In 1991, as often happens when some meteoric little outfit gets lucky and conjures up something promising, a world-class pharmaceutical giant, Sandoz Pharma, Ltd., purchased $10 million worth of GTI stock and promised that it would support the cause by handing over another $13.5 million over the subsequent three years to subsidize various projects as they happened to come along. By the end of 1991, the company reported to stockholders that it had accumulated $20.8 million in cash and marketable securities, all to provide lab services for French Anderson and to keep him from losing his people.

In the meantime, before GTI, working through Blaese, Anderson had joined forces with another Robert Good protégé, Richard J. O'Reilly, chief of pediatrics and director of the bone marrow transplantation program at Sloan-Kettering, for a series of primate experiments the significance of whose results remain debatable even today.

The idea was to see whether they could put a human ADA gene into fourteen rhesus and cynomolgus monkeys via the bone marrow and have it persist. If it did—if the gene worked at all—the team would have taken a giant step nearer to gene therapy. Monkeys are a lot closer genetically to human beings than mice are.

The retroviral vector chosen to do the deed was their house favorite, SAX, and contained both the selectable neomycin resistance gene from *E. coli* and a copy of the protein-coding sequence of the human ADA gene, the expression of which is regulated by a promoter taken from a monkey leukemia virus. With such chemical scavenging, the scientists thought SAX could pull off a miracle in monkeys at a time when nothing was working for them in mice.

The stem cells of the monkeys were the targets. But as the experimenters soon found, to their dismay, the procedures were miserably difficult. Bone marrow samples were removed from the animals and stored, after which the monkeys were subjected to blasts of radiation—more than twice the amount necessary to kill their marrow. The step was mandatory to destroy the resident marrow and give the transplant with the gene-altered cells a chance to engraft. Doing an autologous bone marrow transplant on a monkey, it turned out, is no different from doing one on an infant. "It's a huge deal," Anderson wearily observes, shuddering at the memory of the sterile isolation and around-the-clock intensive-care nursing for the monkeys and costs that skyrocketed to $30,000 per animal. In mouse research, scientists usually gather marrow from several animals bred to be genetically identical, pool it, manipulate it, and put it back. With monkeys, though, the animals are not genetically the same, and the specter of rejection meant that marrow had to be removed from an individual animal, treated quickly—far more quickly than the team had previously suspected, lest it become contaminated—and transplanted back into the monkey whence it came, with the scientists making sure they gave the animal enough active cells that it could reconstitute its blood-making system.

Anderson wasn't pleased that the team had to kill the animals' resident marrow with x-rays. "If you've failed, the animal dies. There's no other option," he points out, adding that few ethicists would have been pleased by the caveat if presented with a human gene therapy proposal. And then Anderson's monkeys, terribly ill, had to be painstakingly rehabilitated and repeatedly tested over six months to see whether the transplant had taken.

The team never expected so many complications. "It was a mess," Blaese sums up today.

The lab turned into a zoo. First off, the primate marrow cells proved very hard to infect with new genes. To begin, Anderson had tried mixing the monkey marrow overnight with the helper virus, because that's the best way to do it in mice. But he found that monkey marrow didn't react like mouse marrow. The cells died. That left too few of them to reconstitute the monkey's blood system. In short order, the first two animals died. The lab was appalled. How could they even think of moving into children? Then two more animals died. How could this fiasco continue? Only later, after a good deal of trial and error, did the team learn to mix monkey marrow cells with virus-infested fluid harvested from the helper cell tissue cultures. That worked better, for some reason. Ten monkeys lived to reconstitute their systems. It was a triumph, of sorts.

But only half of the animals survived in the long term, and *in only one monkey*—Monkey Robert, as the cynomolgus whimsically became known—were the scientists able to detect more than a ghostly wisp of the ADA gene itself. He kept making his own ADA, of course, but levels of the human enzyme seemed to rise briefly to half a percent of his own. In four of five of the monkeys studied, the researchers were able to show that a few cells were producing trace amounts of the human ADA enzyme, meaning some type of infection had indeed occurred. The trouble was, the enzyme expression, besides being faint, was not detected for very long after transplantation. It began to decline after 60 days and had totally vanished in all animals by 170 days (although the *neo* marker could be tracked in T cells considerably longer). It looked as if Anderson had infected some mature immune system cells, which then pooped out as nature had intended them to do when their life was over. Therefore, Anderson really had no evidence that the ADA gene had been inserted in the chromosomes of early stem cells that would repopulate the blood system and cure the animal. Why had the experiment failed? Perhaps the monkeys were too well equipped by nature. Monkeys are higher organisms than mice, live up to twenty times longer, and have evolved more-elaborate defense systems against foreign invaders. Maybe this protection extended to human ADA genes introduced by viruses. Who knew?

Ironically, he later decided that the only reason he got any positive data at all probably was that the two animals that showed the best response (Robert and another simian, named Kyle) actually were mildly ill with undiscovered malaria and other parasites at the time of the experiment. "It is probable that their bone marrow was exceptionally active and, therefore, receptive to gene transfer," Anderson wrote in 1992. "Nowadays, animals always have their bone marrow stimulated [by drugs] before transfer experiments."

Anderson had harbored greater hopes for his monkeys. "I wanted them to make large amounts of human ADA, and to make it for their whole lives," he says wistfully. In point of fact, they made precious little, and did so for only a few weeks.

There are times when William French Anderson seems like a man pulled in two directions at once. There is the circumspect Anderson, NIH section chief, a man who cloaks himself in medical ethics and sound science. He once reassured the authors, "When we have good results in monkeys, when we could sit down with our own sister and say, 'Trust me with your child, we are ready to do gene therapy,' only at that point will we move ahead."

But then there is the impatient French Anderson, rationalizer, eager beaver, harvester, many would say, of half-ripe fields. In spite of his failure with mice and, later, disappointing monkey data, he once again plunged ahead. He submitted to the Recombinant DNA Advisory Committee an application to perform human gene therapy anyway. Only he didn't call it a protocol. He called it a "preclinical data document," sort of a "This is what we would do if we really were going to do it" trial balloon that took up several hundred pages and looked like the Washington, D.C., telephone book.

The picture presented within its bindings resembles the plot of the Kurosawa film *Rashomon,* in that the truth of the data depends on the person interpreting it. The human ADA gene itself, admitted Anderson in the document, was never detected in the monkeys that lived. But its spoor was present in the form of very low levels of human ADA that were produced for a while. These levels measured a maximum of 0.5 percent ADA production instead of the 5 to 10 percent that Anderson sought. The neomycin marker gene, meanwhile, could be detected mainly in *T cells*—a very, very interesting finding, though at the time nobody could figure out what it meant. Most discouraging was that the very low expression of human ADA dropped to zero within five months of transplantation, although in one animal, Robert, expression of the marker gene could be barely detected as late as seven months.

The monkeys that survived their ordeal seemed to have suffered no permanent damage, which was a hopeful sign that reflected the safety of gene transfer, at least to Anderson. But there was no proof that he had managed to infect bone marrow stem cells with new genes or that he could logically expect more than the most fleeting, transient effect in a human subject.

Nonetheless, to the perennially upbeat Anderson, the acceptance and expression of the ADA gene by a few monkeys—particularly stouthearted Monkey Robert—showed that the retroviral-delivery idea essentially worked. Besides, reasoned Anderson, selective advantage couldn't really be tested in monkeys. Monkeys have their own ADA genes, make their own proteins, and wouldn't be sick at the start the way human patients would be. There was no reason to think that transplanted human ADA genes would have any selective advantage in monkey cells.

However, like the Marines, all a *human* body would need was a few good cells, Anderson believed. Selective advantage would take care of the rest. The healthy cells would proliferate, while the sick cells would die. Eventually, they would build a new immune system for a dying child and save him!

To Anderson, "the issue is whether a small amount of expression in a non-selective system [the monkey] could be amplified in a selective system as an ADA patient ought to be." That was pivotal to his think-ing. In other words, monkeys have no need of human ADA, since they have plenty of monkey ADA. But would ADA-deficient children benefit from even a tiny bit of ADA? "That's a question for the RAC to answer," said Anderson, passing the buck. And so, he submitted his 500-page doc-ument to the RAC's Human Gene Therapy Subcommittee, on April 24, 1987.

"They've got it now," he said afterward. "It will be their decision, not mine. I can punt."

In his heart of hearts, though, French Anderson was not yet confident enough to go into a patient. The preceding December, upon hearing the news of Duke University's success in treating ADA kids with PEG-ADA, Anderson asked the members of his team whether they wanted to press on with their gene therapy research. Many did, until Anderson asked whether they would feel the same way if the patients were their own children. "We've taken an informal screening of our own lab people who really know the data," he said. "I asked them, 'If this were your own child, would you be willing to do it?' A third of our lab said, yes, they'd be willing to try it now. A third said they were hesitant. And a third said they didn't know. So we were split."

On reflection, with two-thirds of his own lab expressing hesitation, Anderson decided not to push for permission to undertake a human gene replacement experiment. He called it off once consensus evaporated. Even so, he told anyone who would listen, the personal physicians who dealt with immunodeficient patients and their families were still urging him to go ahead. They kept pressuring him to try, to try anything. "The chances of hurting patients are very slim," he admitted. "But if we do nothing, the patients most likely will die."

There was a method in Anderson's muddle. By putting a halfhearted protocol before the RAC, he certainly took a big chance. But he also was managing to stake out the territory as his own and bring his agenda to public attention. From then on, other proposals from other quarters would have to be compared with his. In fact, pushing forward, Anderson told William Gartland, the RAC's executive secretary at the time, to send the telephone book–sized document to fifteen of his competitors—such as Mulligan, Orkin, Friedmann—and to other experts, such as the retrovirus Nobelist Howard Temin, of the University of Wisconsin, and the PEG-ADA developer Michael Hershfield, whose successful enzyme replacement drug for ADA youngsters was starting to loom as a big rival to gene therapy.

Reaction to the Anderson document ranged from tepid to rabid, but it was almost uniformly negative. The ADA enzyme levels were too low; the risk of viral contamination, too high. PEG-ADA was emerging as a viable alternative therapy. All but one of the reviewers to whom Gartland sent the paper agreed that the time was not yet right for gene therapy. Then came another jolt. To add insult to injury, the prestigious journal *Nature* rejected the scholarly paper reporting Anderson's monkey results, a body blow to the proud NIH team. Only when the *Journal of Experimental Medicine* later accepted the study was Anderson able to save some face. The journal gave the paper a validity among physicians that Ph.D. scientists seemed loath to provide. Anderson's pique must have been keen. Besides, a new field needed its own journal. Hence he started his own, which meant he would never have to depend on another publication. With its debut in 1990, *Human Gene Therapy*, edited by Anderson and published by Mary Ann Liebert, quickly rose to preeminence as the nascent field's sounding board. Even so, Anderson's end run seemed amusingly ingenuous to many in the field.

As he endured his colleagues' withering response to the preclinical data document, Anderson could have been expected to pout. No way. Again, he mounted a fresh horse and dictated an eighteen-page letter to Gartland. "Overall, we agreed with essentially all of the concerns raised by the reviewers," the letter began. But within a few paragraphs, Anderson was making a new case. When there's nothing left to offer, he argued, when bone marrow transplantation and PEG-ADA have failed (even though Anderson had no evidence that PEG-ADA would ever fail), what then? "Would it be ethical to deny the potentially life-saving gene therapy protocol in the face of certain death?"

Anderson said that he knew of three infants that soon would fall into what he considered the "last hope" category. Because of that, he said he felt a limited clinical protocol should be submitted soon.

But none was.

By the end of 1987, after the December 7 meeting of the subcommittee, it became apparent that nobody was ready to let French Anderson do gene therapy. Nobody. Briefly, the embattled lab chief seemed to lapse into a blue haze. He suspended his monkey work. He avoided the bench. He immersed himself in planning for the American tae kwan do team's appearance at the 1988 Seoul Olympics.

"It was a depressing time," remembers Mike Blaese, "because we weren't getting anywhere. Yet I *knew* we could cure these kids, because we'd done it two years earlier in the test tube. I also knew that we had a new research fellow coming to replace Don Kohn in July of 1987. His name was Ken Culver, and I already had his first project waiting for him."

Bearded, blow-dried, energetic, and wearing his heart on his sleeve, thirty-six-year-old Kenneth W. Culver is a quintessential Iowan. He brought compassion, lab skills, and a near-religious determination to the little NIH team. It would be Culver's hands, working with the cells of dead children, that would conduct the laboratory work leading to the first human gene therapy experiment.

Culver has a mystical bent that mirrors Anderson's own. He doesn't believe his relationship with Anderson and Blaese was due to "a chance occurrence." Like the young Blaese, Culver agonized about becoming a doctor. Although he had enrolled at Drake University after a professor friend persuaded him to try premed, Culver kept waiting for some sign that this was what God wanted him to do with his life. He was still drifting when he entered the University of Iowa Medical School. But shortly after his arrival, Culver received his sign.

"I lived a mile from campus, which I used to walk every day," he recounts. "Near my apartment was a small park, with a path that went down into a forested valley and then back up. One day as I was walking down that sidewalk, I suddenly felt very odd, like, somehow everything was off balance.

"As I came out of the bottom of that valley, I felt like God picked me up, and then set me back on the sidewalk. Never again have I questioned why I was there. The sureness of the hand of God was clearly the guidance and reassurance I had been searching for."

Eventually, he decided to become a pediatrician. At first, he wanted to be a well-baby doctor, because it would be easier than caring for sick and dying children. "However," says Culver, "early in my internship, I realized that I really loved the kids who, most likely, were not going to survive."

In 1979, as he slogged through his medical apprenticeship, Culver paused to marry a nurse, Cynthia Alloway. They now have two boys and a girl. When it came time for his residency, Culver moved west, to the crack immunology unit at the University of California at San Francisco. There he took care of a fifteen-year-old girl named Laura who had been fighting leukemia since the age of four. Following three relapses, she was admitted into the UCSF bone marrow transplant unit.

The seriousness of the treatment is often lost on the public, which generally hears only about its high cost in fund-raising campaigns, Culver says. "A bone marrow transplant is much different than usual chemotherapy treatments. First, the child is placed in a sterilized room for about six weeks. During that period, they can't see anyone but their mom, dad, doctors, and nurses all covered with sterile gear. To visit, we all would put on scrub suits, masks, hats, and gloves. This is tough, especially for a fifteen-year-old girl who had her room plastered with letters, cards, and posters from the friends and family she might never see again."

Second, the amount of chemotherapy and radiation the child receives is lethal—the doses are required both to kill the leukemia and to prevent her from rejecting the bone marrow transplant, which in Laura's case came from her brother.

"Following these treatments and the injection of bone marrow cells," Culver relates, "we waited. Gradually, we began to see Laura's brother's marrow grow inside her, and we began to celebrate. But then her liver began to fail.

"The toll of all the chemotherapy over the years was too much, and soon Laura became comatose and died. Before she did, she demonstrated a level of courage that you and I may never reach."

Culver was still grieving over Laura when he met Chelsea, an infant from Alberta, Canada, who suffered from ADA deficiency. "What a beautiful child she was!" he remembers. "Not just physically, but also in personality as she endured the difficulties and horrors of modern medicine."

As Culver worked on Chelsea, "she developed a very severe infection and almost died. We were able to get her feeling better again. Our last hope was a bone marrow transplant from her father, after which we sent her home to Canada.

"But, again, the transplant didn't work. Even though she had little immunity to fight infection, she did have enough to reject a partially matched bone marrow transplant from her father. So we brought her back and gave her yet another transplant, this time from her mother."

But Chelsea also had to undergo chemotherapy that time, and radiation to prevent her from rejecting the donor marrow. That meant her only chance at life was the success of the bone marrow transplant. The night of the procedure, Culver says he received a "powerful urge" to go pray with Chelsea's family. But, not knowing whether that would be professional, he decided simply to trust in God. When his beeper notified him that the cells were nearly ready, Culver excitedly headed to the hospital, retrieved the cells, scrubbed, and entered his patient's sterile Plexiglas environment. He prepared to give her the intravenous infusion of donor bone marrow cells. But then, professional or not, Culver paused and asked the little girl's parents to pray with him.

"I prayed to thank God for Chelsea, her family, and for having given us the ability to do bone marrow transplants," he says. "Then I injected the ten milliliters of marrow cells that would hopefully cure her."

A few days later, despite all the inventions of the physician's art, Culver could not ignore the brutal, agonizing truth: the transplant, once again, had failed. Chelsea was going to die. The family gathered at the bedside of the comatose child who had given her all. They had made their decision to postpone the inevitable no longer. "As I disconnected her from the ventilator and waited for the last heartbeat, we were all in tears," Culver says, his eyes staring emptily into the distance. "I was upset and confused with God."

Hours later, he trudged off from work in a pouring San Francisco rain. He was determined to find Chelsea's parents, but in his despair he'd forgotten to bring along their address. "I didn't know the exact place," he says, "but I stood out in the rain yelling their names at the address where I thought they might be—no answer. Then a nice woman let me through the security gate. There I found them, in a barren apartment still somewhat in shock."

As generally happens in such tragedies, the child's dazed parents simply didn't know what to do next. They didn't have a lot of money, so Culver telephoned friends who agreed to arrange inexpensive embalming. He took the parents to a funeral home to shop for caskets, but they were too costly. Accordingly, Culver brought Chelsea's father and her maternal grandfather to his own house, and there, as the rain rattled on the roof of this appalling night of young death and overpowering grief, the three men constructed a simple casket out of wood.

"This seemed therapeutic to her dad as he labored," Culver recalls, his voice catching. "We next drove around town to obtain the proper papers for her to be returned to Canada. We took our homemade casket down to the mortuary and sealed her body. I will never forget standing in the back corner of that fancy funeral parlor room, where we used a power drill to seal the coffin after placing her precious body inside."

It was very quiet in the car as they drove Chelsea to the airport for her flight back to Canada. "We logged her in to the cargo bay and slowly drove off."

Her father looked back. Would they take good care of her? he asked. Culver was crying.

Since Chelsea died, in January of 1986, Culver has had occasion to visit her parents in Canada. Their intense interest in treatments for ADA deficiency has never wavered. It was they who sent
Culver many articles about gene therapy and, most significantly, the
work of Anderson and Blaese at NIH.

His residency completed, Culver started looking for a new job. One
day he met Don Kohn, a young job-hunting NIH researcher in the UCSF
hospital hallway. Kohn suggested to Culver that if he was interested in
working in Bethesda, he could stay at his house.

"I ended up taking his job, buying his house, and buying his car,"
Culver says with a wistful smile. "When we signed the papers, it turned
out we also had the same birthday. Maybe God had his hand in all of
this."

If so, Culver's first day on the job was indeed providential.

"When I walked into the lab, my boss, Mike Blaese, quietly told me,
'I'd like you to work on a cure for ADA deficiency—the disease that
killed Chelsea.' Somehow he had taken the trouble to find out about her
and about me. I was thrilled."

7

THE TROIKA FORMS

By the time Culver arrived, the equally restless and frustrated Blaese had come up with an idea that would prove to be monumental. Instead of fruitlessly pursuing the fickle stem cell in hopes of curing blood diseases at their source in the bone marrow, another tack was called for, he concluded. Blaese wanted to concentrate on T lymphocytes—fully mature, disease-fighting white blood cells that had already proved their mettle in the body. If for no other reason, Blaese wanted to work with T cells because he understood them.

"I recognized that our idea of simply putting a good gene into bone marrow wasn't going to work," he says. "I didn't know enough about stem cells. I knew I wasn't going to solve that problem—and to date nobody has. But my feeling was that I can try to learn entirely new things in biology, or I can try to use what I already knew."

And, for Blaese, it boiled down to a beautifully simple plan.

Although the exact mechanisms remain a mystery, there is ample circumstantial evidence that T cells somehow possess long-term memories. As children mature into adults, they no longer need regular immunizations, even though they stand the same risk of exposure to infectious diseases that youngsters have. T cells, like all body cells, die at regular intervals, yet immunity to previously defeated enemies somehow gets passed on when the cells divide.

"Let me give you an example," Blaese explains. "I had my last tetanus shot thirty years ago, but I am still immune to tetanus. I tested my blood, and I know this is true—I still can make the antibodies. So lymphocytes, although they are not stem cells, can 'live' a helluva long time. Something happens that causes them to divide repeatedly and carry forth the memory of immunization. My own cells tell me that.

"Because lymphocytes can be so long-lived, it seemed reasonable to believe that if we could get the corrective gene for certain diseases into some of them, it might provide a long and stable source of corrected cells."

That meant the team had to change its thinking entirely about its approach to treating genetic diseases, he admits. "The notion of treating mature, or 'peripheral' T cells, as opposed to stem cells, was a radical idea. A lot of people still haven't come to grips with it."

French Anderson, for one, thought it was a terrible idea. "I just couldn't see it," Anderson says. "It wasn't gene therapy. I was working toward another bone marrow experiment. That was gene therapy."

The two argued for three months, with the normally quiet Blaese pressing very hard and the normally accommodating Anderson flatly refusing to budge. But the shoebox-sized Blaese lab hopped to its boss's new tune right away. Culver instantly saw the possibilities and threw himself into experiments, contributing samples of his own blood for study, savaging his fingers without breaking concentration. The team fought to learn how to culture T cells in volume and how to keep various cell lines growing and expressing ADA.

The plan seemed simple: *temporary gene therapy* for ADA kids by taking their blood, using viruses to pump new ADA genes into the white cells, and then merely giving them back. How odd that no one had thought of it before!

It wasn't a cure, to be sure. But it was a form of gene therapy, even if the treatments would have to repeated every three months or so, to provide immunity for a youngster. A permanent cure probably would not be possible until scientists could solve the stem cell problem, reasoned Blaese. But this radical notion of periodic infusions of healthy genes represented a start for human gene therapy where none existed.

Anderson just couldn't bring himself to accept the idea. It seemed like such a puny gesture when one had worked for so long for the grand slam. Blaese understood that. "It took a long time for me to talk French into it," he says. "Once Ken Culver started working, we made fast progress putting the ADA gene into T cells of ADA kids in tissue culture. It looked good. But French kept saying, 'No, it just doesn't seem like it's going to be something. It's not gene therapy.' He still was very much into stem cells—but at the time the whole field was into stem cells.

"But every Monday morning at nine o'clock, French's lab and my lab would meet, and every week my fellows kept presenting this stuff about making sick T cells healthy in the test tube. We finally just wore down the other lab. And French started getting excited about it as well."

Blaese's brainstorm posed a huge and dangerous adjustment for Anderson. Everyone in the business knew he couldn't get ADA genes to work in mice or monkeys and had been turned down when he wanted to try it in people anyway. He had a reputation as an opportunist and now was being pushed into preempting the field again. *Temporary gene therapy*—that was the ticket? A part-time cure? Gene therapy for a while? His pal Blaese was proposing something that nobody had even considered. And T-cell memory? It was terra incognita, Blaese's tetanus immunity notwithstanding, an area of research that had not been studied, let alone explained. And what about the clinical implications? Anderson's trial balloon touting last-resort gene therapy for kids who had no other options had been shot down, like everything else he proposed. Would anyone buy the justification for "oil changes"—gene therapy periodically performed over a lifetime, much as diabetics give themselves daily injections of insulin, or thalassemic youngsters get nightly injections of Desferal?

"It just wasn't clean," Anderson grouses today. "No matter what happened in the lab, there was no way to tell if it would work unless we actually went into a patient." And he knew there would be hell to pay from Richard Mulligan and the other competitors working on permanent gene therapy (or "real" gene therapy, as they would quickly make the distinction).

But Anderson had no other answers. Despite his massive preclinical data document—which was all but dead—the entire field of human gene therapy had been put on hold. On the other hand, Anderson shrewdly realized, this was not all bad. There was no competing plan out there to stop him if he brazenly tried again to redefine his chosen field.

Finally, Blaese's campaign began to wear Anderson down. After some careful thought, he decided to embrace Blaese's ingenious plan. What did he have to lose? When he started thinking this way, Anderson immediately got a second wind. He would emulate Field Marshal Ferdinand Foch, who, in his darkest hours of the Second Battle of the Marne, in 1918, declared, "My center is giving way, my right is pushed back . . . I am going to attack!" By God, they would try it!

But the methodical Blaese wasn't finished yet. To pull the thing off, he needed somebody who knew how to use T cells clinically. And, mirabile dictu, he knew such a person, a pioneer who had been laboring for years trying to turn T cells into therapeutic agents to vanquish cancer: Steven Rosenberg, whose world-famous lab was now three flights down from Blaese's. But if Blaese needed Rosenberg and his clinical experience with T cells, the canny immunologist thought he had some ideas that might intrigue and tempt Rosenberg as well. Most important, Blaese knew that nobody, but nobody, stood in Rosenberg's way once the government's chief surgeon set his mind to something. It was time to teach him about gene splicing.

Steven Rosenberg's T-cell monomania began in 1968 when he was a surgical resident at Harvard assigned to the West Roxbury Veterans Administration Hospital, in Boston. Rosenberg was about to perform a routine gall bladder removal on a sixty-three-year-old man, when he discovered from the patient's records that there was nothing routine about this fellow. Some twelve years earlier, the man had been admitted to the hospital with a fist-sized cancer in his stomach that already had spread to his lymph nodes and liver. Doctors removed as much as they could, but there was little else that could be done because of liver involvement, and the man was sent home to die. Yet there he was, twelve years later, with a diseased gall bladder for Rosenberg to excise. As the astounded young surgeon numbly went about the task, he could find absolutely no trace of cancer anywhere in the man's body. Rosenberg had just met his first case of spontaneous remission, among the most amazing and rare phenomena in all of medicine—so rare, in fact, that over the next twenty-five years, Rosenberg would take care of more than ten thousand cancer patients and yet see spontaneous tumor regression in only one or two others.

That first case in 1968 might have been passed off by a less imaginative physician as a curiosity. But Rosenberg's mind caught fire. He saw it as a case where the body's immune system had somehow mustered its forces and expelled a mortal enemy. And that inspired him to become a scientist as well as a surgeon, so that he could figure out how such a miracle might happen, and how to make it occur more often.

Rosenberg's career has turned into a frustrating net game with death, yet one enlivened occasionally by the perfect passing shot. And those are what he lives for. He pored over the records of his Lazarus with a bad gall bladder, and something curious leaped out at him: a few days after his cancer surgery in 1956, the patient had developed a virulent strep infec-

tion in his stomach. Rosenberg knew that bacterial infections sometimes reduced the size of tumors. Lab analysis at the time showed that
the man's tumor contained large concentrations of immune cells. Convinced there was something in the disease-fighting white blood cells of
the gall bladder patient that had turned them into anticancer agents,
Rosenberg decided to try a daring clinical experiment. He obtained a
unit of blood from the man and transfused it into another patient suffering from the same inoperable stomach cancer. But to Rosenberg's dismay, nothing happened. The second man's cancer inexorably followed
its course, and he died. Yet Rosenberg was hooked by the possibilities.

Always single-minded, he had gone to Johns Hopkins for both college
and medical school and then moved to Harvard for surgical training. But
now he became as focused as a laser. He interrupted his surgeon's apprenticeship for four years to earn a doctorate in biophysics at Harvard,
followed by another three years to do research in immunology. At the
age of thirty-three, after eleven years of unusually prolonged and specialized training, he felt scientifically prepared to go to war against cancer in
a new way. In 1974, Rosenberg was recruited by one of his mentors, the
cancer specialist Nathaniel Berlin, to become surgical chief at the National Cancer Institute and embark on the animal and human experiments that would lead to the technique that made him famous—adoptive immunotherapy, he calls it—and then to a historic gene therapy
partnership with William French Anderson and R. Michael Blaese.

Steven Rosenberg and his team inhabit an environment
that most people would find too depressing: its watchwords are desperation and agony. "The only patients we ever treat," he acknowledges,
"are those who have been sent home to die by their family doctors." Not
that Rosenberg is a shrinking violet. He vaulted into public view on a
July day in 1985 when he bluntly announced to the world the results of
his physical examination of Ronald Reagan—"The President has cancer"—and then joined the surgical team that removed it. But for years
before he burst into prominence by daring to use the dreaded c word in
reference to a president, he had been dazzling the world of cancer research. Rosenberg is an intense, compactly built, flawlessly tailored surgeon in his fifties whose take-charge attitude, incandescent self-confi-
dence, elongated face, and trademark round, clear, plastic spectacles
framing penetrating dark blue eyes all combine to make him come
across as a hyperkinetic owl. A genuine medical superstar, he is the best-
funded cancer researcher in the world, according to published accounts
(the late Armand Hammer was a prominent private supporter). And he is

noted for developing controversial, audacious, last-ditch treatments that—sometimes—have triggered a terminal patient's own immune system to rise up angry and annihilate cancer cells while sparing healthy tissue. When people seek out Steve Rosenberg, they've run out of options and nearly out of time. The handful whom Rosenberg is able to accept in his oncology service at the NCI are those who have not responded to any other treatment, yet who still are healthy enough to survive his drastic ministrations. That means only six new patients a week, at most, make their way to the cancer establishment's version of Lourdes—office 2B42 in Building No. 10, the Warren G. Magnuson Clinical Center, the clinical hospital of the National Institutes of Health—although, as Rosenberg points out, hundreds of thousands of others deserve the last chance he represents.

"I'm trying to solve an *extraordinarily* difficult scientific and clinical problem, and I think we're making some progress," he explains in the stark, laconic speech patterns of a surgeon who wastes not a word, deflects embarrassing questions, and says nothing he might regret later. "There's a long way to go, but there's also a realistic chance that the thing I am doing can result in better treatments for these patients, if not curative treatments for some of them."

Even though Rosenberg deals only with the incurably ill, his treatments have temporarily shrunk tumors in as many as half of the kidney cancer and melanoma patients he has accepted. And a few, lucky ones have experienced remissions. Breakthroughs he has not achieved, but he has made incremental progress, in the process becoming a hero to the hopeless. His gene therapy partner Anderson loyally professes, "In my opinion, Steve is doing the best cancer research in the world. And I'll tell you, if I were dying and he told me to eat cherry pits, that's what I'd do." Nevertheless, Rosenberg's pursuit of aggressive, highly experimental research—and the inevitable media hype it attracts—have caused other doctors to charge that he raises false hopes and experiments on dying people before answering questions in animals. But Rosenberg, like most surgeons, ignores his critics. The men and women who place their lives in his hands are praying for miracles because the rest of modern medicine has given up on them. Rosenberg alone refuses to let them go without a fight, and, remarkably, not all of them die.

Yet, despite his best efforts, there is no escaping the fact that nine out of ten will eventually succumb to cancer. "About 90 percent of the time, the miracle doesn't happen," he concedes tersely. "But it can happen sometimes, and I won't be satisfied until it happens to all my patients."

"Look around," he says, tilting back in his big leather chair. He gestures at his huge office—grand by NIH standards, with its own adjoining conference room; a messy table scattered with papers and protocols; high walls lined with manuscript shelves, children's watercolors, autographed photos, including several from the Reagan family. His laboratory is just a few steps away. "I spend all my time here. I don't think there have been ten days in the last eighteen years that I haven't been at work. I'm not the kind of guy who has hobbies. This is what I do. I'm a surgeon and I do science. I want to cure cancer in my lifetime."

There now exist three reasonably effective means of treating cancer—surgery, radiation, and chemotherapy. "Those three modes are capable of curing about half of all individuals who will develop cancer," Rosenberg says, "but the incidence of the disease is so staggering: one in every four Americans will develop cancer at some point during their lives, and one in every six will die of it.

"Despite our ability to cure half of the individuals, last year about 514,000 Americans died of cancer—more than those who died in World War II and the Vietnam War combined." Against such a mass killer, Rosenberg is developing what specialists hope will be a fourth major treatment mode—*biological therapy*. It hinges on the theory that throughout life the human immune system vigilantly protects us against cancer and, in extremely rare circumstances, can even reject an erupting tumor as if it were just another germ or transplanted tissue. Most important, immune cells work systemically, ceaselessly patrolling the entire body. That means it might be possible to train them to seek out and destroy errant metastases that have broken off from the original tumor and spread far and wide to seed destruction. Rosenberg's knife may cut out solid tumors and banish them from the body forever. But those "mets," malignant cells that migrate to distant sites to continue their mad dance of wild division, of course, pose the greatest danger.

Rosenberg's strategy is to identify in cancer patients certain immune cells that show anticancer activity. The cells are then "educated" in tissue culture to enhance their cancer-fighting powers, raised in huge volumes, and transfused back into the bloodstream to see whether they can hyperstimulate the body's own immune cells to seek and destroy tumors. Along with white blood cells, or lymphocytes, Rosenberg also administers certain natural communicating molecules, such as interleukin-2, that make lymphocytes grow and divide and that also have been shown to play a crucial role in mediating the immune response. Because these mediating cells can now be produced by genetic engineering, Rosenberg is able to deliver them in immense concentrations never envisioned by nature.

Progress in immunotherapy has been "slow," he admits, "but at least our first biologic therapies have helped a few patients with cancer, and we're getting better at it." Rosenberg is convinced that the body's natural defense mechanism can eventually be educated and superfortified against cancer. His foray into gene therapy is merely a means to that end, he insists. "For me, gene therapy is just a tool, one of many tools I use. I want people to realize that gene manipulating is just one step in a very long, carefully thought-through process to try to develop a cancer therapy.

"I'm particularly excited about the promise of gene therapy. But to my knowledge, nobody has demonstrated that it has cured a single patient yet."

A New York native, married, and the father of three daughters, Rosenberg has published more than five hundred papers. His career has paralleled developments that have allowed scientists to cultivate immune system cells in the laboratory. When Rosenberg was starting out in the early 1970s, there wasn't much going on in this regard, but it was possible to use animals as farms. Rosenberg's animal of choice was the Yucatán minipig, a dog-sized creature preferred by heart researchers because of its convenient size and its quirky physiological similarity to human beings. Trying to stimulate a porcine antibody response to cancer, Rosenberg inoculated his minipigs with cancer cells from six human patients, removed the pigs' white blood cells, purified them, and injected them back into the patients in hopes of immunizing them against their cancer. But, as had happened with the second stomach cancer patient, he could detect no response. He did show, however, that one could administer white blood cells to patients safely—not just white cells but *pig white cells*—an advance that Rosenberg deemed very important. Unfortunately, no one else did, and Rosenberg's four scientific papers reporting the research were all rejected for publication.

The scene now shifts to another NCI laboratory, that of the tumor cell biologist Robert Gallo, later to become famous and highly controversial as the disputed codiscoverer of the HIV retrovirus that causes AIDS. In 1976, while trying to grow leukemic cells in volume so that they could be studied, the Gallo lab happened on a mysterious substance—which it called interleukin-2—that turned out to be a growth hormone that T cells secrete as chemical messengers to turn on other T cells. IL-2 belongs to a class known as cytokines, which regulate and coordinate the activity of the entire immune system. Several powerful cytokines have been discovered and given precise, stilted-sounding names—including gamma interferon, B-cell growth and differentiation factor, and granulocyte and monocyte colony stimulating factor. All of them make blood cells grow exponentially in tissue culture, and all are turning into genetically engineered wonder drugs that, when given in much greater amounts than nature intended, can cure a host of anemias and other ills of the blood and circulation systems.

Now that he had learned of the existence of IL-2, Rosenberg thought that if he could isolate even a small number of T cells from a cancer patient that had reacted against the individual's tumor, he might be able to multiply them in the lab and strengthen them enough to use them for immunotherapy. Was this possible? Tantalizingly so. In 1980, Rosenberg was able to show that incubating a patient's own white blood cells for four days with IL-2 stimulated certain kinds of baby white cells to metamorphose into a special commando army of fierce lymphocytes, which Rosenberg dubbed lymphokine-activated killer cells (or LAKs) and which would stalk and shrink a variety of cancer cells in tissue culture. Much more experimentation followed, and by 1984 Rosenberg was using a combination of IL-2 and LAKs to actually dissolve lung tumors in mice—a formidable achievement. However, it was still difficult to grow the experimental cancer-fighting cells in volume, and when he tried giving patients small quantities of IL-2 and LAKs he had available, he found the doses to be inadequate. A total of sixty-six terminal patients were treated with these new chemicals, and all sixty-six succumbed to their cancers.

At about that time, however, Japanese researchers led by Tada Taniguchi in Osaka cloned the gene for IL-2, meaning that an unlimited supply could be produced by recombinant DNA technology. Rosenberg joined forces with gene splicers at an inventive biotechnology company, the Cetus Corporation, in Emeryville, California, and proved that genetically engineered IL-2 worked just like the natural substance. When administered by itself, however, IL-2 produced an awesome array of toxic side effects, ranging from fevers, nausea, vomiting, diarrhea, skin rashes, fluid retention, and troubling central nervous system symptoms, including disorientation, difficulty in concentrating, and insomnia. Rosenberg began compiling a list of these miseries and set about determining how to manage the side effects in what he hoped would be a lifesaving therapy. Finally, in 1984, he felt compelled to go ahead. He tried for the first time to treat human beings suffering from kidney cancer and the deadly, fast-growing black mole cancer, malignant melanoma, with a blend of LAKs and IL-2.

In November of that year, the frustration suddenly turned into triumph. The patient was a twenty-nine-year-old naval commander from Florida, a nurse named Linda (her last name has never been published) with widely metastasized melanoma that had spread throughout her body, including her arms, thighs, back, and buttocks. Linda had been given only three months to live when Rosenberg became her doctor. The treatment was vicious. Linda, who is five feet two inches tall, retained fifty pounds of water weight during the infusions, and blacked out during most of them. However, within a few days after treatment, her tumors simply began to *melt away*, and in three months, instead of being dead, she was cancer free.

Rosenberg, after helplessly presiding over so much misery for so long, finally had reason to rejoice. His lab exulted. Linda and her family even threw a "dead tumor" party in honor of her rebirth. Today, a decade later, Rosenberg is sure that her remission was in no way spontaneous. She is living proof that he is onto *something* altogether remarkable. But precisely what is not clear.

Soon after Linda's cure came another success story. It involved thirty-four-year-old Kathy Donohue, co-owner of a fund-raising firm in the Midwest, who endured three years of radiation and five different chemotherapy courses for lymph cancer that defied all attempts at removal. Near death, the desperate woman came to Rosenberg in 1985. In just four weeks, his custom blends of IL-2 and LAKs had reduced her tumors by 75 percent. Not that it was easy. During treatment, Donohue's blood pressure plunged like a stone. Her body swelled grotesquely. In pitiful shape when she was wheeled before two hundred NCI researchers making grand rounds, she confessed that the simple act of merely blinking her eyes took all her strength. "Hit by a truck" was her analogy. But her body had prevailed: her solid tumors shriveled away. A handful of malignant cells still lingered in her bone marrow, and the cancer reappeared. But she has experienced prolonged remissions— some for as long as two years—between new bouts of therapy.

When Rosenberg reported such results in a December 1985 issue of the *New England Journal of Medicine*, he naively expected it to attract "some interest." But the furor ignited by his research astonished him. Suddenly, he was on the cover of *Newsweek*. Thousands of desperate patients flooded his office with phone calls and letters, begging him to cure them. Although heralded in a tsunami of headlines in 1985 as the long-dreamed-of cure for cancer, IL-2 has since proven to be a fickle warrior. Sometimes it works; most often it doesn't; but it always is capable of causing liver, kidney, and heart failure, although it may not if carefully monitored. Even so, patients must steel themselves to take it as long as they can. This is very serious therapy.

Summing up his long-term results in 1994, Rosenberg reported on 283 patients with melanoma or kidney cancer. Standard treatments had failed all these patients. Seven percent of them have enjoyed complete remissions. Partial tumor shrinkage occurred in another 10 percent. And, gratifyingly, of those responding, 79 percent have survived for periods ranging from seven months to more than seven years, so far.

In 1987, many of the nation's cancer centers began using IL-2, yet it remains an expensive and often dangerous treatment that many doctors find extremely troubling. Nevertheless, immunotherapy still amounts to the only thing that may halt or slow down certain cancers after ordinary treatments have failed.

"These kinds of therapies can be effective," Rosenberg maintains, "but the response rate is still fairly low, with only about one in five patients responding. Our effort therefore was to try to improve on therapy, which brought us to the second generation of this kind of treatment—gene transfer."

As he explored the bodies of his patients for other kinds of white blood cells that could be more aggressive than LAKs in their anticancer activity, Rosenberg ransacked circulating white blood cells, lymph nodes, the thoracic cavity—any place they might be hiding out. Finally, he decided to look in the tumor itself—"what better place?" Paydirt! Right away he discerned some peculiar blood cells that seemed to have penetrated the tumor deposits he found in patients. These cells, which probably had fought in vain before being swamped by the sheer bulk of the renegade tumor, at first intrigued and then inflamed him. In 1986, Rosenberg gave them a name: tumor-infiltrating lymphocytes, or TILs.

"They're lymphoid cells that infiltrate into solid tumors and can be grown by culturing suspensions of the tumors in IL-2," he explains. "The suspensions consist mainly of tumor cells, with an occasional white blood cell present. But, as we saw to our amazement, the only lymphocytes that can outgrow the tumor cells in IL-2 are those responsible for killing them—TILs.

"Therefore, after three weeks of growth in the lab, we have a pure culture of lymphocytes, free of tumor cells."

Normally, TILs pooped out when a tumor fattened, but Rosenberg now had the ability to grow the microscopic cancer fighters in prodigious quantities. He quickly turned to animal studies. Work on mice soon showed that TILs possess incredible antitumor properties and are fifty to a hundred times more potent than LAKs. Equally important, TILs can go home again.

Or so it seemed. Although Rosenberg couldn't track them, when he rebiopsied animals after injections, he found TILs present that he presumed had returned to the tumors from which they came. Was it possible that TILs could remember their origins and would home in on specific tumors, leaving normal tissue alone?

It sure looked like it. The notion is key to Rosenberg's work, yet it remains a matter of biological debate.

Simply put, although TILs are sometimes found in tumors, Rosenberg can't conclusively prove cause and effect—namely, that the TILs actually home to cancer. They seem capable of zapping cancer cells merely by coming in contact with tumors. But their true effectiveness would hinge on their ability to home. Biologists have contended that it is equally possible that TILs merely get dumped in tumor sites by circulating blood and that their homing talents have been vastly overrated.

Nevertheless, after LAKs, TILs became Steve Rosenberg's new preoccupation. Clinical trials revealed that as many as 43 percent of all terminal melanoma patients (even those who had failed on IL-2 and LAK therapy) responded to TILs in varying degrees.

By 1988, however, Rosenberg was stymied again. Sometimes his TIL cells worked brilliantly in humans, sometimes not. In either case, he had no way of determining why, because he couldn't follow the TILs for more than two weeks once he put them back into a patient's body. Radioactive tracers proved fruitless, not having a long enough half-life to give any conclusive answers. For example, at scientific conferences Rosenberg displays the nuclear medicine scans of a forty-six-year old woman with some forty melanoma nodules and multiple lesions on her thigh. Inside her body were tumors on her chest wall and large, tangerine-sized lumps at the base of her lung. After treatment with massive doses of her own TILs, all the cutaneous cancers disappeared, as did most of the lung metastases.

But then, as if he were watching a movie that abruptly ran out of film, Rosenberg lost sight of his TILs. "We were unable to follow the traffic of those cells beyond two weeks," he explains. "No assay worked. It was impossible." After that, Rosenberg was flying blind. As before, patients either got better or they didn't, and who knew why?

Over several years, as Rosenberg kept struggling to cure dying patients while the rest of the medical community frowned and fretted, sympathetic eyes were watching from down the hall.

Patiently and unobtrusively, Michael Blaese had been following Rosenberg's saga. Blaese thought he could help. With the persistence of an Airedale terrier, he kept trying to arrange a meeting between the surgeon and NIH's house gene splicer, French Anderson, but both men were too busy to make their schedules mesh.

Finally, on March 17, 1988, Blaese brought the senior lab chiefs together. In what would prove to be a fateful meeting, lasting only an hour, in Rosenberg's second-floor library, Anderson's fortunes were to change dramatically in a way that, he hoped, would put him back on the map as a gene doctor.

In fine, there took place a classic marriage of convenience. Rosenberg knew next to nothing about retroviral systems of gene transfer. Anderson knew very little about Rosenberg's brainchildren—TILs. Within five minutes of meeting, the men knew that each had what the other needed: they would mix Anderson's genes with Rosenberg's killer cells.

The scheme was childishly simple, in molecular biological terms. They would genetically engineer a retrovirus, splicing in a bacterial gene that conferred resistance to the antibiotic neomycin. The altered virus would then be cultured with the TILs so that the virus could infect the freewheeling cells and dump its genetic load into their chromosomes. Thereafter, as the TILs replicated in IL-2, so did the bacterial gene that nestled within them. Because the altered TILs alone had the antibiotic resistance gene, those cells—and only those cells—could use it to survive when later doused by neomycin.

A better marker with a longer half-life? Perhaps. But the biggest difference from earlier experiments was that these cells would *not* stay in the test tube. They would be placed in the body of a dying cancer patient. They would be delivered intravenously, along with IL-2, to keep them replicating. When chunks of tumor were removed a month later by Rosenberg, he would have a way of checking for his TILs, a way that would quite conceivably render them distinguishable from all the other blood cells in the body. Only the gene-marked TILs would be able to survive when mixed with neomycin, because of the neomycin resistance gene they could switch on. That meant their fates could be learned. And in that way, when TILs were tagged with a harmless beacon, they could be tracked in tumors. The surgeon could determine how many TILs had survived laboratory engineering and were back on the job. And Anderson could determine whether the gene was still functioning after spending a month in a human body.

The genes could in no way be considered therapeutic—if everything went as planned, they definitely would not *do* anything; they would just be noticeable. But that the idea bore no resemblance to Anderson's previous gene therapy schemes was immaterial. It provided entrée. It was the steppingstone. It was history.

Within an hour in Rosenberg's library, in fact, the entire first human gene transfer scenario had been sketched out. And the sketch would be followed precisely, as Blaese discovered to his amazement several years later when he went over the plan that had been stored in his computer. "I called up my notes of the meeting," he says. "It's remarkable how they mirrored the events that followed!"

As the three scientists talked, they ticked off benefits and were able to rationalize the plan in terms of safety and efficacy. The TILs are already grown-ups. They are differentiated adult subsets of T cells and, theoretically, predictable in behavior, unlike their parents, bone marrow stem cells that must differentiate into all the kinds of blood cells. And because the TILs would be growing in culture between thirty and forty days after being infected with gene-bearing retroviruses, molecular biology's many controls and tests could be deployed to analyze safety, purity, and functioning before the transplant. Besides, if, as the government researchers believed, these cells did seek out and target tumors, the next step might be to "engineer in" actual tumor-fighting genes and deliver them directly to the appropriate battlefield. That would amount to gene therapy against cancer—in effect, the most precise and powerful cancer chemotherapy ever envisioned, because all other cancer drugs indiscriminately kill all fast-dividing cells, healthy ones as well as malignant ones.

First, though, the NIH team would be conservative and begin with simple gene transfer in adults who could give informed consent. Yet in the broader context, what the ad hoc NIH brainstorming session really meant to the clinical development of gene therapy can scarcely be overstated. When Anderson, Blaese, and Rosenberg sat down at a conference table and tossed ideas at each other, the notion of permanently curing an inherited disease was placed firmly on the back burner. Genes would first be used as *drugs*. And like other drugs, they didn't have to last long in the body, which would evade the persistence problem that had held up the entire field.

As for political acceptance, whereas rare genetic diseases like ADA deficiency didn't have much of a constituency or a lot of clout, cancer was the moon. Gene therapy for cancer was very different from gene therapy for a disease that afflicted 1 hapless child in 150,000. By clinging to Rosenberg's immaculate white lab coat, Anderson and Blaese would command a level of urgency and credibility they might never have otherwise achieved. They could make human gene transplants a reality at last. And they could be the first to do it.

Privately, as word of Anderson's latest plans spread from NIH, many gene researchers were floored by the change in the rules. Yet they remained silent—they, after all, had been saying that gene therapy was still years away. Nonetheless, they weren't comfortable with the life-or-death, throw-in-everything tactics of cancer doctors like Rosenberg. It was a cultural thing. If something worked, people in the gene community wanted to know why. They wanted to understand the science fully before handing potent gene pills to patients. To these scientists, genes were not merely drugs but something much more vital. As David Baltimore once excitedly stressed to the authors: "You must remember, *genes are alive!*"

Mike Blaese concedes, "Our gene-marked TIL proposal was culture shock for a lot of gene people. But when I'd go around to cancer conferences and tell them what we wanted to do, the common response was 'What, haven't you already done it? What are you waiting for?' "

For French Anderson, Rosenberg's TILs could allay a lot of fears about gene therapy. "All of us were gun-shy about actually putting new genes into patients," he says. "But if nothing bad happened when we gave them marker genes—no Frankenstein's monsters were produced, no leukemias—that could get us started. And it would show, once and for all, whether you could use retroviruses to deliver genes into human cells and get them permanently planted in people. Even if the gene wasn't going to cure anything, the TILs might. All we had before was animal work. The gene-marked TILs gave us an intermediate step."

Particularly with Steve Rosenberg plowing the path.

Rosenberg is a powerful and astute figure in government research circles and one of the American Cancer Society's biggest stars. When he sets out to do something, there's no stopping him. So it was that by the following June—June 10, 1988, to be exact—the gene-modified TIL experiment was formally submitted to local review bodies at the NIH for consideration.

Since 1974, the government had evolved a nationally controlled mechanism for protecting humanity from poorly designed clinical experiments by mandating the review of such proposals at all institutions that received federal funds. It consisted of three parts. First came review at the local level by institutional review boards (IRBs) and by institutional biosafety committees (IBCs). Then, for a gene experiment, the scene shifted to Bethesda for its toughest scientific hurdle—a mid-level review by the Recombinant DNA Advisory Committee's Human Gene Therapy Subcommittee—followed by scrutiny by the parent RAC (which contained so many members of the subcommittee that it often served as a rubber stamp). If the proposal made it that far, it still had to survive final review by the directors of NIH and the Food and Drug Administration, who would accept or reject the recommendations by the RAC.

The TIL proposal came as a surprise because most experts had been expecting that the first clinical experiment to attempt gene transfer into human beings would involve an immune disorder—probably ADA deficiency—not cancer. Moreover, this proposal amounted to a new diagnostic technique, not a possible treatment.

The approval process dragged on for months as the proposal clawed its way up through the most formidable bureaucracy ever assembled to ponder a medical trial. The delay was probably hardest on Steven Rosenberg, who dislikes few things so much as regulatory committees that cramp his style and, in his view, hurt his patients. Yet he willingly joined the members of the newly formed gene team as, hats in hand, they began making their rounds of the various committees and seeking permission for the gene transfer cancer experiment.

As it moved up the regulatory ladder, the first gene transfer proposal to reach the Human Gene Therapy Subcommittee, on July 29, 1988, was nothing fancy, merely a request to use an engineered retrovirus to insert a selectable marker gene into TILs—the familiar neo-2 (or N2, as it was called). Then, after the TILs had been returned to the patient, Rosenberg would biopsy tumors and Anderson would check for the neo-resistant cells. That way, the scientists could see how efficient the gene transfer had been and how many TILs had, in theory at least, gone back to work.

Until 1992, when the RAC subcommittee was disbanded because so many of its members did double duty on the RAC, subcommittee hearings and committee hearings alike took place in a sixth-floor conference room in the C wing of the cavernous structure named Building No. 31, on the NIH campus. Meetings were organized in a businesslike fashion with an agenda and formal presentations. Committee members, appointed for four years, brought a wide spectrum of viewpoints and skills—gene splicers, doctors, ethicists, clergy, legal scholars, government officials, lay people—to their task of evaluating protocols involving genetic engineering. Over the years, the various committees had earned a reputation for rigor and fairness. As they considered the proposals before them, each subcommittee member would question the presenters, but there was room for informality and interplay, too. The chairman at the time was an ethicist, LeRoy Walters, president of the Center for Bioethics of the Kennedy Institute of Ethics, at Georgetown University, who often set aside parliamentary procedure and sought questions and comments from the members of the public in attendance.

The chairman of the parent RAC, Gerard McGarrity, who served as president of the Coriell Institute for Medical Research, in Camden, New Jersey, worked the same way. Openness and consensus were the goals. They had served the RAC well for fifteen years after NIH had seized the regulatory reins of federally funded recombinant DNA research involving plants, animals, and human beings—years basically free from mishaps and scandals in genetic engineering.

The members of the NIH gene team listed the key things they hoped to learn from gene transfer: How long do the TILs persist in the body? Where do they end up in the body? Does longevity or location correlate with clinical effect? Is it possible to recover the TILs? What functional characteristics of the TILs define their ability to localize to tumors or distant sites? Is there a correlation between localization, function, and clinical efficacy?

This all seemed straightforward, but by the time the proposal reached the RAC subcommittee, several critics had expressed their doubts whether the design of the experiment would provide many answers.

The questioning at the subcommittee hearing was led by MIT's Richard Mulligan, the University of Minnesota geneticist and ADA researcher Scott McIvor, and William N. Kelley, then chief of medicine at the University of Michigan (and now dean of the University of Pennsylvania School of Medicine). In the late 1960s, Kelley, a genetics researcher in the laboratory of J. E. Seegmiller at Johns Hopkins, had played a major role in the discovery of the cause of the Lesch-Nyhan syndrome, that is, the patients' HGPRT deficiency. Later, at Ann Arbor, Kelley inaugurated an important gene therapy research program.

Yet by 1988 Kelley and other genetic engineers had become fearful that the government physicians were crossing a dangerous line with their gene transfer proposal. Every young doctor is taught that risks may be imposed on a patient only if there is the potential for some gain, Kelley reminded the NIH team. Rosenberg's patients stood to gain nothing from the *gene transfer part* of TIL therapy, and the risks were unknown.

The risk issue was thus "exacerbated," Kelley complained. More studies on mice should be done before the technique could be used in human beings.

Mulligan's worries centered on the safety of the retroviruses that would be used in the experiment. Probably nobody in the world knew more about retroviral vectors than Mulligan, and he didn't like Anderson's, and told him so. Scott McIvor also deplored the lack of animal models for the experiment and doubted whether current assays would be sensitive enough to tell the team much about what the engineered retroviruses were doing in human patients when they were infused.

Things went downhill from there. In sessions that ranged from the cozy to the stormy, the Human Gene Therapy Subcommittee challenged Anderson to provide better data on the safety of the experiment. The researchers complied, but they did not submit two important safety studies—an omission the committee took as a slap in the face. Anderson and Rosenberg, it seemed, feared that premature release would endanger their publication in the *New England Journal of Medicine* and in *Science*, both of which demanded exclusivity.

At issue was the controversial "Ingelfinger rule," named for the late Franz Ingelfinger, a celebrated gastroenterologist and former editor of the *New England Journal*, who had demanded that the results of scientific work appear nowhere else—not even at a scientific conference—if they were to be included in the Massachusetts Medical Society's eminent publication. The *Journal* was not interested in old news.

Because careers may hinge on publication in the *New England Journal*, the Ingelfinger rule has assumed the power of law, but many critics, including the *New York Times*'s Lawrence K. Altman, himself a physician, have long been assaulting what Altman regards as merely one journal's protecting of its scoops.

The RAC subcommittee expressed outrage when it learned that Anderson and Rosenberg had been withholding data. James Wyngaarden, the NIH chief at the time, declared that the federal government "would not be held hostage to *The New England Journal of Medicine*." The editors of the two prestigious research journals—Arnold Relman of the *New England Journal* and Daniel E. Koshland, Jr., of *Science*—backed down fast, meekly insisting that their demands for exclusive publication would never apply to a duly constituted committee of the federal government. Relman assured *Science*'s Barbara J. Culliton, "The *New England Journal* doesn't want to hold the NIH or any other government body hostage, and we won't."

In retrospect, Anderson asserts that the real reason the latest safety data weren't presented to the RAC was much more mundane: "We didn't have everything in final form until a week or so before the RAC meeting. All we had were Polaroid slides. But, sure, we were worried that premature publicity might affect our ability to publish—yes.

"But Wyngaarden really was quite clever. He forced Relman and Koshland into an agreement in writing that, in the future, anything could be presented to any NIH committee without jeopardizing publication. That, I think, was the real point of the brouhaha."

The TIL tempest seemed to blow over, especially after it became clear that Rosenberg's *New England Journal* paper reinforced the urgent need for the gene transfer experiment. Rosenberg reported on fifteen patients with advanced, widely metastatic melanoma who had undergone regular TIL therapy (with no marker gene), nine of whom had experienced a 50 percent reduction in the size of their tumors. One even had enjoyed a complete remission. Again, the vexing question for Rosenberg was: Why did nine patients improve somewhat, while six others did not?

Safety loomed now as the main hitch. At an RAC subcommittee meeting, Mulligan lashed out at Anderson, demanding that he demonstrate that the engineered retrovirus destined to carry the marker gene into the TIL cells could not replicate once it found itself inside a human body. Anderson limped home to do some more animal work. When he returned, he flashed a slide on the screen that summed up his safety studies, showing that there were virtually no viral genes left in his vector construct and that there was thus no chance for recombination.

But the others in attendance still didn't buy it. They knew all about retroviral vectors, packaging cells, and safety factors and demanded more evidence than a mere slide. Once again, Anderson's major challenger was Mulligan, who wanted not a slide but hard data he could study. He argued that Anderson's assay designed to detect the presence of potentially dangerous helper viruses wasn't sensitive enough.

Mulligan also noted that a new vector created in his laboratory at the Whitehead Institute was safer than the one Anderson and Rosenberg were proposing. Interviewed later, Anderson admitted that "in theory" Mulligan's packaging cell was probably safer, particularly for a real gene therapy experiment that entailed putting a therapeutic gene into bone marrow cells.

But Anderson added that he had never been able to get Mulligan's packaging system to work in his hands. He said he was willing to stake the fate of the experiment on second-generation vectors supplied by Dusty Miller, of the University of Washington's Fred Hutchinson Cancer Research Center, in Seattle. Miller had rebuilt the retrovirus vector almost from scratch, excising the genes crucial to viral replication and in their stead inserting the *neo* marker gene.

The safety issue became moot after Anderson presented further proof to his colleagues that the packaging part of his retroviral vector had been so changed that it actually contained no genes from the mouse leukemia virus itself, an improvement that several laboratories have pioneered.

The Human Gene Therapy Subcommittee still wasn't happy. Anderson, running out of patience, decided that he wasn't going to get approval from the biologists, no matter what. So he made an end run. On July 29, 1988, the subcommittee voted unanimously against his protocol, whereupon Anderson demanded a hearing by the full RAC, which was his right.

It wasn't until 1993 that Anderson revealed his game plan for the all-important October 3, 1988, meeting of the RAC, which led to the first human gene transfer. His medium was a series of editorials dubbed "Musings on the Struggle" that began to appear in his journal, *Human Gene Therapy*. The recollections had a sly, "how-I-pulled-it-off" quality that may have been more revealing than Anderson intended, particularly to Richard Mulligan, William Kelley, Scott McIvor, and the other molecular biologists who had been his harshest critics. Anderson wrote that the NIH team had agreed beforehand that the biologists on the RAC were never going to allow them to proceed, for reasons that Anderson viewed as ranging from "the most noble to the most base." The noble reasons, he believed, "were founded on a sincere effort to carry out the responsibility of the subcommittee: to carefully evaluate the safety and appropriateness of the initial human genetic engineering experiment. The base reasons, which probably made only a small contribution, were founded on direct personal competition."

To combat scientific opposition, wrote Anderson, the NIH team decided to try to win over the other "presumably uncommitted RAC members." The key to the strategy was to "change the playing field: redirect the discussion onto our strengths and away from the Subcommittee requirements that we could not meet." By the latter, Anderson meant demonstrating that transferring marker genes into mouse TILs with retroviruses was safe and had a chance of working.

Because human experiments throughout medical research are ordinarily preceded by successful animal work, one might have expected the NIH doctors to have proved safety and efficacy before they made the big jump from mice to men. However, the team had strenuously argued before the subcommittee that its laboratory labors had failed miserably in this regard. Mouse TILs would not accept new genes via retroviral vectors, no matter what tricks the troika tried. Consequently, Anderson held, "the Subcommittee had placed on us an impossible burden, but one that sounded reasonable."

The solution? Merely to leapfrog the scientists and play to the non-scientists on the panel by providing answers to what Anderson would argue were the "five paramount questions:

"1. The marker gene could be inserted into human TIL;

"2. The gene-marked TIL would not be significantly altered;

"3. The protocol had a reasonable chance of success based on animal studies in mice and primates;

"4. The risk to the patient was minimal;

"5. The risk to health care personnel and the public was essentially zero."

If the stratagem failed, Anderson was prepared to throw in the towel and base everything on pity. As he wrote, "Our trump card, if needed, would be a carefully planned emotional appeal. We fully understood the risks of calling on emotion in a scientific debate—it could be interpreted as an admission of failure on the scientific front. But if all else failed, we would play our 'dying cancer patients' card."

Tension at the October meeting was palpable, Anderson wrote; the subcommittee members who also held seats on the parent RAC were angry because the troika was even there. "Everyone anticipated we would lose. I think this worked to our advantage in the later stages of the meeting."

Shortly after the proceedings began, Anderson put forth his five paramount questions; then Rosenberg explained the need for the protocol—except that he failed to follow the script. Wrote Anderson, "To set up for the possible 'emotional appeal' later, Steve was supposed to say: '485,-000 Americans died of cancer last year—that means that one person dies of cancer every minute.' Under the stress, he forgot to say the second phrase."

Blaese addressed issues of efficacy, followed by Anderson presenting the latest data on safety. As usual, Mulligan and McIvor challenged the newest offering by the NIH team as inadequate and also complained that the Human Gene Therapy Subcommittee, idle since its July 29 meeting, had never even seen it. Anderson bristled at their criticism. He sensed his team was losing. But then came the opening he had been waiting for. As he wrote later, "one of the molecular biologists made the following we-had-to-destroy-that-village-in-order-to-save-it statement: 'If you plan your animal studies carefully enough, you wouldn't have to go into humans at all.'

"Steve, Mike and I looked at each other in disbelief. I whispered to Steve: 'It is time for Plan B.' "

Anderson seized the moment. He stood up and passionately declared, "I was asked at the break, 'What's the rush in trying to get your protocol approved?' Perhaps the RAC members would like to visit Dr. Rosenberg's cancer service and ask a patient who has only a few weeks to live: 'What's the rush?'

"A patient dies of cancer every minute in this country. Since we began this discussion 146 minutes ago, 146 patients have died of cancer."

Guilt saved the day. "When I sat down, nothing was said for a moment," Anderson recalls. Then, one by one, the RAC members began expressing cautious optimism for the protocol. Finally, it all came down to a veteran RAC member, the Harvard Medical School microbiologist Bernard Davis, described by Anderson as "the grand old man of bacterial genetics," who commanded enormous respect.

Abruptly, Davis announced, "I am troubled by what I am hearing here."

That was the key moment. "If Professor Davis opposed us, we were dead," Anderson wrote, "but if he supported us we had a real chance. He had not said a word all morning so I had no idea which way he was leaning. I could not breathe while waiting for his next words."

Davis, a blunt-speaking, white-haired professor, gazed at his colleagues and made his fateful pronouncement. "I think we are nitpicking these investigators," he declared. Medicine has a long tradition of taking chances, he pointed out, and the volunteer patients who were eagerly waiting for TIL therapy had little left to lose. "It is virtually not possible to have more risk than certain death," Davis reminded the committee.

The troika had done it. "I collapsed in my seat—we would win!" wrote Anderson. "I was so overcome that I had to fight back tears."

After a last spate of hand-wringing and outpouring of emotion by committee members, Donald C. Carner, a Tiburon, California, attorney, made the historic motion that the TIL experiment be approved, with the proviso that it be limited to ten patients, each of whom had a life expectancy of ninety days or less, "and each of whom consented to participation after having a full understanding of what he or she is accepting."

According to Anderson, by that time the mood of the RAC had

shifted so abruptly that "our opponents did not even realize what was happening."

The RAC voted 16 to 5 to let the controversial experiment proceed— over the objections of the molecular biologists. Carner wrote later that he had believed there was a possibility, however remote, that "one or more of the ten patients could benefit from the information gained from the genetic labels they receive." He trusted that the scientific community would be watching the experiment closely and would not let it get out of hand. He added, echoing the sentiments of the exhausted committee members, "After extensive discussion, it seemed we were attempting to pirouette on a molecule with nothing more to be achieved."

Euphoria instantly gripped the Anderson-Blaese-Rosenberg labs. "We had a celebration dinner a few days later attended by 100 people," wrote Anderson. "The lab personnel put on several skits including a take-off of the RAC meeting in which all the key players were caricatured."

To this day, Anderson never talks about the TIL experiments without thanking Dusty Miller, without whose packaging cells the experiment would never have taken place. And Anderson has been just as quick to state that even though Richard Mulligan put him through "hell" during the hearings, the MIT researcher later quietly sent him the latest version of his packaging-cell line in case Anderson wished to try it again.

This single magnanimous act by the cautious, critical Mulligan went a long way to defuse what was widely perceived as a bitter feud between two of the leading scientists in gene therapy. Of course, a cynic might argue that Mulligan's act reflected his grave mistrust of French Anderson and the vectors he was intent on putting into people. But Anderson doesn't look at it that way: "In the end, we all came together," he says. "I was deeply touched."

As the NIH request was picked over by no fewer than fifteen panels of experts, it became clear to longtime observers that the RAC members had been waiting for years for some sign of progress in gene therapy, and they agreed with Anderson that this experiment could go a long way toward removing fears.

Once again, however, the careful Wyngaarden postponed the experiment in hopes that at least some of the five dissenters would change their minds, which, for the record, they didn't.

The haggling over risks versus benefits continued behind the scenes until January, when Wyngaarden felt confident enough to conduct a quick telephone poll with the full RAC. Only after the majority of all twenty-five members decided that, so far as human minds could tell, the experiment was safe did Wyngaarden give it his final blessing.

But even then, an anticlimax. Genetic engineering has a scrappy, self-appointed watchdog, Jeremy Rifkin, president of a Washington-based group, the Foundation on Economic Trends. The author of *Algeny* and other best-selling books that warn (to the delight of a substantial following) about the dangers of the misuse of high technology by what he views as an arrogant scientific establishment, Rifkin is an activist attorney who aggressively uses the courts. After monitoring the RAC hearings, Rifkin at the last minute filed suit to halt the gene transfer experiment. His complaint, which scientists viewed as yet another publicity stunt, involved Wyngaarden's telephone poll of the RAC on the final version of the final protocol. It should have been conducted out in the open, contended Rifkin.

On May 16, 1989, just a few days before the experiment was set to begin, a federal judge dismissed Rifkin's lawsuit after a settlement was reached in which the NIH agreed that in the future every single hearing of the RAC would take place in public. Having made his point, Rifkin declared victory and moved on. He gave the TIL experiment his blessing: "This is a historic precedent," he declared. "It is the first experiment in the world in which a foreign gene is to be placed into a human being. With this experiment we begin the whole era of human genetic engineering."

Anderson and Rosenberg insist that, though they had received official permission months earlier, Rifkin's challenge didn't hamper their preparations in the slightest. They had enough problems of their own, and Anderson asserts that the timing of the experiment, immediately following settlement of the lawsuit, was merely coincidental. "The truth is that his lawsuit, which really dealt with procedural matters, had nothing to do with us and never affected us one way or another. We had our patient, and we were ready to go."

8

GENE TRANSFER IS BORN

It happens like that, sometimes. A fifty-two-year-old man stands in his shower scratching an itchy mole at the base of his neck. He's had the mole all his life, this family man, this grocery truck driver showering for work in his cozy Franklin, Indiana, ranch house this January morning in 1989, realizing that his darkening mole doesn't look good and is bothering him. It starts to bleed. And he starts to worry.

"It was right on the inside of his collar bone," says Sharon Kuntz, unconsciously rubbing the left side of her own neck. "He'd had it ever since I knew him." Cool as a cucumber—that was Maurice Kuntz, the man whom fate was to enlist in May of 1989 as the first human being to have foreign genes systematically introduced into his body. Strong, muscular, five eight, 170 pounds, Kuntz was a good neighbor, yet a private man of quiet conviction. A lifetime member of Teamsters Local No. 135, an active member of the New Life Baptist Church (he drove the bus for the congregation) and the Waldron Masonic Lodge No. 217, Kuntz loved the outdoors and had spent the preceding summer in the sun working in his yard mulching fruit trees. He loved to plant things— if a neighbor needed help in planting fifty spruce trees, Maurice would do it. When he began his time of trial, this rugged, handsome man with the darkening mole, he didn't want Sharon to worry. They'd been married twenty years—she was twelve years younger than he—and were devoted to each other and their two children. "We were very close," she says. "Some people talk about the perfect marriage. That's what we had. He'd talk about it. He talked to everybody." She's crying now, seated alone in her immaculate wood-beamed living room with its beige thick-pile carpeting, its dark brown country-style furniture and hanging plants. The rear sliding window looks out over cornfields, and a little white dog, a bichon frise, yaps noisily. She is slim and attractive, this grieving woman of forty-two, wearing stone-washed jeans, her chestnut hair cut short, her eyeglasses big with green frames.

A religious man, Kuntz turned his worries over to God, but he relied on doctors. And so he nervously entered the customary pipeline: the family doctor told him he had to get the mole taken care of; a plastic surgeon removed it. "We got everything," he told the couple. "I took a wide margin [of surrounding tissue]. But I have the feeling it's malignant." The biopsy results came back: malignant melanoma, the deadliest form of skin cancer.

For reasons yet unclear, the incidence of melanoma has risen dramatically in the last fifty years, and it often develops from harmless moles or in normal skin. The average person's body has ten to forty moles, and often the first sign of cancer of the melanocytes is a change in the size, shape, or color of a mole. Physicians look for what they call ABCD—Asymmetry, irregular Borders, Color variability, and a Diameter of greater than six millimeters. Early, a melanoma is flat or slightly raised and confined within the upper layers of skin. It is quite curable at this stage. Removal of the tumor and a small, one-centimeter-wide margin of tissue will halt the cancer in the vast majority of cases. As a melanoma progresses, though, it becomes deeper and more raised. Once it hits the bloodstream, it seeds the body with clumps of tumor cells that will grow uncontrollably and choke off vital organs. About 32,000 Americans will develop melanomas this year, and about 6,700 of them will die from them, many in less than a year after the fast-growing cancer starts to spread.

"But after Maurice's surgery, even though it had been cancer, there were no more tests done," Sharon says, her voice still betraying wonder. "He went in for checkups. They said he was fine. I don't want to sound bitter . . ." The tears are flowing now and the glasses fogging. Over the next six months, Maurice Kuntz seemed to lose his oomph. He was tired all the time, no matter how much sleep he got. Back he went to the family doctor, who examined Kuntz's neck again but this time seemed to linger on the lymph glands on the right side. They were badly swollen. The doctor said he didn't think it was connected with the earlier melanoma. That had been on the left. But the doctor probably suspected. His fingers told him. The tumor had started to grow.

A CAT scan followed, then a needle biopsy. It was the end of February now, and the Kuntzes made the rounds—tests, pathologists, oncologists, radiologists. The scan results came back. Kuntz had five metastases in his liver—malignant melanoma spreads like wildfire. Then he had to endure a liver biopsy. That was the worst agony, Sharon says. It took the doctors four tries to hit a "met" and withdraw a sample. Cancer. "Fast-growing type."

"We went into shock after that," Sharon remembers. "We'd been hopeful that his neck was fine and he was in remission. Then when they found the tumor, the surgeons had talked about doing an operation. Maurice would lose half his neck, but he accepted that. Now, it had spread to the liver. Suddenly, we didn't have a lot of time." Only about two months, the doctors speculated. But, by chance, Kuntz's Indianapolis oncologist had received a form letter from Steven Rosenberg at the National Cancer Institute offering the possibility of TIL therapy for hopeless melanoma cases. When shown the letter, the Kuntzes didn't hesitate. "We said we'd try anything," Sharon remembers. Here was a ray of hope.

On March 13, 1989, Maurice and Sharon Kuntz made their first trip to Bethesda. "We'd never heard of NIH," Sharon admits. "We initially paid our own way, but after Maurice was accepted, they paid for everything, even the motel. They were very nice. Very caring." The couple's spirits rose after tests revealed that Maurice was the perfect candidate for TIL therapy. A brain scan, in particular, was very important: it showed no cancer there, which was crucial because TIL therapy is unable to cross the blood–brain barrier and do battle against brain tumors.

"They explained about TILs to us in a simplified way," Sharon says. "There was no pressure. They even said Maurice could die from the therapy. But we made friends. They weren't just other people in the waiting room. All the TIL candidates would go back at the same time. It became a family. There was a bond. When we walked into the clinical center, it was like a weight was lifted from our shoulders. Maurice had people with him twenty-four hours a day. He had the best care anyone could have ever wanted." What the Kuntzes didn't know was that, as a candidate for TIL therapy and gene transfer, Maurice Kuntz was viewed by NIH professionals as having no more than three months to live.

On March 20, in a ninety-minute operation, government surgeons removed a cancerous tumor directly above Kuntz's right breastbone. "They told me they couldn't guarantee they got everything," says Sharon, "but they got everything they could see. The tumor was black, like the mole. It was melanoma."

As Kuntz began to recover from his surgery, he had not yet met French Anderson. But Anderson already knew him intimately. Rosenberg's new comrade-in-arms, working in his cramped laboratory, a lump forming in his throat at the thought of the milestone soon to come, had begun extracting TILs' from Kuntz's tumor, which was about the size of a small plum. When he mixed about 50 million tumor cells with IL-2, they all died, being replaced by a rich crop of multiplying TILs, which had been hiding out in the tumor. Not only did IL-2 boost the TILs' numbers exponentially; it strengthened their anticancer tendencies until they were acting like frenzied ninjas. Normally, this would take about a month, but Maurice Kuntz's TILs seemed to be growing extraordinarily well. In the meantime, Kuntz, feeling better, was released on March 24. He flew home and went back to work on the twenty-eighth. In Bethesda, tending to his fast-growing cultures, a marveling French Anderson even started to seed separate batches of Maurice Kuntz's TILs, one without the marker gene and one that he hoped would be given the bacterial gene, by means of Dusty Miller's retroviral vector. A month later, the couple were summoned back for the first round of ordinary TIL therapy. That was on April 26. "Nobody had said anything about the gene yet," Sharon recalls. Scheduled for his first treatment on the twenty-eighth, Maurice was visited the night before by Anderson. The white-haired scientist introduced himself and spent two hours with the couple, getting to know them. He put up his feet and relaxed, and with his customary Tulsa charm told them about Kathy and their cocker spaniels. The Andersons had decided to remain childless and pursue their careers, but over the years the couple has developed relationships with several "surrogate kids" from broken homes, treating them like family, assisting them through college and even medical school, shepherding their careers. Anderson bragged to the Kuntzes about them. He related that he had been fortunate enough to study in England and described where he worked at NIH. No hotshot gene surgeon then, he was as likable as a teddy bear.

"He was humble when he talked about his work," Sharon says. "I don't remember him mentioning the word 'gene,' but I really liked him. Maurice did, too." The next day, Kuntz received ordinary TIL therapy. The couple had been warned to expect severe side effects, but there were some other problems as well. During the first infusion of cells, doctors punctured Kuntz's lung while putting in an intravenous tube. A chest tube had to be inserted to help him breathe.

As she recites the story, Sharon refers to relevant dates in a brown calendar book that she faithfully kept. When Kuntz received the first massive infusion of his own cells, followed by IL-2, he reacted as if hit by a truck. Delirious, he told Sharon he felt he had been in a basement full of caskets containing body parts and that it was his job to tote them off. Another time, he kept hearing the hymn "Amazing Grace" and was astounded that Sharon didn't hear it, too. He had been offered medication to keep down the chills, diarrhea, nausea. But he refused it. Sharon's 1989 date book tells the story:

"April 28—first infusion.

"April 30—he didn't know me.

"May 5—he knew me again."

By Saturday, May 13, Kuntz had recovered sufficiently to be sent home again. In discussing his options, the NIH doctors had told him about gene transfer. "Count me in, if you need me," Kuntz told them. Recalls Sharon, "They told us it had never been done before, but we didn't mind that. Maurice really didn't think it could help him, but he hoped it might eventually help somebody else. So he told them, sure, if they wanted him to go first, he would do it." Back home in Indiana, an urgent phone call from NIH interrupted Kuntz's rest: Steven Rosenberg and French Anderson wanted to talk to him about receiving gene-marked TIL cells in a second round of TIL therapy. Right away. The couple flew back to Bethesda on May 21 and was directly admitted to the intensive-care unit. "I left the kids home," Sharon says. "They stayed with friends—our closest friends knew what was going on. I didn't want the kids with us, in case something went wrong."

That night, Anderson again came to see them and spent three hours this time. More earnest now, he told the couple the details of the experiment, going over every point, expressing his hope that the marker gene would track the TILs to Kuntz's tumors like radio collars on wolves, so that Rosenberg could see where they went and what they were up to. Anderson reiterated to Kuntz that he didn't *have* to do it; that he didn't have to go first; that he might suffer a heart attack from the gene transfer, or even worse. Kuntz nodded. He understood. Anderson carefully read the couple every word of the informed consent. They signed it. Then Rosenberg came into the room beaming with his customary cheer. "We'd met him before," Sharon says, "but we didn't have the same relationship with him as with French. Maurice and French became good friends." At eleven-thirty that fateful night, May 21, 1989, Sharon left her husband to catch a cab back to her motel. Anderson, driving by, saw her waiting and insisted on taking her home.

To Anderson, the gene transfer TIL experiment "was like a moonshot," pervaded by the nail-biting complexities, shattering last-minute delays, and harrowing suspense that NASA had made a trademark. The team had received permission from federal regulators months before to treat a total of ten metastatic melanoma patients with gene-marked TILs. But the "final go," in Anderson's molecular moonshot had come only on May 20, 1989, after a complicated countdown that hinged on about twenty different factors—the delicate preparation of living cells had to be perfect, and the perfect patient candidate had to be in the perfect state of readiness to receive them. Previous countdowns had not made it that far. In fact, Maurice Kuntz was actually the third final candidate. Discouraging laboratory telemetry had spooked Anderson into abruptly canceling one molecular launch twenty-four hours before infusion, and another had been scrubbed with only ten minutes left.

But all systems were go on the morning of May 22, 1989. In the privacy of the surgical intensive-care ward of the 550-bed NIH clinical center, Anderson, Rosenberg, and Blaese hovered over the bedside of Maurice Kuntz, a patient identified in news accounts only as "a 52-year-old man dying of cancer." He was undergoing a massive transfusion, with the first of five plastic bags holding 200 billion of his own tumor-fighting white cells that had been culled from his biopsied cancer and supercharged in the lab. As Kuntz, wearing a white hospital gown with blue polka dots, sat propped up on a hospital bed, bristling monitors like a porcupine, holding hands with his wife through the side rails, a nurse hooked up a plastic intravenous bag containing the milky fluid. "Maurice was not nervous. He was very calm, but I had been praying all morning," Sharon remembers. "We were watching each other the whole time. Smiling at each other. Just looking at each other." Nobody said anything historic, she says. The atmosphere was too tense for that. A male nurse stood quietly in the background holding a syringe filled with adrenaline in case resuscitation was required. But without ado, the IV dripped into Kuntz as the seconds ticked by. And in this intimate setting, the course of medicine was changed forever.

As the doctors monitored Kuntz's blood, Susan Calabro, a surgical resident, continued to withdraw samples as she had before the experiment started, so that crucial differences could be measured. Anderson stood, transfixed. Kuntz's souped-up TIL cells were again being unleashed back into his body. Nobody could say for sure what would happen when alien genes actually were introduced into a human being.

They might just kill him. The persons around the bedside held their breath as the IV dripped and new genes entered the human bloodstream. "You could have heard a pin drop," Sharon relates. Both Anderson and Rosenberg checked their watches: 10:47 A.M. The molecular moon rocket had lifted off.

"Maurice seemed serene," Anderson recalls. "But I was scared to death. My worst fear was that the impossible would happen: that after everything all of us had been through, he would suffer a heart attack and die right then and there!"

Finally, one minute passed. "He's okay," Anderson thought, surprised by his own success. Kuntz just sat comfortably, a small smile playing at his lips.

Three minutes passed. "Absolutely uneventful," noted Anderson. The first stage of the rocket disengaged.

Five minutes. "Perfectly normal." The second stage fell away.

At this point, the patient himself decided to lighten the tension. Alluding to all the mouse research that had preceded this first human experiment, Kuntz cracked, "Well, I haven't grown a tail yet."

That broke the ice. Various jokes about cheese followed as the minutes stretched and the participants relaxed a little. The genes were now in orbit. Steven Rosenberg, as usual, was jubilant. "Today is the first ever!" he obligingly declared for the benefit of the NIH press office, "the first time that a new gene has been introduced into a human." Later that day, a new sign appeared on the wall of Anderson's office: "One small step for a gene, but a giant leap for genetics."

"French was like an excited kid who wanted to jump up and down," Sharon confides. Yet, as she recalls it today, the full import of the moment had not sunk in for her and her husband. "Even though we'd been told what a milestone it was, it wasn't such a great big thing to us. At the time we both were focusing on Maurice. He even later told other patients it was no big thing, just a tracer bullet. He didn't feel it was a foreign gene. He said he knew it wouldn't hurt him."

The first infusion lasted about forty-five minutes. As the cells dripped into a central line on the left side of Kuntz's body—the IV penetrated the subclavian vein that led straight to his heart—Calabro withdrew blood from another intravenous line in his right arm. Left side in, right side out. Input, output. First, the blood samples were to indicate how the TILs and their new genes immediately fared in their big survival test as they passed through the lungs, liver, and spleen—the giant filters that cleanse the labyrinthine tubing of the circulatory system. Together, these obstacles were tantamount to a black hole, so far as gene science was concerned. The cells could easily have vanished forever.

Anderson had spent the weekend before the experiment working off the tension by teaching martial arts classes in the basement exercise center of NIH's Building No. 10. Almost giddy with excitement and worry, he sparred with everyone in the club on Friday night. Sunday night, after dropping Sharon off at her motel, he couldn't sleep. On Monday, he arrived at his lab before dawn to fret over his cell preparations and to add notations to a flow chart that had grown to twelve, highly detailed pages of data even before the team actually started to infuse Maurice Kuntz.

"In case something horrible happened, I made certain that I had written down absolutely everything we had done beforehand," he says. "That way I could hand it to Jim Wyngaarden and say, 'Here. This is what we did.' Whereupon the director would have had to announce the premature death of not only the patient but also of human gene transfer, setting back the entire field of gene therapy for heavens knows how long."

When it became clear that Kuntz was tolerating the treatment very well, the doctors moved on, their faces masking, in a professional way, feelings of overwhelming relief. Recalling his mood after the critical first infusion, Anderson remembers simply thinking, "So far, so good." Quickly, he refocused on the crucial data that the scientific community would demand. As each vial of a total of nine blood samples was drawn from Maurice Kuntz, no participant in the experiment could know what a particular sample represented until it had been analyzed independently. The times and dates of collection were individually coded in numbers and letters. After collecting blood from Kuntz, four different technicians extracted the white cells from each sample. Then they added neomycin, extracted the DNA from the white cells that survived the dousing, and submitted the DNA to polymerase chain reaction analysis (PCR), the exquisitely sensitive laboratory technique that allows scientists to pinpoint, amplify, and study the most minute pieces of genetic material. By targeting a tiny stretch of DNA and adding an enzyme, researchers can duplicate the DNA sequence indefinitely until sufficient quantities are retrieved for further analysis. In effect, PCR allows them to look for a needle in a molecular haystack and, when they spot it, to use it to create whole stacks of needles. The process, discovered by Kary Mullis (which won him a share of the Nobel Prize in chemistry in 1993) and developed since 1985 by researchers with the Cetus Corporation, has swept through science like a cyclone—it is eagerly embraced by AIDS researchers, forensic scientists, genetic disease researchers, even archaeologists. In the TIL experiment, PCR was used to scan billions of normal white blood cells in a sample and to single out the relative handful of TIL cells by spotting the bacterial marker gene.

If the test was working properly, blood samples taken from Kuntz before the infusion obviously should have given no indication of the inserted gene. But as the PCR analyzed the vials of blood taken during the infusion and afterward, matters should have started to get exciting, and they did. A positive PCR—actually a tiny black dot on x-ray film—meant that the marker gene had been detected in the chromosomes of the TIL cells. "The vials, which, remember, were taken from the right side of the body during the infusion, proved to be very positive for the gene," Anderson says. "The lungs couldn't filter out all the TIL cells on the first pass. But the test taken five minutes after infusion was negative. Cells containing the gene had disappeared. The lungs, liver, and spleen had filtered all of them out. One hour later, however, there was a very faint but detectable signal.

"A few cells had been released back into the circulation. By seven and a half hours, there was a strong positive in the bloodstream. An hour later, an equally strong positive. At twenty-four hours, we saw the same thing. And by forty-eight hours, the signal had dropped back to a barely detectable level, which presumably means the cells were migrating to the tumors and piling up. We'd find out when we did the biopsies.

"It's like we sent the gene into orbit, and after the initial communications blackout, it started to send us clear signals. And everything was A-okay."

The TIL cells and their genes were designed to persist, or stay in orbit—for months, it was hoped—and Anderson and Rosenberg would monitor them periodically, depending on the availability of the patients and whether their tumors were capable of being biopsied. "But the early data represented the first time that TILs have ever been detected in the bloodstream," enthused Anderson a week after the first experiment. "Using radioactive tracers, they never were found. Not even immediately during infusion." If, by some incredible fluke, the delivery virus had somehow started to replicate in Kuntz's bloodstream, PCR should have picked up that signal, too. "We asked the PCR to look for the presence of the virus's envelope gene," Anderson said. "It found none. So the virus has done precisely what it was designed to do. It performed perfectly. We couldn't be happier."

A few days after the infusion, Maurice Kuntz was sent home. Every Monday for weeks afterward, he went to his local clinic and had a blood sample taken and sent to Anderson. A month later, he returned to NIH for more tests, at which point Rosenberg biopsied a lymph node for evaluation.

The delicate question in this case was, What would have been the cause of death if Kuntz had died in the interim? "At this point, he would die from his tumor," Anderson said at the time, "because he's clearly not going to die from the gene transfer. We're sure of that, at this stage. Should he die, we would do an autopsy on him and look for virus, look for gene, look for everything."

One of Sharon's most valued possessions is a videotape that NIH made eight months later, on January 2, 1990. It shows herself and her husband chatting with Anderson about the first gene transfer experiment. "I was proud to be part of that program," Maurice tells Anderson. "Everything was thoroughly explained to me. I knew what I was doing." When asked why he had decided to go first, Kuntz replies, "I was glad to do it for some little child, or some old man like me, or even some of those critics of this [procedure], if they ever found themselves in my shoes. I just wanted to help. If everybody had stood still in the 1950s, we'd still have polio. So I did my little part.

"You [French] have been doing this for twenty years. You did the hard part. It was just a little marker gene for me. But it was a big step, too. It could help with all kinds of diseases, not just cancer."

Getting TIL therapy, along with IL-2, Kuntz admits on the tape, "is like being in the ring with Mike Tyson. But you have to take the pain with the gain. I'm getting ready for my next treatment. Then I hope I'll be able to go home and get a little fishing in."

At first, it looked as if the TILs would save Maurice Kuntz. "We drove to Bethesda in June," Sharon notes. "Maurice felt good. He had no pain. Tests showed that some of the mets in his liver, and in his spleen, had disappeared. So he had hope. It looked like it was working." Kuntz fished the summer away, seemed to feel okay. He went back to Bethesda for more TIL therapy, without the bacterial gene, but the regimen seemed to have run its course. That November, Kuntz somehow found the strength to prune his spruce trees. "I want them to look pretty for you in the spring," he told Sharon. A chill seized her heart. "I think deep down he knew," she says. Three metastases had spread to his brain.

After that, things rolled downhill fast. In November, Rosenberg treated Kuntz's brain cancer with radiation and removed his spleen, which again had become cancerous. Anderson grew new TILs from it, and the team tried another bout of TIL therapy on Kuntz. It had worked in the past. At home, Sharon stood by while her husband received repeated radiation treatments—eleven in all—which dehydrated him and made him sicker and sicker. Finally, in desperation, Sharon called NIH and pleaded, "If he is going to die, please don't let him die this way." NIH flew the couple out immediately and pumped Kuntz full of fats and fluids—but no more TILs—so that he could spend Christmas home with his family.

More metastases kept developing. At this point, Anderson took Sharon aside and leveled with her: "Enjoy the time you have together," he gently advised. "How much time do we have?" she wanted to know. "Nobody knows but God," Anderson replied, "but maybe six to eight weeks."

Back home again, Kuntz took to his bed. He began telling Sharon what to do: keep the house, sell the van and buy a car, get the chimney cleaned, trim the pine trees. The family doctor made several house calls. "I want to keep him here," Sharon told him. "No more hospitals." The doctor told her she was doing as good a job as any hospital.

Up until the end, Maurice Kuntz was aware of everything around him. He refused morphine, yet seemed to be in no pain, his wife says. Friends and family came to call. Sharon nursed him through the nights. "We wanted him to know he was with people he loved," she says. "He told me, 'I'm ready to go to see Mom now.' "

Finally, at 6 A.M. on Easter Sunday 1990, nearly a year after he made medical history, Maurice Kuntz quietly died after a long and valiant fight. That night, his body was flown to NIH for autopsy and then home for burial. French Anderson, who had been on vacation with his wife in Hawaii, had tried to get home for the autopsy, but to no avail. "But they did it as carefully and with dignity as if it were surgery," he reassured Sharon. "They did it as if he were going to wake up."

"And I know they did," Sharon softly says, her own love and respect filling her voice. "I feel my husband was chosen to have cancer. He told me, 'I'm glad it's me, instead of one of my brothers.' That was the kind of man he was.

"I only wish he could have known that because of him gene therapy was so soon to come to pass. He would have been so proud to have helped a small child!

"I miss him with all my heart," she says. "But it's time that people know who he was. I want you to tell his story. It is my final tribute to him. Maurice was put on Earth to help other people. I believe that.

"French said he always knew it would take a special person who was willing to go first. And it was. My husband was a very special man."

Periodically, over the months following Kuntz's last-ditch gamble for life, seven of the allotted ten volunteer NCI terminal cancer patients would undergo TIL therapy. All absorbed the marker genes with no side effects, according to the doctors. Despite all the fears and all the emotion-laden committee hearings, the genetically altered retroviruses had, like the reliable Saturn rockets of yore, done their jobs precisely as programmed. They carted in sufficient numbers of the genes for later testing and then stopped cold, leaving the altered cells safe for infusion in the patient. Moreover, enough gene-marked TILs survived to be detected by ultrasensitive lab tests more than three months out—and in one patient for considerably longer. Some genes, at least, were working in their new homes; this suggested that functioning foreign genes circulating in blood cells could indeed be safely implanted into human beings.

Tantalizingly, all of Rosenberg's patients seemed to respond to the opening salvos of TIL therapy. All of them tolerated the treatment well, he says, and were discharged from the hospital after a few days. Patient number five, in fact, seems destined for the history books. She is a twenty-six-year-old woman who had dozens of melanoma nodules throughout her body, as well as lung tumors. She was in desperate straits when she came to Rosenberg on June 21, 1989. A disturbing lesion on her soft palate made it difficult for her even to swallow or eat. Rosenberg harvested under local anesthetic three subcutaneous tumor deposits and for a month grew up her TILs in the lab. He created a virtual army of TILs, tagged them with marker genes, and gave them back to her. The result was phenomenal. The x-rays bring tears to the eyes of all who behold them: in just a few weeks, all the melanoma in this woman simply went away. Thirty days after treatment, her tumors had completely disappeared! She returned home to her husband and eight-year-old child, cured. Rosenberg joyously stamped her records, "NED," no evidence of disease.

Another male patient, number three in the protocol, at first enjoyed a dramatic regression in a large mass in his abdomen, but then it started to grow again. After biopsy, Rosenberg found evidence of substantial numbers of gene-modified cells within the tumor. Worried about another mass in the patient's esophagus that was restricting his swallowing, Rosenberg extracted TILs from the abdominal tumor, grew them up, and reinfused them. The patient's esophagus tumor shrank markedly.

"Of the first five patients," Rosenberg says, "at least three showed some evidence of regression of tumor. That's what we'd expect, after our earlier experience with non-gene-modified TILs." But then came trouble. Despite the last-resort therapy, patient number four died with widespread disease in every organ of his body, including his brain, about six months after receiving the gene-modified TIL. "He died at home in Ohio," Rosenberg relates. "We were not able to obtain autopsy tissue for the kinds of exhaustive analysis and cell distribution that we would have wished."

Summing up what he and Rosenberg had learned over the first year of the experiments, Anderson said, "We've learned that the TILs are detectable in the bloodstream for about three weeks, and then they disappear. But they are clearly not really gone, because if we take out a tumor later and grow up the cells, those marked TILs are still there. We think they're sort of hanging around as memory cells, and if a tumor starts to grow, they go to it." Another important piece of evidence, he said, is that when the researchers remove tumors much later and simply assay them directly for the presence of marked TILs, they cannot find them. The TILs are below the 1-cell-in-100,000 level that PCR can pick up. But when the researchers cull all the TILs from the tumor, the marked TILs are present in about 1 percent. "That tells us when a tumor really starts to grow, it overwhelms the body's immune cells," Anderson explains.

"However, the immune cells that we find in the tumor are more determined to fight that tumor than the other cells are."

The results of the first battery of TILs spotlight the mysteries of cancer and the human immune system that the experiments were designed to help decipher—one terminal patient was sent into complete remission, another was helped, and two others died. (By 1992, seven out of the first ten patients had died.) None, however, had been expected to live more than ninety days. "In general, about half our patients are unaffected by TIL therapy," Rosenberg admits. "I don't know why. There are thoughts . . . lots of thoughts . . . but nothing proven. We're working around the clock to try to answer that question. Around the clock." Nevertheless, a baseline for research was started with the first group of gene-modified TIL patients, and fifteen days after the death of patient number four, on March 30, 1990, Rosenberg, Blaese, and Anderson went before the Human Gene Therapy Subcommittee of the RAC and asked to expand the experimental therapy—with the marker genes inserted—to at least fifty more volunteers.

At that same historic meeting, Anderson and Blaese went for the brass ring and made the biggest move of their own careers: they asked for permission to try to cure ADA deficiency by reprogramming the genetic codes of children.

9

GENE THERAPY MEETS THE PUBLIC

The first human gene therapy document itself, several hundred pages long, awaited the curious in neat piles on tables. Its austere, long-winded title, "Treatment of Severe Combined Immunodeficiency Disease (SCID) Due to Adenosine Deaminase (ADA) Deficiency with Autologous Lymphocytes Transduced with a Human ADA Gene," belied its true significance as marking a sea change in the treatment of human disease. R. Michael Blaese had the lead byline as principal investigator; the associate investigators were listed as W. French Anderson, Kenneth Culver, and (rather peculiarly in this context) Steven Rosenberg, the cancer surgeon-oncologist.

It was a rainy, chilly day in Bethesda, typical for late March, a less than inviting setting for what was to be a historic meeting of the RAC's Human Gene Therapy Subcommittee. An underslept and hyped-up Anderson, elegant in a blue-gray suit, bobbed about the formal stage set like a nervous parent, greeting allies, casting glances at the audience, rechecking the slide projector. As the sixteen RAC subcommittee members took their seats at the enormous, twenty-foot oval table, the scene of so many post-Asilomar battles in the recombinant DNA saga, the crowd leaned forward in anticipation. All eyes seemed drawn to the charismatic Rosenberg, who had seated himself silently on the periphery next to the poker-faced Mike Blaese and earnest-looking Ken Culver, all of them clad in spotless white clinic coats—part of the game plan. The coats conveyed a message—patients were dying, and the time had come to approve this new therapy.

A major stumbling block was to intrude, however. PEG-ADA refused to go away. Back on March 5, 1987, a landmark paper out of Duke University, the University of Nebraska, and Enzon, Inc., appeared in the *New England Journal of Medicine*. The paper, entitled "Treatment of Adenosine Deaminase Deficiency with Polyethylene Glycol-Modified Adenosine Deaminase," heralded a new miracle drug that when administered weekly by injection was saving the lives of ADA-deficient children by replacing the enzyme their bodies could not make.

There were no more bubble boys, a gratifying development, unless your team was trying to get a gene therapy proposal through the RAC. Most of the thirteen ADA-deficient youngsters in the United States and Europe who might have qualified for gene therapy seemed to be doing well on the drug.

Despite this, Blaese and his colleagues were asking the federal regulators to sanction something pretty drastic. First, the gene therapists wanted to splice the human ADA gene into a mutant mouse retrovirus. Then, at the NIH clinical center, blood would be removed from a youngster, and the child's own T cells exposed to this viral concoction. After checking the safety of the gene transfer, the team would, for a week, mix the genetically altered T cells with interleukin-2, which amplifies their numbers and without which they seemed unable to survive. After another week of growth, 80 percent of the cultured cells enriched with healthy ADA genes—about 100 million cells per kilogram of body weight—would slowly be given back to the child in a series of monthly infusions. This sounds like a lot, but it wasn't. By comparison, adult melanoma patients receiving TIL therapy had been routinely given infusions of as much as 400 *billion* cultured T cells in a single day, so the proposed ADA treatment would involve only about 1 percent of the volume of cells needed for TIL. The youngster, who would have been on PEG-ADA for at least nine months, would continue to receive weekly shots—the crucial point.

Gene therapy would be administered over six months and carefully monitored, requiring the child and his family to make more than twelve trips to NIH, first to donate blood for genetic alteration and later to receive infusions. Side effects were possible, and the youngster might have to have a catheter installed in the abdomen to make multiple infusions more convenient. The child's immunity might improve during this period, but the real idea was to see whether the cells were tolerated, how long the genes persisted, and whether what worked in a test tube would work even better in the body because of the selective advantage of the altered cells.

If things went as planned, the protocol would move into its second phase six to nine months later. The patient would continue to make monthly visits to donate blood and receive genes, but the dosage would change. This time, only cells that had been shown in culture to be expressing the introduced genes would be infused into the patient—again, starting with a mere 100 million cells per kilogram of body weight and rising into the billions as the therapy became more aggressive. The point of using only gene-corrected T lymphocytes was to increase the amount of ADA enzyme the child was getting. By the end of the second part of the experiment, which could last between twelve and eighteen months, the patient should be receiving monthly T-cell infusions at a dose between one and three billion gene-corrected T cells per kilogram of body weight—the projected therapeutic dose—which was roughly equivalent to the weekly dose of PEG-ADA (fifteen units per kilogram).

After six months of this regimen, an immunological evaluation would be performed, with the child being inoculated with several different vaccines, including those for tetanus, diphtheria, and influenza, to check the immune response and to estimate the persistence of the gene-modified cells in the blood. If no evidence of expression was found, or if less than 1 percent of the blood cells were shown to be expressing good ADA genes, the intravenous injections would stop and the patient would for six months be treated by means of the abdominal catheter, which allowed a slower, more gradual infusion of genes. But if there was clear evidence that the child's immunity was improving, the periodic infusions of gene-modified T cells would continue indefinitely.

In theory, the healthy T cells should, like an army of mercenaries brought in for the task, start to function as a normal immune system, with their numbers being replenished every month. And only if the repetitive treatments were shown to be conferring normal immunity would the protocol move into its third phase and PEG-ADA be gradually withdrawn—as one might remove the net from beneath circus aerialists.

Normal immunity, to Blaese, meant basically that the child's body was making good use of the modified T cells and that the cells had become fully versatile and therefore equal to the myriad protective tasks they had to perform. Such diversity, or "repertoire," hinged on their ability to divide and produce clones of themselves, each of which would carry on a legacy of vigilance, recognizing and reacting to specific threats—such as tetanus, diphtheria, potential tumors—as the parent cells had in the past, whenever such threats cropped up in the body. Repertoire is something that occurs dynamically throughout life, as T cells encounter different invaders and fashion receptors on their surfaces that fit each invader's topography the way a key fits a lock. One way researchers tell whether someone has good immunological repertoire, in fact, is to study the T-cell surfaces for signs of a wide array of different receptors.

As crucial as repertoire to robust immunity is the issue of balance—balance between the two basic types of T cells. The establishment of such equilibrium is normally the job of bone marrow stem cells, which design the two T-cell types and manufacture them in roughly the same proportion. "Helper," or CD-4, cells alert the immune system that trouble is afoot—that the body is under attack from a pathogen. Then another major type of T cell—the "killer," or CD-8, cell—responds to the warning from its CD-4 brethren, leaps into the breach, and tracks down and gobbles up the pathogen. A child with corrective genes that bestowed normal immunity should be able to balance helper and killer T cells as needed throughout life.

When one studied the NIH gene therapy proposal, it became apparent that a child and its family would be expected to make more than twenty trips to NIH. That seemed like a long—and perhaps unnecessary—haul for a child already getting weekly shots of an effective drug.

So what was the RAC to do? In one corner: treatment by polyethylene glycol–modified adenosine deaminase, the drug that slew the bubble. In the other: treatment with autologous lymphocytes transduced with a human ADA gene.

What was the point of gene therapy? "To try to cure these kids," Anderson asserted heatedly before the meeting. "Not to give them gene therapy instead of PEG-ADA. That is not the protocol. It is to *supplement* the drug and then—only if we've established that the kids are doing much better—to withdraw it."

Despite Anderson's elucidations, the RAC subcommittee had an inflammatory situation on its hands. In the excitement over the possibility of gene therapy, PEG-ADA kept getting short shrift in the press. In fact, the drug had just that week received final FDA approval as the first medicine that could be used to treat an inherited metabolic disease by replacing the missing enzyme, yet this triumph inexplicably continued to be downplayed in reports about gene therapy. For instance, stories in *Science* magazine and the *New York Times* ignored or discounted the lifesaving new drug treatment. It was as if the reporters had not been aware of its efficacy or even its existence. Said *Science*, "After a few doses [PEG-ADA's] effectiveness is often lost, for reasons that are not understood." The *Times* flat-out ignored PEG: "Bone marrow transplants have helped a few patients but only when a perfectly matched donor can be found; otherwise no effective treatments exist."

The oversight, which probably was nothing more than a mistake, had a prescient quality: the picture painted in the press was that gene therapy *alone* offered real hope for children born with the deadly immunodeficiency. Anderson could not have written the script better himself. But the picture simply was not correct.

The PEG-ADA backstop comprised the very heart of the gene therapy proposal. Without PEG-ADA's ability to handle lethal toxins, the ill children would be unable to make any T cells at all for the NIH team to remove and correct with genes. Questions about PEG-ADA's worth were crying out for answers. It should have been routine for the RAC to invite a PEG-ADA expert to attend the day's proceedings to supply those answers.

Unfortunately, nobody had thought to invite one. Correction: one person had. Seated in the audience behind Richard Mulligan was a guest from Durham, North Carolina, none other than Duke's Michael S. Hershfield, the man who had helped develop PEG-ADA. Hershfield, who had paid his own way, was there at the urging of Mulligan. "He [Hershfield] is known to science as 'Mr. PEG-ADA,'" Mulligan said later. "I thought he should be there."

According to the rumors then raging through NIH, the dapper Duke physician, a paid consultant to Enzon, was intent on scuttling the gene therapy experiment, lest it replace the drug. In fact, Hershfield claimed a more altruistic motive. He feared that in their mania to be first to try gene therapy, the experimenters would withdraw PEG-ADA from the children prematurely, placing the youngsters—all of whom he had treated from day one—in mortal danger. Moreover, as the scientist who had figured out the basic biochemical flaw in ADA deficiency and as senior author of the *New England Journal* study that announced the first drug treatment for children formerly doomed, Hershfield believed it was *he* who was the real ADA expert, not Mike Blaese, and certainly not French Anderson, whom Hershfield had not even met. Shut out by the gene therapy team, which had never contacted him or asked to see his PEG-ADA data, Hershfield was hurt and angry. He suspected an end run.

It was indeed Hershfield, not Blaese, who had dealt with the terrified families when their children were desperately ill and facing imminent death. And PEG-ADA had saved them. Hershfield wanted to tell the subcommittee that. He also felt compelled to tell anyone who thought he was a shill for Enzon that, yes, he was a paid consultant to the company but that he considered Enzon an adventuresome enterprise that had taken a big chance in creating an orphan drug for thirteen children in the world. The payoff, he admitted, was that if the preparation worked, PEG technology would serve as a model for other lifesaving enzymes that could be produced commercially and engineered to survive in the body, drugs that might be used to treat leukemia and other diseases. This would certainly be a lucrative development for Enzon, but it was not relevant to the issue at hand.

In any event, Michael Hershfield's commitment to PEG-ADA was total. He had seen what it could do. ADA deficiency is a lethal disease, yet most of the kids were able to live normal lives and even go to school. Hershfield deeply resented being treated like an outcast by NIH's gene therapy team.

Now, in Conference Room 6, he sat bristling like the stone Commendatore in Mozart's *Don Giovanni*, who, as every operagoer knows, eventually rises up in fearsome majesty and hauls the profligate Don Juan down to the fires of hell. Would Hershfield do the same to gene therapy?

Other problems loomed. Not enough time had been allotted for questions of any complexity, because NIH planners had optimistically summoned both the subcommittee and the parent RAC members on the same day, on the slim chance that the gene therapy proposal might sail through and gain full approval. Furthermore, there were nasty overtones of conflict of interest because some questioners, particularly Mulligan, were generally considered to be active competitors.

Perfectly attuned to this volatile atmosphere, Anderson had prepared thoroughly for the subcommittee meeting, as he always does. "It'll be very emotional," he predicted. "You really have to be sharp and follow the flow of the thinking and try to direct the subcommittee, even though you can't look like you're doing it. You have to be very careful. I don't talk to anybody twenty-four hours before the meeting except those people who are part of the game plan. It's too important. There's too much at stake."

For an essentially shy and socially awkward man, Anderson is amazingly adept at the political arts when he needs to be. This quality stems from his childhood, when he was detested by grade school classmates for his intellectual arrogance. Anderson, the class brain, looked down on anyone who wasn't as smart as he. He remembers walking home with a classmate in the fifth grade and being told, "You're the most unpopular boy in school." At the time, Anderson recalls, he considered that a badge of honor. But as the school year progressed, his social problems grew worse. Eventually, the school psychologist was brought in on the case.

Even Anderson began to sense the need to change. "I realized that if I was ever going to be successful in life, I had to learn to get along with people," he recalls today.

But he did it with his usual all-out intensity. He began to press the flesh, as if working on a five-year plan to become the toast of Tulsa. He even changed his name. He no longer used his given name, William, but began going by "W. French," assuming it would give him some cachet. Over time, what started as an expediency has become another eccentricity. Today, he turns on his heel if anyone dares to call him William or Bill. Letters so addressed make him see red.

Nevertheless, his schoolboy crash program to become popular seemed to work. By the seventh grade, Anderson was so well liked he was elected class president.

He had learned to manipulate people with the same flair with which he later handled genes and retroviruses. Now, with the stakes infinitely higher, he was still approaching matters like Eisenhower at Normandy, plotting his campaign down to the last detail.

The game plan began with a straightforward cancer research request from Steven Rosenberg and Anderson that mysteriously turned into a big deal. First LeRoy Walters called the meeting to order. Anderson strode to a lectern and began flashing slides on a large screen, briefly tracing the history of the gene-marked TIL experiments and concluding that there was a "slow but steady acquisition of data." Then he turned the microphone over to Rosenberg.

The surgeon, a veteran presenter, matter-of-factly reported the results for the first six patients and asked that he be allowed to treat at least another fifty volunteers with gene-marked TILs, instead of the mere ten he had been allotted by the RAC. Chafing at bureaucratic restrictions, Rosenberg asked the subcommittee to remove the patient limitation at once, pointing out that 150,000 people would die of cancer in the four months between each regularly scheduled RAC meeting.

That was a bold, if planned, move. Rosenberg's posture suddenly seemed aggressive, judgmental. From then on, he played on feelings of guilt, with allusions to paper shuffling and foot-dragging.

With passion in his voice, Rosenberg argued that with the need for prior notice to interested parties, publication in the *Federal Register*, and appearances before the various committees, "a year easily could pass from the time an important clinical clue to cancer was discovered in the lab to when it actually could be put into practice." Rosenberg's patients didn't have that kind of time, and everybody in the conference room knew it. So, after the lengthy process that had preceded approval of the gene-marked TILs, the cancer doctor now wanted carte blanche from the RAC. He believed he could show that TILs worked by directly invading tumors; the experiments had demonstrated that it was safe to put genes into humans; and the time had now come to move ahead.

Then Rosenberg shrewdly played his hole card. He confided that in the laboratory he had already inserted the gene for a potent cancer-killing molecule called tumor necrosis factor into TILs—and that these TILs were "making large amounts of TNF!"

The statement stunned the audience. All those present knew that TILs sometimes were supposed to flock back to tumors, but now the specialized white blood cells were being characterized as cruise missiles that could deliver cancer-blasting drugs into the very heart of the enemy. As the subcommittee digested the news, Rosenberg threw in the clincher: "We soon hope to come before you with a proposal to give *these* cells to cancer patients."

That meant gene therapy for *cancer* was imminent, although it was a bolt out of the blue. Rosenberg wasn't espousing classic notions of cancer gene therapy, in which tumor-controlling genes would be replaced, or perhaps switched on or off. He was proposing, in effect, an incredibly precise form of cancer chemotherapy. Tumor necrosis factor, produced by the immune system, is known to kill cancer cells by choking off the blood vessels that feed them. Discovered in the early 1970s by researchers at Sloan-Kettering, the powerful cytokine is produced by the body during bacterial infections and has been shown quickly and magically to destroy cancerous tumors in mice. Its discoverer, the tumor immunologist Lloyd J. Old, who also delineated the surface antigen structures that characterize normal and malignant cells, describes TNF as a "central regulator of inflammation and immunity, the intertwined processes that limit and repair injuries and fight infection." TNF, in fact, can switch on the patient's entire immune system and make it shove the pedal to the metal.

However, the body normally uses TNF with great care, summoning it when needed and then disarming it quickly when its task is done. Under intense study since 1977 when Sloan-Kettering announced that several of its top researchers had been assigned to unveil the secrets of the substance, TNF made even bigger news in 1984 when the gene that encodes it was cloned. Since then, the chemical has always been viewed as highly promising, yet extremely dangerous. It is likely to kill an animal as fast as it can cure it, if given in too great amounts.

Nonetheless, at least three times during his presentation, Rosenberg mentioned, almost teasingly, how he might come up with a gene therapy proposal for cancer within a month or so and hoped it wouldn't have to wait for RAC meetings, because great numbers of people would die in the interim.

Yet, curiously, since his name was on it as a coauthor, at no time did Rosenberg call for a similarly quick approval of the existing ADA proposal that Blaese and Anderson were putting on the table. Obviously, ADA, although deadly, could not be compared to cancer in terms of enormity. But even before the Blaese proposal was formally presented, some observers were starting to view it as something of a sacrificial lamb.

In his impatience to proceed, Steven Rosenberg seemed to be challenging the RAC to redefine its very role in clinical research—to get out of the way—a move that by rights should have put him in direct conflict with his newly acquired partners, especially French Anderson. The latter, despite his own reputation for impetuosity, had always brought his ideas before the RAC. Moreover, Rosenberg's new tactic of accusing the RAC of fiddling around bureaucratically while cancer patients died wasn't destined to sit well with the advisory body that had been protecting the public from misuse of recombinant DNA since long before Rosenberg appeared on the scene.

Nevertheless, when Rosenberg showed his slides of the first gene-modified TIL patients (in which the added genes had no therapeutic effect whatsoever), an impressed subcommittee watched raptly. He flashed x-rays on the screen. There were gasps as the cancerous tumors melted away in x-rays of patient number five, the twenty-six-year-old woman who had enjoyed a complete remission.

"This is a biopsy taken from this patient to illustrate how the treatment works," Rosenberg explained, his voice rising to its characteristic cheery tenor.

He flashed a pathology slide that showed chaos—a mass of weird, scythe-shaped cells. "Notice the monotonous melanoma cells . . . prior to treatment."

The next slide depicted an invading army of pastries—TIL cells, as seen by the electron microscope, look for all the world like coconut puffs—consuming the cancer cells.

Rosenberg's tone underscored the wondrous thing happening on the screen: "At day three into the treatment, one begins to see infiltration of lymphocytes into the tumor."

As he clicked the slide projector again and again, a pattern of order replaced chaos. The puffs predominated. The weird cells were vanishing.

"By day five, we see more of them infiltrating, some of which contain the marker gene. By day nineteen, you see the tumor almost completely replaced by lymphocytes.

"This patient had biopsies out beyond thirty days of residual deposits that contain no viable tumor. They disappeared completely."

Astounding! The audience had just seen the rarest phenomenon in all of medicine: cancer that gave up and just went away. Rosenberg had the committee members eating out of his hand.

It was his hope, he quietly explained, to use gene modification to design cells with new properties that would increase their effectiveness manyfold. He wanted to expand the number of his patients so that he could treat more skin tumors, do further biopsies, and study the functional properties of the TILs, now that the marker genes gave him a way of tracking them.

But there was more. By later removing the TILs from patients who were responding to the therapy, Rosenberg hoped to be able to identify the specific cell populations responsible for the greatest amount of anticancer activity. Those were the cells he wanted: the ones that killed cancer. And it wasn't beyond the realm of possibility, he ventured, that the patient's own tumor could tell him precisely what population of TIL cells could be used to kill it.

Now, here was an ingenious scheme: "We can use the marker gene to have the patient's own tumor select for us those cells which traffic to tumor deposits [and have] antitumor activity," Rosenberg outlined.

When removed from the body and multiplied billions of times in the laboratory, those identical cells would then be given back for a reprise treatment that could really be awesome.

Eventually, he said, it might be possible to give back to the patient IV bags full of smart molecules that had already proved their cancer-killing prowess. "It's the second growth of TILs that hopefully will have more profound antitumor activity than the first population," he pointed out.

"That means we could take that subpopulation and expand it to hundreds of billions of cells—that is, the total white blood cell population in the entire patient.

"Then, when we give it back, every lymphocyte in the patient's body would have antitumor activity."

Here was yet another of his great notions. In theory, the entire immune defense system would be brought to bear on the cancer, led by swarming legions of cells that the patient's body had already armed and primed. Here, if he would pull it off, would be the ultimate biological therapy for which Rosenberg had been questing for twenty-two years.

"But you must realize that it's entirely speculative," he cautioned reporters later. "This is at the cutting edge of technology. It's not something we're going to be able to do tomorrow."

Although stirred by Rosenberg's presentation, the subcommittee members nonetheless harbored doubts. Leading the way was the nemesis who had kept dogging Anderson and Rosenberg throughout the TIL saga—the University of Pennsylvania's William Kelley. Despite the excitement over TILs, Kelley continued to view the experiments skeptically. To him, they didn't appear to prove much. Precisely how many TILs were reaching the tumors? he asked. How long did they persist in the body? What happened to the rest of them? "I assume the [TILs] replicate [in the body]," Kelley intoned. "Do you have any idea what the functional half-life is? We've seen data that at least some of the lymphocytes persist for a long time. But how long?"

Kelley kept probing the same weakness. Rosenberg sighed. "That's the crucial question," he admitted—then, he dodged it.

What was really important in the killing of cancer? he asked rhetorically. The number of TILs? The kind and amount of chemicals (cytokines) they secreted? The actual mechanism of action? "We know they can make TNF and other things," Rosenberg offered. "But we don't know exactly what kinds of functions to look for."

But did it matter? Everybody had seen Rosenberg's slide show. TILs apparently had the capacity to kill tumors, and that was Rosenberg's job.

Kelley, however, was not mollified. He wrinkled his nose. It would be "helpful," he said pointedly, to know *scientifically* what was really going on with the TILs, not just how long some of the cells persisted in patients' bodies. Of the TILs infused, what percentage actually went to tumors, rather than got lost elsewhere in the body?

Rosenberg shrugged. "A very tiny percentage," he acknowledged. He did not bother to add for the uninitiated just how poor the showing really was—that according to the best guess it was less than 1 percent of the cells that were infused back in a patient. The point was that some of them made it back.

But with the numbers so low, what really was going on? Kelley's stubborn interrogation again highlighted the central difference between research scientists and physicians—scientists treated diseases, doctors treated patients, and TILs sometimes helped Rosenberg's patients; that was what mattered most to him. What works, works, confound it!

After more back and forth, Rosenberg conceded, finally, that the TILs that made it to the site of tumors represented only the smallest fraction of the total cells infused. But, he told Kelley, the meticulous animal studies that could have answered these basic questions had proved fruitless. Rosenberg was unable in any consistent fashion to introduce the marker genes into mouse TILs, no matter what vector he used. They just wouldn't go. "That makes it extremely difficult, if not impossible, to perform therapeutic studies in mice," he said.

But another subcommittee member didn't buy that. Studying Rosenberg's evidence, the University of Minnesota geneticist Scott McIvor, whose RAC role seemed to be that of the loyal opposition, wasn't sure that TILs were even attacking tumors, at least directly. "You can't really *prove* the TILs are going to tumor, can you?" queried McIvor, a quietly individualistic New Age type who is in his thirties and sports a ponytail. The TILs could have gotten there randomly, he mused. "Maybe they were merely passing through" and got dumped off by accident.

In fact, McIvor went so far as to venture the heresy that the TIL experiments had revealed nothing certain "beyond the fact that the cells can be marked and found in the bloodstream for a certain period of time. It's uncertain as to whether anything else is being learned." Slides or no, from a scientific point of view, the experiments had shown nothing.

Stunned, Anderson and his teammates sat poker-faced. Kelley's and McIvor's questions had stumped them. Without persuasive animal data, Rosenberg was still flying blind. He could interpret only what seemed to be happening in the bodies of his patients, even though the gene manipulation had allowed him to spot a few TILs in tumor deposits, and he could see—in some patients, anyway—a positive response.

Earlier Rosenberg had argued that, at the very least, the experiments proved the safety of gene transfer. But did they really? To get a bona fide scientific judgment of whether there were ill effects, it didn't seem enough to merely say that the dying patients didn't look any worse. You really had to figure out ways to look in detail at those infused cells. You would need to know with precision how many cells survived after infusion. If 99 percent of them disappeared instantly into the filtration system of liver, lungs, and spleen—as seemed to have happened—all Rosenberg could reasonably say was that the *remaining* cells didn't appear to cause any difficulty.

But what would happen, say, if he transferred tumor necrosis factor genes into the lymphocytes and they similarly vanished into the body's big filters? This time, instead of harmless *neo* marker tracer bullets, the TILs would be armed with the molecular equivalent of nuclear warheads—genes that were pumping out tumor necrosis factor, something nature had never intended. Could the lungs and liver cope, or would they themselves be killed? There was no way to say, if you didn't know what was really happening in the body.

Suddenly, observers found themselves in the TIL approval battle all over again. Landmark experiments that afterward didn't tell one much of anything were useless, even if one did manage to cure a patient or two—marvelous as that was.

Richard Mulligan was growing restive. He decided to make what he viewed as a game stab in behalf of science. The bearded MIT researcher had sat through the morning proceedings twiddling a pencil and trying not to smirk. It all was too slapdash for him. He was particularly angered during the questioning of Rosenberg when he heard a piece of potentially important data on a gel referred to as "this big blot over here." To Mulligan, blots mattered.

In fact, all the evidence before him looked sloppy and unscientific; it reeked of having been assembled at the last moment. When the floor came around to him, Mulligan quickly proposed that a scientific panel be set up to examine all the safety data from the TIL experiments, to try to get a handle on what really was going on inside those melanoma patients. Mulligan was pushing, in effect, for a subcommittee of the subcommittee to analyze the data before allowing more human experiments to proceed.

"My point," Mulligan vehemently told Rosenberg, "is that the data are really more sophisticated than you might think. One person's blot on a gel means everything to somebody else. I'm worried about safety."

French Anderson's poker face contorted into a scowl. He had to squelch this line of discussion fast. He interjected, "Safety data are maintained by the FDA and the NIH review committees. To have yet another subcommittee of the subcommittee to examine all the clinical data would be burdensome."

Burdensome beyond belief, he implied.

Rosenberg politely chimed in, "I would welcome a review of the safety data."

But seated in the audience, the FDA representative at the meeting, Henry I. Miller, acted as if he thought things were getting out of hand. Miller, a fan of Anderson who repeatedly uses terms like "saintlike," "brilliant," and "fanatical" when describing the latter, raised his hand and was called upon to speak. The FDA had been diligently monitoring every single move of the TIL experiments, he declared, calling the safety data "impressive." All Rosenberg wanted was to be able to treat more patients. "We should vote to approve, and move on," Miller urged.

In an attempt to make peace, Chairman Walters, the ethicist, suggested that Mulligan or some other expert leaf through the huge books of raw patient data before the afternoon RAC meeting, three hours away.

"Nonsense!" reported Mulligan. "You just don't leaf through this stuff. It will take careful examination and explanations by these guys. . . ."

Walters called for a vote on Mulligan's proposal that the subcommittee should further review safety aspects of the TIL experiments. Many members collectively shrugged, accepting the view that such matters should be decided before a protocol ever reached the RAC. The motion was voted down by a majority, 11 to 5.

"So much for science," an onlooker muttered off to the side.

Finally, after two hours of rancorous debate over what should have been a simple matter, the subcommittee voted on whether to remove the patient limitation in the TIL experiments. The motion was passed unanimously—16 to 0—with Mulligan, McIvor, and Kelley voting for it, too. No one, Mulligan later admitted, was willing to block Rosenberg's patient-volunteers from their last chance at survival. Indeed, the change in procedure sailed through the RAC later in the day. Rosenberg clearly had won round one.

Tempers were on edge, however, and only about forty minutes remained for the ADA gene therapy proposal. The subcommittee would officially dissolve when the full RAC convened for its meeting after lunch. French Anderson could do nothing but plunge ahead. In this familiar room and before these regulators he considered his colleagues, he began the most important presentation of his career as if he were about to hurriedly show vacation slides to a group of pals.

Anderson reminded the subcommittee how, back in 1985, he and his colleagues had begun their attempts to use gene therapy to treat ADA deficiency in bone marrow. He invoked the now deceased preclinical-data document, calling it "a dry run," and said it had been put together to answer the various "points to consider."

"Everything you will hear today builds on that background," Anderson said.

The "points to consider"—the RAC's trademark federal guidelines—were devised in 1985 when it began to appear that some team would soon request permission to try gene therapy. Essentially, the government forbade any experiment by any institution receiving federal research money without NIH and FDA approval. First, a proposal had to pass the appropriate committees at the home institution. The points are largely in the form of questions indicating what kind of preliminary research the RAC considers necessary ("What laboratory studies have been done, with cells and live animals, that make researchers hopeful that gene therapy will help patients rather than harming them?")

As spelled out by the RAC, all research studies must "demonstrate that the gene therapy does not harm laboratory animals and, in fact, demonstrates that the desired biological effects occur." Even then, the researchers must explain the probable benefits and harm of the proposed treatment and deal with two potential problems: "Transmission of altered genes to a patient's offspring, and viral infections of persons who come in contact with the patient."

Anderson, an RAC member in 1985, was instrumental in the creation of the points to consider. Having submitted the dry-run gene therapy protocol in April of 1987, and now following up with the real thing three years later, he was about to play the endgame like a grandmaster. "You just whittle them down," he once remarked about the points. "Eventually, you've answered them all."

When that occurs, when all the quibbles have been settled, the advisory panel has few choices, it would seem. It can collectively throw up its hands and declare, "This whole thing is ridiculous; let's forget it for now." Or it can give its approval. Anderson, of course, was banking on the latter.

On the other hand, his contention that the current gene therapy proposal built on the 1987 preclinical-data document was wishful thinking, from a scientific point of view. The preclinical proposal had bombed because Anderson could not get genes into the bone marrow stem cells of either mice, primates, or humans, as he himself acknowledged. Afterward his team had stepped back and reconsidered its situation.

"Rather than go for the home run, [we] decided to use gene transfer in another way," Anderson explained to the subcommittee. They would try to make base hits instead, and transfer genes as markers in cells that could be taken out and studied—"namely TILs." And in place of children, they would enlist terminally ill adult volunteers who could give informed consent.

On the basis of those experiments and the safety data obtained, Anderson said, the team now proposed "to slightly refine the protocol we gave you two years ago, only instead of going into bone marrow, we'll go into T cells. It builds on the original ADA protocol and the TIL protocol. This is the next step."

Anderson didn't point out a few obvious things. Without the juggernaut of favorable national publicity generated by Steve Rosenberg, without Mike Blaese's ingenious T-cell strategy, without Ken Culver's laboratory wizardry, without the presumed safety of the confusing TIL experiments, and without the unexpected support provided by PEG-ADA, which made supplemental gene therapy possible for youngsters no longer in mortal peril, this ADA bid would never have been born.

Undaunted, Anderson smiled to the subcommittee and turned the microphone over to Michael Blaese. The immunologist began by acknowledging that the new experiment did indeed mark a compromise with what previously had been envisioned as gene therapy for severe genetic disease. These new genes would probably have limited life spans; they would die when their T cell hosts died.

Nevertheless, Blaese said his team had to prove only four things to the evaluators: that the introduced gene corrects the biochemical defects in cells; that the transplanted cells can persist and express in animals; that the approach provides an acceptable level of risk to the patient; and that it provides no risk to the public health in general.

The new protocol, he said, had the same basic design as the TIL cancer protocol, the same basic target type (T cells), and the same retroviral vector. Only the disease had changed. After discussing the history and nature of ADA deficiency and flashing slides of pitiful young victims, Blaese got around to PEG-ADA, calling it "a very important advance for the therapy of this disease."

Having talked to several of the primary-care physicians before the meeting, Blaese summarized what he said were the clinical results in the cases of the thirteen children treated so far: "Some patients have had relatively little improvement, although in general this has been a very useful treatment for most patients." The majority have shown improved immunity, he said; many had gained weight and enjoyed a "general improvement in overall health, and a significant decrease in number of opportunistic infections."

However, none of the patients could really be considered normal as a result of PEG-ADA therapy and would throughout their lives be at risk for opportunistic infections, virulent pathogens, and cancer, Blaese said.

They reminded him, in fact, of Wiskott-Aldrich children (his specialty), in that they had some immune function but could in no way be considered normal. Moreover, the drug was expected to cost upward of $60,000 a year per patient, Blaese said. He couldn't have known that he was actually understating things. A year later, Enzon would raise the estimated cost to $205,000, citing rigorous quality-control standards imposed by the government as the reason for the increase.

As Blaese continued to press his case, he next reviewed T-cell data from a dozen patients and animal studies and argued that it was possible to correct the biochemical defect. His team had shown that altered cells could persist and express ADA in animals for various amounts of time, he pointed out—although in fact the results would be challenged later. But he was right in saying that the proposed human experiment had several things going for it. "The cells are easy to culture, easy to administer in transfusions, and can be [frozen] for repeated administrations. . . . These are generally a mature population, so the inserted gene is less likely to be shut off during the multiple steps of differentiation, which may be occurring when trying to introduce the gene into bone marrow stem cells."

Blaese wanted to try to supplement PEG-ADA slowly and to end the need for it, by transplanting the healthy genes themselves. Once in the body, the gene-altered cells should have a selective growth advantage; the body, when given the finished product in the form of healthy T cells, would opt for health over sickness and make billions of clones of the genuine article. That was the key.

Yet, as Blaese talked, the long-term future of such a child seemed hazy. The injected T cells had somehow to produce enough ADA to protect the youngster; otherwise there was no point to the treatment. But mature T cells are believed to live only several months, meaning that periodic transfusions of new genes for the rest of a child's life were a definite possibility. The only advantage to the patient would be to reduce the need for doctoring, to every few months from every week, as is the case in PEG-ADA injections. And while the gene infusions probably wouldn't hurt as much as the PEG shots, they sure were a lot more complicated.

On the other hand, T cells can develop vast repertoires to antigens and are believed to endow their progeny, known to biologists as "daughter" cells, with those memories for years and years (remember, Blaese's favorite example is his own immunity to tetanus, although he has not been inoculated for thirty years). If some of the gene-altered cells turned out to be memory cells, they could perhaps pass on their immunity to their daughters by normal cell division and permanently reconstitute the youngster's immune system.

However, in putting forth his proposal, Blaese neglected to mention that PEG-ADA also showed early signs of allowing at least some memory cells to develop in some children. For example, according to Michael Hershfield, chicken pox, which formerly could kill ADA kids, was survivable now, and the immunity seemed to forestall a second, a third, or even a fourth episode, which used to routinely befall ADA-deficient children.

Nonetheless, Blaese did seem correct in arguing that if the NIH team got lucky, and the memory cells kicked in, the child would conceivably retain a healthy immune status indefinitely and, in effect, be cured. There didn't appear to be any way to tell without actually trying the therapy in a human patient.

But there was another problem. The only known way that cancer-fighting TIL cells could be kept alive in the body for more than a few days was in partnership with IL-2 and its toxic side effects. Lab-grown TILs, and presumably other T cells, seemed to be exquisitely sensitive to IL-2 and needed it to survive. Yet who wanted to give that much toxicity to children?

As Blaese cited in the protocol the lab results that he considered significant, most of the work backstopping him had been performed by Ken Culver. Working night and day, Culver had taken blood from three ADA-deficient children who were surviving on PEG-ADA therapy, grown up their T cells in the laboratory, and inserted the ADA gene. He then had tested the cells for a number of properties.

For one thing, Culver believed that he had shown that the gene-altered T cells were able to help B cells make antibodies. For another, if he took the altered cells out of IL-2, they died, indicating—to him—that they were not on their way, as some feared, to mutating into cancer cells that would grow wildly on their own (T cells in the body depend on naturally produced IL-2 for growth). When Culver put the gene-corrected cells in deoxyadenosine, they cleared the toxin. The cells expressed the *neo* marker that had been added. And Culver measured ADA production that, gratifyingly, was about four times greater than normal. In tissue culture, at least, the therapy seemed to work: it cured ADA deficiency in the cells of affected children without apparent harm.

To Blaese, the lab data indicated that placing the new gene into the deficient T cell "probably gives a survival advantage to the cell in tissue culture, and we'd think, perhaps, as well in the body."

Moreover, he had some immunologist tricks up his sleeve to tilt the odds further in favor of the new cells. Although they are not stem cells, T cells can be coaxed to proliferate by exposing a youngster to specific antigens, Blaese told the subcommittee. Work with mice had indicated that T cells juiced up in this way had continued to express for four weeks and longer—in one animal the expression could be traced after eighty-nine days. But that was in mice. Attempts to look for expression of the

ADA gene in monkeys, Blaese said, "have been very frustrating, we frankly admit."

Culver had tried to do in five rhesus monkeys the same experiment that he did in mice. He would remove blood from a monkey, put the human ADA gene into the cells, and shoot them back into the animal. Then he checked the monkey periodically to see whether the gene was still there. But he encountered a big problem: assays were not sensitive enough to reliably distinguish the human ADA in the monkey from its own monkey ADA. Moreover, Culver was unable to grow up enormous amounts of monkey T cells to put back. Monkey cells just didn't grow as well as human material.

However, another surprising lab finding had emerged that couldn't be analyzed in animals but that would prove crucial: when Culver inserted the ADA gene into deficient human T cells, they lived substantially longer in culture than did cells without the new genes.

Nobody knew why. Nor did anyone know whether the same thing would happen in the human body. How long would the cells live? The researchers could say only that faint expression of a *neo* marker gene could be detected by PCR in one monkey's T cells as late as five months after insertion.

But what did that prove? It was not an ADA gene. Anyway, was the NIH team pinning its hopes for human gene therapy on one monkey, yet another Monkey Robert, on whose fragile shoulders the preclinical-data document had rested? It appeared so.

Blaese took the high road. As he told the subcommittee, in the cancer patients who received gene-marked TILs (and TILs, he stressed over and over, are a kind of T cell), persistence of the cells in the bloodstream—where they would be needed for ADA therapy—had been spotted three weeks out, and in one patient for as long as two hundred days. But such cells were rare, to say the least, and the ability to track them might be a tribute more to the technology than to the biology behind the experiment.

Nevertheless, Blaese summed up as follows: "These initial observations, with about 45 patient months accumulated, suggest that this procedure is relatively safe. No side effects at all, related to the gene insertion, or the giving back of gene marked cells."

Safety first, safety above all. Blaese had done his best to focus the committee's attention on the probable safety and possible efficacy of gene therapy. But the committee members were like children who will not be deterred from noticing a fat man on a bus. They quickly started asking about PEG-ADA. It was unbelievable that the gene therapy request came the same week that the FDA had approved PEG-ADA.

"I just attended a press conference," said Judith Areen, dean of the Georgetown University Law School, "and six of the 13 kids on the drug were in that room. I talked with their parents. Some of the children have gotten chicken pox and recovered. Many have gotten over colds. It seems to have restored their immune systems."

How, she asked Blaese, if the test subjects were receiving both gene therapy and PEG-ADA, would the team know which effects were due to the drug and which to the new genes? It was a very good question.

Skin tests would show a lot, Blaese replied. "Some of the patients on PEG have not developed reactions that will show up in skin tests. Their ability to [react] over time would indicate that the gene therapy was helping."

The answer seemed a little pat. Response to standard allergy tests might also be due to the growing power of PEG-ADA in these patients. The drug was, after all, allowing them to develop immune systems for the first time in their lives. The point was, as several members noted, could something be done for these kids that even PEG-ADA hadn't been able to achieve?

It was now five minutes to noon. The meeting could not extend past twelve-thirty, because the full RAC was due to convene at one o'clock. Aware of this, the hard-pressed subcommittee members soon turned away from Blaese at the lectern. They looked expectantly at Michael Hershfield and began questioning him directly.

"What's the data on the immune reactions?" barked Dr. Robert Cook-Deegan, of the Office of Technology Assessment. He wanted to know whether the children on PEG-ADA who had some small immune function had developed antibodies to the drug. There were reports circulating that several kids had, one of them Alison Ashcraft. They were indeed true. At first, PEG-ADA had made Alison sick, instead of helping her. However, by interrupting her treatment for eight weeks, doctors were able to induce her immune system to calm down and tolerate the drug, whereupon it was successfully resumed.

With the spotlight on him, Hershfield reacted petulantly. "I'd be happy to present them [the data] if there was time," he said. But he complained that he hadn't seen or had a chance to study the Blaese protocol, so he didn't know what charges he was supposed to rebut. "I think it's unfair in the last few minutes of this meeting to have to deal with the presentation I've heard, or the actual clinical experience with PEG-ADA."

But Blaese cut in. He quickly tried to reassure the panel. "Since these patients are getting their own cells back, there should be no rejection," he cooed. "One patient we've studied developed antibodies to PEG-ADA and was treated with immunosuppression, and she became responsive."

Hershfield interrupted. "Do you want to see the data on that patient? Or do you want to take it second-hand?" he angrily asked Cook-Deegan. Then he swiveled to face Blaese.

"When you say you studied the patient, you mean you studied the patient's cells?"

"Absolutely," replied Blaese calmly.

"I studied the patient!" boomed Hershfield. He turned to the panel. "If you feel it's important to know what the experience with PEG-ADA is, I suggest that the subcommittee formally ask me to present it. I may not sound very diplomatic at the moment, and some people may be concerned that because of my connection with PEG-ADA and that I serve as a consultant to the company, I can't be objective. But if you want to know the data, I think I'm the only person outside the company who has studied all thirteen patients."

The tension was palpable now. But nobody pounded the table. Blaese seemed to sense he could make points by appearing to be above the fray. Even so, he grimly refused to hand the meeting over to Mike Hershfield.

"In the past week, I've been in contact . . . ," he started to relate his conversation with the hometown doctors.

Hershfield cut him off. But the dapper Duke physician seemed all at once to recognize that he was losing his audience as a result of his pique. Abruptly, he did an about-face. He oozed charm. Rising, he exclaimed, "If you have an overhead projector, I'd be happy to show you the data. Is there one here?" He was purring now. "If not, I'll just pass it around [the table]. . . ."

"Just to respond . . . ," Blaese went on icily, noting that he, too, had been a paid consultant to Enzon, had studied the data over the preceding two and a half years, and had been in contact with five of the primary-care physicians just the preceding week. His conclusions, apparently different from Hershfield's, had come from that background.

Was the drug working or not? Who could tell? Obviously, it was keeping most kids alive and letting them make some T cells. But Blaese implied it wasn't working well enough and he could do better, while Hershfield's data presumably said it was working superbly. The trouble was, he had not published his findings.

At this point, the veteran subcommittee member Charles Epstein, a wry, low-key geneticist from the University of California at San Francisco, put his foot down. "I think we're in a crossfire that nobody understands," Epstein complained. Then he slickly diverted attention away from Hershfield and PEG-ADA.

First, he praised the "very clever" protocol, adding, "Until now, everybody's been locked into bone marrow and all of a sudden, here's a different way to go."

Next, he asked Blaese about the balance of the T cells he hoped to give back. The vital issue centered on the quality of the disease-fighting arsenal. Obviously, if Blaese and his team were going to return T cells to enhance immunity in a child, they would want to give back as optimal a selection of clones as possible. How would Blaese do this?

His data—at least as stated in the graphic numbered Figure 5, 13.2, of the protocol—indicated that when a child's T cells were given new genes in culture, the clones that resulted were mostly helper (CD-4) cells, instead of the fifty-fifty balance of helper and killer (CD-8) cells that a normal body would make.

"I'm not uncomfortable with giving back CD-4 cells," Blaese said. "We can't really answer whether we're going to unbalance the [CD-4/CD-8] system, in any model we have."

With the clock ticking, the subcommittee members seemed to be getting antsy. William Kelley intervened and returned to his principal theme: How long would the gene-altered cells live in the body?

"Mike, have you looked at other animals, with and without ADA, getting a real definition of the half-life?" he asked. "It seems to me that a half-life of circulating cells that have been genetically modified is a key factor here. I'm feeling a little uncomfortable. . . ."

Was the NIH team merely presenting another, gimmicky way to give the children the ADA enzyme? Like any other enzyme, plain old ADA, when infused in the body, swiftly decomposed in the blood—that's why enzyme replacement therapy had never worked before. Encapsulating the ADA in PEG prolonged its half-life to a week, long enough for it to be therapeutic. In contrast, how long would gene-altered T cells stay alive? "If you can show me that the net half-life of lymphocytes that express the ADA protein is substantial, then terrific. It should represent a viable approach," Kelley declared.

Blaese argued that only human studies could show that. But he was willing to hazard a guess. All he needed, he said, was a mere 1 percent of the new T cells to express ADA, and the child would be cured.

No, Kelley pressed, that wasn't the point. Would Blaese reach that 1 percent in six hours or six months? Blaese shrugged. The gene-marked TILs in cancer patients sometimes reached the equivalent amount of detectable protein product in two weeks. And sometimes they did not.

But, again, Blaese's answer seemed slightly askew. TILs were highly specialized T cells that supposedly were designed solely to fight cancer. Where was the hard scientific evidence that they behaved like ordinary T cells, let alone like gene-bolstered T cells whose role in life was to fight immunodeficiency?

Nervously watching the give-and-take, Anderson again interrupted. "Clearly, this issue is going to come back again," he remarked, seeming all at once to capitulate to the clock and the realization that they were not going to get the proposal through that day. "Tell us what you want to see by the next meeting. That would be helpful to us."

Jaded onlookers now suspected that Anderson was not banking on immediate passage—that he was using the meeting to gauge what the committee members wanted, so that he could give it to them the next time out. All in all, there was a wheels-within-wheels quality to the meeting, a sense of many layers of reality.

Indeed, the protocol appeared to be growing as it was presented. In fact, it had become what Mulligan and other critics would derisively label "a moving target," a constantly shifting proposal based heavily on the TIL experiments. But the national advisory panel seemed confused by Anderson's strategy. Was the protocol genuine? Or was it a work in progress?

"This was amazing," Mulligan was to observe later in the first of many critical analyses. "What were they really trying to do? It seemed to me they were really saying: 'We want to do this thing any way we can. Just tell us *how* you'll let us do this."

The next questioner, the bone marrow transplant pioneer Robertson Parkman, of UCLA, who had been instrumental in eliminating the Lesch-Nyhan syndrome as the prime candidate for gene therapy's debut, now turned a cold eye on ADA deficiency as well. Where, he wondered, was the NIH team's animal data to support its hypotheses—particularly the basic data showing that ADA expressed *inside a cell nucleus* by means of gene therapy was better than ADA expressed *outside* in the bloodstream by PEG-ADA?

"You must state your hypothesis," Parkman lectured the senior NIH researchers, as if they were undergraduates, "then submit preclinical data supporting the hypothesis."

Blaese and Anderson grinned, nodding politely. To them, Parkman was flying the flag of bone marrow transplantation, which was not an option for these kids, because they had no compatible donors.

But Mulligan wasn't smiling. Sickened by the way the NIH clinicians had been orchestrating the entire meeting, he took a final shot. "All in all," the biologist asked plaintively, "why do we need this treatment? Why is it better than PEG-ADA?" Mulligan couldn't tell from the protocol. The document didn't deal with PEG-ADA at all.

More than that, Mulligan was flabbergasted that these other doctors would not defer to Michael Hershfield. Hershfield was the central repository of data about PEG-ADA. Why didn't that matter?

To do things right, Mulligan knew in his bones, especially when one was proposing to experiment on children, one summoned all the major players, threw probing questions at them, and let the chips fall where they might. How else could one arrive at an informed scientific judgment about something as important as human gene therapy? But here was the panel ignoring a prime source of information.

The public, Mulligan said later, didn't understand how real science should work. "Instead, people see those two NIH scientists [Anderson and Blaese] in nice white lab coats, and they'd been in the press and stuff, and obviously they must be very good scientists, or they wouldn't be up there giving this presentation.

"Then they go through their data, and the public doesn't understand exactly what they're talking about, but it sure looks like real data. And gee, good scientists can disagree about things, can't they? And why should they listen to guys like me who are obviously competitors just trying to pop the other guys' bubbles?

"But, dammit all, we weren't nitpicking!" Mulligan's voice roughened in frustration. "These were fundamental, basic points. They hadn't even demonstrated that any T-cell balance remained after they cultured the cells. They misrepresented PEG-ADA therapy. They misrepresented the science behind the persistence of the cells in TIL cancer patients and monkeys: they give you the impression that the cells lasted some amount of time; but they don't talk at all about only the *tiny fraction of cells*—the absolute numbers that lasted—and that was below 1 percent!"

There were ways to do this experiment, Mulligan had told the RAC subcommittee that blustery March morning in Bethesda; ways to find out what happened to the T-cell ratio between CD-4 and CD-8 cells, as well as to their repertoire, that is, their ability to respond to new threats, when you grow up cells; ways to attain "the minimum standard of science." You could take the T cells before expanding them in culture and, by examining their receptors, determine their disease-fighting potential.

"Then," noted Mulligan, "you could mark each type of cell with genes transported by retroviruses and culture them up to the quantity you'd want to give back to a patient and ask the question again: Would you be giving a strong army back to the kid?"

Mulligan was willing to bet that the repertoire would be severely distorted, because the cells had been grown to huge quantities. But he didn't *know*, because nobody had done the experiment.

Part of the problem, Mulligan later explained, was that he and the other subcommittee members had been too polite, too kind, to the gene therapy troika. He snapped out the words sharply: "We haven't said, 'This is shitty science. You're misrepresenting the data.' Instead, we've said, 'These are our concerns. Help us out here.' "

In the meantime, the clock had run out. It was 12:30 P.M. in Conference Room 6. The subcommittee was forced to adjourn, having decided nothing—except to hold another meeting. Tactfully summing up the session, LeRoy Walters ordered Blaese and Anderson to do more homework and to answer several central questions raised by the subcommittee. Before the next meeting, RAC reviewers would study the effectiveness of PEG-ADA. Just how good was the stuff?

As a concession to Rosenberg, the subcommittee agreed to move up its next meeting to June 1, 1990, from the scheduled date at the end of July. On that sour note, the historic human gene therapy meeting ended with observers confused and disenchanted. After all, the proposal had been approved and sent along to the subcommittee by the review boards of the National Cancer Institute, the National Heart, Lung, and Blood Institute, and the NIH Institutional Biosafety Committee. Its failure to pass even the most basic muster by outside experts raised serious questions about what was going on inside NIH and its own, vaunted regulatory procedures.

"We never intended to get approval first time out," Anderson said defensively afterward. "We knew if we presented evidence on twenty mice, they'd want forty. If we did two monkeys, they'd want four. We went before seven committees before reaching the RAC subcommittee, and each one wanted additional clarifications. So the point here was to familiarize everyone with the proposal so they could consider it, ask questions and give us a chance to respond."

Reporters clustered around Michael Hershfield, who reacted to the attention with the nervousness of a deer caught in a headlight beam. He managed, however, to express grave doubts about what had just happened.

There was no need for gene therapy, he insisted. There was no sign that PEG-ADA would ever fail. The kids on PEG-ADA were not "normal" and probably never would be, he explained, because the defect was programmed in their genes. Only bone marrow stem cell therapy might permanently change that condition, not the T-cell therapy Blaese was proposing.

As he spoke, it seemed cynical to reporters that the NIH team had ignored Hershfield. The official reason, expressed later by Ken Culver, was that the Duke doctor, though a biochemist, rheumatologist, and molecular biologist, was not "a card-carrying immunologist." But that didn't make sense, because Duke undoubtedly had its quota of immunologists for Hershfield to talk to. It seemed more likely that Hershfield got steamrolled by the gene therapy troika because he posed a serious threat.

Nevertheless, Culver contributed some relevant new information: although most of the children on PEG-ADA were doing better, apparently not all of them were. "We didn't go into detail at the meeting, because we didn't have the time, but that very week one of the kids on PEG-ADA was admitted to a hospital in the Northeast for a severe infection."

Culver speculated, "Maybe this boy needs our gene therapy now, while other kids don't. We didn't want to get into a big fight and make problems at the beginning."

Regardless, in the confusion of the moment, there was no denying that the quick-tempered Hershfield had played right into the NIH team's hands. Reflecting on the proceedings later, he said he had run into trouble with the gene therapy lobby before. Once, after presenting unpublished PEG-ADA data at a genetics meeting, he was standing in an elevator when a prominent gene therapist (not from NIH) told him, "Well, you really stuck it to us, you son of a bitch."

"The man seemed to be only half kidding," Hershfield recalled, with some awe.

10

VICTORY

Perhaps it was jet lag after a long flight from Milan. Maybe it was the wilting heat and humidity of midsummer Washington. But when Claudio Bordignon stepped forward to save the day for his old friends French Anderson and Mike Blaese, the Italian researcher seemed jumpy and ill at ease.

His discomfort contrasting sharply with his suave, well-tailored presence, Bordignon strode to the lectern before the RAC's Human Gene Therapy Subcommittee on Monday, July 30, 1990, betraying one thing above all: he did not want to be there.

As recently as the subcommittee meeting of the preceding March, Bordignon hadn't even been listed as a contributor in the original protocol submitted by the NIH doctors seeking to try gene therapy for ADA deficiency. However, by the time of the second RAC subcommittee review hearing, on June 1, that omission had been rectified in spades.

There he had merited costar billing. Claudio Bordignon, of the Department of Laboratory Medicine, at the Istituto Scientifico H. S. Raffaele, in Milan, Italy, suddenly found himself a "major collaborator" on a hastily revised ADA protocol. The Milanese hematologist was about to back into immortality because he possessed unpublished data that the NIH team considered dynamite. Working entirely on his own, Bordignon had conducted a straightforward experiment that Blaese and Anderson had not, and probably would not have been able to pull off if they had, if only because they lacked a suitable patient. The NIH people had cured ADA in cells but not in animals; they had put genes into people

(TILs), but they were the wrong genes. They had not been able to cure *anything* in a living system, whereas Bordignon apparently had. One of Bordignon's patients was a very special little boy whom he was treating for ADA deficiency with PEG-ADA. The youngster (known only as G.B.) responded beautifully to the medication. In fact, he could have been a model for the drug's safety and efficacy as his body obligingly churned out T cells for his doctor to remove and tinker with. Having genetically altered those cells, Bordignon proceeded to conduct animal experiments that he hoped would someday lead him to the human equivalent. And in immunodeficient mice he obtained results so striking that he quickly found himself hauled across the sea to rescue the Anderson-Blaese consortium.

Professional courtesy aside, Bordignon, deep down, considered himself a competitor of the NIH team, not a collaborator. He was rushing to perfect a gene therapy technique he could try in Italy, where, despite that country's bad experience with Martin Cline a decade earlier, Bordignon would never have had to undergo the daunting bureaucratic scrutiny that bedeviled Blaese and Anderson. Bordignon emphasized his independence on July 30, the third time the RAC subcommittee chewed over the ADA protocol that, if approved, would ensure Anderson, Blaese, and Culver their place in history. "I am not a close collaborator in this study," Bordignon declared in accented English, "nor was my work done with a view of answering some of the questions raised during the review of this protocol."

Moreover, Bordignon said he feared that his ability to publish his own data in an important American scientific journal might be jeopardized by his disclosing it prematurely. "But I have confidence in the fairness of the [American] system," he noted, adding that in his country "a process like this would be unthinkable."

The last-minute, third-act entrance by the dapper Italian rescuer served to deepen the confusion and heighten the improvisatory air surrounding history's first authorized gene therapy experiments. By the spring of 1990, the entire scientific debate had boiled down to one basic question: What evidence was there that ADA enzyme as expressed *inside* T cells—via gene therapy to the nucleus—would work better in the body than the enzyme floating free in the bloodstream, where it would act like cellular dialysis, detoxifying poisons by bathing the cells with PEG-ADA? Bordignon had the answers. And, thanks to an old relationship with the American researchers, he was there to divulge them. In 1985, he had served a fellowship with Richard O'Reilly at Sloan-Kettering during the frustrating period when O'Reilly was performing bone marrow transplants on monkeys for Anderson and Blaese, facilitating their attempt to hit stem cells with new genes in order to cure ADA deficiency that way. After the venture failed dismally, Bordignon returned home, where several years later he seized on Mike Blaese's lymphocyte idea as a clever way to do gene therapy. Luckily for Bordignon, he had two aces in the hole: little G.B., who made T cells in volume, and a colony of SCID mice, mice born without immune systems, a man-made species that NIH did not yet have. It was a simple matter to piggyback new ADA genes into G.B.'s cells with retroviruses, so Bordignon set about designing an experiment that might answer the pivotal question about the superiority of gene therapy over PEG-ADA. Providentially, he bumped into Blaese at an immunology conference in Egypt in the fall of 1989 and let the NIH researcher in on his plans. Blaese recalls that he did little more than nod sagely, masking his excitement, while making a mental note to follow the results very closely.

In essence, Bordignon took T cells from his young patient, armed them with healthy ADA genes, and squirted them into several SCID mice. He then analyzed the mice for survival of the human cells, efficacy of ADA gene transfer and expression, and the degree of reconstitution of immune functions. Eight mice received G.B.'s cells bearing new genes. Four other mice were given "negative control" cells that had been manipulated in tissue culture in exactly the same manner, but without the introduction of ADA genes ("mock infection," this is called). Twenty animals received normal human blood cells alone, to act as "positive controls" of the efficiency of the model. All the mice were sacrificed from six to ten weeks after treatment. When Bordignon peered at their organs, he found human cells in significant numbers (1 to 10 percent) in the spleens of four of the eight animals reconstituted with the ADA-engineered cells. Similar efficiency was discovered with the control mice treated with normal human blood cells, which, of course, contained normal ADA genes.

Not surprisingly, G.B.'s mock-infected lymphocytes did not persist in the four mice injected with them. In order to work in a SCID mouse, human T cells needed functional ADA genes in their chromosomes, and retroviruses would deliver them in a very capable manner. Bordignon concluded that intracellular expression of ADA was necessary for survival. Was this indeed the crucial animal model that the NIH researchers had been dreaming about for so long?

Skeptics later pointed out that the test mice were not given PEG-ADA, which might have allowed the mock-infected T cells to survive in mice as they do in kids, so that crucial variable could not be weighed.

Nor could Bordignon demonstrate expansion and proliferation of human T cells in the mice—their ability to respond to new threats, something that Blaese would dearly want if his children ever developed anything resembling normal immunity. On the other hand, safety was shown to be little cause for concern. Bordignon's mice did not seem to be at all harmed by the gene-altered human lymphocytes. Moreover, expression was excellent for a couple of months, and the Italian was able to detect normal mature T-cell function, including some specific immune responses to threats (mainly tetanus) in the animals that received the human genes.

Bordignon was to write in 1991 in *Science,* "We show that synthesis of vector-derived enzyme confers to the positively transduced lymphocytes a significant advantage over non-transduced cells in vivo." In plain English, the cells with transplanted genes quickly elbowed out T cells without ADA, giving living mice workable human immune systems, at least for a while. Bordignon's conclusion was delivered in the steamy conference room: "For patients failing therapy with PEG-ADA, the gene therapy approach may represent a safe and efficacious alternative." ADA-deficient children, like a man wearing both a belt and suspenders, might benefit the most from new genes to go along with the drug. Bordignon killed his animals soon after they received human genes, and the short duration of the experiment was troubling. How long would the healthy cells remain in mice? Would some of them exhibit memories to specific antigens and pass them on to progeny, as Blaese hoped?

Ken Culver said not to worry. "The mere fact that it worked mattered; and that the T cells had some sort of function, such as the ability to react to antigens that the mouse didn't have before. The duration becomes less important, and it sure made a big difference in terms of how the subcommittee thought."

Word of Bordignon's experiment had reached Bethesda while the gene therapy troika was still licking its wounds after the March meeting. Press reviews of the hearing had been critical and disastrous. Typical was the reaction of *Science:* "One had the feeling of watching an existential play. Was the RAC imposing bureaucratic delay or exercising good scientific judgment?" Then the author segued to a quote from William Kelley: "If we were dealing with impeccable science here, there wouldn't be any delays."

Impeccable science or not, it began to appear that some ideas were just too exhilarating to be denied, and frustrating as the March meeting had been to Anderson and Blaese—"It was just plain crazy!" in Blaese's words—a seed had been planted. After fifteen years of fruitless hopes, a real human gene therapy experiment had been presented to the national advisory panel, and over the weeks from March to June a reversal seems to have occurred. Whether because of a wave of optimism among individual subcommittee members or as a result of collusion—for which there is no evidence—a pact was made. It occurred so subtly that most science writers skipped the June 1 meeting in Bethesda, feeling that nothing of import would emerge, just more wrangling. But the NIH gene therapists brought a different agenda.

Anderson and Blaese knew that the Milan data would impress the RAC subcommittee. In fact, it could turn the tide. After several pleading phone calls from Blaese, who can be very convincing when pressured, Bordignon finally faxed a letter summing up his findings. In Bethesda, the scientists raptly studied the fragile piece of paper and then exploded in ecstasy. This was it! Anderson immediately set about making arrangements to fly Bordignon in for a subcommittee meeting, although today he hints darkly that jealous former collaborators, whom he declines to identify, were not averse to trying to prevent the flight. "A number of people tried to keep Claudio from coming," Anderson says. "They tried to push him. At one point, he said he was just sick of it. He was tired of being in the middle. I told him, 'Fair enough. If you're really distressed, don't come.' He finally gave up and asked me what he should do. He worried about the best thing to do for patients, and decided that to get a protocol through that he believed in would be the best.

"Also, if we could get approval, it would be easier for him later. And it was. When it came time for him to try his own experiment in Italy, I was there for him. It was the only trip to Europe that I was planning for. I would support Claudio's protocol whenever he was ready."

Such mind games typified the pressures on the gene therapy players as the stakes started to rise. Even when it looked as if the NIH team was grasping desperately at any wisps of justification, Anderson, Blaese, and Culver refused to be embarrassed or deterred. Characterizing their efforts during this period, a prominent RAC member noted, "The guys always tried to push as far as they could with what they had. But that was okay. We were onto them." The team dutifully kept appearing at the RAC, limping away to do more homework, then coming back again, whittling down the objections—pressing forward no matter what—while the rest of the world's gene therapy community stood pat and sullen, content to let the NIH clinicians commit what might be professional suicide.

Concerned about the fuss that Michael Hershfield was making and fearing derailment by PEG-ADA, the NIH team had spent the time since the March meeting editing its protocol and scrambling to come up with the additional data demanded by the RAC. By then, the team had already bowed to the inevitable. Chewing over ethical worries expressed by RAC members who reviewed the first version of the ADA protocol— namely, Robertson Parkman; Paul Neiman, of the Fred Hutchinson Cancer Research Center, in Seattle; James F. Childress, of the Department of Religious Studies at the University of Virginia, in Charlottesville; and the Howard University medical geneticist Robert F. Murray— Blaese and Anderson made an extraordinary move: they voluntarily hacked off the crucial last phase of the three-part experiment.

That was the fateful decision.

Formerly the proposal had read, "In the final part of this protocol we will *withdraw* [italics added] PEG-ADA treatment while continuing periodic T cell infusions. In this part we will determine whether the infused ADA modified . . . T cells have become sufficiently established to provide adequate . . . function."

Not any more. No longer would the ADA children be weaned off PEG-ADA to see whether gene therapy could save them. They would continue to receive the lifesaving drug indefinitely. Despite efforts to keep Michael Hershfield out of sight like an embarrassing uncle, the idea of removing PEG-ADA entirely was just too risky, and Blaese's expedient deletion had the added benefit of making the proposed experiment seem safer: chances were that gene therapy wouldn't hurt the kids, and it just might help them.

True, the concession dealt a body blow to efficacy—as proposed, the experiment was unlikely to reveal, for sure, whether gene therapy alone was better than PEG-ADA therapy. That information, which after all was the point of the exercise, would have to wait, perhaps for years. In a further bow to RAC pressure, the NIH doctors agreed not to administer toxic IL-2 to the youngsters in hopes of prolonging the survival of the gene-altered T cells in their bodies; nor would the kids have to endure permanently installed catheters in their abdomens for gene drips. Blaese and Culver would be limited to simple monthly blood infusions.

Yet even these dramatic concessions probably wouldn't have mattered without the precious gift that became known as the Milan data. Suddenly, the controversial NIH plan showed great promise, because Bordignon had made a strong case for intracellular ADA in a immunodeficient animal model that seemed to reproduce what the NIH team had found in all the model systems it could offer.

In the meantime, since April, Culver and Blaese had been beating the bushes for the perfect patient for their own experiment. "Up to that point, I didn't really want to focus on specific patients," Blaese explains. "But if I was going to be able to write a protocol, I was going to have to design it more carefully for some specific patients, and we were going to have to come back to the RAC with specific stories." Although clinical data on PEG-ADA had not yet been published, the drug was being used at several centers—among them, Duke University; University of Kentucky Medical Center, in Lexington; Albert Einstein, in the Bronx; Rainbow Babies and Children's Hospital, in Cleveland; UCLA Medical Center; and Children's Hospital of Philadelphia. In addition, the drug was being administered in Europe at the Hospital Necker in Paris, at the Hôspital des Enfants Debrousse in Lyon, and, of course, at H. S. Raffaele in Milan, Bordignon's own hospital.

Culver's and Blaese's travels had taken them to Philadelphia and Cleveland, where they extracted blood samples from several promising ADA-deficient patients who were being treated there, and met with their families. An earlier plan to rely on Michael Hershfield to pick a patient had been abandoned, even though Hershfield, as the dispenser of the lifesaving PEG-ADA serum, knew all thirteen children on the drug quite intimately and had an extensive set of vials containing each one's blood. But, says Anderson, tartly, "after that March 30 meeting, we decided we shouldn't depend on Hershfield. We resolved to conduct our own interviews with patients."

Among the first patients whose mailed-in blood Blaese examined was the teenager Alison Ashcraft. Alison's stock as a potential gene therapy patient had soared unexpectedly when she began receiving PEG-ADA shots, in August of 1988. For a time, it looked as if she would not be able to tolerate PEG-ADA. But an infusion of antirejection drugs helped quell the reaction, and by mid-1990 Alison was doing well on the preparation. That fact, and her family's long-standing wish that she not be the subject of the initial gene therapy experiment, ruled her out as the first patient. Another prospective subject was an eighteen-month-old Amish boy from a community near Philadelphia. The toddler was ideal in one sense. Culver met with the parents, examined the lad, and reported to Blaese that he had "no improvement in lymphocyte count, no reactive lymphocytes, no positive skin tests, no antibodies, no nothing." In fact, his doctors were leaning toward the idea of trying to find a bone marrow donor for him. Sicker than the rest of the youngsters on PEG-ADA, he was the child the team wished to treat most immediately. But social factors impinged. The parents, having spent a lifetime in a sequestered religious setting where modern machines are anathema, expressed strong reservations about the high technology of a contemporary hospital. They were not even sure they could allow themselves to make a trip to the relatively low-key suburban environment of Bethesda, let alone commit to something like twenty visits as outlined by the protocol. Curiously, they were not opposed to gene therapy itself. But they wanted it performed in their hometown medical center. This alone was not a hindrance—the Anderson-Blaise team was willing to fly its highly mobile operation virtually anywhere—but the wealth of problems associated with the case made it less than perfect for a first audition of gene therapy.

Their attention now centered on the two young girls from Ohio, raven-haired Ashanthi DeSilva, of North Olmstead, and tow-headed Cynthia Cutshall, from Canton. Both were patients of Melvin Berger, an immunologist who had worked at NIH until 1981 and was a friend of Mike Blaese's. For two years, Berger had been sending Blaese blood samples from Ashanthi and Cindy, whom he saw routinely at Cleveland's Rainbow Babies and Children's Hospital, where he was chief of immunological and infectious diseases. Berger was pleased with the way the girls were faring on PEG-ADA, but he saw a pressing need for a treatment that was more effective and dependable in the long run. "I have nothing but the best things to say about PEG-ADA," Berger noted. "It's a true wonder drug. It's taken kids who otherwise have a fatal disease and made them clinically normal. But there are a lot of unknowns about PEG-ADA.

"Our patients have made antibodies to it that so far haven't impeded the drug, but who can say about the future? Such a pernicious antibody can form any time, spontaneously. The way I see it, it's only prudent to develop the next level of technology.

"Personally, if it were my child at risk, I'd like to know that you all have something up your sleeve before this drug stops working. I don't want you to wait until the drug fails and then have a crash program." To Berger, Anderson's gene therapy scheme embodied that long-range promise. "I think there's a chance that this offers a cure," he said in a candid opinion of the NIH team's protocol. "I don't think it's going to happen in a few infusions, or in a few weeks or months, but down the road it carries a possibility for cure."

Cleveland looked good to Culver and Blaese. They took copious notes and wrote case summaries, but decided not to pass them out at the June meeting, in order to keep the patients' identities confidential. The doctors did, however, agree to read the summaries to the subcommittee as representing "reasonable candidates" for this revolutionary treatment.

"This was very important," Blaese said later. Speaking of Ashanthi, he explained, "We could say, 'This is the patient we want to treat. This is what she's been going through. She's been on PEG-ADA for a year and a half. She still has a lymphocyte count of only 1,000.' We had real examples with specific numbers, and I think that was very important for the committee. It took it away from the abstract."

While his partners were searching for gene therapy patients, Steven Rosenberg, of course, had no shortage of the desperately ill. Rosenberg's lab—with Culver's help—was working full-time trying to insert tumor necrosis factor genes into tumor-infiltrating lymphocytes—gene therapy for cancer. Seeking more expertise, Rosenberg turned to a source with which he had maintained a close relationship since the early days of recombinant IL-2—the Cetus Corporation, in California. Rosenberg was working with the TNF expert Michael Kreigler at Cetus, who had made a potent vector that delivered TNF genes into TILs. A few days after the March RAC subcommittee meeting, Kreigler phoned Rosenberg with great news: the TILs indeed seemed to be secreting TNF in therapeutic amounts. In an interview, Rosenberg excitedly passed on the glad tidings: "We can show the human TILs now make 20 to 100 times more TNF than non-gene transduced TILs make, and this would be safe to do in humans," he said, though it wasn't clear how he knew that. Then he dropped a bombshell: he planned to submit a proposal at the June meeting to do gene therapy for cancer. That meant the RAC would have to ponder not one but two historic gene therapy proposals.

Actually, as Rosenberg pointed out, his fascination with TNF was not new. He had long been experimenting with the substance clinically. However, merely giving cancer patients TNF by injection had proven to be a bust. Rosenberg had published the results of such therapy in thirty-nine patients, to whom he administered escalating doses until he either reached the limit in terms of toxicity or saw antitumor effects. Unfortunately, the patients couldn't stand enough of the drug for it to have a chance to work.

"It works in mice, but not in man," Rosenberg said. "If you give TNF to a mouse, you see dramatic necrosis [death] of the tumor within an hour. An hour! It's really remarkable." But mice can tolerate a lot more TNF than men can. "They can handle about 500 micrograms of TNF per kilogram of body weight, whereas humans are much more sensitive. They can tolerate only 8 to 10 micrograms per kilogram. Larger doses cause severe problems, such as a sudden drop in blood pressure."

However, because TIL cells re-injected into the body often return to the tumors whence they came—at least according to Rosenberg—the concentrated production of TNF by gene-altered TILs at those localized sites might have profound effects in such a "micro-environment." If the RAC allowed him to proceed, his ability to regulate the TNF genes themselves would be rather primitive, he acknowledged. TILs bearing transduced TNF genes could live only in the presence of IL-2, so Rosenberg figured that by merely stopping injections of the cytokine in patients he would kill the TILs, thus shutting off the new genes. Moreover, if he needed a quicker rescue, he had it. Steroids had been shown to mitigate the effects of TNF, he said, adding that he didn't think such emergency measures—rather like Alice's having one bottle to make her larger and another to make her smaller—would be necessary.

"We'd measure the amount of TNF the cells are producing before we give them to the patient," Rosenberg promised. "There are many ways to do this safely."

In light of what had happened to Blaese and Anderson in their attempt to initiate the ADA proposal at the March meeting, it seemed unlikely that the RAC would quickly give Rosenberg permission for a much more ambitious experiment fraught with many unknowns. But the surgeon refused to be discouraged from trying in June. "We're going ahead," he vowed, grimly. "I'm not going to let them sit on this one."

As he often does, Rosenberg said he drew continued inspiration from a quotation from the Belgian writer and philosopher Maurice Maeterlinck that he keeps in a gold frame on his office wall: "At every crossway on the road that leads to the future, tradition is placed against each of us—ten thousand men to guard the past."

"This stuff is new," he explained. "It's understandable for people to be worried by what I want to try. But what can I say? I'm a doctor.... My patients are dying."

Rosenberg's daring proposal ran into trouble right away and never got to the RAC that June. At its first hearing before the NIH Institutional Biosafety Committee, it was voted down unanimously. Rosenberg's fellow scientists and citizen-neighbors sent him back to the lab to do more animal experiments before even asking permission to move on to human beings—even though those human beings were at death's door. Ironically, however, Rosenberg's decision to collaborate with Cetus instead of with Genetic Therapy Inc., Anderson's company, set off warning alarms among the ADA team. Although no words were spoken—indeed, laboratory members talked to interviewers only on condition of anonymity—Anderson and Blaese woke up one morning and found themselves in an apparent race with Rosenberg to be the first to do gene therapy. Why did Rosenberg need Cetus to provide his viral vector, when he could have obtained the LASN vector from GTI, by merely walking down the hall and asking Anderson for it? Here was a rift as obvious as yellow snow. Many thousands of dollars were involved, but that sum paled next to the prestige that would fall to the biotechnology company that, if Rosenberg was correct, had a role in curing cancer.

On the other hand, what did Rosenberg care? When it came to getting results, he had never been particularly concerned with niceties and sometimes even pitted two members of his own lab against each other on the same experiment because creative tension might make good things happen quicker. Still, the ADA researchers were stunned. "There certainly has been brutal competition between us and Steve," said a team member. "He's been rushing to beat us, and we've been rushing to beat him. It's been an outright race." Obviously, it was a race that all cancer patients and their families would want Rosenberg to run. But the NIH gene therapy troika suddenly found itself in turmoil.

None of this was evident, though, at the second of the three RAC subcommittee meetings, that of June 1, 1990. Like Arctic musk oxen threatened by hungry wolves, the anxious NIH gene therapy team formed a tight circle against its critics. Fifteen of the sixteen subcommittee members were present; only Richard Mulligan, widely viewed as the principal antagonist of ADA experiment, had stayed away. In a letter dated May 31 to the RAC's executive secretary, Dr. Nelson A. Wivel, Mulligan had pleaded that a "pressing commitment" at MIT precluded his attendance, although he presented several objections in his letter for official consideration.

The meeting opened with the familiar complaint from a few committee members that they had received the revised ADA protocol only two days before having to leave for Bethesda. Anderson explained that, because of the many issues raised by the subcommittee, it had been necessary to obtain new data about possible risks and efficacy, and that had pushed the NIH team to the limit. Next, Robertson Parkman was asked to present "a short historical outline" of his specialty, bone marrow transplantation, as a background to the gene therapy protocol. Parkman did so in rather agonizing detail and then turned his attention to the protocol itself.

The original plan presented the preceding March, he said, was "ambiguous" about how gene therapy was expected to work. He posited a hypothesis: the T cells into which ADA would be introduced would permit a greater efficiency in detoxifying deoxyadenosine metabolites than was achievable with PEG-ADA.

But there was a paucity of data supporting such a hypothesis, Parkman asserted, a slap in the face to the NIH team. However, Parkman went on, Claudio Bordignon in Milan had recently addressed this issue and shown that when gene-altered T cells were put into SCID mice, they appeared able to call up antibodies. Without such an animal model, said Parkman, issues of function and persistence of the ADA gene in T-cell populations would have been unclear.

Parkman then admitted that "one very important issue" was that the parents of the two children who might be receiving gene therapy had decided not to opt for bone marrow transplantation. But he wondered about the untested powers of PEG-ADA in such children. Perhaps merely increasing the frequency of dosage could constitute a treatment as effective as gene therapy?

Standing at the lectern, Blaese politely countered that doubling, or even tripling, the dose of PEG-ADA had not seemed to change the immune response in children, according to studies. He also said that he had talked to Michael Hershfield about the current studies of fourteen youngsters treated with the drug. Seven of them had developed antibodies to it, and two of them had to be taken off therapy and treated with immunosuppressants. Weeks later, PEG-ADA was resumed and the children were responding, though there was no way of knowing whether raising the dosage might precipitate a new round of rejection.

After a coffee break, Chairman LeRoy Walters asked Parkman to summarize the changes in the protocol that the NIH researchers had made between the March and June meetings. Parkman said that the hypothesis of inserting the ADA gene was better spelled out and that additional data had been presented. However, the level of detoxification activity to be gained by gene therapy had not been expanded. But he admitted that the Bordignon finding that transduced T cells appear to be capable of doing their job in mice represented "a major step forward."

Walters then did something important. Like a judge instructing a jury, the chairman explained that the third part of the original protocol had been abandoned, thus leaving the subcommittee to ponder only the first two parts. What this really meant was that the children would not be losing anything—they would still be maintained on PEG-ADA—but the subcommittee would be voting on an additional therapy that offered the possibility of a cure.

When Walters opened the floor to discussion, William Kelley again claimed that two questions were critical: "Can the experiment work? Can it be determined if it's working?"

Like a pit bull that refused to unlock its jaws from Anderson's calf, Kelley still wanted to know what the half-life of the implanted cells was and whether the presence of PEG-ADA would muddy the results when it came time to evaluate gene therapy. But oddly enough, Kelley's fire, after having blazed so long, now seemed to be dimming. This time, the frequent critic Scott McIvor also seemed to be keeping his own counsel. It was only when Walters asked the members of the subcommittee whether they had read Richard Mulligan's letter that a whiff of trouble wafted into the room.

Mulligan wrote that he had read the reviews of the protocols and that they covered most of his concerns. However, he believed that the issue regarding the functionality of T cells cultured in IL-2 was "a major issue needing much more discussion." Talks with "experts in the field" led Mulligan to fear that the method described to expand the T cells would lead to cells very dependent on IL-2. "Therefore, the issue of the need for and risk of IL-2 administration may be quite important . . . ," he wrote. The point had already been made, though; the team had by then agreed not to give IL-2 to the kids, because it would be cruel to subject them to so much toxicity.

But then Mulligan turned to something that Charles Epstein had briefly mentioned at the March meeting. It had been eating at Mulligan for months.

"Also from the data presented," he wrote, "the very important parameter of CD4/CD8 ratios appear to be altered in a major way after T cell expansion.

"Accordingly, even if the cells would persist in vivo [in the body], the procedure may paradoxically lead to an 'induced' immunodeficiency state. I am very disappointed that you [Wivel] and LeRoy [Walters] chose not to solicit reviews by a collection of expert T cell immunologists, as I had suggested. They could have been quite helpful in the review process."

Judging by the response of his fellow subcommittee members, no one really understood what Mulligan was getting at. But he kept pondering that single piece of data in the protocol—Figure 5, 13.2—suggesting that the ADA-deficient children might be given back an imbalance of T cells by gene therapy. Blaese had said that he wanted to return a full complement, at least a fifty-fifty balance of the two major T-cell subtypes—CD-4 (helper) cells and CD-8 (killer) cells—that would be necessary to confer immunity. But Mulligan didn't think that would happen. In fact, the NIH team's own data indicated the opposite after Blaese and Culver had tried to investigate what would happen to the ratios when they cultured gene-altered T cells from an ADA patient. Thanks to PEG-ADA, the T cells they had removed from the child had contained CD-4 and CD-8 cells in nearly equal proportion; and, indeed, they would want to preserve that balance to reconstitute a child's immune system. When Blaese and Culver cultured cells from an ADA patient in IL-2, they started out with a balance of 41 percent CD-4 and 33 percent CD-8. After forty-five days in culture, though, the cells had changed their character. As the eagle-eyed Mulligan noticed, nearly 100 percent of the cells had evolved into CD-4 helper cells, while the proportion of CD-8 killer cells, which the CD-4s would call on to joust with an invader, had plunged to a scant 1 percent.

They were gone, most of them. There were no soldiers. Only helper cells had survived that genetic manipulation. What did that mean? To Mulligan, in terms of therapy, it meant that if that cell population were given back to a child, the cells might not be able to mount an immune response, because they would now have an imbalance of helper T cells and killer T cells. In short, they could order a hit, but there were no assassins to take them up on it.

If the confusion displayed by his RAC colleagues at the June meeting (which he attended only by sending a letter) was any measure, Mulligan alone was worried by what that piece of data showed. In his letter, he hypothesized that, if one were to infuse a patient with a distorted ratio of T cells, the patient's already weak immune system might be further compromised.

At the meeting, the RAC member Paul Neiman, who had been assigned to address the issues brought up by Mulligan's letter, said that the possibility Mulligan had raised was considered "purely theoretical" and that he was not sure what the background for the statement was, or whether there were models to suggest that this could occur.

In the letter, Mulligan also complained that "the involvement of commercial concerns in this protocol needs clarification." His impression, he wrote, was that Genetic Therapy Inc., Anderson's company, would be involved in the preparation of the virus. "This was not pointed out in the proposal," Mulligan noted. "If this is true, the public may want to know the relationships between the people involved in the protocol and the company."

Now, here was a nasty little innuendo. When asked about it at the hearing, Anderson calmly assured everyone that there were no commercial concerns. Genetic Therapy Inc., he said, is a company "which is performing official business of the U.S. government under a CRADA agreement from the National Heart, Lung, and Blood Institute." That meant it was all on the up and up; technology was being transferred from the public to the private sector, as the government had instructed; and if anyone thought Anderson was making any money out of his affiliation with GTI, he was dreaming. Of course, the explanation held water only so long as Anderson was a federal employee. Should he ever leave NIH and head off to the greener pastures of, say, academe, all bets were off. Then he would be in a position to profit from research done completely at taxpayers' expense, if he so chose.

So much, though, for Mulligan's objections. There was no further discussion about them. Before breaking for lunch, LeRoy Walters summarized the questions that remained to be answered by the NIH team. Who are the patients? How were they chosen? Is the therapy safe? Should animal models be used to ascertain the possibility of cancer? Will catheters be required, as opposed to intravenous infusions? Is the protocol likely to be effective? "If so, how will we know?" What differences exist between the cell cultivation technique used in Italy and those being used in Bethesda? How long are the gene-altered cells likely to survive? How will long-term follow-up be communicated to the patients? What rules will there be for stopping the experiment? What needs to be included in the consent form?

But then Walters paused . . . and just stopped. The questions seemed to sum up the remaining objections!

As the members headed for lunch, Anderson, Blaese, and Culver could smell the bacon. They raced for the RAC's Office of Recombinant DNA Activities, commandeered the aide Becky Lawson's computer, and feverishly began answering the questions Walters had raised. By the time the session reconvened, at one-thirty, Blaese was ready. He distributed a list of the criteria for a patient's inclusion in the experiment; it limited the candidates to children who were already on PEG-ADA but who had opted against a bone marrow transplant. He then presented case studies of Ashanthi DeSilva, Cindy Cutshall, and the Amish boy. The subcommittee listened with interest.

Then Blaese called on Anderson to answer the question "Is it safe?" Anderson mentioned that it was always possible for any retroviral vector to land in the wrong place on a chromosome and produce cancer but that, so far in all the monkey work, this had not occurred. No problem from the subcommittee on that one.

Blaese next took up the issue of whether the therapy would be effective. Gaining confidence, he said that by the time the doctors had treated three or four patients, they would know whether the therapy was worthwhile. He promised to return to the RAC every six months or so and keep everyone apprised of the progress. He vowed that if two patients died of treatment-related problems, he would stop the experiment. Nobody disagreed.

Continuing to press the case for efficacy, Blaese cited the work of Bordignon, presenting a table in another handout summing up the Milan data. The differences between the NIH protocol and Bordignon's were "trivial," Blaese argued.

He then distributed a page that sought to address the issue of half-life of the transduced gene in the absence of IL-2. Blaese guessed that it was about five days, but who knew? The only way to determine whether gene therapy would work was to do gene therapy.

By this time, the subcommittee's mood had turned decidedly warm and cozy. The experiment seemed to have assumed a momentum all its own. Following further collegial discussion, there didn't seem to be much more to say. In all the years since its founding, the RAC for the first time had run out of questions.

With an exquisite sense of timing, one of the few subcommittee members who was a bona fide gene therapy researcher seized the bureaucratic reins and broke the silence: the San Francisco geneticist Charles Epstein, an impish, diminutive man who favors preppy clothes and wears clear plastic spectacles.

It was Epstein who earlier had saved Anderson and Blaese by fending off the challenge from Michael Hershfield. This was an understandable action on the part of a fellow gene therapist. In Epstein were blended all the talents the NIH team could have wanted in an ally: he is a brilliant man, a professor of pediatrics, an expert in the creation of transgenic mice, an active explorer of chromosome 21 in pursuit of the genes responsible for Down's syndrome. Nor was the four-year RAC veteran any stranger to committees. He had served on more than he could remember, to his occasional vexation. The time had come, he decided, to get this show on the road.

Without further ado, Epstein made a motion. He moved for "approval of the submitted protocol of May 29, 1990, with the following provisos":

1. That the consent form be revised, and accepted by the RAC at its next meeting.
2. That a stronger warning about the possibility of causing a malignancy be inserted.
3. That a "stop-criterion" of two therapy-related deaths be included.
4. That abdominal infusions of genes not be done on the children without RAC approval.
5. That the team not proceed on its own volition to the second part of the protocol—selecting only gene-altered cells to give to the children in order to supercharge the therapy. If they even thought about doing that, the gene therapists would have to go before the IRB.
6. That full data from the Milan experiments be provided to a subcommittee review team prior to the subcommittee's meeting on July 30, 1990. Although the committee members could not command Bordignon's presence, they privately told Anderson that they would appreciate his appearance so that they could question him directly.

With that, the motion was on the table. Anderson and his teammates found it hard to breathe. "It looks like it's gonna finally happen," Anderson thought. After years of work, they were going to be allowed to do it! Only a few trivial provisos stood in the way of gene therapy.

After several more moments of discussion and "friendly amendments" from subcommittee members, Epstein, like a lawyer tidying up the details after a complicated settlement, added two more technical provisions, ordering that a specific protocol be provided for the evaluation of the patients' immune and clinical status as the experiment progressed.

Then Walters asked for a second. The previously critical Parkman chimed in with his wish that the protocol be approved.

Walters quickly called for a vote by the Human Gene Therapy Subcommittee. The motion passed, 14 to 0, with no abstentions.

And that was how history would record the birth of human gene therapy. More approvals would have to follow, but this was the big one. A flushed and exultant French Anderson thanked the members of the subcommittee "for their extremely conscientious and superb review," but he also expressed his concern that the full RAC would not meet again until October 7, 1990. He suggested asking the parent body to move its meeting ahead to July 30, the same day the subcommittee was meeting, because he felt that the rest of the review would not take long.

Ever agreeable, Walters asked Blaese to have the final revision of his protocol in the hands of the subcommittee by July 2. Then it was agreed to move up the date of the next RAC meeting.

Walters adjourned the subcommittee meeting at 3:20 P.M. More bureaucratic niceties were in the offing—final approval by the RAC, the FDA, and other officials—but the major obstacle, approval by the Human Gene Therapy Subcommittee, had been virtually overcome.

"I feel great! I'm relieved and elated," Anderson told an interviewer after the room emptied. "But I must admit, I'm not surprised. I didn't poll ahead of time: you don't do that. But I know everybody *so well.*"

Looking back on his historic motion a year later, Charles Epstein said he had done nothing special, nor had he been co-opted. "There were no deals, no arm-twisting, no politics that I'm aware of," he said. "I think I wanted to ensure the integrity of the process."

The problem with the ADA gene therapy proposal, Epstein felt, was the lack of a good animal model. "My position was that in a situation like this, one could do a fair amount of work in a test-tube system, and because they already had the TIL experiments out of the way, it didn't seem that they needed to have an absolute animal model to go ahead."

Claudio Bordignon turned the tide, Epstein admitted. "He had a good model showing that the cells could survive in SCID mice. So people were willing to accept that as reasonable.

"My bottom line: it's an experiment. There was every reason to believe it would be reasonably safe. The committee had put in the stipulation about keeping the kids on PEG-ADA. The team made a reasonable case with Bordignon's stuff. And I think there was a lot of neatening up of the protocol as time went on. Editing, I mean. The reason it looked like a fiasco so often was that they kept coming forward without all the data. But we'd been yelling at them about that all through the TIL stuff."

The outcome of the July meeting was a foregone conclusion, with Bordignon as the guest star and Mulligan sure to attend this time. By then, gene therapy was in the air—even cancer gene therapy, as Steve Rosenberg found out to his delight. There was a feeling among the nonscientific RAC subcommittee members, people like Abbey Meyers, director of the National Organization for Rare Disorders, that the time had come to try gene therapy, especially on patients who had no other hope. Thus in July the subcommittee seemed content merely to tie up loose ends of the ADA protocol.

Mysteriously, though, by the time Richard Mulligan started to prepare for the July RAC meeting and checked the revised version of the protocol the NIH team was to resubmit, the troublesome piece of data about the ratio of CD-4 and CD-8 cells had vanished. Not only had it disappeared; the pages had been renumbered, rather clumsily, by hand. In place of the missing data, Figure 5, 13.2 (as reproduced on page 92 of the July version of the protocol) now presented a vector study entitled "ADA Levels in LASN Transduced T-Lymphocytes."

For all intents and purposes, the CD-4/CD-8 experiment had never been conducted. The disappearance looked suspect to Mulligan. Where was the old version of Figure 5, 13.2? Why the renumbering? If the experiment had been repeated, and different results obtained, where was that data? Had some tidying up occurred? Some adjusting of the moving target?

This just wasn't done—it wasn't the way Mulligan was trained to do science. Caginess was expected if you were great lab workers, but when you did go public, you were expected to show good faith by presenting your data, warts and all, and then try to figure out what it meant. Clinical-research types, Mulligan thought, in their zeal to make discoveries, customarily played fast and loose. But this seemed beyond the pale. Could the NIH team have been intellectually dishonest enough to dump the data? Or was it merely a mistake? A clerical error? Mulligan couldn't say. But the bottom line was plain: the only piece of data that presented a serious challenge to the NIH protocol was no longer in it.

Mulligan tried to bring it up. "This data doesn't address repertoire, except to tetanus," he told his colleagues, after studying the latest version of the protocol. "I want to get back to a piece of information from the initial protocol looking at T cell subsets. And the fact that there is data that suggests when the cells are cultured, the important ratio of CD-4 and 8 is quite different after culture."

The main thing Robertson Parkman had addressed, Mulligan went on, was whether or not gene transfer affects the capacity of the T cells to be transplanted. "That's absolutely the key," Mulligan agreed. But if the RAC looked at the current version of the protocol, he pointed out, "you won't find what I thought existed, which is the T cell subsets that resulted."

"I was just curious to get some response. Is there anything about the data that now makes it less important?"

No one responded.

"This is probably the only data that I think deals with the issue of repertoire," Mulligan continued. "When one takes blood from these patients and looks at the ratio of T cell subtypes, you get a reasonably typical ratio. After culture, conditions are quite different. That raises the possibility that by virtue of culture, one alters the ratio."

"My greatest concern would be that this is true," he said, and that if the NIH team gave children an overbalanced ratio of subsets of T cells, "that would be potentially harmful.

"Is there something about this data that isn't important?" he asked again. "Or anything in addition that gives us a sense of what happens after culture?"

Again, no response from the NIH team.

Weeks later, Blaese gave his version of what happened to the data in an interview. Essentially, he said he couldn't recall having removed it, but if he had, no matter. Life was a madhouse during that time, with the researchers having to come up with new data to answer the subcommittee's objections. "I considered the graphic insignificant when we wrote the first draft," Blaese said. "I may have removed it because I considered it irrelevant or misleading."

That was because when the team repeated the experiments on the blood from ADA-deficient children, it usually got different T-cell subsets in culture. "One time CD-4 would predominate, the next time CD-8," Blaese said, noting that such variability happens all the time in immunodeficient patients. "I'm not concerned, so long as we do get a representative sample.

"Rich Mulligan I have the highest respect for as a scientist. He's had an enormous impact on gene therapy. But he's not a card-carrying immunologist. He's not a clinician. I feel very comfortable with the decisions I've made with patients. We resubmitted the protocol so many times that it wasn't an intentional omission. The thing was getting so thick, and we'd Xeroxed it so often as the body of the manuscript was revised. I honestly don't remember."

Blaese emphasized that the NIH team had submitted all its data to the RAC subcommittee—"It certainly was not an attempt to mislead"—although the hand renumbering of the graphic indicated that someone knew that it had been removed.

But at the July meeting, Blaese had not been pressed to make such an explanation. After Mulligan aired his misgivings, the subcommittee chairman, LeRoy Walters, asked him, on behalf of the lay people on the committee, to explain the major subsets of T cells and what he was worried about. Mulligan briefly defined the roles of helper and killer T cells and gave an example of how the ratio could be crucial. "In HIV patients, you have a selective lack of one of these subsets. So even though they have some CD-8 cells, a depletion, a change, in the ratio of the subsets is very important, and not good."

Oddly, however, his point didn't seem to impress the RAC subcommittee. The matter was summarily dropped. "It fell on deaf ears," he noted later. "It went right past them."

In fact, generally speaking, no one responded to Mulligan's satisfaction to any of the questions he raised about the science behind the ADA experiment. He had argued yet again that the protocol didn't adequately address four major points: whether the repertoire and proper ratio of the T cells could be maintained in the body by means of added genes; whether tissue culture conditions were adequate for preserving T-cell function and a clearer understanding of the role of the growth factor IL-2; whether intracellular expression of the ADA enzyme was more beneficial than direct infusion of PEG; and whether it was possible to introduce enough modified cells in the patient to detoxify the blood.

By this time, however, Mulligan's had become a lone voice. Charles Epstein and then LeRoy Walters checked off the issues raised on the list of provisions from the June meeting, leaving only the presentation of the Milan data. It was as if Mulligan had become invisible.

At this point, Bordignon marched briskly to center stage and described the results of his experiment. After more polite discussion, Epstein said that in light of the history of the ADA proposal and the scrutiny with which it had been reviewed, the only way to finally test the hypothesis was to go ahead with the protocol.

The motion passed by a vote of 12 to 1. The only no vote came from Richard Mulligan—expressionless yet defiant, unconvinced to the end.

It was amazing, given its gravity, how quickly the tumor necrosis factor protocol received the blessing of the RAC. With no break after the ADA protocol approval, Scott McIvor presented an overview of Rosenberg's latest proposal, which, oddly enough, was being proffered not as gene therapy but as a drug safety study—a phase I cancer therapy protocol for desperately ill patients with very short life expectancies. Rosenberg himself called TIL/TNF "a classic Phase I study which has been used for evaluation of a very large number of chemotherapeutic agents in patients with advanced cancer."

In the three-hour discussion that followed, as much time was spent on revising the informed-consent form that patients would be required to sign as on evaluating the hard science behind the Rosenberg experiment. Rosenberg presented what he knew about TNF and why he thought inserting it into TILs would be beneficial to terminal patients. He emphasized that all he wanted to do at first was put a few genes into TILs and see what happened. The initial dose of TNF made by the gene-bolstered TILs would be 0.7 micrograms per kilogram of body weight, he predicted, which was only about a hundredth of the TNF dose that human beings can tolerate, according to many studies. Reiterating that normal TILs already had their own TNF genes, he pointed out that, in his earlier experiments, laboratory supercharged TILs making three times as much TNF had been given to patients with no ill effects.

Rosenberg next presented data in a few monkey studies showing that no adverse reactions occurred when the animals were given TIL/TNF cells in concentrations equal to those now being proposed for people. He went over his safety data and explained how, in an emergency, he could switch off the infused TNF genes by stopping IL-2 or by injecting steroids.

In reviewing the protocol for the RAC, Scott McIvor pointed out the obvious, calling it "more complex" than the ADA protocol, because it sought therapeutic benefit from the insertion and expression of a very potent gene whose mechanism of action was not completely understood. McIvor noted that TNF had proven toxic in cancer patients, but also showed some efficacy, though he wasn't sure about the safety of the experiment as evidenced by the animal work that the Rosenberg lab had done. McIvor wanted more. But even he admitted that Rosenberg's prediction of the low dosage that the genes would produce seemed safe, though McIvor doubted that Rosenberg could get enough TNF genes delivered by TILs to tumor sites to do any good. McIvor never had been satisfied with the TIL experiments.

By way of rebuttal, Rosenberg reiterated that efficacy was not the issue here; safety was. He wanted to begin with only a hundredth of the TNF that humans could tolerate. What was wrong with that?

Well, he would be working with live genes, and that was still an "unknown unknown," as researchers like to say. Moreover, the differing philosophies between cancer researchers and genetics researchers kept surfacing during the discussion.

Even the veteran gene splicer Charles Epstein said he was "surprised" to see the protocol being described as a Phase I study. In two years of discussing human gene therapy protocols, the emphasis had gone from diagnosis to efficacy and, now, to safety.

Rosenberg said that the efficacy stage, when he worked his way up to it, would begin at a dosage that had been tolerated by 80 percent of the patients in the safety study. He promised to come back to the subcommittee before beginning Phase II. He noted that the institutional review board of the National Cancer Institute had finally approved the protocol and that extensive in-house review, and review by the FDA, would follow.

Abbey Meyers then noted that if this were an AIDS protocol, "the room would be filled with patients demanding that it be approved. The people who will be entering the protocol are terminal cancer patients. The risks involved are more than outweighed by the possible benefits. These people should have their chance."

The idea was reinforced by Brigid Leventhal, a pediatric oncologist from Johns Hopkins. To her colleagues who wanted more animal work, Leventhal declared, "Waiting for data on three more monkeys is pointless since data on TNF toxicity already exists on eighty-six human beings." Apparently, that made sense to Epstein, because, as he had with the ADA protocol, he once again cut directly to the coda, as musicians say.

Epstein moved that the protocol be given provisional approval, pending final approval by "all the institutional committees" that would later monitor Rosenberg. Epstein made his motion contingent on changes in the informed-consent form to make sure patients knew what they were getting into. The motion carried by a vote of 13 to 0, with even Richard Mulligan voting for it.

This marked a bizarre moment in the history of American medicine: cancer gene therapy had been grandfathered in. ADA deficiency had broken the ice, a new field of treatment had been born, and cancer was going along for the ride.

The next day, as expected, both protocols were rubber-stamped by the parent RAC. It was now up to the Food and Drug Administration.

A few hours after the subcommittee meeting, in the lounge of the Bethesda Hyatt, Richard Mulligan looked depressed and exhausted. As he settled down to a vodka martini prior to catching a few hours' sleep, he pronounced it a black day for science. A bad beginning for gene therapy.

Why had he voted for TNF? "My reasons for having a certain opinion are little related to the committee's reasons," the MIT professor said bitterly. "Frankly, the committee would have passed that proposal in any case." Mulligan said he had voted yes, even though he was no great fan of the TIL experiments. "There are so many unanswered questions. We don't really know anything about the biology governing how those cells behave. Despite what you're told, we don't even know if they really *do track to tumors*. We don't know that. Nevertheless, it's a cancer treatment that has some success. And it can be fantastic. I mean, it takes things that wouldn't go away by themselves and makes them disappear."

Mulligan sipped his drink. Adding TNF genes just might work better, he guessed. Why not see? "Rosenberg may have to sift through a bunch of these chemicals like TNF before finding one that's effective," and immunotherapy might indeed work someday. "But not because it brings TNF to the tumor by virtue of the lymphocytes," Mulligan predicted. "It may work by triggering other changes in the system that cause other cell types to release lymphokines to fight the cancer."

To Mulligan, the questions about Rosenberg's TNF protocol boiled down to two: Was it safe? And did it have a prayer of success? Mulligan felt the answer to both questions was yes. "It met the minimum scientific standard," he said. "The ADA protocol didn't. ADA has no chance of success. I could be wrong, but from all the data presented, there was no support that it would work. In fact, I felt the data often supported that it *would not* work." Clinical research played by rules different from those of basic science, Mulligan said. Nonetheless, the NIH team was playing in its own park with its own ball.

"If an MIT graduate student was sitting and watching what went on at the hearings, it would be appalling for that student to think he was watching a good scientific discussion, or that the points they were making were good scientific points," Mulligan said. "They misinterpret all the time." Anderson and his colleagues and the nonscientific members of the RAC saw only what they wanted to see.

To show what he meant, Mulligan composed an analogy: "It's like saying, 'I'm almost blind, and I see a yellow half circle there. Therefore, it's a lemon peel.' That is the NIH way of doing science. As opposed to: 'I'm looking at this thing. It's curved around. It has all the appearances on the outside of a lemon peel. It looks like one on the inside. It is likely one.' "

Scientists are supposed to reach conclusions solely on the basis of hard data, even in dynamic and often fickle biological systems, insisted Mulligan. "And they're stringent about it. They don't overinterpret data. They don't underinterpret. They interpret right. In this case, from virtually start to finish, they have tremendously misinterpreted data to their own benefit in terms of support for their proposal.

"Whether that's conscious or not, I don't know," Mulligan admitted. But he did hazard a guess. "I can't imagine, frankly, that they're that thick. I really can't."

Mulligan was still stewing over the piece of data that had disappeared after the early versions of the protocol. "When I saw that originally, I snickered," Mulligan recalled. "I wondered why they would include that particular piece of data? Helper and killer T cells are supposed to be kept in balance. When it changes, as in AIDS patients, you're very sick." And, sure enough, that piece of data had vanished. Was this the result of sloppiness, embarrassment, dishonesty? Mulligan wouldn't say what he thought. He merely finished his drink, frowned, and stalked off to bed.

11
SEPTEMBER 14, 1990

Green and yellow, the highest ratings French Anderson can bestow upon himself, vividly highlight early September 1990, on the color-coded personal-achievement calendars he faithfully maintains year after year. It is surely only coincidence that those are the colors of the Green Bay Packers, whose legendary coach, Vince Lombardi, once said, "Winning is the only thing," but in Anderson's peculiar symbology, the colors also signified scientific performance of the highest order. By his own estimation, Anderson did an outstanding job in that momentous period.

As September arrived, Anderson and his cohorts, Blaese and Culver, had become convinced that FDA approval was at hand for their pioneering attempt to use gene therapy to treat ADA deficiency. The team decided to summon to Bethesda the small child who earlier that summer had been selected for the inaugural experiment. The purpose of the trip: a dry run of the procedure.

And so it was that young Ashanthi DeSilva, a chubby, poker-faced three-year-old with huge brown eyes, shiny sable hair, and tea-colored skin, arrived for the first time at the imposing city within a city, the mile-square complex of eclectically designed medical buildings that make up the National Institutes of Health. It was here, along a gently curving, tree-lined boulevard, that a tenacious team of molecular cryptographers had first unraveled the genetic code in the mid-1960s. And, fittingly, it was here where the first attempt to remedy a mistake ingrained in the DNA of a living human being was about to be made.

By early August, less than two weeks after winning RAC approval to conduct the first human trials, Anderson, Blaese, and Culver had all but agreed that one of the Cleveland girls—most likely Ashanthi—would be their candidate. When word hit Rainbow Hospital, the public relations machine kicked quickly into gear. In a matter of hours, the hospital put out a press release naming the two youngsters and quoting Cynthia Cutshall as having said, "My mother told me if the therapy works, I will be healthy for the rest of my life." Mel Berger was in the act, too, enthusing, "Gene therapy is something I barely dreamed of when I was in med school. Now this gene breakthrough potentially offers a cure for this disease and in the future, for others such as cystic fibrosis and sickle cell anemia."

How could the press resist such sumptuous fare? The story had everything: kids, suffering, and miracle cures. The only thing missing was a talking dog. And who knew where this genetic business was going? It might produce *that*, too, someday, editors muttered to themselves. When Rainbow Hospital called a news conference, reporters turned out in droves. Photographers snapping their shutters made a sound like a flight of arrows as little Cindy Cutshall capered for the media, at one point peering through the lens of a television camera held to her eye by an obliging technician.

Notably absent was little Ashanthi and her family. The reason behind this sudden retreat from the spotlight had much to do with events of several days before. A medical writer for a local newspaper, having been tipped off to Ashi's identity by a friend at the hospital, showed up at the DeSilva's house requesting an interview. The DeSilvas—who months earlier had willingly appeared in a promotional video made at Enzon's headquarters ballyhooing PEG-ADA—let him in, and Ashi's father, Raj, agreed to answer questions. Later, a television reporter from a network-affiliated station appeared, also asking for an interview. He, too, was granted entry. But when the tape began to roll, Ashanthi would not face the camera, let alone speak. Every time the cameraman moved, the child turned away. Her mother, Van, also shunned the lens. When the television crew members returned some hours later, asking to shoot the background footage known in the trade as "B roll," the family refused them entry. Shortly afterward, Berger received a call from Raj, extracting a promise that the hospital would do nothing further to identify the family. Berger complied, even though the issue had been rendered academic with the initial release of the DeSilvas' name and hometown.

Raj DeSilva later explained why he was so desirous of confidentiality. "My philosophy was, if the procedure worked, we owed it to the rest of society to go public," he said. "But if it didn't work, I didn't want to put our daughter through the harassment that she would inevitably be subjected to."

In Bethesda, Blaese and Culver assiduously tested and retested the girls' blood samples. Which child would they do first? Both youngsters fit the overriding precondition of the experiment, that is, they were in as good shape as their illness allowed. As a researcher, one naturally sets out to treat a sick patient, but, to state the case bluntly, one wants the healthiest sick patient available. At first glance, this might seem paradoxical and self-serving, an insurance policy against embarrassing failure. The healthier the child, the better the chance of success. But, in fairness, the issue is more complex. If the new procedure generates an adverse effect, it is imperative to have a patient strong enough to withstand it.

In the case of Ashanthi and Cynthia, that meant having good lungs. Many ADA deficiency victims have ravaged respiratory tracts, as a result of their almost incessant bouts with bronchitis and pneumonia. The lungs were of paramount importance to the NIH team because when the first gene transfer experiments were performed on Steven Rosenberg's melanoma patients the year before, the patients experienced serious, albeit temporary, respiratory problems. One could argue that the complications were due to the IL-2 the test subjects received in their cancer treatment. But one could make a similarly persuasive case that the TIL cells impregnated with new genes were responsible. No one really knew. But since the ADA patients were to be pumped full of white cells closely resembling TILs, it was best to come down on the conservative side and use patients with the best chances of survival.

Just as vital, however, was the number of T cells that were recoverable from the patient. Even though they were going to be multiplied ad infinitum in the laboratory, one needed a good supply in the first place to prime the pump. T cells are precisely what ADA deficiency victims don't have. Only those who had fared well on PEG-ADA were able to produce T cells in any volume. The question boiled down to whether Ashanthi or Cynthia was better in this regard. In the end, it was Ashanthi who met the criteria best.

Thus it was that on the morning of September 4, immediately after the Labor Day weekend and two days after Ashanthi's fourth birthday, the DeSilvas arrived at NIH for the first of what would be many visits. They had just completed a long-planned vacation in the Poconos, which made it a convenient time for them to come. Although Culver had met the family before in Cleveland, this was Anderson's and Blaese's introduction to the plump, thirty-five-pound little girl.

As the DeSilvas stood in the lobby of the Children's Inn, a dormitory at NIH for sick children and their families that is patterned after the Ronald McDonald Houses, Anderson did what he always does on his first meeting with very young patients. He dips down to the child's eye level and tries to make contact. The very shy Ashi immediately turned away.

Fortunately, as the morning wore on, Ashi came to accept the gentle, unpretentious gene therapist who wore his heart on his sleeve. She spent the day under his wing, being taken on a tour of the various parts of the clinical center where she would be during the actual procedure. All the while, Anderson was explaining to her what she should expect. "She doesn't understand what a gene-transduced cell is," he admitted, "but she knows what *pain* means, and with a child it is absolutely imperative that you never lie. You say, 'Here is what's going to happen,' and that better happen, just the way you said it would."

The next day, the NIH doctors got down to business, withdrawing white cells from Ashi in a process known as apheresis. As the afternoon wound down, Anderson, in a burst of tenderness, took the little girl to the hospital gift shop and told her she could pick out any stuffed animal she wanted. His magnanimous gesture was met with giggles from Ashi's parents. Mystified, Anderson scanned the furry inventory and, spotting the object of their mirth, forced himself to laugh, too. One of the animals was enormous and cost over a hundred dollars. He breathed a small sigh of relief when Ashi chose a small, inexpensive animal. "I think she was primed by her parents," he said afterward.

There now began an agonizing period of expectancy. All of the regulatory bodies whose approval is required for a gene therapy experiment had checked in with their assent, save one—the Food and Drug Administration. Anderson had strong reason to believe that the FDA would say yes. But according to the protocol submitted by him and Blaese, the procedure had to be carried out within twelve days of the apheresis, giving the FDA only until September 17 to grant consent. Of course, Ashi's T cells could always be frozen and stored for the day when the FDA finally *did* approve. But to Anderson, consumed by the desire to get on with the world's first gene therapy attempt, it was nail-biting time.

The DeSilvas returned to their Ohio home, near the greater Cleveland airport, to wait. They are people of great patience, having had to surmount many severe misfortunes in the course of their parenting. DeSilva, a chemical engineer, is a senior investigator in the research and development department at the B. F. Goodrich Company. Ashi, the middle child, was born during the family's previous residence, in England, and she alone among the children can talk or walk normally. But her fragile immune system places her in the most precarious position of all the siblings.

"We've had our share of problems, but we manage fine," says Raj.

When the DeSilvas went home, Anderson, Blaese, and Culver wasted no time. On Thursday and Friday, and once more on the Saturday following the family's visit, they exposed Ashi's T cells to a swarm of retroviruses bearing the normal version of the gene for adenosine deaminase. In a matter of moments, 200 million viruses went to work on 200 million white cells—"a liter of soup," Anderson called the cloudy broth in which they interacted. In the ensuing days, Ashi's T cells grew quietly but lushly in the brand-new incubator that had been installed as part of a major renovation of the C wing on the first floor of Building No. 10. Their new home was a set of plastic tissue culture plates with six buttonholes in them, each holding ten milliliters of fluid. Swimming in the fluid, carrying one extra gene in their bellies, the cells were alive with the promise of good health as they divided, subdivided, then subdivided again.

For its part, the FDA was trying to live up to its reputation as a stickler for documentation. After wading through the research team's voluminous protocol, not to mention its answers to a mound of questions that the agency had already posed, FDA officials sent Anderson a note stating that the team's replies were too brief. Out of the envelope fell a new vexation: twenty-three more questions to be addressed. Anderson and Blaese breathed hard.

The next seventy-two hours were a blur. Working straight through the night, Anderson, Blaese, and Culver put together a three-inch-thick stack of data that they hoped answered the FDA's questions in exquisite detail. "When they got those replies, I think they were so taken aback they were afraid to ask us any more questions," says Anderson.

Whatever the reason, the FDA put aside any further cavils. On the following Wednesday, the agency granted the team what it had been seeking: preliminary approval to perform the experiment. Immediately, the team put through a telephone call to the DeSilvas. Anderson fairly tapped his shoe through the floor as he waited for the family to pick up. When Van answered, the message was brief: come to Bethesda right away. It looked as though the FDA was poised to give the go-ahead, and the team was shooting for a Friday target date for Ashi's first treatment.

The DeSilvas needed no prompting. They had been ready for days.

The next morning the entire family piled into their large van and began the trek to Bethesda. They arrived at the Children's Inn around seven o'clock Thursday night, road weary and apprehensive. On hand to greet them, wearing the requisite business jacket and tie—the inn's policies prohibit lab coats and any other articles that might remind guests that they are there for serious doctoring—was Ken Culver, gentle, reliable, Eagle Scout Ken Culver. Notably absent were Anderson and Blaese. As Anderson tells it, they were "tying up loose ends. There were so many things to do. It was critical that nothing go wrong. You have no idea how many people are involved, how many services. There's the blood bank, the intensive-care unit, x-ray, the clinical labs, the labs responsible for safety studies. There were in the range of nearly one hundred people all told. Thursday I went to each place in the clinical center that would be affected to make sure that everybody knew Ashi was coming." Also consuming his time, he notes, was the infusion the same day of the ninth cancer patient to receive marked TIL cells in the collaborative effort with Steven Rosenberg.

Anderson's crowded schedule on Thursday included a press conference. Called on the shortest of notice, the briefing was designed to alert the media to the coming experiment while allowing Anderson to take direct command of the publicity effort, something he would not be available to do on Friday. Interestingly, Steven Rosenberg was not present. More than fifty reporters crammed the conference room. They represented primarily the influential East Coast media, whose coverage the media-hungry Anderson most coveted. The journalists were taken aback by the suddenness of the experiment, having been assured by Anderson himself that the procedure would not be attempted before the late fall. For example, among the five networks present was CBS, which by coincidence had been scheduled to spend Friday shooting B-roll material for telecast later in the year when the experiment was expected.

By this time, Ashi had become an anonymous person, her identity ferociously protected by the NIH team. With unusual restraint, the media respected the family's desire for privacy—and continued to do so from then on—despite the feeding frenzy of publicity ignited just a month earlier by the Rainbow Hospital's public relations department, which had made the secrecy surrounding the little girl's identity moot.

Thursday night, Anderson was unable to choke down dinner. He had not eaten a full meal since Monday, surviving chiefly on chocolate. Thursday morning, he awakened to find himself breaking out in pimples, so he cut out the chocolate as well. Only the intervention of his wife caused him to down a salad at Thursday's lunch. Sleep was as out of reach as food. Anderson observed the formalities of bedtime, climbing under the covers at one in the morning, but his brain was still on fast-forward. When Kathryn rolled him out of bed at five-thirty, he had not slept a wink.

Anderson threw on his clothes, made a revulsive sound at the suggestion of breakfast, and flew out the front door of the couple's newly remodeled, two-story, white brick rambler house. He hopped into his gray Chrysler LeBaron, a hand-me-down from his wife that replaced the Volkswagen Beetle he had driven from 1963 to 1985, and made the three-mile drive to the hospital in five minutes. It was 6 A.M. on an overcast day when he walked into the intensive-care unit to check on TIL patient number nine.

Some three hundred feet away, in their suite at the Children's Inn, the DeSilvas arose at five-thirty. They dressed the children carefully, slipping Ashi into a pair of turquoise trousers and a white top with Mickey Mouse effigies on it that her mother had sewn from scratch. "Are you going to talk to anybody today?" Van asked Ashi. "Are you going to smile?"

"No," said Ashi, with conviction.

As the gray dawn broke over the sleeping campus, the DeSilvas strolled along the winding cement path to the hospital and rode the slow elevator up to the clinical laboratory on the third floor. They were almost two hours early; not surprisingly, the apheresis unit, where they were to meet the research team, was not open. A yawning orderly told them the facility would not unlock its doors until seven-thirty. The news was oddly welcome. The delay would give the jittery family a chance to relax. They retraced their steps to the laboratory's admitting room and settled into the deep chairs of the lounge.

But the calm in the empty room proved only the more unsettling. The mundane setting and night-shift quiet were at odds with the gravity of the occasion, seeming to mock the enormity of the ordeal to come. From somewhere, the electronic song of a telephone punctuated the silence. A woman's intemperate laugh—a nurse's?—was heard. Long moments later, a distant door slammed. More time passed. The wheels of a trash dumpster groaned outdoors, and there was the whap, thunk of an elevator door being held open. All the while, Ashi stood in a stolid posture, her elbows on her mother's knees. She might have been waiting for a school bus.

At a bit past eight, Anderson and Blaese completed their routine morning rounds of the TIL patients' ward and headed for the apheresis lab. With them was Melvin Berger. At eight-thirty, the appointed hour for the family's arrival, Anderson set out for the admitting area. On the way, he ran into Culver, and the two of them checked the lounge area without success. They then hurried to the blood bank. Still no DeSilva family. Suddenly, the members of the domestic entourage appeared. Anderson and Culver had walked right past them in the lounge without seeing them.

Anderson considered the anxious tableau made by the DeSilvas. In spite of their fresh clothes and impeccable grooming, the parents looked drawn and uneasy. Van, a short, dark woman with a sweep of black hair and an engaging, earth mother smile, in particular looked tense. Their two disabled children sat impassively in their wheelchairs, a living admonition for the researchers to exercise extreme care. Anderson groped for words. "Well, it's the big day," he said in an attempt to be hearty. But his bleary eyes betrayed his own condition.

DeSilva, a barrel-chested, ruggedly handsome man with luxuriant eyebrows and curly black hair, graying at the temples, asked whether the all-important FDA approval had been granted. Anderson replied that it hadn't actually come through yet, but that it was all but in the bag. With that, the phone rang just to prove Anderson wrong. He took the call privately.

It was Jay Greenblatt, the FDA's liaison with the gene therapy laboratory. Apologetically, Greenblatt revealed that the powers that be at the FDA *still* weren't satisfied. They wanted the results from a few more tests conducted by Anderson's laboratory, including tests for contamination by a microbe known as mycoplasma. Anderson slapped his forehead in exasperation. His face turned gray as he sensed that his careful plans were about to be chopped into splinters. "My God, Jay, we didn't do one," he said desperately. He silently mouthed the word "mycoplasma" to Blaese, who groaned. "It's nonessential. You know that."

Greenblatt agreed but said that it wasn't his decision. The FDA wanted the test results in the record. The test would have to be done.

"There isn't *time*," Anderson pleaded.

His brain worked feverishly to come up with a solution. Dialing an extension, he reached the clinical center's mycoplasma specialist. A few minutes later, the expert was on the phone to his counterpart at the FDA, begging the FDA man to perform the test himself. Apparently not willing to stand in the way of history, the FDA technician agreed.

At 8:55, Anderson's Bellboy pager went off. On the liquid crystal readout was the number of Jay Greenblatt. Anderson picked up the phone. His hand was trembling.

"All right, French, you've got the green light," said Greenblatt.

Anderson was so benumbed, his mind could not absorb what Greenblatt was telling him.

"What's that, Jay? Repeat, please."

"I said, you've got the green light."

Anderson rushed back to the blood unit. He was out of breath by the time he got there. "We've got FDA approval," he managed to blurt out as he came bursting through the door. Blaese's impassive mask relaxed, and he smiled broadly. Culver exclaimed "Yes!" as he whapped his palm with his fist. The DeSilvas exchanged glances. "So then this is go?" said Raj DeSilva.

"It's go," Anderson replied.

The first order of business was to perform another apheresis on Ashi, to give the team a large supply of her blood, to be used as a baseline if it became necessary to know what her blood values were before the gene infusion. The quarry were white cells—in fact, no less than half of those circulating in her body. Even though Ashi's white count was excessively low—the average individual has 400,000 white cells per deciliter of blood, but Ashi had only 750—the medics did not fear depleting her body of the essential cells. That is because the body replenishes white cells in a short time. Within minutes, Ashi's count would be back up to 750.

She showed no emotion as the machine, with a soft whirr, took her blood, separated out the white cells from the red cells and platelets, and then returned the red cells and platelets to her. Her attention was riveted on the movie *Dumbo*, which was playing on the VCR in the small room. In time, Anderson's overheated gaze fell on the screen and stayed there.

Minutes passed. Ten, twenty, thirty. Walt Disney's poignant little elephant was being forced to perform humiliating clown tricks in the circus, while his mother lay in chains outside in the dark. Berger slipped urgently to Anderson's side. He spoke sotto voce. "I just figured out what you're doing," he said.

Anderson looked at him blankly. "You've been looking off into space for the longest time," Berger said with an accusing smile. "I wondered if you were passing out, having a seizure, or just thinking great thoughts. Then finally it came to me. You were watching *Dumbo*."

Anderson's face reddened. "It's a great movie, Mel, one of the greatest of all time."

Berger cackled.

"No, really, I happen to love it."

Berger cackled some more.

After forty minutes, the apheresis was complete and Ashi was disconnected from the machine. There now remained just one final step, an official signing of the informed-consent form. Informal consent had been granted by the DeSilvas the week before, but NIH wanted a formal meeting with the events recorded on videotape. The group, which now had grown to include Ashi's sisters, Berger, and Anderson's chief nurse, Lois Knapp, crowded into Blaese's private conference room, leaving scarcely any room to move.

Anderson conducted the lengthy proceedings. It consisted primarily of asking the parents a series of Miranda-like questions: "Do you understand that a malignancy might develop if the new gene happens to come to rest in the wrong place among your daughter's genes?" DeSilva looked a little green, but he nodded valiantly. So did his wife.

Each time that Anderson raised a daunting possibility—most of them related to cancer—the DeSilvas seemed to shudder, but they, like Sharon and Maurice Kuntz before them, had long ago decided to throw their lot in with Anderson. It was no longer appropriate to back out. They continued to nod.

There is still some of the Oklahoma cornball left in French Anderson. With a flourish, he produced a handful of government issue ballpoint pens and handed one to each participant. He had carefully applied a small ring of adhesive tape to his so that he wouldn't slip and use it again. "Historical occasions demand a little pomp," he explained in his boyish manner. "We'll all sign with different pens, just like a presidential ceremony."

The last person to sign was Culver. He put the pen back in his pocket in the midst of a cluster of identical pens.

"Ken," reprimanded Anderson. Culver looked up, furrowing his eyebrows. "How are you going to tell which pen is which?" Culver's gaze fell on his pocket. "*Jeez*, French, I dunno," he said.

The clock hands reached high to clasp the number twelve, then began their long descent. It was time to head for the treatment room in the tenth-floor pediatric intensive-care unit where the historic gene infusion would be performed. Anderson, moving athletically, maneuvered Ashi's wheelchair expertly down the halls. Culver was dispatched ahead to make sure no reporters were lurking in their path. Already, Barbara Culliton of *Science* and Natalie Angier of the *New York Times*, reduced to the juvenile connivance required of journalists covering live news events, had slipped past security in an effort to get a look at the "gene girl" and had to be sent back to the first-floor waiting room with reprimands.

The infusion site is one of a handful of enclosed treatment rooms feeding off of the pediatric intensive-care unit, which is located inconspicuously behind the adult ICU on the D wing of the tenth floor. The room is small and, for the initial gene therapy experiment, is jammed with state-of-the-art monitoring equipment. A soft thrum stirs the air. The indirect wall lights bathe everything in a soft, pinkish glow that suggests a cheap romance novel instead of epochal medicine. Ashi, clad in a billowing yellow hospital gown to allow quick access to her body in case of emergency, climbs with help onto the tubular metal bed, where she allows herself, rag doll–like, to be arranged on a clump of pillows. Her expression and body language remain reserved as ever as a dark blue blood pressure cuff is wrapped around her plump upper arm, followed by brightly colored electrodes, which are attached to her chest to register her heart rate. A nurse places a small black box on the bed sheet and molds Ashi's right hand around it. The box is intended to measure how much oxygen she has in her system. If the percentage begins to drop, one of the ICU nurses will swiftly clasp a mask over Ashi's face and start feeding her oxygen from the large torpedo-shaped tanks behind the bed.

There is a rustle at the door, and Culver enters bearing the transluscent vinyl bag containing the one billion T cells that will be slowly instilled into Ashi's arm. He suspends the pendulous bag from a hook projecting from the spine of a shaky stand and begins hunting for a vein on the back of Ashi's left hand in which to insert the intravenous needle. His first attempt to sink the needle fails—for all of its miracles, medicine still has not improved the arduous procedure for getting IVs in. Only after five minutes of eye-straining toil does Culver succeed in getting the small needle to stay put. He places a piece of tape over it to hold it down. All the while, Ashi remains unmoved, sitting darkly on the bed like some tiny, enigmatic potentate. Even when the needle slides in, she barely winces, her features frozen.

The deadpan look persists while the nurse performs a brief physical exam. A digital thermometer is stuck in her mouth. A stethoscope alights on her chest. An illuminated cone penetrates her ears. Her throat is plumbed. Her abdomen palpated. Her reflexes elicited involuntarily. Everything seems in order.

At 12:52, Anderson, more to calm his own jumpiness than to satisfy any clinical purpose, plants himself next to Ashi on the bed and places his fingers on her pulse. He sets a small pile of colored index cards next to her on the top sheet and writes down the time, the pulse rate, the volume of cells, and a number of other data. "We've got to be precise," he counsels Ashi, as if she were a junior colleague. "I don't trust these machines." He is rewarded with the flicker of a smile. Ashi likes Anderson. He is the only one on the research team who seems to be able to coax a reaction out of her.

She reaches over and places a rabbit sticker on Anderson's lab coat. For good measure, she affixes one to Blaese's and Culver's, too.

"Look at her," says Anderson admiringly. "She's as calm as a Zen monk."

The moment of truth is at hand. Blaese nods his head at Culver, who reaches carefully to turn the clip on the IV. Liquid begins to travel through the tube in a thin column at the rate of 250 milliliters an hour. For thirty seconds, the T-cell solution disappears into the back of Ashi's hand, while Ashi plays with a yellow sponge ball, the kind used by donors when giving blood. She squeezes it rhythmically between her chubby fingers. In that brief interlude, she receives 5 percent of the gene-altered cells that will eventually be placed inside her.

"Cut it off," Blaese directs Culver. The younger man reaches out and turns the clip, stopping the flow. Raj and Van look at each other. They have been thoroughly grounded in the steps of the procedure, but, for a moment, they have become confused. Why were they stopping?

"We are going to check for any reaction," says Blaese, sensing their anxiety. "Remember, we told you about it."

"Oh, yes," says DeSilva, relieved.

Seven minutes, the prescribed waiting period, begin to tick off. Ashi, bored, moves restlessly on the bed. Sensing the girl's need for something to break the tension, Anderson, ever the pediatrician, picks up three of the squeeze balls and begins to juggle them adroitly. Ashi's eyes suddenly grow wide, her brown irises shining in the pinkish glare.

"If you want, I can teach you," Anderson says. Beguiled, Ashi nods. Her eyes follow the balls in trajectory.

The child's attention now shifts to the television bolted to a raised metal rack on the wall. If she has any notion that hundreds of thousands of her own cells, reconstituted with the addition of the absent ADA gene, are now flowing through her veins, she gives no sign.

Again the intravenous flow is started. Ashi settles back and watches a situation comedy on television, while Anderson continues his hold on her wrist and her gently throbbing pulse. He has scarcely let go, except to juggle. Moving around the crowded bedside now, like some familial paparazzo, is Raj, snapping pictures as if he had motor drive. By the time the infusion is over, he will have taken 130 photos of the two-day procedure.

Throughout the infusion, Raj has promised Ashi a treat as a reward. Toward the end, he asks her what she would like to have. She is silent. Raj begins to think she has not heard him. But then she replies softly that she would like some M & M's, her favorite candy snack. "You got 'em," Raj says breezily.

At 1:20 P.M., exactly twenty-eight minutes after the infusion began, the dregs of the T-cell concentrate run into the tube and down into Ashi's arm. The vinyl bag is at last empty. A cosmic moment has come and gone, with scarcely a sign of its magnitude. To the small group of humanity jammed into the stuffy room, the procedure has seemed unsatisfyingly mundane, with none of the majesty one might expect to attend such a watershed event, an event whose intellectual roots lead back to antiquity. All the drama has had to be supplied by the observers themselves. Yet the thought that this young girl on the bed is now teeming with a billion infinitesimal particles, not of her own flesh, and that these particles are speeding to her aid, is more than the mind seems able to absorb.

Ashi, still shackled with monitors, looks over at her father. "My M & M's," she says softly.

DeSilva does not hesitate. He leaves the room in full stride and heads for the cafeteria, where he empties the candy machine like a man possessed. Visitors and hospital staff alike stop to gawk at the sight of the strange, grinning dark-complected man frantically draining the vending machine of bag after bag of M & M's.

"It's over," Anderson whispers, as much to himself as to Michael Blaese at his side. For weeks, Anderson has been haunted by the specter of disastrous failure. In his nightmares, T cells become trapped in Ashi's lungs, causing her to expel a blood clot, which would stop her heart. "I saw her gasping and dying on the bed," he confesses later. "The Inder Vermas and Richard Mulligans of the world would have come down on me like a ton of bricks. Gene therapy would have been set back years, perhaps indefinitely."

But it was not really over. Many months of subsequent infusions lay ahead, with the number of T cells per infusion rising each time into previously uncharted territory. Months more would be needed before the success or failure of the experiment could be evaluated.

Yet, for a day's work, it was not bad. Virtually forgotten was the carping criticism that the experiment was mere drug delivery. A breach of DNA had been filled in.

Close to nervous exhaustion, almost overcome with relief and exultation, Anderson stretches his weary body. Culver helps put away the infusion equipment, a broad smile wreathing his face. The immediate task is to sit and wait in case Ashi shows any late signs of trouble. By three-thirty, it is plain that she is out of the woods, at least for the moment.

Later, a nursing supervisor was to put her hand on Anderson's arm and say, "You were exuding so much emotional energy in there, it was filling the room."

Anderson would answer, quite rightly, "That's the way I've done everything in my life that was really important. My way is to exert so much energy it forms a protective cloud so nothing bad happens. I guess I believe in karma."

It is an improbable procession that wends its way back to Ashi's third-floor room. The child's sisters, on their backs and slightly sunken in their wheelchairs, their heads tilted at an angle. A woman in Eastern garb and her husband, his coat pockets stuffed with yellow bags of candy. Anderson still wheeling Ashi in her chair, although she has begun to protest the means of transportation mightily. As the small band passes in the hallways, few people pay attention. Says Anderson, "It's like people say in downtown Manhattan. Jesus Christ himself could walk by and nobody would notice."

At various times, insiders acknowledge Anderson and Blaese as they trundle by. Anderson points to Ashi. "Number one," he says.

They are rewarded with a thumbs-up sign. If people are looking for two-headed monsters, they are disappointed.

And so, French Anderson was first. He had ridden to glory, not on the flowing steed he had once imagined, but aboard a patchwork mount, its parts concocted to please everyone, its legs hobbled so as never to reach its destination, and its body oxbowed from the weight of the three collaborators Anderson had found it expedient to hitch up with—a fourth if you count Claudio Bordignon and his fine Italian hand. Nevertheless, within hours, the press had picked up the message. William French Anderson, not Michael Blaese or Ken Culver or Steven Rosenberg, had been the man on horseback, the white knight who had seized gene therapy from the land of reverie and made it come alive.

12

ONE ERRANT CELL

To succeed in the world, we do everything we can to ap
pear successful, La Rochefoucauld once noted, and the egos of the NIH
gene therapy troika, muted so long for the common good, were now col-
liding like thunderheads. Steven Rosenberg worked in the big world of
cancer and suddenly found himself upstaged by Anderson and Blaese,
whose patient load numbered no more than fourteen, if one charitably
included all the SCID youngsters in the world on PEG-ADA. No one
could gauge the disappointment that seemed to be burning in Rosenberg's breast, but during the summer of 1990 it had been obvious to the
members of the ADA group in Building No. 10 that their own partner
was racing to beat them to the punch. What had particularly shaken
them was the RAC's instant approval of his TNF protocol and its endless agonizing over giving them the green light to treat little Ashi
DeSilva. Speaking privately, they frequently grumbled about Rosenberg's style, particularly his rumored acceptance of a six-figure publisher's advance to write his autobiography, centering on his work leading to gene therapy and its use to fight cancer. The huge advance was the
result of an auction at which Rosenberg sold his story to the highest
bidder, which turned out to be G. P. Putnam's Sons.

In his own account, *The Transformed Cell: Unlocking the Mysteries
of Cancer*, written with John M. Barry, Rosenberg takes no credit for the
ADA experiment and has nothing but good things to say about his
friends French, Mike, and Ken. There Rosenberg explains that he would
have loved to have been present for Ashi's first infusion, but he was only

an associate investigator, hence not responsible for treatment, and felt he should fulfill a promise he had made months earlier to take his daughter, Rachel, a high school senior, on a tour of New England colleges.

Much of a physician's life involves preserving life and then dealing almost the next instant with death. Late Friday, after that momentous event with Ashi, Anderson went down to a room on Two East, one floor below her hospital room. Two East is Rosenberg's cancer wing. The object of Anderson's trip was a failing TIL patient—number three of the ten—who had been drifting in and out of a coma since the preceding Tuesday. "He just squeezed my fingers; he couldn't talk," Anderson relates. "Everybody thought he would die on Wednesday, but he was still there on Friday after the ADA infusion. I went down and told him everything had turned out well . . . and he just squeezed my fingers again and died. It was like he had waited. I mentioned it to other people on rounds, and they just said, 'Yeah, sure.' "

On Saturday morning, Anderson got to the hospital early although still physically exhausted and emotionally in turmoil. Suiting up in the anticontamination outfit he calls a "monkey suit," he performed a postmortem on TIL patient number three. Usually, he and Rosenberg did the autopsies together, but that Saturday morning Rosenberg was not in evidence. Anderson had to work alone.

Completing the autopsy, Anderson returned to the Children's Inn to say good-bye to Ashi and her parents. It would be Monday morning before he saw Rosenberg again. Anderson, who can be almost sappy in his magnanimity at times, reportedly approached the oncologist and seized him by the shoulders. "Steve," he said earnestly, "I just want to tell you I dedicate myself to helping you get cancer cured!" Rosenberg could not bring himself even to look at Anderson, as the story is told. It would be days before the surgeon would again interact with him. According to those who know the participants, the reason was simple: Rosenberg felt beaten.

In the aftermath of triumph, even the tightly knit Anderson-Blaese partnership was to pop a few stitches. Suddenly, Anderson began to assume center stage in the media. For instance, a Sunday magazine piece in the *Washington Post*, entitled "French Anderson's Genetic Destiny," unequivocally portrayed Anderson, not Blaese, as the leader of the ADA project, as well as the spiritual father of gene therapy. His home base, NIH, kept referring to him as the father of gene therapy, too. And the *New York Times Magazine*, giving its profile of Anderson the whimsical, Willie Wonka–esque title "Dr. Anderson's Gene Machine," crowned him as no less than "the nation's leading genetic surgeon." Even *Science* magazine, which ought to have known better, made Anderson sound like the Lone Ranger when it called him "the heart institute clinician who recently began the first federally approved gene therapy trials in children."

"Children" referred to Ashi and Melvin Berger's other ADA-deficient youngster at Rainbow Hospital, Cindy Cutshall, the nine-year-old gamine from Canton, who began her regimen of monthly treatments in Bethesda on January 31, 1991.

Blaese gulped at the media pieces but refused to be jealous. Soon, however, it would be Anderson's turn to become infuriated. Critics of the NIH team had been strangely silent since RAC approval was granted. Had they become converted to the teachings of Chairman French? Were they willing to see, or at least resigned to seeing, him and his colleagues propel the human race into its biomanipulative future? Well, the opposition was about to come roaring back.

Coincidentally with the infusion of new genes into Ashi, a September conference was to be staged at Chicago's dowager queen—the stately Drake Hotel. The conference dealt with a technology that, in its own way, was every bit as cutting edge as gene therapy. Some 250 of the world's leading in vitro fertilization obstetricians and reproductive geneticists had gathered in the hotel's grand ballroom for the First International Symposium on Preimplantation Genetics—the alteration of sperm and egg *before* fertilization that would permit such oddities as the diagnosis and discarding of defective reproductive cells and, someday, the ultimate piece of sideshow medicine, human cloning. Arthur Bank, a professor of genetics at Columbia University's Health Science Center, had agreed to attend the conference months before and interpret for the assembled researchers the recent spate of headlines datelined Bethesda.

Leaning heavily on the podium, he looked drawn and tense. An owlish man, with large eyeglasses and hair slicked to one side, Bank began straightforwardly enough. "The outlook for gene therapy over the next few years is favorable," he advised the audience. "The original barriers, which were social and political, have essentially been overcome." Bank continued in this vein for a time, outlining, in an academic drone, the arcane methodology involved in replacing genes. All at once, without warning, his speech took a most unscholarly turn.

Waxing waspish, the Columbia researcher declared that French Anderson's experiment upon Ashi had been premature, if not irresponsible. It was driven by "personal ambition," he continued, "not by good science or good medicine." The ballroom filled with the sound of bottoms moving uncomfortably in vinyl chairs. "The main impetus for this procedure," charged Bank, glaring through his spectacles like a stern preacher, "was not science, but the need for French Anderson to be the first investigator to perform gene therapy in man."

At this, there were outright gasps. One of the rules of the scientific community is that one does not lambaste one's fellow scientists in public—though this gentility has often been abandoned. Men as august as Sigmund Freud and Louis Pasteur endured public tongue-lashings from their colleagues in their day. A young lady who announced that she worked in Anderson's laboratory arose to defend her mentor. But Bank made short work of her. Her passionate feelings quickly caused her to become tongue-tied, and she soon took her seat again in vexation.

After his speech, Bank sounded the bell for round two, holding forth in an anteroom of the ballroom. Far from backing down on his charges, he repeated them as reporters took it all down eagerly on tape. Hearing of Bank's charges from these reporters an hour later, another Anderson detractor happily jumped aboard. He was Stuart Orkin, professor of pediatric medicine at Harvard Medical School, a sometime collaborator of Richard Mulligan's and full-time critic of William French Anderson's. "I think it's highly unlikely that [the ADA experiment] is going to work," Orkin told media callers that afternoon. "I think it's a long shot. Many people who are dealing with research in this area would prefer to see very careful studies done in primates . . . before going ahead to the human experiments." This was a clear slap at Anderson's poor results in his monkeys.

You get to be first only once, Orkin noted, when asked to assess what he believed was the prime motivation for the experiment. "In this case, I think NIH wanted a first strike," he said. "But there are a large number of scientists who believe that this kind of experiment is really not well founded scientifically as yet. If Mulligan gets up now and opens his mouth, the NIH group can say it's sour grapes. But I'm really surprised there hasn't been more vocal response, more criticism from completely objective scientists."

A phone call to Richard Mulligan found him more than willing to speak up. "You know what I think," he said. "The experiment is a sham. It ought to be scuttled." For a reporter, the harvest was bountiful: with just three phone calls, respected researchers from three of the world's leading centers for gene discoveries—Columbia University, Harvard, and MIT—were now on the record lambasting the federal government's ADA experiment. Informed of Orkin's remarks, Anderson quickly dismissed them. "Stu Orkin is entitled to think what he wants," he said. Mulligan's remarks he expected. However, Bank's comments were a different story. Anderson and Bank had known each other for a quarter century, Anderson remarked, adding, "We've been the closest of friends. His son stayed at our house last Christmas. We've been coeditors and cochairmen for the Cooley's Anemia Foundation. Two months ago, at a Cooley's Anemia meeting, he smiled and took my shoulder. Isn't that a friend? I don't understand why he is suddenly doing this." Anderson almost seemed to be bleating. "The ironic thing," he said, "is that he called me three months ago and wanted to put a gene therapy protocol in. He wanted me to fax him all my material that I sent to the RAC. . . . I never heard any more from him."

Anderson's only explanation for Bank's behavior was that perhaps illness had made Bank crotchety. Though he was at first inclined to take the remarks in that context, and accept them with some equanimity, his mood began to darken when reports of Bank's diatribe began to appear in the national media. At the height of his fame and glory, French Anderson, the grammar school outcast who had made himself into the most popular boy in school, was being slugged from behind, and it hurt. When he should have been riding on his colleagues' shoulders, he was even now being torn down. It wasn't fair. By the end of the week, French Anderson was no longer full of grace and equanimity. He was pissed. What would Ashi's and Cindy's parents think? That they'd entrusted their children into the hands of cynics or madmen? Their French, an opportunist?

The media setback soon fizzled out, and the lionization of Anderson resumed. But as it continued, Blaese, ordinarily a man of few pretensions, began to chafe. When the New York Times referred to "Dr. French Anderson, who runs the [first gene therapy] project," Blaese's disgruntlement grew. And then, early in 1991, Anderson was nominated as a sole candidate for the prestigious Albert P. Lasker Prize, an honor that often precedes the receipt of a Nobel Prize. The nomination letter had been written by Anderson's boss, Claude Lenfant, director of NIH's Heart, Lung, and Blood Institute, who might be expected to try to reserve the spotlight for his own man. But the letter, which all but ignored anyone else's role, amounted to a major revision of the facts, and a clear injustice. Inadvertently, Blaese saw a copy of the letter, as it was about to enter the pipeline that ultimately leads to the Lasker Prize. He immediately called Anderson and demanded to see him. It was the last straw.

Interviewed that morning, Blaese was still fuming. "The letter's

written in such a way as to intentionally minimize anybody else's contribution," he said. "I don't know if it was aimed at shutting Rosenberg
out of the prize or what. But it bothers me tremendously, and I have to
talk to French about that.

"He didn't write the letter, but I have to let him know now that
people have to know, from my scientific point of view, what really happened. I had to talk French into this whole T-cell strategy."

Anderson had never claimed otherwise and had always gone out of
his way to credit Blaese with having moved him away from his obsession with bone marrow stem cells, at least long enough to bring gene
therapy to fruition. Within days, the Lasker omission was rectified.
Blaese seemed appeased, but beneath the surface his indignation remained.

"Philosophically, I would hate to have a major award made that
didn't include me," Blaese conceded shortly afterward, in Chicago. He
was speaking at a gene therapy conference. Around his neck he wore an
elastic pink cloth neck brace to soothe the muscles and ligaments he had
strained months before while sleeping in a cramped position on an Italian airliner bound for Milan and Claudio Bordignon's lab. "I decided a
long time ago that I love what I'm doing," he said, "and I don't want
anything to spoil that. So it doesn't bother me that the *Washington Post*
has an article about French.

"But I feel very close to everything that's happened, because it really
was my idea. French got involved because I invited him. He was all hot
about this last-hope thing [for patients with no other options] and was
ready to go with that, and I said *not* with *my* patients, you're not. I just
thought it was a stupid idea."

Despite all the wrangling, they would not go on to win the Lasker
award.

Ken Culver, for his part, was annoyed because of ongoing friction
with Rosenberg, who he believed simply didn't like him and had tried to
get Culver's name dropped from the ADA protocol. Another incident,
petty to be sure, nettled the youngest member of the gene therapy team.

One day, around the time the TIL experiment began, Culver was
asked by Anderson to carry some papers to Blaese. As he walked down
the hallway, Culver noticed that among them was a document he
quickly recognized as a patent application.

"I'd never seen one," he says, "so I decided to flip through it, and I
wound up in shock." The patent was for the therapeutic use of TIL cells
containing genes to enhance their anticancer potential, in essence a pa-

tent for cancer gene therapy. The names on the patent application were those of Anderson, Blaese, and Rosenberg.

"They did this even though they had attached all my data on the back," Culver remarks, reacting as underlings in labs headed by heavy-weights have always reacted. "This isn't connected with GTI. There's no CRADA. Just those three guys will have the patent. They're never going to make any money on it. But, still, it's my data."

Having lost the battle but still in the thick of the war, Steven A. Rosenberg was dueling with the Food and Drug Administration. Even though the RAC, in that heady outburst of enthusiasm the preceding July, had quickly approved the TIL/TNF protocol, the regulators had timidly admitted their concerns about Rosenberg's paucity of animal data by attaching several caveats. For example, the RAC required that Rosenberg not inject gene-modified TILs into patients unless the cells were expressing at least 100 picograms of TNF per milliliter (ten million TILs) for twenty-four hours. As it turned out, he would have difficulty in reaching even this modest threshold—and the disheartening fact was that 100 picograms is at least one order of magnitude beneath what it probably would take to shrink tumors. Even two years later, as reported in the December 3, 1992, edition of *Nature,* TNF expression in the clinical trials reputedly varied from 100 to 1,200 picograms, with only a few patients showing the 1,000-picogram level that mouse work suggested would be therapeutic in a human being.

No matter what NIH had decided, it quickly became clear that the FDA was not going to lie down with the lambs and approve Rosenberg's cancer protocol unless he provided more safety data. TIL was one thing, TIL/TNF quite another. Extraneous issues, such as interminable political wrangling over the retroviral vector, would also emerge in the coming months.

To the ADA team, the Cetus vector approved for TNF wasn't much different from the trusty GTI vector used for ADA—the genes were nearly identical—except that the vector didn't seem to be working for TNF. The FDA had approved Blaese's ADA protocol with the GTI vector in September, so in an apparent attempt at peacemaking Rosenberg began using the GTI vector as well as the balky Cetus vector in his experiments—what did he care? But in November the FDA approved the Cetus vector for TNF, but not the GTI vector, because most of the safety tests had been conducted on the Cetus vector. To Rosenberg and Anderson, this was ridiculous. In the first place, the two vectors were so similar that only a patent lawyer could tell them apart. In the second, as a routine part of the gene therapy regimen, every viral preparation would undergo a dozen stringent safety analyses before being administered to any patient.

Unimpressed, the FDA next tightened the screws on the TNF dosage Rosenberg could use. He had planned to begin with a negligible amount from a tiny number of altered TILs—only a hundredth of the amount of exogenous TNF that had already been shown to be tolerable in people— and to work his way up from there until toxicity was determined or he reached his standard TIL therapy dosage of 200 billion cells. *Whoa!* cried the FDA. Nobody knows what these TNF genes might do. Take even tinier baby steps, officials ordered, forcing Rosenberg to decrease tenfold the number of cells he would infuse—in effect increasing the safety factor from a hundredth to a thousandth of the known tolerable dose. As if that weren't cautious enough, the FDA next placed a catch-22-style restriction on Rosenberg, refusing to let him give IL-2 to the first three patients who would be receiving TIL cells armed with TNF genes. "Let's see what happens without IL-2," the FDA said. But Rosenberg already knew what would happen. Without the cytokine to keep them growing in the body, the TILs had no chance to survive, and since he was being allowed to give only a few of them anyway, the treatment in no way constituted valid TIL therapy. Rosenberg had dying patients who were counting on him. He flatly refused to comply with the latest FDA dictum. A stalemate ensued, with both sides digging in their heels.

Unbeknownst to the FDA, the argument over precisely what Rosenberg would be allowed to do was academic. Behind the scenes, his lab had its hands full trying to get patients' TIL cells merely to accept the insertion of TNF genes and subsequently to express the protein. Cells are not stupid. Human TIL cells have their own TNF genes and are very careful about how much they make. Given additional TNF genes in culture, either the TILs were dying—presumably because the TNF was too toxic—or they quit producing it before they had increased in numbers enough to be given back to a patient as a therapeutic dosage. This is what happened with cells from the first four patients who had volunteered for cancer gene therapy—their last hope—and Rosenberg couldn't work his way around it. Other NIH scientists had found the same thing. When they stuffed TILs with various cytokine genes—for IL-1 and IL-2, and for TNF—the TILs would rebel and shut themselves off in self-defense. But Rosenberg's problems became a grotesquerie. As he wrote in his book, he almost grew afraid to go to the lab. One day, he would find out that the titer, or number of viruses, in solutions provided by both

GTI and Cetus was too low, making the preparations useless for trans-ducing TNF genes into patients' TILs. The next day, a contaminated shipment of solutions would arrive, and Rosenberg would be told, apologetically, that fixing the production glitch that had ruined the batch would take months because of FDA safety regulations.

Then, though he doesn't report it in his book, when things seemingly couldn't get any worse, his mice started dying when injected with TIL/ TNF cells. Rosenberg had instructed the members of his lab not to talk to the press, but some of them tattled. "They're dropping like flies," confided a source about the hapless mice. "No one knows what the hell is going on."

Rosenberg, frantic now, blamed the plague on a hepatitis epidemic. Chaos threatened to swamp TIL/TNF, history's second adventure with gene therapy and the first for cancer. The net result was that despite having won blue-sky NIH approval in July, Rosenberg did not receive FDA approval for another six months (on January 8, 1991). Even then, TIL/TNF was to be continually plagued by the problem of introducing the vectors into the patients' TILs and getting them to express proper amounts of TNF. Although Rosenberg had been granted permission to work with fifty patients, in the first year of the trial only four eventually received the new genes. By monitoring Rosenberg so closely, the FDA tacitly admitted its own misgivings about the experiment.

At the end of 1991, Mike Blaese finally broke what had become an embarrassing silence and articulated what Rosenberg was loath to admit: "We've had major problems," Blaese said in a publication cir-culated throughout NIH, "in getting human TIL cells and human T cells to express cytokine genes."

Despite everything, however, medical science's second authorized gene therapy experiment finally got under way, under a vir-tual news blackout, at 8:49 A.M. on the cold Tuesday morning of January 29, 1991, when Rosenberg, working alone, infused into a twenty-nine-year-old woman and a forty-two-year-old man about 100 million TIL cells (without IL-2), only about 5 percent of which bore added genes for TNF and neomycin resistance. He refused to identify the patients fur-ther, but said both had tolerated the low-dose, twenty-minute therapy well and were "resting comfortably." Although the infusions lasted only minutes, Rosenberg ironically noted that "it took ten years of work to get here." He restated his hope that TNF would throttle cancer cells by choking off the blood vessels that fed them and that, if successful in attacking malignant cells by sparing normal tissue, the treatment could open the door to the most powerful and precise cancer chemotherapy ever envisioned by medical science.

He refused to go further. "This trial is the first to apply gene therapy to cancer which affects millions of people," Rosenberg said, stating the obvious. "We must determine if it's safe by gradually giving escalating amounts of gene-altered cells over the next several months. Then we must see if it works."

The brouhaha over vectors from Cetus and GTI had changed the construction of the experiment somewhat, starting it at a lower dose and increasing the dosage twice as fast until the maximum tolerated dose was reached. Rosenberg, who seemed nervous, refused to say any more, except that because about 40 percent of his patients showed some improvement when given TIL cells alone, it would take at least a year for any benefits from TNF to become apparent. "And then, the proper forum for reporting would be a scientific journal," he said, grumpily.

From then on, the TIL/TNF experiment seemed to vanish behind a fire curtain. Rosenberg wouldn't take repeated calls from reporters, citing patient confidentiality and the scientific method as the reasons. In his book, published in July of 1992, the details suddenly emerged. At that time, although he had yet to publish anything in a journal, Rosenberg provided the only known account of what may be one of history's most important medical experiments. Both metastatic melanoma patients, whom Rosenberg calls Robert Antrim and Suzanne Marotto, showed no toxic reactions to the tiny amounts of gene-altered cells in the first several hours after infusion. The following Friday, February 1, 1991, they received a second infusion, of 300 million cells (still minuscule in cellular terms), and were doing so well that Rosenberg sent them home the next morning. At midnight, however, Marotto was back in the hospital, complaining of abdominal cramps, and was treated and released the following morning. On Tuesday, February 5, both patients received another billion TIL/TNF cells without adverse side effects. However, by then, their remaining cells were not growing well in culture, reports Rosenberg, and he was afraid he soon would run out of cells to infuse.

The following Friday, February 8, both patients received another three billion cells each, but by Tuesday only Marotto had any cells left to support another infusion (5.4 billion cells). Rosenberg's TIL surplus was exhausted. He had never even made it to the point in the protocol where he would have been allowed to administer IL-2 to his patients and help his TILs survive. "The number of cells they received was pitifully small," he writes. Follow-up examinations on both patients ominously revealed that their tumors were progressing, which came as no surprise to Rosenberg. He decided to operate on Antrim and Marotto again, excise more cancer, try to grow new TILs from the tumors, and attempt further therapy.

Antrim's TILs grew well in culture this time, but would not accept TNF genes. In April 1991, Rosenberg treated him with untransduced TILs; gratifyingly, several of Antrim's newer melanoma nodules began to shrink, and a "huge" tumor in his groin seemed to soften, a sure sign of regression. On May 2, Antrim returned for another round of TIL therapy, but, to Rosenberg's dismay, his tumors were erupting again. "The April TIL treatment had killed most of the tumor cells," he writes. "But the few cells not killed were so virulent that in two weeks the tumors had grown back to their former size. I worried that these cells would be very likely to be invulnerable to a second round of TIL." This proved sadly true. As melanoma claimed Antrim's body, Rosenberg offered him another experimental therapy, which he didn't really believe would help but which would at least let the man know the team would not abandon him. Despite everyone's best efforts, Antrim died on August 5. "I liked him very much," Rosenberg writes. "I wished I had been able to give him TNF-transduced TIL in large quantities with IL-2. Perhaps it would have helped."

When she received TNF genes, Suzanne Marotto had already been treated with ordinary TILs and had enjoyed a seven-month remission. But then things went downhill. A month after she had received TIL/TNF gene therapy, Rosenberg was able to palpate sixty tumors on her body. He surgically removed a third tumor and tried to grow TILs from them. He was unsuccessful. As the months passed, the treatment drew no response. Rosenberg performed a fourth operation, removing tumors and trying to grow TILs, giving them back to her. But nothing seemed to help. In fact, she nearly died when he administered IL-2 alone. "Her cancer was mocking everything we tried," Rosenberg writes, "and mocking my efforts to offer some hope." He does not explicitly say what happened to Marotto, except that "neither Suzanne nor Mr. Antrim responded."

But then, they had not received the therapy that Rosenberg had wanted to give, he writes. This occurred in part because their cells had not grown well and in part because FDA restrictions had limited him to administering only tiny doses of cells, and then without the nurture of IL-2.

What had the FDA limitations, though well meaning, allowed

Rosenberg to prove? Only that if one undertreats dying patients with TILs, the patients will have no chance whatsoever.

However, with Rosenberg's third TIL/TNF patient, a fifty-two-year-old woman he calls Barbara Spengler, the story brightened. She had failed to respond to a large dose of regular TILs (300 billion cells) accompanied by eleven doses of IL-2, but by May, when Rosenberg gave Spengler TIL/TNF, he had received FDA permission to select for gene-altered cells.

That meant the woman would receive a more powerful treatment. At the time, Rosenberg found thirty palpable melanoma tumors on and inside her body. To fight them, fully half the TILs she received had new genes in them. In the earlier experiments, Rosenberg had not been allowed to select for gene-altered cells to administer to his first two patients, presumably to set a baseline. This time, he could. Moreover, Spengler's cells grew much better in culture than those of the previous two patients had, and Rosenberg was able to administer gene-altered cells twice a week, tripling the number each time. In July, he gave her 100 billion transduced cells, but still without IL-2. Like a racehorse chafing at the gate, Rosenberg couldn't wait to get going. But even at this point in the trial, the FDA wanted him to be very careful about adding the growth factor to his gene-altered cells. It ordered him to drop back the cell numbers by a factor of ten and slowly add one-quarter the IL-2 doses he usually administered with TILs.

On July 5, Spengler received 11 billion cells and the low-dose IL-2, whereupon Rosenberg was ordered to wait three weeks and watch for side effects. There were none. So, on August 5, he was granted permission to give Spengler 28 billion cells and IL-2; then he was told to wait again. Nothing adverse occurred. Finally, on September 23, Rosenberg gave her most of the remaining cells he had—90 billion, along with low-dose IL-2.

Even before she received her last treatment, Spengler's tumors had begun to shrink. "None had grown," Rosenberg writes in his book, "and no new ones had appeared." It looked as if the miracle that Rosenberg had witnessed so rarely had finally materialized. He excitedly biopsied the woman's tumors. Pathologists pronounced them dead. How long would they stay dead? At the time when Rosenberg finished his book, the woman's tumors were continuing to shrink, and she seemed heading for a complete remission.

But was the remission due to TIL? Was it due to TIL/TNF? Who could tell? Apparently not the National Cancer Institute, which gave indications that it was getting nervous about the human experimentation. Rosenberg's TIL/TNF travails were exposed to the light in December of 1992 when an NCI scientific oversight panel, the Division of Cancer Treatment, suddenly froze funding of a $3.9 million contract with an outside lab to grow TILs after blasting the cancer specialist for not providing adequate evidence that the trials were succeeding and should continue.

Not only was the board worried about gene expression; its members seemed to be having doubts about the fundamental premise of the experiment: that TIL/TNF cells were indeed flocking to tumors and not ending up in the liver, where with escalating concentrations, they might kill the patient.

The biology of TILs—whether they indeed homed back to the cancer or were merely dumped there by circulating blood—was the objection that RAC subcommittee molecular biologists had kept raising earlier. Their doubts and the absence of a firm answer had been crushed by the euphoria over a cancer cure and the steamroller acceptance of an iffy clinical protocol.

Two years later, at the October 1992 meeting of the National Cancer Institute board of scientific counselors, its chairman, the Stanford University oncologist Ronald Levy, said that Rosenberg's unpublished data showed that the numbers of gene-modified TILs gathering at tumor sites were "about a quarter of a magnitude lower than was projected." On the basis of that meager showing, Levy said, he could not in good conscience go along with further experiments. "I think . . . that it is premature and unscientific to proceed with a clinical trial based on the pre-clinical and preliminary data that has been developed."

The feuding had apparently started the preceding February, when Rosenberg seemed rather cavalier about an on-site visit from members of the board. "There was a lack of preclinical data for virtually every one of the issues that [Rosenberg] was proposing, and it would help if Steve provided the preclinical data to us," a board member, the University of Washington oncologist Philip Greenberg, told *Nature*, which highlighted the showdown on December 3, 1992, in a major story headlined "Gene Therapy Researcher under Fire over Controversial Cancer Trials."

Instead of greeting his cancer colleagues with new data, Rosenberg seemed to deflect them. "What we got," complained Greenberg, "were a series of reprints (of earlier studies), none of which really answers those questions. His response to the site visit was not adequate."

Two years earlier, when Rosenberg would consent to sit down for an interview, he also fended off reporters' demands for specifics by handing out reprints. Apparently, his peers were insulted. Rosenberg has always been a controversial figure within cancer circles, and these colleagues worried that the line between clinical research and human experimentation was getting blurred. They were also concerned about Rosenberg's tendency to continue treatments of dubious value until ordered to stop. In *Nature*, for example, the University of Chicago oncologist Ralph Weichselbaum complained that despite the disappointing results of trials that compared high-dose IL-2 and IL-2 plus LAKs (lymphokine-activated killer cells), Rosenberg was going to continue them. The board recommended that the trials be stopped, and Rosenberg grudgingly halted the study the following August.

Although the board generally expressed support of Rosenberg and his quest, several of its members told *Nature* they didn't like his methods and were withholding some of his funding until he cleaned up his act. Rosenberg, not very contrite, later explained that a misunderstanding had occurred and that he was adequately funded anyhow.

TIL/TNF had barely gotten off the ground, however, before it was eclipsed by an even more encouraging strategy euphemistically known as cancer "vaccines." In April 1991, instead of removing TILs from the body and arming them, scientists had begun treating the tumor itself in such a way that the body might spot it as foreign and expel it. That way, the immune system might be roused to track down and kill free-floating metastases of the original cancer. If it weren't for mets, cancer would be a very treatable disease—in fact, it wouldn't be cancer at all, but an "in situ" tumor that could be removed.

Evolution has admirably trained the human immune system to recognize countless viruses and bacteria. Invaders have chemical keys on their surfaces called antigens. These keys fit into locks—receptors—on cells, enabling the virus to gain entrance and cause infection. But they also provide targets for the immune system's Y-shaped antibodies to latch onto and neutralize. The discovery of an antigen, or something foreign, is done by the system's frontline guards, the white cells called macrophages, which ceaselessly roam amoebalike throughout the body and the brain as sentinels in search of invaders to scavenge.

But cancer rises from within. The surveillance system falls apart with cancer, perhaps because the difference between a tumor cell and a normal cell—both of which the body sees as "self," instead of "non-self"—is relatively slight. Immune sentries often miss cancerous cells until it is too late and the tumor has grown too large to be killed. The new immunotherapy strategy has been to devise ways to treat antigen proteins genetically on the surface of cancer cells and turn them into flashing marquees for immune defenders to aim at. But scientists working on the problem had long been stymied by their inability to determine which of the myriad proteins that stud the surface of a tumor cell—the so-called tumor antigens—would be most likely to trigger a potent immune response if the gene that produces it is altered in the lab, and thus be most useful when inserted into a cancer vaccine. Most antigens that have been discovered on tumor cells are also found on normal cells, which may be how the tumor fools the system. However, in November 1991, researchers in Belgium announced the discovery of the first set of three tumor antigens that are unique to tumors and incite killer T cells, the immune cells responsible for destroying malignancies. When HIV-infected people, for instance, lose their CD-4 cells, they become susceptible to a host of cancers. The Belgian work involved melanoma, and—an important point—the trio of antigens identified by Thierry Boon and his colleagues at the Ludwig Institute for Cancer Research, in Brussels, were found in about half the melanoma patients studied. That suggested that the antigens might be common in the series of mutational events that occur when a harmless mole, its genes perhaps damaged by exposure to excessive sunlight, evolves into a deadly, fast-spreading black mole cancer.

Another strategy was unveiled that month when scientists at Johns Hopkins working with Richard Mulligan at the Whitehead Institute used gene therapy to rid mice of an existing cancer by stimulating the animals' own immune systems to destroy tumor cells. Although he had voted for it, Mulligan had never totally approved of Rosenberg's TNF strategy, because of problems in understanding the basic biology of TIL cells. The alternative concept pleased Mulligan more: flag the cancer cells themselves, and let the whole body rise up angrily, instead of trying to trick a few TILs into becoming guided missiles.

Writing in *Science*, the Hopkins/MIT team, led by Drew Pardoll, took tumor cells from a mouse kidney cancer and engineered them to secrete large dosages of another cytokine molecule, interleukin-4 (IL-4), that T cells normally release when activated to notify other cells that there's work to be done. Boldly, Mulligan's Boston-bred viruses inserted the gene that makes IL-4 into the tumor cell nucleus itself, thus turning renegade cancer cells into little IL-4 factories, which then were injected back under the skin of the mice, whereupon the animals produced an immune response to the cancer that actually cured the disease!

"This is the first time that this type of gene therapy has been shown to cure an animal of an already established tumor," said Pardoll. "Until now, most studies have sought to immunize animals against developing subsequent cancers." The added benefit of the new treatment, Pardoll said, is its ability to destroy cancer cells not only of the primary tumor but also at a distance from the site, without apparent damage to normal cells. When cancer slips by T cells, Pardoll explained, the inactive T cell may be likened to an army general asleep at his command post and IL-4 to the signal he fails to give his soldiers. "The body's soldiers may recognize the enemy but need orders from the general to kill them. In our study, we bypassed the general and signaled the soldiers with orders to kill. Once activated, the T-cells function like smart bombs that kill the specific target but leave little collateral damage."

Pardoll's colleague Paul T. Golumbek said it would be feasible to make this method work in human cancer therapy. "But a number of additional studies need to be done to establish the effectiveness of this strategy and develop its full potential." Scientists had to find a better way to introduce genes that make cancer-fighting chemicals into tumor cells removed during surgery. "We then have to perform many animal experiments, including those that study the effect of tumor-fighting chemicals besides IL-4 to determine which tumor fighting chemical or what combinations of chemicals work best."

As usual, Steven Rosenberg wasn't about to wait around. He had been working in cancer vaccines, of course; in fact, he had earlier shown that when he administered to experimental animals the cytokine alpha interferon along with TILs and IL-2, he could "up-regulate" the histocompatibility antigens on the surfaces of tumor cells—these are the antigens that determine one's tissue type—to make them more susceptible to the killing effects of TILs. The laboratory maneuver, he said, improved the therapeutic aspects of TILs about fivefold. He had already started doing that in patients (with full FDA approval) in hopes of boosting the success of ordinary TIL therapy.

On October 8, 1991, a few weeks after his patient Barbara Spengler had received her final dose of gene-modified TILs, Rosenberg treated his first patient with a cancer vaccine. In his book, he describes the patient only as a forty-six-year-old surgeon. Rosenberg removed a tumor from the man, inserted a gene into his cancer cells, grew them up in culture, and then injected them into the man's thigh. Three days later, he repeated the treatment on a thirty-year-old woman. He reports no results in his book, but writes that he hopes by the end of 1994 to have treated fifty melanoma patients in each of his two new protocols—the first group with TIL/TNF and the second with gene-altered cancer cells.

In the meantime, a dramatic announcement of success with cancer vaccines was provided by one of Anderson's old mentors, the molecular biologist Eli Gilboa of Memorial Sloan-Kettering Cancer Center, in New York. Gilboa, working with Bernd Gansbacher, a cancer specialist at Sloan-Kettering, and Michael Feldman, chief of cell biology at the Weizmann Institute of Science, in Rehovoth, Israel, reported encouraging results in animals suffering from a variety of common cancers—sarcomas, carcinomas, melanomas—that were given genetically engineered cancer cells. The cells were irradiated first to destroy their ability to cause cancer. New genes that produced a substance called H2K, which normally makes tumor antigens, were inserted into them to "hypersensitize" the immune system to the invaders lurking in its midst. When the altered cells were returned to the animals, and their immune systems tipped off, they rallied and stopped the tumors from spreading.

Marveled Gilboa, "The ability of these tumor vaccines to stop the spread of cancer in mice is amazing." H2K, which belongs to a family of immune-system genes that make histocompatibility antigens, is normally expressed very weakly in metastasized cancer cells, said Feldman, so very little antigen is made and the cells escape detection by the immune system. Feldman's idea was to supercharge H2K expression so that the immune systems of patients would become much more sensitive to the low level of antigens on malignant tumor cells—rather like giving hearing aids to elderly people. Two other Sloan-Kettering researchers, Carlos Cardon-Cardo and Zvi Fuks, had made a discovery that seemed to answer the burning question of how cancer cells hid out from the patrolling T-cell sentries that were created by nature to seek them out. The scientists showed that tumors contained some cancer cells that had literally pulled in their antigens—rather like retracting a radio antenna in a car—thereby making them invisible to the immune system. Cardon-Cardo also found some of these cells floating freely in the bloodstream. "Ghost cells," he called them, "that the immune system can't see and attack." Evidence was thus mounting that the primary way cancer cells avoid detection is to simply hide their antigens, or protein identification cards, which normally sit on the surface of cells.

Not stopping there, Gilboa and colleagues had produced similar antimetastatic results in mice with another cancer vaccine. Instead of inserting an antigen gene into irradiated cancer cells, they this time used a modified retrovirus to make IL-2. The strategy worked like a charm, and the next step will be to insert genes for both the H2K antigen and IL-2 into the same irradiated cancer cell to see whether that will produce a stronger cancer-killing response.

What would come next? Gene-based immunotherapy, that belated but very crucial part of the saga, had entered the picture on May 22, 1989, when Maurice Kuntz placed his trust and his life in the hands of four men he had barely met (crediting them in alphabetical order, French Anderson, Michael Blaese, Kenneth Culver, and Steven Rosenberg) and permitted the U.S. government doctors to transfer foreign genes into the body of a human being for the first time. In the brief span of just three years, the revolutionary therapy, so heartbreaking yet so singularly promising, had swept out of their hands into the mainstream of medical research. Kuntz, the truck driver from Indiana, was no longer around to see it. But he would have loved it. It was all starting to happen.

13

BELT AND SUSPENDERS

Another grueling workweek has come to an end on the NIH campus. The stifling workout room in the basement of Building C rings with the brute sounds of martial arts—piercing cries, body slams, the scrape of bare feet on vinyl mats. It is here on Friday nights where William French Anderson, fifth-degree black belt, has for many years taught classes in the ancient Korean discipline of tae kwon do.

On this night in the late summer of 1991, Anderson is conducting a drill, taking on his students one by one. Most are NIH researchers like himself. For the drill, Anderson has purposely adopted a defensive mode. He offers a seemingly inviting target, blocking with forearms and knees while keeping his chin back. The posture gives him a weak, almost sissified, aspect as he allows his white-clad disciples to flail at him with hands and feet. But his passivity is deceptive. No one seems able to land a decent blow. And just when the pupil thinks he has Anderson totally cowed, the master takes sudden control, either scoring with a lightning set of combinations to the midsection or grasping the other by the shoulders and crying, in the voice of the happy camper that he is, "Good. Very good. You're coming along great."

There is a curious analogy between Anderson's demeanor on the mat and his image in scientific affairs. To the world, he often seems inept and marginal, an intellectually shaky figure who is ever on the ropes scientifically. Yet, at the very moment when his competitors are most disdainful, when they are about to inflict the coup de grace, Anderson abruptly makes his move. Too late the competitor finds that while he was busy underestimating Anderson, the cagy Oklahoman has passed him on the outside to win. And win French Anderson has. Make no mistake about that.

It is an altogether relaxed man who wipes the sheen of perspiration from his forehead with the baggy sleeve of his robe and orders his class to take a breather. Long forgotten are the slurs and barbs of those who took French Anderson lightly. Now, at almost a year's remove from his epochal experiment, Anderson is secure in the knowledge that when the history of human gene therapy is written, his name will head the list of those who pioneered the science. Even detractors must grant him grudging credit for championing the technique when no one else would even speak its name, for navigating the choppy regulatory waters to win approval for a trailblazing effort, and for being first, along with Mike Blaese and Ken Culver, to try the procedure in a clinical setting.

Where critics felt sure the Anderson team would stumble was in its results. Few believed that the experiment, riddled with so many uncertainties, could possibly succeed. But the outcome, while falling well short of the team's most optimistic predictions, has been heartening, with the data seeming to indicate a qualified success. All things considered, the debut of gene therapy has been promising.

The initial goal of the therapy, as envisioned by Anderson and Blaese, was to elicit a modest amount of ADA production in Ashi and Cindy. All he needed, Blaese speculated before the RAC, was 1 percent of the normal human production to cure them. And *cure* was what Anderson called the point of the exercise. In terms of ADA production, what was actually achieved far exceeded expectations. Several months into the regimen, Ashi's cells were manufacturing 20 to 25 percent of the normal volume, an amount well beyond that necessary for a reasonable level of immunity. In effect, she was making 40 percent of what a carrier of one bad gene, such as her mother, ordinarily makes. Cindy's levels were well below that. Where an operative ADA gene could be found in 30 percent of Ashi's T cells, only 2 percent of Cindy's cells had taken up the gene. But she was still making ADA in amounts sufficient to confer immunity, the lab team insisted.

From a purely functional point of view, other lab findings were even more encouraging. A healthy immune system depends on the production of antibodies. The classic way to tell whether someone is making antibodies is to expose him or her to various substances, known as antigens, that cause a reaction in normal people. Shortly after the start of gene therapy, Ashi began undergoing skin testing, first with a preparation of tetanus bacteria and then, as time passed, with extracts of *Can-*

dida, Diphtheria, and *Hemophilus influenzae B.* To the delight of her doctors, she responded to these agents provocateurs with a series of little bumps at the site of the shots, a sure sign that something was afoot in her immune system. When Cindy was also tested, she performed even better. Besides reacting to the same stimuli as Ashi, she showed a marked response to polio virus and cytomegalovirus (CMV). Such positive reactions in girls who a year earlier could muster no response seemed to indicate that they might be developing a repertoire of white cells with memories for specific intruders. If that was true, the cells could be relied upon to attack these trespassers on sight the way a mongoose bristles whenever a cobra comes on the scene. T cells work by recognizing marker proteins on the surface of various antigens. Thus, one T cell is programmed to mobilize when the tuberculosis marker is sighted and another when ragweed appears, still another responds when it spies scarlet fever, and so on. On swinging into action, the T cells activate awaiting B cells, which crank out the needed antibodies. Cindy was able to produce more antibodies on skin testing, because being older and having received the usual childhood immunizations, she had been exposed to more antigens, including polio and CMV. Her B cells were primed to manufacture antibodies, but they could do nothing without the signal from T cells—T cells she did not have. Now, thanks to gene therapy, she had the T cells and thus the means to protect herself. "Our purpose is to give our patients as thick a mug book for potential antigens as we can," said Blaese shortly after the favorable skin tests. "One of the reasons we have all been seeking a way to put genes into stem cells is because such cells are nearly immortal, assuring virtually lifelong immunity. But these gene-altered T cells we are giving back to patients have long lives, too—at least three months and possibly more. They also appear to have a selective advantage in the body. The altered T cells, each carrying some memory hopefully, appear to be dividing at the expense of the nonaltered cells, which will boost their numbers all the more."

By way of demonstration, Blaese motioned toward two trays of Ashi's T cells, which had been drawn a fortnight earlier in preparation for her next infusion. Nature is wont to provide exquisite aesthetic experiences if one knows where to look, and cell division is one of her visual feasts. When cells are dividing healthily, their elixir takes on a bright yellow hue, a golden cream color resembling that of good chardonnay. The yellow tinge is imparted by acid secreted by the cells as they grow. Cells that are pink have a higher pH and are not growing appreciably. One of the two trays contained T cells from Ashi that had been altered with new ADA genes. Their elixir was as yellow as a field of wheat. By contrast, the cells in the other tray had received no new genes. Their broth was a sickly, pale pink.

"Before we started this protocol, Ashi's T cells were always pink. They just refused to grow," said Blaese, still wearing a cloth brace months after straining his neck during an overseas flight to visit Claudio Bordignon's laboratory. Now frayed and soiled, the brace stood as mute testimony to the physical hazards of the scientific life.

Cell division in a laboratory dish is one thing. But replication must ultimately occur in the body or the game is lost. The all-important measure is the T-cell count. Here again, the early signs seemed to indicate that gene therapy had reversed Ashi's condition. Prior to therapy, when she was being sustained by PEG-ADA alone, Ashi's cell count had stood at 582 per microliter of blood. In the months following infusion of the gene-altered T cells, her count soared to 2,546, well beyond the threshold of normalcy, which is 1,000 to 1,500. Not only had her T cells stopped perishing; in the absence of toxic deoxyadenosine, they seemed to be flourishing. Not surprisingly, Ashi's natural levels of immunoglobulins, the broad class of antibodies that fight viruses and bacterial pathogens, began to climb, too.

There was one more very tangible indicator that gene therapy had succeeded. A hallmark of a humming immune system is a robust pair of tonsils—though this may come as a shock to members of the baby boom generation whose pediatricians were bent on making the structures all but extinct. Tonsils form a base camp of immunity, a place where B cells are bred by the body like rabbits. To the disbelieving eyes of Anderson, Blaese, and Culver, the girls began growing tonsils as their therapy proceeded. Ashi who possessed no visible tonsils when she began treatment, now sported small ones. Cindy, who had small tonsils to begin with, now developed large ones, as well as "two nice-sized lymph nodes in her neck," according to Culver.

Despite these positive signs, Blaese and Culver remained cautious through the summer of 1991 about the significance of what they were seeing in the girls.

Their wariness stemmed, in the main, from their limited ability to make conclusions based on the evidence. "Saying we only know the tip of the iceberg about genes is an understatement," says Culver. "In medicine we have an old adage that the dumbest kidney is smarter than the smartest intern. The same thing is true of the scientist. The dumbest cell is smarter than the smartest molecular biologist."

The acid test remained. There was only one way to determine

whether the reversal of immunological fortunes was indeed due to the infusion of new genes. The two young patients would have to be withdrawn from gene therapy for a period of months to see whether any of the salutory signs disappeared. If they did, it would strongly suggest that the treatment was responsible for the upswing and not PEG-ADA. On the other hand, if the girls maintained their good immune status, the team could declare a victory and move on. They were in a no-lose situation, as a matter of fact.

For the time being, Ashi's numbers were so unexpectedly high that the team could scarcely conceal its elation and was actively contemplating allowing longer intervals between her gene infusions. "If we can keep her at the level of what we've seen in the last few weeks," said Ken Culver, on the occasion of Ashi's seventh infusion, "it will be enough to give her cells every six months. It's impractical to keep giving them to her once a month as we've been doing."

Blaese also was beginning to waffle on the issue of withdrawing the girls from PEG-ADA. For months, he had insisted, in conformity with the protocol, that he and his colleagues would not lobby to remove their young patients from the drug, on the grounds that it would be unethical to step out on such a dangerous limb when the treatment was already working. But now the idea of taking the safety net away was proving seductive. "We're at least thinking about it," Blaese said pregnantly one afternoon in the blood lab as he watched the NIH team's blood-growing wizard, Charlie Carter, manipulate the vinyl bags in which he was growing up Ashi's cells.

"We couldn't be happier," he gushed. "Everything is going the right way. I couldn't have designed her response better had I scripted it myself."

Nonetheless, Blaese admitted that the full returns on gene therapy would not be in for some time. "People ask me, 'When are you going to claim success?' " he noted, clicking a ballpoint in and out reflectively. "I think this therapy has been successful. But it's new and untried. If, God forbid, one of them should develop a lymphoma next year due to us, could we call the procedure a success?" There is an introspective quality to Blaese, a sense of irony and existential futility, uncommon in his profession. It is one of his most endearing qualities, and the more he articulates the ultimate limitations of medicine, the more parents contemplating treatment find him reassuring. He who respects the dark side can be trusted.

French Anderson made no bones, however, about his conviction that the therapy is a success. "There are not really any other conceivable reasons why Ashi's ADA level has risen so high," he noted one morning in June, tilting back in the broken recliner chair in his office that only he seems able to manage without slipping a disk. "Mike is continuing to be rather conservative because he is the one talking to the press right now, and when he says something, he's the one who has to answer for it. He doesn't want to continue to be leveled by Mulligan and Kelley. He'd rather just lay back, and that's fine. But I think what has happened is unmistakable. I'm satisfied it's because of the gene, and that's why I am planning to move away from the experiment and on to something else. I've made my case."

Given the belt-and-suspenders approach, however, couldn't one argue that PEG-ADA is an indispensable part of the mix? How much of the improvement is due to the drug and how much to the gene? Do the two work in synergy? No one, not even Anderson, could really say. "There is for sure some synergy between PEG and the gene," Anderson allowed, "in the sense that the drug keeps the deoxyadenosine levels low so that even if there isn't enough ADA being made in the T cells to do the job, the PEG does it. We'll need to test that. It's possible that without PEG-ADA, the gene wouldn't work as well. But once it's up to where she's making 50 percent of the amount of ADA that a carrier makes, she might not need PEG-ADA any more."

Anderson admitted that while carriers produce only half the normal amount of ADA, they do it in every one of their cells, while the children produce it only in their T cells. "So, no, they haven't been given a complete immune system, and that's the concern."

The summer wore on. As July arrived, bringing with it the killer heat that enervates the Washington-Bethesda region, the Anderson team undertook to steal a page from Hollywood by hosting a gala party to mark the first anniversary of gene therapy's approval by the RAC. For several weeks, Ken Culver took on the assignment of impresario, hiring an entertainer, arranging for refreshments, and sending out invitations to more than two hundred people, most connected in one way or another to the first experiment. The guests were asked to pay homage to Ashi and Cindy and the infant technology that had rendered them more nearly normal than they had ever been in their lives.

The party was slated for July 25, a date that coincided with the regular visit of the girls to NIH for their monthly treatments. What was to be Ashi's eighth and final infusion of new genes was scheduled for the day of the party. Under the protocol approved the year before, the treatment would wrap up phase one of the experiment; in the ensuing hiatus of several months, the efficacy of the procedure would be evaluated by means of a battery of tests. Cindy, in the meantime, was at the halfway point of her regimen. She was to receive her fourth infusion the next morning.

Shortly before seven o'clock that night, the guests began to gather in an auditorium on the top floor of Building No. 10, a space with all the charm and character of a grammar school assembly hall. The crowd was composed primarily of top NIH brass, clinical staff, and laboratory personnel, accompanied in some cases by their families, but it also included some members of the national press who, in attending a historic occasion that they could not write about because of the DeSilvas' desire for privacy, were a little like dogs obliged to hold a sirloin steak in their mouth without biting in. President George Bush was invited to attend the affair but declined. Conspicuously missing was Steven Rosenberg, an absence that could be explained, of course, by simple group dynamics. More mysterious, however, was the behavior of Mike Blaese, who, after making a perfunctory appearance, quickly disappeared, pleading that he had a sick patient elsewhere in the hospital.

From the outset the party had the provincial flavor of an Odd Fellows dinner, perhaps because it had been staged by Culver, who, for all his scientific expertise, retains some of the makeup of a Des Moines booster. Hence, two sheet cakes resting on a long buffet table near the door, their legends spelling out, in pink frosting, "Congratulations Ashi DeSilva, Sept. 14, 1990—Cindy Cutshall, Jan. 31, 1991," and, as a salute to the doctors, "Great Team Work, ADA Deficiency." "ADA Deficiency" had a line drawn through it, as in a "no parking" sign. Huge bowls of caramel corn and vegetables with dip filled out the repast. As for the entertainment, the guests watched in bemusement as a performer in military garb, calling himself General Foolishness, presented a vaudeville-style act that once was a staple of such programs as Ed Sullivan's TV show but is seldom seen any more. He rode a unicycle, climbed an unsupported ladder, which he jumped up and down upon as if it were a pogo stick, and spun dinner plates on a row of dowel rods. At one point, he prevailed upon a game but awkward French Anderson to join him in his act. There is a side to Anderson that seems to enjoy burlesque performance. Earlier in the day, during Ashi's infusion, he had startled onlookers by seizing Blaese's neck brace and slipping it on. Letting his head droop forward, he affected a grotesque shamble reminiscent of Quasimodo. It was out of character, like Richard Nixon attempting stand-up comedy, and it made his retinue laugh nervously.

The entire evening had a strained aspect, a sense of false hilarity. Few could escape the nagging notion that there was nothing to celebrate yet. The girls were still not cured. Ahead of them, by any reckoning, lay an ordeal of many months' duration. Nor was gene therapy in its present form anything but an interim treatment best suited to bridging the gap until something better came along.

It is later that night, much later. Eleven o'clock has come and gone, but the DeSilvas and Cutshalls, badly in need of winding down, have set up shop in the playroom of the Children's Inn. The room, brightly lit and decorated in primary colors, is a whimsically furnished space that doubles as a library. Its shelves are chockablock with young people's books, especially those "quality" works that have won the Caldecott and other self-congratulatory medals with which the publishing industry promotes juvenile literature. There are also magazines, including, inexplicably, a copy of *Archaeology*. One wonders whether it was placed there by French Anderson.

The Children's Inn itself is a spectacular place built on the lines of a great hunting lodge. It is designed around a central atrium whose focal point is a huge, three-story chimney of glittering flagstone. At the chimney's base is a cavernous but somehow charming fireplace, kept lit even in summer, as if to warm the haunches of the pair of ten-foot stuffed pandas that flank the hearth. A balustrade encircles the atrium and girds an overlook that houses the living quarters of the inn—comfortable two-bedroom apartments with all the amenities—as well as a recreation room with pool tables, pinball machines, and advanced video game technology; an entertainment room featuring a giant-screen TV; an aquarium; and a kitchen of oak cabinetry, chrome range tops, and cedar timbers. The kitchen is a nosher's fantasy, with big plastic sacks of popcorn resting next to bowls piled high with Teddy Grahams and sugar wafers.

In the playroom, the three DeSilva children are still awake, along with the two Cutshall youngsters. Ashi is seated with her legs crossed. Cindy is right beside her, while Laurie Cutshall, Cindy's younger sister, is seated in a chair shaped like a bear. They are watching an Inspector Gadget video on the television. Cindy is eating a late snack from the kitchen. She holds the last of a plate of cold cuts between her thumb and forefinger and smacks her lips loudly as the morsels slide into her mouth. She turns to address her mother, Susan, who is deep in conversation with Van DeSilva. The two families, which met several years earlier, during an Enzon press briefing announcing FDA approval of PEG-ADA, have become fast friends and talk frequently on the telephone. But their visits to NIH have not dovetailed until now, and in their medical isolation, they seem to hunger for the contact. "We're sort of in this thing together," says Van.

"Mommie, I need a drink," wails Cindy. Ashi, dressed in green sweatpants and baggy T-shirt, yells, "So do I." Her loud voice is the more startling for its departure from her customary shyness. With their parents' permission, the girls run out of the room to get a drink of water, with Laurie pushing Dilani DeSilva's wheelchair. The wheelchair careens through the doorway, as in a slapstick movie.

Raj DeSilva is sitting in an armchair alongside Cindy's father, William. Bill Cutshall is a diffident man with dark red hair and a neatly trimmed red beard. He is wearing a polo shirt, gray Dockers, and deck shoes. His wife, Susan, a large and striking blond-haired woman, is likewise dressed informally, in blue elastic shorts and tennis shoes. Sitting across from her is Van DeSilva. It is a peculiar happenstance that both sets of parents involved in these first gene therapy trials have professional links of one sort or another to science. Both mothers are registered nurses. Bill is a former respiratory therapist—he met Susan in 1980 at Altman Hospital—who now teaches respiratory care at Stark Technical College, a two-year institution in Canton. And Raj is a researcher with B. F. Goodrich working in the area of plastics, particularly in the development of polyvinyl chloride dissipators, the nonconducting sheathing that surrounds electronic equipment. His brother, coincidentally, is an immunologist and clinical allergist in New York. It was he who suggested that Ashi's troubles after birth might be immunological in origin.

The four parents are discussing how their children are faring under gene therapy. Raj DeSilva takes the lead, describing Ashi's progress in glowing terms.

"I'm really elated," he says, his thick arms folded behind his head. "When you get a minute to think about it, our world has changed so much more for the better in the past year. A year and a half ago, Ashi could never have been playing out there with these other kids. We didn't even dare to hire a babysitter for fear of infections. With PEG, we'd seen a tremendous improvement. But even with the drug, she had runny noses and a constant cold, was on antibiotics all the time. But by the second infusion of genes, in December, it began to change."

The first sign was the runny nose drying up. "We noticed because we weren't using up so many boxes of tissues," Raj says. "Then all of a sudden, she wasn't getting fevers and colds as often. If she *did* get them, she'd recover normally."

Now Ashi is so improved that she is able to go swimming in public pools, says Raj. "She can even keep her head down in the water. No nose plugs, no ear infections. She mixes with other kids. We think we've hit the jackpot. You can't know what it means to our family to see her begin to lead a normal life."

Have the treatments been a burden? "She's got no conception of what's going on here," DeSilva points out. "We told her she can still get sick, that she's not all well, and that everybody here is trying to make her healthy. That seemed to do it. She doesn't really mind this. It's the PEG shot every Friday that she hates. That's the worst thing to her."

"Now we'll have a little break," he says of the hiatus the protocol calls for. "That pleases me. It's been tough, lots of traveling, lots of work. This will give us a chance to sit down, relax, and think."

Cindy's story parallels Ashi's. "Cindy's just now starting to get better," Sue Cutshall booms in a strong voice. "It's not a big improvement yet. But to give you an example, she just got over a cold. Usually her colds end up in pneumonia. This one didn't. Moreover, it's the first time since January she's even had a cold. That's a breakthrough for her."

Bill Cutshall picks up from there. "I feel fortunate we're able to profit from this treatment," he says in his painfully shy manner. "We were under the impression, we had been told outright, that this wouldn't happen for ten years or more. In fact, when they came to us and said they were suddenly able to do it, it made us nervous. How could the estimates have been so far off?"

Nervousness was only one reason they viewed gene therapy with caution at first. "We had mixed feelings when it was proposed to us," notes Sue. "Cindy was doing okay on the PEG. The question for us was, did we want to rock the boat? We were concerned about where the new gene would land on the DNA. Would it strike a cancer gene? Only after we talked to our immunologist, Mel Berger, did our fears subside. He told us she was much more likely to get a tumor if she *didn't* have gene therapy because of her lack of immunity.

"We've become real enthusiasts. I think we'd be upset now if we'd have been turned down. The possibility of a total cure is what made it palatable to us. Also liberating Cindy from her weekly PEG shots. That's what she ultimately wants."

The girls come running back into the room brandishing candy bars. Van peels off the wrapping for them, and Ashi seizes hers exultantly. Van shakes her head in wonder. "This is more like we see her at home," she says of her daughter. "There, she's boisterous, cheeky, the earth mother of the world. Very bossy and very stubborn. But whenever she comes here, she refuses to smile. I try to make her, but she won't do it. I guess it's a defense mechanism."

Says Raj, "We were dreading the first infusion. If she froze, it would make it hard. But fortunately, that hasn't happened." He studies his daughter, who is wolfing down the candy bar while Cindy slips a new tape into the videocassette machine. "We had our doubts about the procedure," he says pensively. "It's hard being first. You don't know how many sleepless nights we had. It was a hard decision, but what else can you do? One thing that helped was the kind of people doing the experiment. They're conservative, and I like that. I trust them. Mike Blaese, I trust him. And French, he never leaves Ashi's bedside. He's there 100 percent of the time. He really cares about kids, I have no doubt about that, despite what some people say about him. Oh, I was skeptical about him initially. I'm skeptical of all scientists, including myself. I've seen too much fudged data. Everybody's got to write papers and produce in order to survive."

Says Sue, "We feel close to these men. Dr. Anderson is the one who, when Cindy gets wound up, or out of line, like when they put the huge apheresis needle in, he can calm her down. He can just sit there and put his hand on her wrist, and all of a sudden she relaxes. Nobody else can do it but him. It's very strange. And Dr. Blaese. He's very informative, and matter-of-fact. He considers you a partner. Both kids love him." Cutshall's eyes well up with tears. By the time she gets around to talking about Culver, her makeup is streaked. "Culver, we're closest to because we see him all the time." The Cutshalls have become personal friends with Ken Culver's family, and the Culvers' ten-year-old boy is one of Cindy's best friends. "We're really sold on this team. They really care."

The subject of the high cost of PEG-ADA comes up, as it often does when the two families compare notes. "When we got our first bill from Enzon, it set us back on our heels," says Sue Cutshall. "We were never told to expect this kind of money."

"They told Mel Berger it would only cost $1,000 a shot," says Raj. "We thought even that was high. But now it's $2,200 a vial, and each kid gets two vials a week. That's $16,000 a month. They sent us a letter saying they wouldn't do anything to harm their patients' well-being. Well, I want to know if reducing someone's insurance to nothing is in a patient's interest. The funny thing is they are hardly hurting. Their stock was at $4 per share and now it's $9." Raj is a seasoned investor himself, having made a fair amount of money dabbling in penny stocks, particularly those of biotechnology companies such as Genentech and Collaborative Research. Ironically, he bought Enzon stock in 1984, long before PEG, long before Ashi was even born. "I liked their management, and the fact they were a start-up company working in cancer," he once explained.

Says Sue, "I understand that they have to recoup the money they spent on research and development of the drug. But there are not very many people taking this stuff, so the full cost falls on a very few. They'll never recoup it. A lot of these kids are on Medicaid already. There will come a time when Cindy will be on Medicaid, too. We have no choice. When your insurance is used up, it's used up."

It is nearly midnight. Ashi, who has been up since five in the morning, when the family left Cleveland, curls up on her mother's lap, burying her head in Van's bosom. Cindy has turned off the television and is rather desultorily reading a book called It Was a Dark and Stormy Night. She is wearing a pink ribbon in her hair. A T-shirt with a bubble design on the front and torquoise shorts over her slender legs complete her attire.

Yawning, she closes the book. A fifth-grader, she is an excellent student, whose favorite subjects are math and biology. At the moment, her aspirations are to become a veterinarian, although because of her respiratory problems, the family has no pets except for a blue-green parakeet named Mango.

Cindy professes to be angry with her treatments because they are interfering with her softball practice. "I'd rather be home playing ball," she complains. "Instead, I have to come here." But her disgruntlement is short-lived. She grudgingly admits that it is the treatments that have allowed her to play softball at all. "I couldn't do it before."

"I need these treatments," she affirms. "I understand my condition. At first, it made me feel funny to be getting new genes. But I'm used to it now." Asked if she knows what a gene is, she screws up her nose, one of her characteristic gestures. "No, not really. Well, a little bit. Sort of."

This night, with its relaxed, almost euphoric reflections on gene therapy, was to be the end of the beginning for the first experiment. As the next few months passed, Ashi was taken off T-cell infusions to see whether the beneficial effects would continue or whether there would be a decline. It would be the proof of the pudding—or so everyone hoped—for Blaese's theory that in spite of the limited life span of T cells, some cellular memory and immune repertoire could be permanently instilled.

Right from the start, it looked as if the gene therapy had benefited Ashi. On the PEG-ADA alone, without the addition of new genes at regular intervals, her T-cell count started to descend, dropping to 1,240 per microliter before finally leveling off at about 1,300. The gradualness of the decline, however, and the fact that the count never plunged to pre–gene therapy levels, were sources of satisfaction to the Anderson team, suggesting, in Culver's words, "that the half-life of these cells is quite long." The result was auspicious enough that when it was time to put Ashi back on the T-cell therapy, in early 1992, the team felt sufficiently comfortable to spread her treatments out. Infusions would now come at three-month intervals, instead of every six weeks. Then a strange thing began to happen. Her numbers started rising again. By August, when the gene treatments were once more suspended, the T-cell count was fluctuating between 1,300 and 1,600. Nearly a year later, though she had not received any new genes since the preceding summer, her cell count had leveled off at around 1,600.

It was roughly the same story for Cindy. Taken off gene infusions in September of 1992, she was still maintaining a T-cell count of approximately 1,600 nine months later.

By any interpretation, their T-cell counts had risen substantially. Moreover, Ashi's ADA levels remained encouragingly high. But while it appeared that the gene infusions were responsible for the improvement, there remained the possibility that the gradually increasing power of PEG-ADA over the long term was the cause. There also remained a third explanation, that the gains were due to the interaction of the genes and the drug.

For his part, Culver was convinced the gene infusions, not the PEG, had done the trick. "It wasn't until they got the altered T cells that they had clear and persistent improvement," he said. He was seconded by Raj DeSilva, who noted that before the gene treatments, when Ashi was on PEG-ADA alone, her T-cell count "had hovered around 500 and her ADA levels never came up to 1 percent. I'm absolutely satisfied it was the gene therapy, no question about it. It's like a magic wand."

Still the issue seemed far from closed. So what is one to make of this research event? Was it worth the effort? Was it a triumph, a disappointing exercise, or something in between? Was it so muddled by variables—as Richard Mulligan and other scientists insisted—as to yield little if any useful knowledge?

Any innovative treatment is always judged on the basis of two criteria—safety and efficacy. The first question is, Does the new procedure carry with it any overt or hidden dangers? Here, the team seemed to have demonstrated that this particular form of gene therapy, lymphocyte alteration, posed little short-term risk. Genes were shot into Ashi and Cindy by the billions, and yet neither child, to use French Anderson's words, "exploded or turned green." As Mike Blaese maintained, no one is in a position yet to evaluate the long-term dangers. Nonetheless, the experiment silenced those who feared that the introduction of new genes into the body might invite biological disaster.

From the point of view of effectiveness, however, things remain murky. Even though the girls were taken off the infusions in the summer of 1992 in preparation for a new form of gene therapy that had been approved by the RAC, and were able to go for many months without losing immunity, the research team knew full well they would ultimately have to receive more genes. "These altered T cells won't last forever," admitted Culver. Moreover, not even the staunchest proponents of the protocol would argue, on the basis of the initial experiment, that the gene infusion should replace PEG-ADA. The PEG safety net remained firmly in place with everyone's blessing. With only a fraction of their T cells producing ADA, it seems likely the youngsters would continue to need indefinitely the intercellular protective bath that the drug provides. Michael Hershfield was right: such children were not normal, and T-cell therapy did not make them so.

In the light of this, one might legitimately question as a matter of public policy the value of an inconvenient, very expensive therapy that only augments the action of a very expensive drug. Unless one's own child is involved, of course. And unless the very expensive therapy was designed to take medicine into a new age and has a gallery stomping its feet and urging that it be tried.

Viewed that way, the experiment takes on a more magisterial quality. Even if its strongest justification was clinical improvement, not scientific advancement, that is sometimes reason enough to try an experiment. Edward Jenner proved the worth of his smallpox vaccine by testing it on young children more than one hundred years before scientists could explain why it worked. That said, it is still likely that NIH was the only place the experiment could have been attempted, as several RAC members confided privately. Cutting-edge research is the reason NIH exists, although the limits that even government scientists should be allowed to go can be questioned. Remember the "moving-target" protocol with those mysterious renumbered pages. It is likely that the experiment could never have passed muster at a university or a major hospital institutional review board. The liability issue alone would have squelched it.

In the final analysis, however, the experiment was less about treatment than about the making of history. The day the RAC gave its blessing for the ADA experiment, the RAC's chairman, Gerard McGarrity, neutral up to then, gave away his true feelings: "Doctors have been waiting for this day for a thousand years," he said, deeply moved. He was, of course, correct.

By the spring of 1992, however, the value of the gene-altered T-cell experiment was on the verge of becoming moot, for the Anderson team now proposed to take gene therapy to the mountaintop. With a flair honed by several years of courting the media, it announced to the world it was about to attempt the ultimate dream, stem cell therapy.

Whether the protocol approved by the RAC on February 11, 1992, could properly be called true stem cell therapy was in doubt, though. The team was contriving not to directly target stem cells but to vastly improve the odds of striking stem cells by casting a highly selective net. The plan called for a new technique, featuring monoclonal antibodies that would sniff out stem cells as they float through the peripheral blood. These antibodies recognize a marker on the surface of blood cells called CD34, which is present in some 1 to 3 percent of all corpuscles. This percentage of cells seems to include some, or perhaps all, of the blood-making stem cells. This is known because of a series of bone marrow transplantation experiments carried out by a biotechnology firm, CellPro Inc., of Bothell, Washington, in collaboration with the Fred Hutchinson Cancer Research Center, in Seattle. The researchers have found that on its own, with the addition of no other cell types, a population of CD34-positive cells is capable of fully replenishing the bone marrow, cranking out the complete repertoire of blood cells. While having the CD34 marker does not guarantee that a cell is a stem cell, enough stem cells do possess the marker that it can be considered a calling card. The technique had come along at the right time. Even though Irving Weissman and his group at Stanford had recently asserted that it had perhaps found a way to home in on human bone marrow stem cells, the claim had yet to be conclusively verified. CellPro's breakthrough offered a way to circumvent the time-consuming follow-up on Weissman's work.

The Anderson-Blaese-Culver strategy involves shotgunning the ADA gene into a community of CD34-positive peripheral blood cells taken from the patient's body. The expectation is that, in the process, enough stem cells will be infected with the ADA gene to cause the patient's immune system to become born again. Once more, the Anderson team seemed to be leapfrogging ahead, bringing the future to life in the here and now, even if the particular form the advance took would be something short of the real thing. CD34 therapy might prove to be better than T-cell gene infusion, but it was not exactly stem cell gene therapy, any more than monkfish is lobster.

Not surprisingly, the first candidates for the stem cell trial were to be Ashi DeSilva and Cindy Cutshall. However, in a reversal of the original order, the team was planning to perform the procedure on Cindy first. There was a good reason for this. Quite simply, the researchers could learn more from Cindy than from Ashi. Because Ashi had responded so well to the T-cell infusions, her body was teeming with gene-marked T cells. It would be difficult, following the CD34 stem cell procedure, to tell how many of her ADA-equipped T cells were already there, and how many came from the activity of a stem cell.

Cindy's T cells, meanwhile, were making substantially less ADA than Ashi's. If, following CD34 therapy, a big bump in her levels of ADA was suddenly seen, it would be pretty reasonable to infer that it was due to the action of stem cells. "We would get quicker information that way," Culver noted.

For a time, it was hoped that the clincher would come from trying the procedure on a third girl, a teenager from Omaha. The oldest survivor of ADA deficiency, next to Alison Ashcraft, she represented one of the failures of PEG-ADA therapy. Although she was the second child ever given PEG-ADA, and had been on the drug for six years, it had resulted in no appreciable T-cell production. Meanwhile, constant infection had ravaged her lungs to the point that the team at NIH, where she had been hospitalized for many months, was seriously considering a lung transplant. If what few CD34 cells could be harvested from the girl's peripheral blood suddenly sired hosts of T cells in her body, it would provide instant evidence that stem cells had been activated. There would be nothing to complicate the picture.

By the end of 1992, though, the Omaha girl's lungs finally gave out, and she died. By default, either Cindy or Ashi would have to be first.

"We're going for cure this time," vowed Ken Culver, who as usual had been working in the lab around the clock. "It's only fair. We owe it to these kids."

But in spite of the best of intentions, months of delay followed. Target dates kept getting postponed because of technical problems involving the vector. In the meantime, the Europeans used the occasion to leap ahead of the NIH team. In Italy, Claudio Bordignon performed stem cell procedures on two children, including G.B., the boy whose T cells had originally convinced the RAC that gene therapy was possible. Two children were treated with CD34 cells in France, as was one youngster in England. Not until early May 1993, however, was the experiment tried in the United States. But then all hell broke loose.

On Thursday, May 13, Cindy Cutshall was brought to NIH for CD34 cell therapy. First, the necessary cells were collected through apheresis, considerably more arduous this time around than it had been in basic T-cell therapy. Seven liters of her blood had to be processed, so that Blaese and his associates could glean enough CD34 cells. The extraction of so many blood constituents caused Cindy to become anemic. But by Saturday, when the actual infusion was scheduled, she had bounced back. Over several hours that day, hordes of gene-altered stem cells were dripped into Cindy Cutshall's arm.

As luck would have it, the same weekend found stem cell therapy busting out on the West Coast as well. But there, although the research team was closely associated with Blaese's group, the circumstances were radically different.

At Los Angeles Children's Hospital, where French Anderson's wife, Kathryn, was now chief of surgery, Anderson's former postdoctoral fellow Donald Kohn announced to the world on Sunday, May 16, that he had just tried stem cell gene therapy on a five-day-old.

Months before, using amniocentesis, Kohn's team had learned while the child was still in the womb that it would suffer from ADA deficiency. Thus, when six-pound twelve-ounce Andrew Gobea was born that week, Kohn extracted "cord" blood from his mother's placenta and proceeded to isolate CD34 cells from it (the umbilical cord is regarded as a rich source of such cells). On Saturday, the very day Cindy Cutshall was being infused in Bethesda, Andrew's cells, which had been cultured for many hours with ADA genes, were shot back into his body in a two-minute procedure. His ordeal was far from over, however. For the next six months, he was to be kept isolated in a germ-free plastic bubble until it was determined whether the treatment had been successful.

On Monday, May 18, doctors at the University of California at San Francisco reported they had performed the same procedure on another newborn. If the treatments worked, both children would grow up with complete, functioning immune systems. It would be as if they had never had a genetic flaw in the first place.

Watching reports of the procedures on the national news, Raj DeSilva could only marvel at a technology that had come so far in less than three years. His own daughter, now a first-grader going on seven years old and weighing a robust seventy pounds, was scheduled to have the stem cell treatments in August. According to DeSilva, Ashi was now so immunologically normal that she was regularly having sleep-overs at girlfriends' houses. "It's kind of scary," he said. "If we didn't have to give her a [PEG] shot, we wouldn't even know she had a problem."

Of the two babies treated in California, DeSilva exclaimed, "This is thrilling. These children won't have to go through what we've gone through. I know all too well how much heartache there's been for others and us, too."

As the year wore on, however, initial optimism over the stem cell treatments turned guarded. The efficiency of the gene transfer was proving to be disappointingly low, as measured by the number of cells with new genes to be found in the children's bloodstreams. In Cindy's case, the results were almost negligible. The two California babies showed slightly better numbers, but the outcomes were still far from clinically useful. "Either we're not hitting stem cells or the efficiency is just too poor," grumbled Culver.

By July of 1994, the ADA teams had gone back to the drawing board, trying to learn how to improve their stem cell marksmanship. In the meantime, Cindy was returned to her regimen of altered T-cell infusions. Both she and Ashi, as well as the West Coast infants, were to be on hold for a number of months, with all of them continuing to receive weekly shots of PEG-ADA. But Culver did not consider the delay to be a major setback.

"I think the stem cell therapy will work," he insisted. "It may take a year or two, but we'll get it to work."

14

AFTERMATH

Even as what remained of the NIH gene troika was gearing up to try its version of stem cell therapy, William French Anderson was doing a slow disappearing act from active participation. Professing to feel suffocated by the tedium of wetnursing the ADA experiment in the clinic on a day-to-day basis, he declared that it was high time that he returned to his first love, basic research.

Anderson cultivates within himself the sense that he is a molecular Captain Kirk, going boldly where no scientist has ever gone before. When he finds a McDonald's wrapper in a once pristine galaxy, he knows that it is time to move on. In his view, the ADA experiment was becoming a familiar star system with few new worlds to conquer.

"So I am starting over," he confided one afternoon, sitting deep in his blighted little desk chair, his hands locked behind his head. "I'm creating a totally new laboratory within my laboratory, just like I did in 1984 when I switched over completely to gene therapy."

This time, he said, the game was going to be a new vector—a vector that would not sit passively in a petri dish waiting to infect bone marrow or T cells that had been extracted from a patient. It would be a vector that could be introduced directly into the patient with a hypodermic, a vector that once set loose in the body's freeway system would pick its way to the very street, the very house, the very tissue, one is seeking to correct. In other words, it would be the brass ring of molecular biology, the "targetable, injectable vector."

"This, I believe, is the direction in which gene therapy should go,"

Anderson said, "and I think we can get there in a couple of years."

As Anderson was contemplating this quantum leap, gene therapy was already taking off in flight. And the nexus was shifting away from NIH to other institutions in the United States and Europe—a sure sign that the technology was taking hold. By mid-1992, some thirty gene therapy protocols had been submitted to the RAC and were either working their way up the ladder toward approval or had been authorized outright. By mid-1994, the number of approved protocols stood at 59, and nearly 150 patients had been treated, including a quarter of all the ADA-deficient children in the world.

For NIH, the shift of the epicenter away from Bethesda could not have come at a worse time. The great biomedical campus had been brought to its knees by a series of political and public relations disasters. Its star virologist, the AIDS pioneer Robert Gallo, was under intense investigation, having been accused of biological piracy. In an old dispute that had resurfaced, Gallo was alleged to have filched a copy of the HIV virus from the laboratory of his French competitor Luc Montagnier and then passed the discovery of the virus off as his own work, a charge he denies. The affair, which remains unresolved, reflected a gathering mistrust of the scientific establishment that had Congress questioning the integrity and veracity of many big-name scientists. Even the Nobel Prize winner David Baltimore had felt the taste of the lash. He had been forced to resign the presidency of Rockefeller University in a continuing flap over work that had been done in a collaborator's lab. But a special place of torment had been reserved for NIH, which was being keelhauled by a committee chaired by Congressman John Dingel, of Michigan, who was scouring the landscape for fraud and waste. In one of the more notorious episodes, the travel budget of NIH's parent agency was drastically slashed by Congress in 1992, in the wake of reports that more than fifty federal scientists had flown to Florence, Italy, for the preceding year's International AIDS Conference, in what some lawmakers considered a massive boondoggle. As a result of the cutbacks, and the repressive ethical climate that had descended on the once laid-back agency, a number of angry researchers were bailing out of NIH, seeking new jobs.

In the fall of 1992, NIH even lost French Anderson, although the move, in the works for many months, had nothing to do with any dissatisfaction Anderson may have felt with respect to NIH or Congress. Personal considerations dictated that he accept a full professorship of biochemistry and pediatrics at the University of Southern California Medical School and the directorship of a $30 million gene therapy institute to be housed at USC's Norris Cancer Center, in East Los Angeles.

The reason, Anderson declared, was that his wife, Kathryn, had been rudely passed over for the job of chief of surgery at Washington's National Children's Hospital, even though, as acting chief, she was directly in line for the post. In the ensuing ugliness, with charges of sexism poisoning the hospital's working atmosphere, Kathryn Anderson succumbed to the blandishments of recruiters from USC's Children's Hospital and accepted a job as surgery chief. There was never any doubt that French would go with her. "For twenty-five years, Kathy has subordinated her career to mine, and now it's payback time," said French. Cynics, though, pointed out that French's departure could, as a by-product, benefit him financially. NIH's squeaky-clean ethical machinery, with its cumbersome CRADAs and stiff-necked rules, hampered Anderson's ability to profit from either his scientific discoveries or his connection with GTI. Once at USC, no longer a government employee, he would be freer to deal commercially with the company.

But NIH's problems were not gene therapy's. The revolutionary new technology had established beachheads at an impressive number of medical institutions. For example, at St. Jude Children's Research Hospital, in Memphis, Malcolm Brenner and his colleagues were the first researchers outside of NIH to win permission to perform gene transfer. By an eerie coincidence, Brenner's game plan virtually mirrored the experiment that Richard Mulligan had considered doing on his friend Richard Parker several years earlier. A caring, contemplative Englishman, Brenner set out to use bacterial genes to mark cells taken from children suffering from acute leukemia and an abdominal cancer called neuroblastoma, the object being to learn why patients frequently relapse after transplants using their own bone marrow. Marking the cells would, he hoped, explain whether the relapse was due to residual tumor cells that had not been killed when the marrow was purged before reimplant or to systemic disease in the patient that was not eradicated by chemotherapy. Brenner could soon demonstrate that the purging was at fault, and he was working on ways to make it more effective.

Similar marking experiments, for everything from TIL cells to liver cells, were meanwhile under way at the University of Pittsburgh, M. D. Anderson Hospital in Houston, and Baylor University in Houston. In a flurry of acceptance of gene transfer, the RAC approved five such experiments at one extended, steamroller of a meeting in May 1991. Most of these protocols were not sexy, and as marker procedures, they certainly did not involve gene therapy per se. But they showed the power of the new gene transfer technology to greatly enhance medicine's understanding of the disease process. And they showed something else as well: the power of French Anderson. Shortly after the first experiments on Ashi and Cindy, Anderson made it his personal mission to shepherd through the RAC a second wave of gene therapy procedures. Accordingly, he conducted a whirlwind tour of some eleven medical centers around the United States, offering his expert counsel in the drafting of gene transfer protocols. Then, when the applications came before the RAC, he painstakingly guided them through the gauntlet toward approval.

It was "a momentum thing," he explained.

As wave after wave of protocols rolled into the RAC in 1992 and 1993, Anderson was enjoying his patriarchal role in California—"I love it here," he summed up. He and Kathy had bought a house in San Marino, near Pasadena, leaving him with a twenty-minute commute that skirted any and all freeways. He still didn't have his California license to practice medicine, but that was in the works. In the meantime, he had recruited twenty scientists and technicians for his lab while working constantly with Genetic Therapy Inc., which contributed several million dollars to the institute ("Now that I'm no longer a government employee, I'm involved in all aspects of GTI," he affirmed.) Unlike those in his tomblike quarters in Bethesda, the windows of his small sixth-floor office at the cancer center were always open ("Beautiful sunlight all the time," he marveled. "How could I have lived east for so long?") Open, too, was his door. He had put out the word that anyone with a gene therapy proposal to put before the RAC was welcome to pay a call and pick his brain. On one occasion, the physician-molecular researcher Gary Nabel from the University of Michigan had come to discuss his new idea for treating melanoma with gene therapy. For Nabel's benefit, Anderson conducted a sort of mock trial, in which Nabel was able to practice his upcoming presentation to the RAC while Anderson and his colleagues fired questions as if they were RAC members. On another occasion, he arranged for the cancer investigator Jack Roth of Houston to do the same thing at GTI. Anderson was equally accessible to his new colleagues from various local hospitals. One by one, specialists from USC, UCLA, the City of Hope, and other institutions came to chat and try out potential genetic therapies on him. Within a few months, Anderson had helped set up sixteen clinical protocols in the Los Angeles area, most of them marker experiments, to give doctors practice in working with genes.

Did that make him Mr. Gene Therapy in Los Angeles?

"Well, I would say it more modestly," he joked. "I'm Mr. Gene Therapy, period. Of the thirty-seven gene therapy protocols worldwide, I'm involved in twenty-six of them."

As usual, Anderson made no small plans, but from now on he would not be interested in small—or at least rare—diseases, either. For instance, he was extending marker technology to autoimmune disorders such as lupus. It would be easy, he concluded, to remove self-attacking T cells during flare-ups of the connective-tissue ailment, mark them with genes, and give them back in order to study the disease process. During subsequent bouts, one would be able to tell what caused the T cells to flare up. The answer might determine a potential cure. For cystic fibrosis, his wife had suggested it might be fruitful to treat patients who were about to undergo lung transplants by supplying the CF gene to the resident lungs and later, when those diseased lungs were excised, to examine them directly and determine how and where the new genes had worked. "That could tell us where the defect operates," Anderson speculated. He was also developing advanced gene therapy delivery systems for the eye (to treat retinal degeneration), for bladder cancer, for muscular dystrophy, and for brain tumors. He wanted to study the efficacy of bone marrow transplantation for leukemia and breast cancer to see whether bone marrow stem cells or peripheral blood stem cells were more effective in reconstituting immunity.

His biggest clinical push, however, was in organ transplantation. *Xenografts!* The magic word to transplanters worldwide. Grafts of animal organs offer humanity a way out of the organ shortage that kills thousands of people each year while they wait vainly for human donors. Anderson thought he could eventually alter genes to make animal organs work in people. *Donor-specific tolerance* this is called. By switching off certain surface antigen genes that cause rejection, the human body could be induced to ignore the presence of a baboon liver, heart, or kidney, he believed. What a blessing that would be! In fact, confided Anderson, his new dream lab in the planning stages at USC would ultimately occupy an imposing 250,000 square feet, and half of it would be devoted to xenograft research and the other half to neuroscience.

If the tag-along gene therapy marker experiments that followed in the wake of the first NIH experiments seemed to lack the grandeur that one hoped for from the new technology, far bolder concepts awaited. By the summer of 1992, two more trials of full-blown gene therapy were under way, this time at the University of Michigan's Howard Hughes Medical Institute, which was fast supplanting NIH as the hub of the emerging science. As the year closed, NIH reclaimed some of its lost glamour with an audacious experiment designed by none other than Ken Culver.

One of the Ann Arbor experiments took gene therapy to a higher plane. It involved the first transfer of genes to an organ other than the bone marrow, in this case the liver. The experiment was the brainchild of James Wilson, a protégé of Richard Mulligan and a leader of the second generation of gene therapists now seizing the reins. Wilson, a Michigan alumnus, had followed up a medical residency at Massachusetts General with postdoctoral work under Mulligan at the Whitehead. His consuming passions there were the perfecting of vectors to carry the ADA gene and the trucking of new genes into the liver.

Wilson's introduction to ADA deficiency had come in 1987 when he and Mulligan flew down from Boston to a meeting of the RAC, to hear Anderson propose to treat the disease. "On the plane back," he recalls, "we said, 'We've got to fix this. There's something wrong here if this is the state of the art.'" Thus began a period during which the tall, strapping Wilson, who bears a resemblance to the actor Christopher Reeve, down to a pair of Clark Kent–style glasses, tested more than eighty vectors in trying to solve the problem of getting genes into bone marrow stem cells. Like everyone else, he came up empty.

The liver work was another story. Wilson admits he got into the liver field for reasons that were "not totally noble." It was an area where there appeared to be little competition, and Wilson figured he would be able to make a name for himself. Not that the liver was boring. "It's a nice organ," he says.

From the first, Wilson worked with Mulligan on a peculiar problem involving the liver: how to cure, using genetic techniques, a deadly disease in patients who lack the necessary receptor for clearing cholesterol from the bloodstream. Victims of the disease, familial hypercholesterolemia, are born without either of the two genes necessary for making the receptor, which is produced in the liver. Without these tiny vacuum cleaners, which allow the liver to process what are called low-density lipoproteins, or LDL, the body cannot effectively dispose of cholesterol, and it begins building up in the arteries in prodigious amounts.

In 1991, Wilson used a unique strategy invented by a colleague, George Wu, to treat the disease. He injected rabbits with LDL receptor genes that had been tightly bound with sugar, which the liver soaks up eagerly. He thus tricked the liver into absorbing their gene package, without having to resort to viral vectors. Within minutes, the genes had gained entry to liver cells and, before long, were manufacturing LDL receptors in quantity. The results were astonishing. The rabbits' LDL cholesterol dropped by 50 percent.

Although a good friend of Mulligan's, Wilson worked closely with French Anderson over the course of the next year to gain RAC acceptance of an attempt to move the strategy into human beings. By late spring 1992, approval in hand, he was ready to move. On June 5, he and his colleagues went into the first of three patients with familial hypercholesterolemia so severe that their cholesterol readings were greater than 700. The two other patients soon followed. The plan of attack involved surgery. Up to a third of each patient's liver was removed, and after extraction, the liver cells were exposed to retroviral vectors that piggybacked in the gene for the LDL receptor. The corrected liver cells were subsequently injected into the liver's portal vein, where they were expected to implant themselves.

The second experiment at Michigan's Howard Hughes Medical Institute focused on a different target—cancer. In the wake of Steven Rosenberg's experiments with TNF and cancer vaccines, and having survived Anderson's mock prosecutorial grilling, Gary Nabel, a Wilson colleague, won easy approval from the RAC for a more straightforward kind of immunotherapy for advanced melanoma. He planned to inject genes directly into the tumors in hopes of immunizing the body against its own cancer. The key would be the gene for HLA-B7, one of the antigen proteins on cell surfaces that allow the body to recognize homegrown tissue as its own and that of others as foreign. "The antigen serves as a flag for the immune system," says Nabel. "When the system sees a foreign version of an HLA antigen, it responds with killer T cells and eliminates any cells with the protein on it." HLA-B7 was chosen, by the way, because only a small fraction of the population has that particular antigen, making it a foreign protein to the great majority of patients.

The idea was simpler, and more direct, than anything tried by Rosenberg. Instead of taking out the tumor, treating it, then putting it back, while draining a patient's lymph nodes to obtain TIL cells, Nabel proposed simply to inject the antigen into the tumor and see whether the body zeroed in for the attack. Eventually, he planned to treat twelve patients, each suffering from melanoma. As a precaution, the test subjects were required to have been rendered sterile already—either infertile or postmenopausal—for fear that the injected DNA might break free from the tumor, migrate through the bloodstream, and find its way into one of the patient's sperm or egg cells. There it could conceivably be passed along to future generations, a remote possibility but one that had to be dealt with.

On June 4, 1992, beating his colleague Wilson by a day, Nabel treated his first patient, a sixty-seven-year-old Michigan woman whose body was riddled with melanoma. The genes were wrapped not in a virus but in tiny envelopes of fat called liposomes, which allow DNA to pass through the outer membrane of cells. It was the kind of protective sheathing that the inventors of PEG-ADA had used. The Nabel team injected the genes directly into a tumor nodule about an inch wide under the woman's left arm, performing five infusions before standing back to let things happen. "What we were essentially trying to do was to mimic the process of organ rejection in an artificial way," says Nabel. "We're asking the immune system to do what it does but do it better."

Once inside the tumor, the liposomes were to penetrate some of the malignant cells and off-load the DNA, which would proceed to manufacture the HLA-B7 protein. As in normal cells, the protein would then make its way to the membrane and begin protruding from the tumor cells, signaling the immune system to begin killing the cells. Ideally, the immune system would not only massacre cells in the main body of the tumor but also track down metastasized cells everywhere in the body, so long as they, too, bore the antigen. There was strong reason to think that the killer cells might even be stimulated to go after tumor cells that had not picked up the HLA gene. "In our studies in mice, we were able to induce the immune system to recognize even unmodified tumor cells and respond vigorously," Nabel notes.

Nabel, a Harvard product who did his postdoctoral work under David Baltimore, moved a week later into his second patient, a sixty-eight-year-old man. Over the next few months, he would do three more. Early results were inconclusive. In only one patient was anything resembling a positive response noted. But Nabel had a ready explanation. The number of cells altered by needle injection was too small to generate much immune activity. To improve the result, Nabel announced, he was switching to a new delivery system for the next round of patients. Thereafter, the HLA antigen genes would be shot into the tiny blood vessels of the tumor via extremely thin catheters. "This ought to improve the technology pretty substantially," Nabel predicts.

Also on the drawing board were plans to extend his novel immunotherapy approach to other sorts of cancers, most likely those of the colon, breast, and kidney. "In some ways, this strategy should be universal," Nobel says. "It could work on any tumor we can get at."

Nabel, whose wife, Elizabeth, is also a gene therapist at Michigan, experimenting in cardiovascular diseases, is an unabashed booster of the new technology. "This is the golden age of biology," he says. "In the same way that Einstein and Planck and Heisenberg were around for the golden age of physics earlier in this century, we are present at a very special point in time."

But it was Culver's experiment that caught the fancy of everyone in the field. Conceptually, it was daring, improbable, and unlike anything that had come before. For one thing, it involved the first direct injection of retroviral vectors into the bodies of living patients. Before, the eccentric vectors, whose safety remains a diminishing but still-real question mark, had been allowed to infect only tissue removed from patients for return to the body later. Of arguably greater long-range significance, the experiment marked the first attempt to perform gene therapy on a disease involving that daunting frontier—the human brain, a place where even Anderson feared to tread. It thus stood to blaze a new trail for the infant field, one of huge and untold potential.

The disease Culver had set his sights on is known as glioblastoma multiforme, a tenacious cancer that accounts for 55 percent of all brain tumors. Largely incurable, it kills three-quarters of its victims within one year of diagnosis. Radiate it, and it resists. Remove it by surgery, and it grows back a few centimeters away. Attack it with chemotherapy, and the omnipresent blood–brain barrier slams shut like a portcullis, refusing the anticancer drugs entry to the brain by constricting the tiny capillaries of the neck.

Culver's strategy for treating the dread disease, developed with help from his mentor Michael Blaese, bore a superficial resemblance to Nabel's immunological approach to melanoma. Like the Michigan experiment, it contrived to set the tumor up for a contract hit. But there the similarity ended. Culver's rather antic idea was to harvest skin cells from mice and equip them with viral vectors containing a gene native to the herpes simplex virus. Herpes simplex is the bugaboo that causes genital herpes and cold sores. Of the virus's small array of genes, the one Culver wished to put into the mouse cells was its version of thymidine kinase, a viral protein. Once outfitted with this new cargo, the mouse cells would be shot directly into a brain tumor by means of a special instrument that fits on the head and facilitates the computer-guided drilling of a tiny opening in the skull. The device lets neurosurgeons target specific sites in the brain with an accuracy of one millimeter.

As the tumor grew, reasoned Culver, the mouse cells would multiply and begin seeding the tumor with vector virus. These vectors would, in turn, insert their payload, the herpes simplex gene, into the very heart of the tumor cells. Critical was the fact that no healthy brain cells would be affected. Being retroviruses, the vectors would penetrate only dividing cells. And it is a hallmark of neurons and other brain cells that they do not divide. Tumor cells, of course, divide exponentially. That's what makes them tumor cells.

It was at this point that the craft behind the strategy became apparent. Inserting a herpes gene into a tumor's network of cells has the effect—in theory, at least—of making the tumor genetically similar to the herpes virus. For all intents and purposes, it transforms the tumor into a giant cold sore.

But why should anyone go to such absurd lengths? Because the thymidine kinase gene is the DNA equivalent of a Trojan horse. It is known in the gene therapy trade as a "suicide gene," one of a number of genes in nature that queerly invite ruin upon their owners. The thymidine kinase gene makes the herpes virus biochemically vulnerable to the potent antiviral drug ganciclovir.

The implications were awesome. By infiltrating the tumor with the suicide gene, the researchers just might be able to treat brain cancer with a drug normally used to cure genital herpes. Assuming the hypothesis was correct, ganciclovir would mistake the tumor cells for herpes lesions and start annihilating the cells. Culver and Blaese could only giggle insanely at what they were trying. It was a long way from trifling with some of Ashi DeSilva's T cells!

The idea, far-out as it seemed, nevertheless had a certain logic. And so, very quietly, Blaese and Culver and a few National Cancer Institute associates began trying it out on rats. No sooner had they treated a few animals than it became apparent that their wildest hopes might be fulfilled. Large brain tumors, which had been induced artificially in the rodents, began to shrink. As if that hadn't been enough, the team members straightaway got another shock. It was not necessary, they discovered, to infect all the cancer cells with the suicide gene to get a high mortality rate among the cells. Even tumor cells that had not been imbued with the viral vectors began unexpectedly shriveling up and dying. This mysterious phenomenon, which left healthy tissue unscathed, added enormous leverage to the treatment, doubling and tripling its effectiveness.

It is one of the perks of being a molecular biologist that you get to name things, as astronomers do the heavenly bodies that they find. After combing around a bit, Culver and Blaese came up with a catchy moniker for the new phenomenon. They christened it "the bystander effect."

By any name, it seemed a godsend. Marveled Culver in the days that followed those first experiments, "We're actually only able to infect about 30 percent of the rat tumor cells, but we can destroy the whole tumor anyway. It's the bystander effect that does it. It sees to it that surrounding cells without the thymidine kinase gene also die."

Using the new technique, the Culver/Blaese group could completely eradicate tumors in eleven out of fourteen cancerous test animals it treated, an astonishing 80 percent of them. The three remaining rats were left with only small pockets of residual tumor. What's more, the cured rats were still alive three months after treatment, equivalent to a five-year survival in human beings. Untreated control rats, on the other hand, all died within four weeks.

So much for rats. It was time to try it in people. Joining the rush of applicants seeking RAC approval for human gene therapy trials, the Culver/Blaese team leaped to the front of the pack. The RAC, impressed by the startling animal research, expeditiously granted the group approval for an experiment in twenty brain tumor patients. Ten of the patients were to have tumors that had originated in the brain. Ten more were to have primary tumors elsewhere in the body that had traveled to the brain. Culver and Blaese's allies for the experiment would be Edward Oldfield, chief of neurosurgery at the National Cancer Institute, and his colleagues. Oldfield's expertise with brain tumors was needed for the delivery of the genes and evaluation of the treatment's progress.

On the eve of going into the first patient, Culver commented, "This strategy could make an enormous difference. If it works, it could change cancer therapy forever." Indeed, Culver already had plans for adapting the treatment to other cancers, including those of the liver, breast, and kidney. Nevertheless, even he was not blind to the bizarre connotations of what he was trying to do. "It sounds crazy," he admitted. "We've got government approval to put mouse cells that ooze infectious virus into people's brains. It's the craziest thing alive."

Crazy or not, within a week after the first patient—an elderly man with kidney cancer that had spread to the brain—received the herpes gene, followed a week later by ganciclovir, his brain tumor began to shrink. Down, down it went, by as much as 90 percent, as the two weeks of ganciclovir treatment continued. For a few days' time, it looked as if the field of oncology might be witness to the Second Coming. Alas, early optimism had to be tempered when the mass inside the patient's skull began to grow once again. Magnetic resonance imaging and PET scans suggested that the growth was caused not by a resurgence of the cancer but by a swelling response to the treatment and by the accumulation of dead cancer cells. Indeed, it soon went back down, and several months later the patient was doing "really well," according to Culver.

But when five more patients were tried over the next few months, only one of them had a favorable response. In the four others, the experiment showed no signs of working. A somewhat chastened Culver returned to the drawing board.

As others were concentrating on bringing gene therapy into more frequent, if far from routine, practice, French Anderson was still looking at the big picture: his targetable, injectable vector. Coming up with such a vector would constitute a grand slam home run, for the prospects of gene therapy would be enhanced a thousandfold by the ability to aim at specific organs. Of all the hurdles that must be surmounted before gene therapy realizes its full potential, the most challenging is the inaccessibility of most disease sites to gene delivery. It is this that renders the technology impotent against all but a handful of genetic diseases. Practitioners are currently limited to those conditions whose milieus are the extractable tissues—the blood, the marrow, the cells of the liver. Dangling beyond reach are the in situ tissues, those homebody structures that may not be removed, even temporarily, without deleterious consequences: the brain, the muscles, the heart, the kidneys, the stomach, lungs, eye, and spine, to name a few. Even if one could safely collect a piece of, say, brain for exposure to a gene-carrying virus, and then somehow put it back in its space, as one would reshelve a book, the resultant improvement would be confined to that small theater, with no effect on neighboring tissues also in need of remediation. One would have to excise hundreds of contiguous sections of brain, sewing them back in like quilt pieces, in order to make a dent in the illness, a Sisyphean task indeed. Likewise with the muscles, that vast network of elastic matter. A muscle disease, such as Duchenne's muscular dystrophy, would preferably be attacked globally in order to achieve a total cure— although some researchers are claiming success with a painstaking, piecemeal approach.

What a boon it would be if it were possible to send in the clones, as it were—to just shoot genes in aboard vessels that instinctively know how to take the genes to their ultimate destinations! But how would one accomplish such a bewildering feat?

Several means have been suggested. The most frequently mentioned is the so-called tropic virus, the magic bullet of beasties, a creature with a natural predilection for certain body parts. The gene therapist, in this scenario, is cast in the role of a molecular sommelier, choosing viruses for the occasion as if they were dinner wines. Thus, if one were interested in delivering genes to the brain, to treat Parkinson's disease, for example, one might consider using the rabies virus to convey genes for dopamine, the protein deficient in the ailment. Rabies virus is neurotropic—that is, it gravitates to the tissues of the central nervous system. Or if one needed to target genes to the gastrointestinal tract, in the cause of treating, say, colon cancer, it might be feasible to employ the poliomyelitis virus. The polio virus is ingested orally, and enters the body via the digestive system. Treating respiratory conditions, such as cystic fibrosis, is expected to call for vectors that favor airway passages— namely, adenoviruses, those blackguards responsible for the common cold, or their brothers-in-misery, rhinoviruses. There are viruses that favor the myocardium, or heart muscle, and others that congregate in mucosa, the mucous membranes. A viral cargo vessel suitable to almost any task can be envisioned. The approach also offers a bonus: many of these viruses are DNA viruses, much larger than retroviruses and able to contain large genes that retroviruses cannot—such as the gene for the protein deficient in muscular dystrophy.

The idea of using the likes of rabies or polio virus to mitigate disease understandably causes most people to shudder. It is difficult to imagine organisms responsible for some of the most horrible afflictions known to medicine being deliberately introduced into patients. Yet it must be kept in mind that, like any viral vector, these viruses would presumably be rendered innocuous in the process of outfitting them to carry therapeutic genes. Theoretically, they would have lost the ability to induce and sustain infectious disease. We say "theoretically," because there is some evidence to suggest that many of us carry around fragments of certain viruses in our systems, bits too small to cause any harm, bits that may have allowed us to gain immunity to the parent virus. Then, too, almost all of us retain viral genes in our cells deposited there during previous bouts of influenza, colds, and the like. Because these viral odds and ends are swimming around inside us, the possibility exists that they may supply the raw material needed to reconvert crippled vectors into actively infectious ones. The prospect is frightening enough that researchers are proceeding slowly.

Anderson has been examining a different strategy. His approach calls for isolating the region or regions on a virus that make it favor one variety of cell and not another. These sections of virus are called the host range and are arrayed along its outer surface, or envelope. No one has yet identified the precise location of this host range, but Anderson, after several years of study, claims to be on the verge. He believes that it will eventually be possible to use one universal, all-purpose virus as a vector, but creatively tailor that vector to the needs of therapy by slapping in different host ranges in modular fashion, as one might change the lenses of a 35-mm camera. It won't be quite that simple—a number of regions may have to be modified simultaneously—but conceptually the task is similar.

Up to now, he has been investigating which portions are pertinent, which portions to snap in and snap out. To test his theories, he is using a mouse retrovirus, removing a region that makes it grow in rodent cells and replacing it with a region that makes it thrive in other cells, but not in rodent cells. The question to be answered is whether by modular refitting one can modify a virus's receptivity to its environment. Once Anderson is certain that he has the region that controls viral adaptability—and he claims he is already there—he will then be in a position to slot into his retrovirus, the Moloney murine virus, two new host ranges. One of these is from the hepatitis B virus, which makes its home in human liver cells. Thus, Anderson hopes it will confer an affinity for the liver on his vector. The other is from the HTLV-1 virus, which has a taste for T cells. If his theory is right, the vector will turn away from liver cells and begin infecting T cells.

Tinkering with host ranges is only half of Anderson's self-imposed mission. The other half is making his dream vector injectable. Instilling this quality is not a mere matter of loading the virus into a syringe and pressing down the plunger. In most cases, a virus that is cast adrift in the bloodstream is very quickly lysed by the immune system, specifically by an agency of the blood called complement. In one experiment, 22 percent of a monkey's blood volume was replaced with pure, live virus. "The monkey never even blinked," says Anderson. "He lysed it all."

But, if you know the trick, then . . . . Anderson contends that he has discovered a surface marker on the viral envelope that can be switched off, making the virus effectively invisible to the immune system and therefore unlysable. Stealth technology applied to vectors. "We think we know exactly where the marker is," he says, confidently. The phenomenon already exists in nature. The HIV virus that causes AIDS somehow evades the immune system, traveling through the human system incognito. "The system doesn't even know it's there," says Anderson. He has learned to change the protein sequence in that site by one amino acid such that complement will no longer bind to it.

The term "targeting" as it applies to gene delivery does not refer only to the ability to steer the vector to a given type of cell. It has also come to embrace what is known as "site specific integration": namely, the capacity of the vector to deposit a therapeutic gene at the exact site within the nucleus where it is supposed to go—that is, at the appropriate location on the appropriate chromosome, the precise place where the gene occurs in a healthy individual. A virus that the body cannot fend off, no matter how well it has been engineered to be safe, would still be potentially dangerous if it were not tooled to home directly to its proper site within the genome. Otherwise, the potential for triggering oncogenes or disrupting the activity of other vital genes would be substantial.

French Anderson says he is approaching his vector research in the same right-brain way he has attacked all previous challenges. "What I do is I try to get inside the retrovirus intellectually, to get a feel for its three-dimensional structure," he says. "I just go through all the previous studies of viral envelope, asking specific questions about envelope function, how it behaves, and I put all this information in my subconscious, and my subconscious just keeps working on the problem until I get inside that molecule, and I'll feel how it works. This is the way I've done things before."

What he is stuffing into his subconscious is formidable stuff. He has accumulated stacks and stacks of computer printouts of several different viruses, and different mutations of each virus, showing their envelope (encapsulating) membranes and various regions along those membranes—regions with technical names such as the "leader" region, the "cytoplasmic tail," the "polypropylene hinge." These structures are depicted in linear profile, with ups and downs reminiscent of a Dow Jones chart. Anderson's purpose is to get profiles for as many different regions as possible and then to match all the printouts together, line them up, as it were, to see whether there is any similarity among the viral strains indicating where the host region might be. He taps one of the printouts, which are spread out on a long table in his office like blueprints. And in a way that is what they are. "What you notice is that parts of these molecules are very similar. Some are very different. But if you live with these things long enough, you get a feel for what this part, or this part, is doing, and how this interacts with this. I've become convinced," he says, pointing to a specific sequence of membrane, "that these pieces are homologous in all these variants, even though they don't look it. And that tells me how to go about putting pieces of this molecule into one of these other molecules in order to turn a mouse host range into a human host range.

"It's a matter of living with this data and just having it all inside the computer up here"—he touches his forehead. "I wake up at night and think about this. And now I need to build up a group of assistants who are doing this, too, and GTI is going to convert more and more people to doing it. I think with a little bit of luck I'll be able to get into a monkey with an injectable, targetable vector in six to twelve months."

Anderson is defensive about his scientific skills and his methodology, to the point of incessantly explaining himself and his offbeat, lonewolf ways. He tends to romanticize himself as a sort of beatnik scientist, drifting from problem to problem. "This is pure science, and it's the kind of thing I do. But I don't get up and talk about it. And when I finally accomplish it, I'll simply have something that works, and the Mulligans will quite legitimately say, 'Ah, he didn't do anything.' That's because I don't talk about it. I don't go to meetings. I don't talk to people. I don't do what scientists are supposed to do. I just do the science.

"I apologize if this sounds arrogant, but I just don't learn from other people. They're still worrying about yesterday's problems. It upsets people very much that I don't do all the things politically that one is supposed to do. And then I come up with an end product and people say, 'How'd he get that?' And then as soon as it's working, I leave it, and they say, 'This guy's gotta be crazy.' "

One of his visitors points out that he is not a virologist. "I am now," he protests. Then, contradicting his assertion that he never learns from other people, he relates how he has become a viral expert through the offices of Eli Gilboa, Dusty Miller, Ed Skolnick, Janet Harlet, Alan Ryan, and Robert Gallo.

So you are the Lone Ranger of genetics?

"Yeah, but I occasionally find people I can learn something from, and then I interreact with them. Rosenberg I still learn from, because he's great. Blaese I learn from, so I maintain that. Most people don't really learn anything, because they're still thinking about today's problems. Today's problems are already solved. What I want to do is open up new areas, and once it opens up and other people get into it, I'm happy.

"My ultimate goal for the past thirty years has been to develop gene therapy and revolutionize medicine. I've only got about twenty years left. I'm in my midfifties. I'm not going to go on forever.

"Regardless of what the Mulligans say, I'm a pure scientist. I don't care about getting feted. I don't really care about the Nobel Prize. If I get it, I'll go to Stockholm, but I don't really care one way or the other."

Anderson's right-brain kind of science has been a part of him since childhood. "I don't know how I got this way. All I know is my conscious mind is fairly dumb, but my subconscious is really very good. And that's why it doesn't do me any good to talk to other people, because they're not talking about the things I'm thinking about.

"I've done everything with my subconscious mind since I was little. I drift off. It used to drive my wife batty because we'd be in conversation and I'd be gone. She'd have to sort of get me back. It used to get me mad when she'd do that, and then finally I recognized I had to make some accommodations.

"See, if a person is musical it's okay for them to practice the piano twelve hours a day and have no other interests. People say, 'Bravo.' But if a thinker does something on the same order, he's some sort of nut."

But is he a nut? Anderson thinks about the possibility. "There is certainly an element of not intentionally going with the flow," he decides. "Like, I don't drink coffee, or cocktails at parties, you know. I finally realized in the last few years that some of that was just being obstinate for the sake of being obstinate, so I taught myself to drink coffee and take cocktails. I don't drink them, actually, but I sort of hold them.

"But eventually, when I'm doing what everyone else is doing, I start thinking what everyone else is thinking, and then I realize that I'm not thinking. That means I'm not doing anything, and when that happens, I start a revolution.

"That's what happened in 1975, when I realized I was thinking like everybody else, and just churning out papers. And so I just quit."

Anderson, who was born into a Presbyterian family and still attends church, professes to be religious, but in the "internal" sense, "not in any public religion sort of way." He elaborates, "I believe in God. He basically guides me, and that's why I'm not worried about what humans say; I do what I'm supposed to be doing." Anderson believes, at some level at least, that he is acting as God's instrument. "My rational mind thinks that's all hogwash, but in another part of my mind, yes, that's exactly what I'm doing."

His earthly idol Immanuel Kant is never very far from his thoughts. Anderson is currently working on a reinterpretation of Kant's categorical imperative, about which he started to read in his student days. "When I publish a pure philosophy paper, it's gonna blow people away," he vows. "They're going to say he's got no business writing about philosophy, but I've thought about Kant now for more than thirty years, and he can be reinterpreted in a way that provides a real basis for what we do as human beings."

Anderson sees no inconsistency in a scientist whose muse demands a kind of mystical union with the cosmos, a free-floating, Buckminster

Fuller–style unconsciousness, finding an icon in a philosopher most associated with the idea of natural, inflexible moral obligations. "That's apples and oranges," he says. "That's a misinterpretation of Kant. He wasn't rigid. He's just interpreted as being rigid. I've read and reread him so many times now. You have to reread him, you see; it's like the Bible. You have to put Kant into his framework. His grounding in metaphysics and morals was published in 1785. You have to put it into the framework of what he knew about the world back then. Kant had this awe about the universe. There were two things that overwhelmed him. The universe above him and the moral code within. That's how I feel."

The week that Anderson and his colleagues performed their first gene experiment on Ashi DeSilva, Anderson tacked on his bulletin board the news account of his speech in 1968 in which he predicted that gene therapy was just around the corner. Sort of a mind game with his confreres. Only two people in his entire contingent noticed that the date on the paper was June 1968. Everyone else thought that the speech had been made only a few months before. One of those who did detect the time difference was his secretary. But rather than praise his foresight, she fixed the World's Greatest Gene Doctor with a quizzical look and said, "You mean, you haven't had a new idea in twenty years?"

PART II

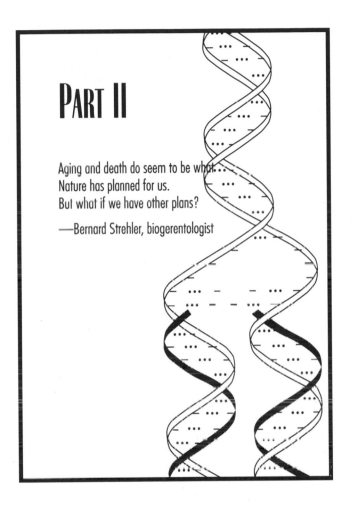

Aging and death do seem to be what
Nature has planned for us.
But what if we have other plans?

—Bernard Strehler, biogerentologist

The early gene therapy experiments were merely prelude to what will come—indeed to what already has come.

The second part of this book will explore the ways in which gene therapy is refashioning the practice of medicine, extending its power against a huge array of diseases affecting both mind and body. It will also consider how gene therapy may ultimately extend the human life span and perhaps even influence the course of human evolution.

15

A MATTER OF THE HEART

Jan Rapacz's laboratory at the University of Wisconsin is literally a pigsty. To see this spare, bespectacled Polish émigré clutching one of his prized porkers against his neat, lambswool sweater, his affection unabashed, is to swear he was once one of those countrified veterinarians who are forever writing poignant books about the Lord's wondrous bestiary. Rapacz is anything but that. He is, in truth, one of the world's most sophisticated researchers into cardiovascular disease. And his pigs are far from ordinary pigs. They are the rarest of the rare, genetically speaking. They are mistakes of nature, and within their pink loins lie answers to a murderous clan of ailments that are the primary killers of human beings.

For fifteen years, Rapacz painstakingly scoured the most distant corners of this planet and tested the genes of more than fourteen thousand swine, from domestic breeds to wild pigs caught in the Polish forests, before he found what he was looking for: a mutant strain with a unique genetic defect that makes the hapless pig a wallowing time bomb. Because of a defective gene, the pig is unable to clear cholesterol from its blood. Consequently, deadly fat deposits inexorably build up in arteries throughout its body in ways that can tell investigators such as Rapacz how the same process leads to heart attacks and strokes in people with similar genetic flaws.

Any farmer who knew ahead of time what Rapacz's pigs were like would instantly destroy them if they showed up in a litter. But Rapacz treats his animals, which were originally found on a European farm, like royalty. They receive around-the-clock care on a spotless research farm in Arlington, a community outside of Madison, where they live out their brief destinies in cleanliness and comfort, all the while aging at quadruple the natural rate. Whereas the normal life span of a hog is twelve to sixteen years, Rapacz's charges die of sudden, catastrophic heart attacks in only three or four years, even when fed a low-fat, zero-cholesterol diet. One of his boars, "ancient" at the age of five, turns blue every time he mates. Rapacz, in fact, willingly helps the winded Lothario do the deed by holding him in place: the researcher wants those mutant genes passed on.

Rapacz believes that his pigs offer a direct window to understanding the relationship between cholesterol, genes, and cardiovascular disease. "Pigs and ourselves are not really all that different genetically," he notes with his habitual high-pitched giggle, gently cradling one of his squealing piglets, whose cholesterol reading is a formidable 400. "Pigs can tell us a lot."

The role of genes in heart disease and stroke is by now universally accepted, thanks to a variety of studies. But the precise way that one's hereditary profile dictates susceptibility or nonsusceptibility to these ailments is still shrouded in mystery. A number of genes, most of them as yet unidentified, are involved. Some affect the processing of cholesterol and other fats and are probably of paramount importance. Others determine one's proneness to such risk factors as high blood pressure, diabetes, and the tendency to develop clots. All of these genes unquestionably interact with lifestyle—how one eats, whether one smokes, how often one exercises, and how well one deals with stress. Sorting out these interrelationships is a complex task. The muddle has forced a shotgun approach to disease prevention. Unable to tell who is genetically likely to get heart attacks and strokes, researchers must conduct studies retrospectively examining behavior patterns—one study actually tried to link the failure to get regular haircuts to a type A personality and the risk of heart disease. A series of rather wacky and inconclusive studies attempted to associate cardiovascular problems with visible clues—the incidence of male pattern baldness, for example, or the presence of creases in the lobes of the ears. In the meantime, doctors must caution everybody to behave as if he or she were a prime candidate for a fatal heart attack. As the Nobel Prize–winning heart researcher Michael Brown says, "Every disease is an interaction between genes and environment. Sunburn is a genetic disease. Two people can get the same exposure, and which of them becomes burned is a genetic trait. Black people, for instance, are born with a sunscreen value of four. Some are going to

fall prey to its effects, and some are not. But whereas we know beforehand who is sunburn prone, we don't yet know in any broad sense who is
cholesterol prone. So we have to treat everyone as if they are, and warn
them all of the dangers of eating too much fat."

The paradoxes here are plentiful; the numbers tell the story. Only 50
percent of all people with high cholesterol levels—defined as more than
200 milligrams of cholesterol per hundred cubic centimeters of blood
plasma—will ever get coronary heart disease. On the other hand, one in
four people with low cholesterol readings—that is, a count under 200—
is also destined to have a heart attack. Such persons can eat oat bran
until they are brimming with it, and it will not help. But who is in these
subgroups it is at present impossible to tell. Clearly, genetic factors are
at work, "and probably five or six major genes will explain a lot of it,"
says Peter Kwiterovich, chief of the atherosclerosis research unit at
Johns Hopkins.

Jan Rapacz is concentrating on a key gene, the gene for a
substance known as apolipoprotein B, or apo B, which squires most of
our cholesterol around the body. As Rapacz was the first to prove beyond
dispute, defects in the apo B gene are directly associated with high cholesterol levels and heart attacks. Rapacz's discoveries were made in pigs,
but they have since been found to apply to human beings as well. In fact,
there is reason to believe that apo B–related problems may be a primary
cause of atherosclerosis in the general population.

Rapacz's career has been an uphill battle against war, politics, and
the skepticism of his scientific colleagues. The son of a famous Polish
horse breeder, his idyllic childhood on the family's farm was cut short
by World War II. When the war ended, in 1945, Rapacz was fifteen years
old and could neither read nor write. "My teachers told my dad to take
me back to the farm and let me be a shepherd," he recalls. "That's all I
was good for, they said." Instead of meekly pursuing a career in she-
pherdy, Rapacz fought his way to the university, studied like a demon,
and boldly and presciently entered the infant field of immunogenetics.
When one of his early papers on the genetics of the immune system of
rabbits was published in *Nature* in 1960, it was spotted by the dons at
the University of Wisconsin, who invited him to their campus as a visiting professor. They weren't so much interested in rabbits as in young Jan
Rapacz. "At the time, it was extremely difficult to leave Poland," he
points out. "The government would only select a few people to leave,
and I was not even a Communist party member. So my old professor had
to put his signature on a paper saying I would visit the U.S. and then

come back." He winks. "Yeah, sure. I was coming back."

Two years later, he was extended a standing invitation to join the Wisconsin faculty in Madison, having impressed his hosts by doing something that was as commercially gratifying as it was offbeat: he discovered the cause of a serious blood disorder in minks that was ravaging the mink-ranching industry. After a brief return to Poland, he went back to Wisconsin and embarked on a three-year study of immunity in cattle. It was then that a colleague broached the idea of working in pigs. "I said, 'No, they smell like hell,' " Rapacz recalls. "But he said, 'Look, as a geneticist, you'd have whole litters to work with. You'd solve your scientific questions faster.' That was the convincing factor. He was right. I couldn't say no." Thus was born an intimate relationship with the porcine world that has endured for more than two decades. Rapacz has worked with every kind of pig, from the largest wild boars to the Yucatán minipig. He has turned hog genes inside out in an effort to find mutations linked to cholesterol accumulation.

Interestingly, not all of the genetically aberrant hogs have an inability to clear cholesterol. Rapacz's global quest has also yielded a "superpig" with a gene that makes it the polar opposite of his pigs on death row. The superpig's body chews up cholesterol with such remarkable efficiency that it is invulnerable to heart ailments no matter how much fat it consumes.

The study of such animals is eventually going to determine why some people can smoke, drink, and eat anything they desire and never have a problem, while other assiduously observe all the dietary rules, quit smoking, exercise resolutely, reduce stress levels, avoid sodium, and nevertheless succumb to heart disease and strokes. The arbitrary nature of it all is illustrated by the case of Richard Lewis, a retired New York junk dealer. Every month, Lewis routinely consumed three dozen eggs, a half pint of salt, fifteen pounds of sugar, thirty pints of wine, innumerable cups of coffee, slabs of fatback, and bread fried in a bed of grease. When he died, at the age of 105, his total cholesterol count was an admirably low 165. Like Rapacz's superpig, Lewis seems to have been blessed with genes that enabled him to eat with impunity prodigious quantities of foods that for other people would constitute a death warrant.

Cholesterol is, as Michael Brown and his collaborator and fellow Nobelist, Joseph Goldstein, have put it whimsically, "the most highly decorated small molecule in biology." No fewer than thirteen Nobel prizes have been bestowed on scientists who have concentrated on this waxy, white, fatty substance, which was first isolated from gallstones in 1769 by the French chemist Poulletier de la Salle. Cholesterol is no idle passenger in the body, visited upon us as a biochemical pest without which we could do very well. It is a vital substance, necessary for life. So important is it to survival that we make our own supply. Up to 75 percent of it is manufactured internally by the liver, the rest coming from the food that we eat. The adrenal glands and sex organs use cholesterol to manufacture steroids, hormones that control key processes of the body. It is also a primary component of bile acids, metabolic cutting tools that break dietary fats down so that they can be absorbed. Above all, cholesterol is the stuff from which cells make and repair their membranes.

Because all cells need this raw material, cholesterol is forever on the run, flowing through the bloodstream to the remotest outposts of the body, where fresh supplies are taken on by cells in exchange for "slag" cholesterol, whose services are no longer needed. This used cholesterol then goes back to the liver, where it is either recycled or eliminated. Structurally speaking, cholesterol is ideal for making cell membranes. It does not dissolve in water. Hence, in the world of blood plasma, which is largely water, it acts as a perfect barrier for the cell, an impermeable wall broken only by the series of receptors, or doorlike structures on the cell surface that allow essential chemicals to pass.

But what is a virtue in the production of cell membranes can backfire disastrously if cholesterol somehow winds up in an inappropriate spot, say, along the wall of an artery. Then the molecule's insolubility works against its host. The body has no storehouse of solvents to break up this fatty gunk. Blood plasma only washes against it impotently. So there the cholesterol chunk sits, difficult if not impossible to dislodge. Eventually, more and more of it may accumulate, in a process called atherosclerosis, creating the deadly deposits called plaques that constrict arterial blood flow. When the clogging occurs in the network of vessels that feed the heart, the so-called coronary arteries, it can dramatically cut the supply of oxygen to the heart muscle, often without the individual's awareness. Should a blood clot then happen by and get wedged into the partially blocked passageway, blood flow to the part of the heart muscle served by that passageway can be completely shut off, causing the sudden oxygen loss and intense chest pain that signal a heart attack. A similar mechanism in the carotid arteries or cranial blood vessels underlies a stroke.

The association between cholesterol and these deadly ailments was not well understood until the middle of this century. To be sure, for more than a hundred years, autopsies had turned up the presence of cholesterol in choked arteries. But it took a long time for the sinister connection to be recognized. The earliest suspicions that cholesterol was not just a benign bystander date back to the early 1900s, when German researchers noted that heart tissue that had been affected by atherosclerosis contained much more cholesterol than normal tissue. Later, a Russian investigator named Nikolai Anitschkow established the dietary link when he fed cholesterol to rabbits and observed resulting fatty buildups in their arteries. In the 1940s, scientists began to close in. The discovery that such factors as blood pressure and smoking were related to atherosclerosis caused a number of researchers, among them Ancel Keys of the University of Minnesota, to suggest that hardening of the arteries, as atherosclerosis is sometimes called, is not the inevitable consequence of aging. Instead, the disease might be brought on by environmental agents, with cholesterol at the top of the list. It was around this time that epidemiologists, alarmed by a surge in the number of deaths in Western countries from heart attacks, began to find that blood cholesterol levels were far higher in Westerners than in Orientals, among whom the incidence of heart attack is quite low. The majority of adult Americans have cholesterol readings ranging from 200 to 230, while the Japanese, who enjoy the world's lowest death rate from coronary heart disease, have an average cholesterol level of only 150. Medical sleuths did not have to look far for the reason. Paralleling the startling increase in mortality from heart disease among Americans and Europeans that peaked in the 1960s was a radical change in the diet. Perhaps because of growing prosperity, people were eating more than ever, and gravitating to foods that were high in fat. Whereas in earlier times ordinary persons in the West had eaten mainly vegetables, fruit, and legumes, they now consumed large quantities of animal fat in the form of meat, cheese, eggs, butter, and highly processed foods. Orientals, on the other hand, had stuck primarily to their traditional diet of fish and vegetables.

Soon, long-range studies such as the well-publicized Framingham, Massachusetts, study, demonstrated a striking correlation between blood cholesterol levels and heart disease, and a similar risk association with saturated fats. Two later studies, in which populations were given cholesterol-lowering drugs over a period of time, showed reductions in coronary heart disease of up to 34 percent. By the mid-1980s, cholesterol had been positively implicated as a primary culprit in cardiovascular disease, even if it was impossible to tell who was genetically most vulnerable to its depredations.

More than 93 percent of the cholesterol found in our bodies is housed safely inside cells. It is the remaining 7 percent circulating in the bloodstream that is dangerous, particularly to certain people who, thanks to their genes, cannot seem to dispose of that floating surplus by absorbing it into their cells. What is it about our bodies that makes a substance so necessary to our well-being potentially lethal? To grasp the answer, one must first understand how our systems attempt to protect us. Because cholesterol is not soluble in water, nature has devised an elaborate method for shipping cholesterol from place to place. The method relies on a diverse fleet of transport vehicles called lipoproteins, which swim freely through the plasma and carry cholesterol in a kind of cargo hold. These same lipoproteins do double duty, conveying from the intestines to the liver the fats that we ingest every day as part of our diets.

There are four major lipoproteins: very-low-density lipoprotein (VLDL), intermediate-density lipoprotein (IDL), low-density lipoprotein (LDL), and high-density lipoprotein (HDL). Each is really a different phase in the continuing metamorphosis of lipoprotein, and each has its own role in the intricate sequence of events by which cholesterol and other fats circulate constantly through our systems. VLDL is the starting point. These lipoproteins contain not only cholesterol but also triglycerides, the predominant form of saturated fat that we take in each day in our diet. VLDL ships out from the liver, laden with cholesterol and triglycerides, and along the way drops triglycerides off at various way stations to be burned as energy or stored as fat. During an intermediate stage, the transport vessels, having been issued new registry as IDL, contain much reduced levels of triglycerides. Some of this IDL now returns to the liver for uptake, a process that is enhanced by surface molecules on IDL that allow IDL to bind to liver cells with great efficiency. Meanwhile, the remaining fleet, renamed LDL, contains only cholesterol, and it is this form of lipoprotein that carries cholesterol to the 100 trillion cells of the body for making membranes. At any one time, 75 to 80 percent of the body's cholesterol is contained in the LDL compartments. When cells need to get rid of used cholesterol, the final form of lipoprotein, HDL, takes over, barging the excess material back upstream to the liver for disposal.

From this simplified description of a very complex process, one can appreciate how LDL and HDL have developed their different reputations. LDL is the "bad" form of cholesterol that one hears so much about, bad because it is associated with the outbound journey of cholesterol and with the potential for the accumulation of cholesterol in the arteries. HDL is the "good" form of cholesterol because it is the means whereby cholesterol is taken out of the loop. As medical science has discovered, the more HDL we have in our systems, the less susceptible we appear to be to atherosclerosis, because more cholesterol is making its way to the liver for removal. Genetics can enter this picture in a number of ways, as we shall see. But the role of genes was very much terra incognita in the early 1970s when two farsighted researchers at the University of Texas Health Science Center, in Dallas, uncovered, for the first time, the way that a specific genetic defect can directly cause coronary heart disease.

Joseph Goldstein and Michael Brown had known each other for a long time. Goldstein, who grew up in the small resort town of Kingstree, South Carolina, where his family owned a clothing store, and Brown, a native New Yorker educated at the University of Pennsylvania, first met in 1966 as medical residents at Massachusetts General Hospital, in Boston, where they discovered they had a mutual taste for medical research—and for contract bridge. After completing their training, the two friends joined the National Institutes of Health. Goldstein, who, like Brown, was fascinated by the new science of molecular biology, became part of the NIH team that was finishing up the work of cracking the genetic code. At the same time, he took on a clinical assignment in the National Heart Institute, where he helped a senior researcher study people with excessively high levels of cholesterol, a condition called hypercholesterolemia.

Some patients had a genetic form of the condition and cholesterol was completely out of control, a fact that piqued Goldstein's interest enormously. He discussed these cases far into the night with Brown, who, as a biochemist, was himself working at the National Heart Institute, studying the digestion of food and disorders of cholesterol metabolism. Goldstein's preoccupation with cholesterol only grew when, as a fellow at the University of Washington, he led a study of diseases of lipids, or blood fats. In 1971, he persuaded Brown to join him at the University of Texas, Goldstein's alma mater, where the two of them resolved to use burgeoning molecular techniques to probe the genetics of heart disease.

The vehicle they chose was the one that had originally caught Goldstein's attention at NIH—a rare inherited disorder called familial hypercholesterolemia, or FH. First described in 1938, by the Norwegian scientist Carl Müller, FH runs in certain families and occurs in two, dominantly inherited forms. Some people, about one in every five hundred, carry a single copy of the defective FH gene, which gives them blood cholesterol readings that are two to three times normal—in the 300 to 500 range. Such people begin to have heart attacks at the age of thirty and forty. Male heterozygotes, as those with single copies of disease genes are known, have a 75 percent chance of suffering a heart attack by the age of sixty, compared with a 15 percent chance in the general population. Females have a 45 percent risk. A more devastating form of the disease occurs when two carriers of the single bad gene mate. Usually, neither parent knows she or he has the gene, since the condition does not ordinarily rear its head until late in the reproductive years. When two heterozygotes conceive a child, the youngster has a one-in-four chance of inheriting the bad FH gene from each parent. The youngster who does will be homozygous for the disease—that is, he or she will have two copies of the defective gene. The condition, which strikes only one person in a million, is tragic. Such children have blood cholesterol levels that soar six to ten times higher than normal—to 1,000 or even higher. And the cholesterol is almost exclusively the "bad," LDL, kind. The ravages of atherosclerosis begin in early childhood, and victims frequently experience their first heart attack before the age of ten. Few survive beyond twenty years of age. Only liver transplants can possibly save them.

When Goldstein and Brown began researching FH, medicine had no idea how the disease wrought its devastation. It was known only that the liver is the major site of the production and metabolism of cholesterol. But it was impossible, from an ethical point of view, to study the livers of living FH patients. Nor could one easily work with test-tube cultures of their liver cells, since it was rather intrusive to obtain such cells and since they stubbornly refused to live outside the body. So the two investigators started prying into the mysteries of FH by taking a gamble. Being shrewd card players, they bet that the defective FH gene would also be expressed in human skin cells, which were easy to get and readily grown in culture. Like all cells, skin cells need cholesterol to make their membranes—and it turns out, when the liver doesn't supply it, they can make it themselves. In late 1972, the Texas researchers began culturing skin cells from three groups—children with two defective FH genes, individuals with one bad gene, and normal people. The skin cells were placed in a solution that approximated plasma, complete with cholesterol in lipoprotein packets. Brown and Goldstein tagged the cholesterol with radioactive markers and then settled back to see how the three types of skin cells differed in their handling of cholesterol. In a short time, they observed that normal cells stopped making their own cholesterol whenever the LDL cholesterol content of the simulated plasma solution was high and started making it again whenever the lipoproteins were removed from the solution. But FH cells were oblivious.

They went right on manufacturing internal cholesterol, regardless of whether the LDL cholesterol level in the solution was high or extremely low.

Brown and Goldstein first thought that something was haywire in the gene for HMG-CoA reductase, an enzyme that controls how much cholesterol the cell produces. But they discarded this theory when they discovered another difference between normal and FH cells. Whereas normal cells were able to seize cholesterol from the nutrient solution and absorb it, FH cells seemed abysmally deficient in their ability to take up external cholesterol.

It now became apparent what was really going on. The heart of the problem, Brown and Goldstein realized, must lie with receptors on the surface of the cells. These receptors, they reasoned, must serve to bind LDL cholesterol so that it can be ushered into the cell for processing. A secondary job is to participate in a type of feedback mechanism. When enough LDL is present in the bloodstream, the receptors carry a signal to the cell, which immediately takes two steps. First, it stops making its own cholesterol. Second, it ceases to make new receptors, since it does not need them for the time being. When there is a shortage of LDL, on the other hand, the cell is alerted to start making receptors again—it takes only a few seconds for the single-stranded receptor to be generated—and to synthesize cholesterol until the optimal level of LDL in the bloodstream has again been reached.

In 1974, Brown and Goldstein unveiled their theory to the scientific community. The defect in FH, they announced, involves the gene for the LDL receptors, which, as they later verified, dot the surface of liver cells. The healthy person has 250,000 of these receptors constantly at work binding LDL and taking it out of the circulation. But people with one defective gene for the LDL receptor have only 40 to 50 percent of this normal complement. They are much less successful in removing cholesterol from the bloodstream, and their feedback mechanism is impaired. Homozygotes have the worst luck of all. They make no LDL receptors, or very few at best. As a result, liposomes containing LDL cholesterol float eternally through their blood like the Flying Dutchman, never putting into port. To exacerbate matters, with no feedback, their liver cells go right on making cholesterol, most of which is dispatched into the bloodstream and adds to the glut.

Goldstein and Brown's findings were momentous. Not so incidentally, they greatly advanced the fledgling science of receptors. Much more than that, they established for the first time how a genetic defect could bring about coronary heart disease, in the bargain demonstrating beyond doubt a direct relationship between LDL cholesterol and athero-sclerosis. For their achievement, they were rewarded with the Nobel Prize in medicine in 1985. Since then, the gene for the LDL receptor has been identified, and five types of mutations have been pinpointed. Brown likes to point out that the cholesterol receptor system of man-kind is woefully unsuited to the modern-day diet. The system is essen-tially the same as that of the frog, an organism that greatly preceded man in evolution. "Think of it," he says. "The system was in place 400 mil-lion years ago. There are parts of the gene that are different, but the key working parts are 90 percent the same in a frog and a human being. I can't keep things on my desk for three hours without losing stuff, but nature has kept this gene the same for all those millennia. The trouble is the frog eats a few grubs and some flies. He has no cholesterol problems. We're using a gene meant for grubs and flies to try and handle steak."

Brown and Goldstein have been richly rewarded with accolades. Their collaboration is so intense that they are often called, collectively, Brownstein, and it spills over into their private lives. Their bridge part-nership is legendary. But while the importance of their findings can hardly be overstated, FH nevertheless accounts for only 5 percent of all coronary heart disease. What, if any, genetic factors are involved in the remaining 95 percent of all cardiovascular disease? The answer lies with the other half of the cholesterol equation—the lipoprotein carriers themselves.

Each lipoprotein has a broker that manages its affairs. These brokers are called apolipoproteins. Two of them, known as apo E and apo B, travel along and introduce and bind their lipoprotein to cell receptors, which then take the cholesterol or triglyceride cargo on board. They correspond to a particular receptor the way a key fits a lock. Two other apolipo-proteins, known as apo A and apo C, serve to intercede on behalf of the lipoprotein with certain vital metabolic enzymes. Just to illustrate, the job of apo C, which rides along on VLDL, is to interact with an enzyme called lipoprotein lipase in order to tell the VLDL that it's time to unload its consignment of triglycerides. As one might imagine, mutations in the genes for these apoprotein brokers can make the brokers unrecognizable to receptors or enzymes and prevent the critical handoff of cholesterol and fats. For example, during the second phase in the life of a lipoprotein, when it is called IDL, apo E sees to it that IDL binds to fat storage cells and muscle cells so that triglycerides can be transferred cleanly. Apo E is very adroit at this, and one of the reasons is that it is generally present on the lipoprotein surface in many, many copies. But, as a group headed by Robert Mahley, of the University of California at San Francisco's Gladstone Laboratories for Cardiovascular Diseases, discovered, there can be up to five different mutations in the gene for apo E. These mutations make apo E look like a stranger to the fat or muscle cell receptors, and the cargo transfer never takes place. Triglycerides, unable to be off-loaded, start building up in the bloodstream at alarming rates. Problems may also arise with a form of apo A called apo A-I. This broker's job is to escort the "good" HDL as it lugs used cholesterol from the cells back to the liver. Mutations in the gene for apo A-I would hamper, or even inactivate, one's HDL, leading to serious difficulties in clearing cholesterol.

A unique case of HDL insufficiency was jointly uncovered by the geneticist Robert Norum, of Henry Ford Hospital in Detroit, and Jan Breslow at Boston Children's Hospital. Norum was treating two sisters in their twenties who had advanced coronary heart disease, even though their cholesterol levels were normal. The only noteworthy abnormalities he could find to explain the women's problem were a lack of HDL in their blood and a shortage of apo A-I and a related protein called apo C-III. Interestingly, when he tested members of the sisters' immediate family—including their five children, their parents, and their brother— he found that these kin had only half the normal levels of HDL and apo A-I. This was an unmistakable sign of a hereditary pattern. Inasmuch as most members of the family apparently had one defective copy of a gene, reflected in their blood levels, they had escaped any serious consequence. But the sisters had been unfortunate enough to receive a double dose.

What the underlying genetic problem was, however, baffled Norum. That is where Breslow came in. Coincidentally, Breslow's laboratory had recently cloned the genes for apo A-I and apo C-III, which, in a quirk of nature, lie side by side on the same chromosome. When the Breslow group studied blood samples from the women, it found that the DNA in the two adjacent genes had become mixed up sometime in the family's past. A segment of the apo A-I gene had exchanged position with a segment of the apo C-III gene, rendering both genes nonfunctional.

Still another example of how genetics can lead to heart attacks is provided by a substance called apo (a), whose function in the body is not known. A strong association between apo (a) and a high risk of coronary heart disease had been noted for some time. But no one could explain why. Then, in 1986, joint studies between the laboratory of Angelo Scanu, at the University of Chicago, and Genentech Inc., provided an answer. There is a striking similarity between the DNA sequence of apo (a) and another plasma component, plasminogen, whose job is to break up clots and keep traffic in the blood vessels moving freely. A team at Cornell University, headed by Katherine Haijar, has proposed that apo (a) brings on heart attacks in some people by competing with plasminogen for cell surface receptors on blood cells and arterial walls. Because the two look so much alike, the receptors become prey to mistaken identity, with disastrous results. In order for plasminogen to disrupt blood clots, it must bind to those cell surface receptors. If it can find no free receptors, because of a squeeze-out by apo (a), blood clots will form at a much higher rate and eventually find their way to the coronary or the carotid arteries, setting up a potential heart attack or stroke.

But by far the greatest scoundrel among the apolipoproteins appears to be apo B, the consuming passion of Jan Rapacz.

Rapacz's high-cholesterol pigs all suffer from flaws in their genes for apo B. It is the defect that makes them die. From the outset of his career, Rapacz has had an interest in polymorphisms—those slight variations that exist from one person to the next in the makeup of a gene or protein. As early as 1970, Rapacz observed that there were at least two distinguishable variants in the apo B protein. Although apo B's relationship to high cholesterol was then unknown, these variants intrigued Rapacz, and he speculated that such polymorphisms might represent mutations that led to a breakdown in function and thus to disease.

"That observation about apo B really created my life's work," he says. "I was perplexed how there could be two variants, two different proteins, really, encoded by one gene. I didn't realize that apo B is such a monstrosity and has so many parts that could mutate." In looking for polymorphisms, Rapacz came at the problem with the tools of the immunologist. He immunized pigs with pig plasma, reasoning that the most variable proteins in the plasma would stimulate the most active antibodies. By working backward from these antibodies, he could then identify the most mutant proteins. Apo B proved to be especially polymorphic, indicating it could be rife with dangerous mutations. He decided to concentrate on it. Throughout the 1970s, Rapacz tested pigs all over the world, eventually distinguishing eight different variants in pig apo B. One of these mutations was found in a particular strain of pig with very high cholesterol. When he cut the pigs' coronary arteries open, he discovered strong evidence of atherosclerosis. Rapacz was ecstatic. He had found a mutation that appeared to be associated with heart disease in pigs. And since pig apo B is very similar to that in people—70 percent of the DNA is exactly alike—he conjectured that a like mutation was responsible for human heart disease.

Furiously, he began inbreeding the mutant pigs in an effort to learn more about them. But when he published his results in 1978, the response from the scientific community was dead silence. "Nobody accepted it. I never got any credit," he says, with more than a trace of bitterness. "The biochemists told me my methods had been immunological and that I would have to use biochemical techniques to prove my case." He grins wickedly. "So now, all these years later, I have been shown to be right."

Indeed, once the gene for human apo B was isolated, in 1986 and 1987, it was soon determined that the gene is as variable in people as in pigs. Even a small mutation in a critical part of the gene could well render its protein inoperable. Several studies have now shown that, in certain people, such defects contribute to a greatly reduced rate of uptake of cholesterol by liver cells. In one of these studies, at the University of Texas, a third of a population of people with elevated cholesterol were found to have an impeded uptake, and in one of them an apo B mutation was positively found to be at fault. "These findings suggest that apo B mutations may be a relatively common cause of . . . hypercholesterolemia in the general population," write the heart researchers Margaret Prescott, of the CIBA-GEIGY Corporation, and Alan Attie, a frequent collaborator of Rapacz's at the University of Wisconsin.

Why is apo B so nefarious? Because it is the broker for "bad" cholesterol—LDL. Without normal apo B, LDL cannot bind to the LDL receptor, whether or not an individual has the normal set of LDL receptors. The key will not fit in the lock. Hence, LDL's cholesterol burden does not get unloaded, and the LDL is destined to ride around the circulatory system indefinitely. But even without a defect in apo B, one can still be in trouble.

If a person habitually consumes a diet high in cholesterol and saturated fats, the liver's feedback mechanism turns sour. Like a man sticking his hand out the window and concluding that the trickle of water he feels from an upstairs air conditioner is really rain, so the cell takes a reading of the passing cholesterol level and decides there is sufficient cholesterol floating around in the bloodstream that the manufacture of new LDL receptors can be halted. This creates a distressing situation. With too few cell receptors, the apo B–guided LDL finds no place to dock. Again it begins an odyssey around the body.

At this point, another disturbing characteristic of apo B comes into play. Apo B is a tenacious, insoluble material that, when not attaching itself to receptors, has an unfortunate affinity for hooking onto arterial walls. When it latches onto an artery, it won't let go. And where apo B goes, there goes its boon companion—LDL cholesterol. Anchored fast, unable to swim away, LDL begins to collect.

Yet a third possibility has been offered by Brown and Goldstein. They suggest that when LDL spends too much time in the bloodstream, it begins to degrade. The changes in its structure make it no longer recognizable to the LDL receptor, but suddenly an admirable fit with another receptor—that on the surface of scavenger cells of the immune system called macrophages, which avidly take up the modified LDL. Adding weight to this theory is the fact that cholesterol-laden macrophages, called foam cells, are a conspicuous part of arterial plaque accumulations, especially in the early stages.

Still more evidence: cigarette smoke is known to be one of the prime modifiers of LDL that make it more attractive to scavenger-cell receptors, and smoking, as was noted earlier, is correlated strongly with atherosclerosis.

As scientists close in on a full understanding of the roots of cardiovascular disease, the possibility of gene-based interventions suddenly looms. "We have a long way to go," notes Robert Mahley. "The trick will be to get healthy genes into liver or adrenal cells because that's where they will function most effectively. It certainly would be possible, and it is a goal that all of us think about. It's very exciting." Rapacz sees a "fantastic" potential for such therapies. "Once we know which amino acids in proteins such as apo B are responsible for the delivery and metabolism of fats, we can substitute a correction in the gene very easily. Then it is a matter of getting it to the liver, which is being done already."

By 1989, various scientists, including Richard Mulligan at MIT and his former postdoctoral fellow James Wilson, currently in charge of the gene therapy effort at the University of Pennsylvania, had succeeded in getting a gene for the LDL receptor into the liver cells of Watanabe rabbits, a strain that Yoshio Watanabe of the University of Kobe in Japan discovered to be genetically deficient in these receptors and, hence, fated to develop abnormally high cholesterol levels. Wilson tried two techniques. In the first, he removed a chunk of about 30 percent of the rabbit's liver, infected it with virus containing a human LDL receptor gene, and then regrafted the piece of liver back into the rabbit. The new gene proliferated as the liver cells divided. In the second method, Wilson simply dripped the altered virus into the liver's portal vein with a catheter, and the virus transferred the gene to liver cells directly, without surgery.

It was a combination of these approaches that, while at the University of Michigan, Wilson used in the May 1992 Ann Arbor gene therapy experiment for familial hypercholesterolemia. It turned out to be a marathon effort, much more difficult than he had anticipated. Surgeons routinely removed 15 percent of the female patient's liver. Then, working straight through for the next twenty-four hours, Wilson and his associate Mary Ann Grossman extracted the hepatocytes and transduced them with new receptor genes. Immediately thereafter, they delivered them via catheter to the recovering patient's liver in hopes of lowering her cholesterol levels.

The patient not only survived the procedure; her cholesterol levels dropped 25 percent, and stayed there for the next two years, although she never was able to stop her medication and, admitted Wilson, the experimental technique probably had failed to stop the progression of atherosclerosis caused by her disease. By mid-1994, Wilson had performed the procedure on five patients, some of whom had "fared better" than the first patient "and some worse," he reported, declining to be more specific, pending publication of his results. Although he had not cured FH, he seemed upbeat. "None of our patients had any problems. We didn't hurt anyone, and no one is taking any more drugs than they were before," he said. "I want to get the first series of experiments published soon, so we can close out this phase."

Wilson hoped soon to be able to inject new genes directly into the liver circulation of such patients by using adenovirsuses as vectors, thus avoiding surgery. "We may hit some roadblocks," he said, "but right now, it's full steam ahead with adenoviruses."

An intriguing by-product of this line of research in Richard Mulligan's lab—a direction that, incidentally, he has always credited totally to Wilson—eventually led to another important discovery. Like all other liver researchers, Wilson was having a devil of a time keeping liver cells alive in culture and getting retroviruses to infect them with any kind of alacrity. His cultures always seemed to be contaminated by endothelial cells, the single layer of cells that come from the thin inner lining, or endothelium, in blood vessels and arteries. Endothelial cells secrete proteins and control constriction of the vessels, the formation of blood clots, and the proliferation of cells themselves. After fighting a fruitless battle to get rid of them, Wilson began to grasp how wonderfully willing his viruses were to infect endothelial cells.

Wilson suddenly became aware of a piercing light.

At about the same time, early in 1989, French Anderson got interested in endothelial cells, too. "If you want cells which are closest to the bloodstream to secrete chemicals, it's the endothelial cells that line the blood vessels," he figured. Anderson pursued this line of inquiry because endothelial cells are, in effect, ready-made stem cells, although of the vascular system, not of bone marrow. "They divide when they're damaged, so they are the vascular equivalent of stem cells."

Working with colleagues at GTI, Anderson became the first to report the successful insertion of genes into endothelial cells in vitro, whereupon the cells obligingly produced large amounts of the compound known as tissue plasminogen activator, or TPA, the natural human enzyme that dissolves blood clots. A very nice thing for endothelial cells to do.

Wilson and Mulligan, in high gear by then, had moved on and were testing endothelial cells in dogs. In June of 1989, they reported that they had successfully seeded endothelial cells bearing genes called *lacZ* into the lining of a dacron graft, a synthetic blood vessel, and could detect the gene for as long as five weeks after they put the grafts back into dogs. They chose *lacZ* as a marker, or "reporter," gene because it produces the E. coli enzyme beta galactosidase, which mammalian cells do not make but which is readily identifiable when stained in cell culture. The cells containing the genes expressing their beta gal product turn a pretty shade of robin's egg blue.

The point is to design new biological drug delivery systems—genes as drugs—and they should soon provide the first gene therapies against heart disease.

Despite better prevention and treatment, the results of atherosclerosis—heart attacks, strokes, and amputations resulting from vascular disease in the legs—are the major cause of death in our society. One of the best available therapies is to replace blocked vessels with a bypass graft. Sometimes it is necessary to use a synthetic graft, made of dacron or other material, for the bypass. But man-made grafts create clotting problems. "Our idea," Wilson says, "is to line them with endothelial cells genetically modified to promote a 'natural' lining, prevent clotting, and possibly secrete therapeutic proteins. This is a first step toward improving graft performance."

At the same time that Wilson, Mulligan, Anderson, and colleagues were tinkering with endothelial cells, Gary Nabel in Ann Arbor implanted similarly modified cells into the odd little creatures called Yucatán minipigs, which are a lot closer to humans than dogs are, in terms of vascular structure. In this experiment, the enzyme-producing cells were placed directly onto previously stripped artery walls by means of a catheter. The cells attached themselves to the vessel and produced pretty blue B galactosidase, leading Nabel to theorize that the technique might be useful not only for heart disease and other circulatory disorders but also to secrete therapeutic proteins directly into the blood.

"It may have application to the management of cardiovascular disease in several ways," he says. "For example, by introducing cells modified to release anticlotting agents, we can prevent clots. Or we might induce the formation of new blood vessels, or to dilate blood vessels and keep them open."

In addition to dacron prosthetic grafts, Mulligan and Anderson were also competing to perfect the use of tiny surgical sponges onto which they have transplanted cells with new genes that could be placed in the circulation system to produce high levels of therapeutic proteins, such as the clot buster TPA. About 350,000 coronary artery bypass operations are done each year in the United States, and 100,000 of them run into problems because of clotting. Tiny fibrous sponges or Gore-Tex fibers containing anticlotting genes or factors crucial to blood vessel growth and patency could be placed at graft sites, or lining the inner surfaces of prosthetic devices, to provide protection until healing occurs.

In animals, the cells are placed in the portal vein of the liver or in the belly; when treated with chemicals, they will develop their own blood and nerve supplies—a remarkable thing, if you stop and think about it— while secreting a crucial enzyme or other protein that the animal needs to survive. Anderson, who says he got the sponge idea from his wife, Kathryn, showed that the sponges can be used to cure the bizarre yellow rodents called Gunn rats, whose color comes from their inability to eliminate bilirubin from the body. When the yellow animals received sponges and corrective genes, they soon diverged into the murine version of Joseph's coat of many colors.

It is Gary Nabel's wife, Elizabeth, who has quietly been making her way into the lead to win the race for gene therapy for heart disease. Associate professor of medicine at the University of Michigan Medical Center and director of the cardiovascular research center, she describes herself as an interventionist cardiologist, a doctor who does heart catheterizations and balloon angioplasties.

"That's how I got interested in gene therapy," she explains. "I've treated a lot of people and been frustrated that so many of my patients would come back in with what we call restenosis of the artery after we'd break up a chunk of plaque with the balloon. There was nothing I could do about it. It seemed to me that the only way we were going to get a handle on this was to change the genetic programming of the cells that cause this to happen—the cells that proliferate after angioplasty."

As a clinician, Nabel is looking toward two major approaches for gene therapy for heart disease: treating the heart muscle itself and treating veins and arteries. For treating the heart itself directly, several groups have demonstrated that "naked," or plasmid, DNA can be injected directly into heart muscle cells (myocytes), with a subsequent expression of the recombinant gene. Beta gal is the marker, and all one need do, apparently, is mix liquefied DNA in a sugar solution and inject it directly into the heart of a mouse by means of a small-gauged needle. The essence of simplicity. The reason this seems to work is that heart cells are terminally differentiated cells—they're not proliferating like, say, tumor cells—hence their requirements for the uptake of DNA may be quite different from those of cells that are still dividing. Once inside the nucleus, the DNA seems to replicate as what is known as an episome, a circular gene piece that lives quite independently of the nearby chromosomes but, like many new neighbors, is a borrower—utilizing the enzymes that the resident genes employ to control cell division.

What might one want to inject into the myocardium, and into whose myocardium would one choose to inject it? The first heart patients to undergo gene therapy probably will be those with end-stage heart disease who are awaiting heart transplantation. From an ethical perspective, these would be good candidates: they already have exhausted all available forms of therapy—without a new heart, they are going to die—yet very few such patients (only about 25 percent) on transplantation lists ever actually receive hearts. Three out of four expire from their disease while waiting, because there just are not enough donor hearts around. The goal would be to give these patients genes that create new blood vessels in their heart muscle to improve blood flow—either temporarily or perhaps even permanently.

The manufacturers of vessels are a family of powerful chemicals known as angiogenesis factors, growth factors that work to promote the proliferation of endothelial cells, which line all blood vessels. One group, fibroblast growth factors, particularly FGF-5, has been shown to induce angiogenesis in the very small capillary blood vessels within an organ—precisely where one would want them to establish collateral circulation in a dying heart. However, a number of stumbling blocks line the road to gene therapy: scientists must tweak the fibroblast growth factor gene to express its protein in high enough levels within myocytes and tiny blood vessels; once new blood vessels do form, it must be demonstrated that they are intact, that they are not leaky or prone to breaking down, and that they can tolerate blood flowing through them; next, scientists must show that the new pipeline actually is able to increase blood flow to the diseased portion of the heart muscle. Obviously, safety and lack of toxicity must be demonstrated, because growth factors have the nasty potential for transformation of a proto-oncogene, and the last thing one would want to do is to give cancer to a dying heart patient. Delivering a recombinant growth factor to the human heart would surely be permitted only after the most stringent monitoring by the Food and Drug Administration, and that could lead to further delays.

Nonetheless, Nabel estimates that the first such clinical trials could begin in from three to five years. She also is excited by studies showing that immature muscle cells, called myoblasts, apparently may be willing to serve as vectors, perhaps to perk up ailing hearts. When injected into muscles, the little cells can deliver serum proteins, and perhaps one day can be trained to do so in sufficient quantities to treat systemic disorders. In December of 1992, Nabel's friend Jeff Leiden, now chief of cardiology at the University of Chicago, and the Stanford University myoblast specialist Helen Blau showed that myoblasts transfected with the human growth hormone gene can produce HGH for several weeks when injected into the leg muscles of mice. A promising finding.

In Ann Arbor, though, Elizabeth Nabel has systematically been tackling an even more complex scientific problem: gene therapy for restenosis, the major complication after balloon angioplasties to clear clogged cardiac arteries. Since 1979, when it was imported from Switzerland, balloon angioplasty as a palliative procedure for angina has spread like wildfire—even most community hospitals have gotten into the act. About a half million such procedures are now done each year in the United States. However, one out of every three patients will require a second treatment for restenosis, or renarrowing of the artery at the site of the injury caused by the balloon when it fractured the plaque deposit. The failure of the treatment is attributed to the buildup of repair-minded smooth-muscle cells at the site. In most people, the repair by these cells causes no problems. But sometimes the response is so vigorous (or "exuberant," as Nabel describes it) that the blood vessel closes and requires emergency coronary artery bypass surgery.

The biology of smooth-muscle cell proliferation is only beginning to be understood. Several growth factors probably are involved, including platelet-derived growth factor (PDGF), transforming growth factor (TGF), and fibroblast growth factor (FGF). "We think growth factors are produced at the local site by smooth-muscle cells, endothelial cells and platelets," Nabel says. "These factors bind to receptors on smooth-muscle cells. The receptor binding them then initiates a signal to the nucleus, whereby cell division proteins get stimulated and probably stimulate other growth factors to continue the process." Nabel seeks to determine at which point she should intervene genetically to block the process.

In 1989 and 1990, the Nabels published papers in *Science* showing that genes could be introduced into vascular cells by means of several different methods. "So, our next step was to try and introduce a growth factor gene into an artery and see what happened," she recounts. By late 1993, she had published a series of studies in which she introduced three growth factor genes—PDGF, TGF-beta, and FGF-1—into arteries and noticed differences that helped sort out the growth factor puzzle. PDGF induced a very intense cellular proliferation—it actually mimicked stenosis. TGF-beta produced proteins that act as a scaffolding for atherosclerotic plaques—a good thing to know. FGF-1, on the other hand, induced growth within the wall of the vessel.

"These studies told us we can introduce a gene for growth factor and study the biological effect of that gene, which should help us learn how to block part of the process leading to the artery injury," Elizabeth Nabel says. "We want to devise inhibitors of these growth factors.

"I believe now that restenosis will be a very amenable condition for gene therapy, especially because we could introduce a recombinant gene through the same catheter right at the angioplasty site to keep the artery open and prevent smooth-muscle cell proliferation."

What she has in mind is a product, a kit consisting of a catheter and a gene therapy solution, so it would be simple.

Diffuse atherosclerosis, she concedes, "is going to be a little bit trickier—the biology is even more complicated than the biology of restenosis. Superimposed on smooth-muscle proliferation is infiltration of lipids into the artery wall. That's going to take us a while.

"But it would be wonderful if we could target a gene therapy treatment aimed at plaques. Maybe all we'd have to do is to inject a solution which would home in and bind to a particular receptor in plaque." That would make angioplasty and bypass surgery obsolete.

Michael Brown is more guarded than some of his fellow scientists about the prospects of gene therapy's ability to correct cardiovascular illness. He notes that coronary heart disease is basically a chronically debilitating condition, "and chronic by definition means it isn't so severe that it can't be treated by manipulations other than gene therapy—drugs, changes in diet, organ transplants." Furthermore, Brown points out, the genes ultimately responsible for atherosclerosis may not be the obvious ones, such as those for the LDL receptor or the various apoproteins. "Rather, there appears to be a class of regulator proteins that supervise such things as the LDL receptor. They tell the cell how much receptor to make, for example, and they do this by binding to the receptor gene. This activity is at a much subtler level, but we are actively trying to find these proteins and their genes. There are two kinds of genes in the body—working proteins, and regulator proteins—and I along with many others think that minute differences in the regulators may determine the major differences in people, such as height, blood pressure, and so on."

It may take years to isolate these proteins, Brown says, although he, along with most researchers, thinks that the map that will be produced as part of the Human Genome Project will accelerate matters. "Once we have a genetic map, we can take families where one child differs from another in, say, hypertension, and by comparing maps of both children, we will find most of the DNA will be the same, but the differences will lead us to the genes for hypertension, or whatever we are seeking."

Brown believes that a more indirect form of gene therapy is more likely to be employed against cardiovascular disease. This scenario involves understanding the control mechanism that turns such genes as that for the LDL receptor on and off. "Maybe we can learn to fool that mechanism," he says. "My feeling is that in the next several years we will learn how to activate the LDL receptor gene to maximize LDL activity without adding a gene. If we could get our own LDL receptor genes to work as fast as they can, we would have LDL levels as low as that in mice."

While gene therapy is a long-range goal, most experts agree that in the near term, the fruit of today's research into the genetic underpinnings of cardiovascular disease will be primarily diagnostic. We will be able to distinguish those at risk for premature heart attacks long before that eventuality comes to pass. "I think the potential definitely exists to identify early in life—in childhood and even earlier, perhaps at the fetal stage—individuals who are going to be at high risk for premature coronary heart disease. Then we can adjust their lifestyles," notes Peter Kwiterovich. "Genetics has the potential to really change our thrust in medicine toward prevention, rather than treatment of the complication after it develops." Brown agrees: "As it is, the death rate from heart disease has gone down more than 25 percent in the United States in recent years. It is not unreasonable to speculate that within a generation, heart disease as a general problem will be a thing of the past."

Befitting his role at the leading edge of that era of diagnosis, Jan Rapacz has already tested his own apo B genes and compared them with those of his pigs. He came out favorably. "I have two mutations in my apo B," he says, "and they cause a slightly increased risk, but not a serious one. I can have a cholesterol level as high as 220 and still be okay."

Rapacz, as big a fan of bacon as he is of pigs, believes he can eat bacon and eggs twice a week and still have good odds of living as long as his grandfather, who died at 101. But he is still not as well off as Michael Brown, who during a buffet lunch was observed to pile on his plate roast beef, potato salad, and a number of other items proscribed by the American Heart Association as killer foods. When he was called on it, Brown shrugged. "My cholesterol is in the 160 range," he said, happily biting into his beef. "Just the genetic luck of the draw."

16

JEKYLL AND HYDE

If all goes according to plan, Jack Roth may soon become known as the man who made it safe to smoke. It would be a description, ironically, that Roth would abhor. As one of the nation's leading cancer surgeons, a specialist in lung tumors, he wages constant war on the use of tobacco. Yet he is the mastermind of a gene therapy experiment that could bring smokers and other malignancy-prone individuals back from the brink of lung cancer.

For sheer daring, not to say novelty, his approach to treating cancer rivals Ken Culver's "crazy" strategy of treating brain tumors with herpes genes.

Walt Whitman once marveled that a mouse "is miracle enough to stagger sextillions of infidels." Granting Whitman's assessment, Jack Roth's mice must be beyond miraculous. His little rodents shrug off cancer as if it were heartburn. By giving them this ability in the laboratory, using a technique that is unprecedented, even in the brave new world of gene therapy, Roth has managed to restrain one of nature's loosest cannons, a cancer-causing gene.

Early in 1992, Roth, who is chief of chest surgery at Houston's M. D. Anderson Hospital and Tumor Institute, injected human tumor cells into the airways of a population of nude mice, so called because they have been deliberately bred to lack a functioning immune system. Such creatures are prized by researchers because they will readily incubate almost any human ailment, allowing the testing of putative therapies for diseases for which there is ordinarily no animal counterpart. By flooding the mice with tumor cells, Roth and his colleagues seemingly wrote out their death warrants, for nude mice treated with such cells generally die within four weeks, their lungs eaten up by cancer. But the M. D. Anderson researchers were not finished. They believed they had an antidote.

All cancer is the result of a security breach. Throughout our lives, a rigidly disciplined police force composed of regulatory genes, messenger proteins, repair enzymes, and watchdog lymphocytes protects us from developing malignancies.

One might conceive of this defense system as a sort of Maginot line, were it not for the fact that the enemy is within us as well as without. A better analogy might be a gigantic industrial complex manned by a workforce of powerful but easily influenced toughs. These toughs, not unlike a group of prison trusties making license plates, do the chores of signaling cells to divide, to differentiate, and so on. They are held in check by an equally hard-nosed crew of guards, who tell them when to knock off work and correct them when they get out of line. But the situation is in delicate balance. If the surveillance is relaxed for even a minute, the toughs may be lured from the straight and narrow by corrupting influences, in which case they will riot and take over the premises. Should these hooligans gain control of even a single cell, the insurrection will spread like wildfire, as they force more and more cells to divide, until the body is choking on them.

In the case of the lung cancer running rampant in Jack Roth's mice, the roughnecks belong to a class of cancer genes called *K-ras*. What the Roth team has come up with is a way to gum up their works, to lock them down, in effect, to stop them cold.

Soon after the treatment was begun, Roth's team noted an enormous change. Ordinarily, up to 80 percent of the nude mice infected with *K-ras* develop lethal lung tumors. But Roth's gene therapy strategy caused the number who fell prey to tumors to plummet to just 20 percent. Moreover, among those few who did acquire tumors, the cancers were much smaller than usual. Almost immediately, Roth started petitioning the RAC to let him try the treatment on human beings, and very quickly he was granted approval. By mid-1994, he was ready to attempt the procedure in end-stage lung cancer patients who had not responded to any other form of treatment.

According to Roth, his goal is to eliminate his own job. "It would be a wonderful thing if I didn't have to treat cancer patients any more and I could go into another field," he says, only half joking. "If my services weren't needed any more, wouldn't that be great?"

Modern cancer research began with, of all things, the untimely death of a farmer's chicken. The year was 1910, and the bird, a Plymouth Rock hen that succumbed to a breast tumor, fell into the hands of a young pathologist named Francis Peyton Rous, of New York's Rockefeller Institute. It was Rous's idea to use the hen's tumor to study whether cancer might be caused by certain infectious motes called viruses that had been discovered just a decade earlier. Grinding up the tumor and adding water, he strained the blend through a filter whose holes were too tiny for objects the size of a cell to pass through. Then he injected the result into a flock of healthy hens. Soon large numbers of normal fowl came down with breast tumors. Something in the brew was obviously infectious, something smaller than a tumor cell or bacteria. Rous had little doubt that the something was a virus.

But it took no genius to see that cancer was not communicable in the way that other viral diseases, such as mumps and the common cold, were. Rous correctly assumed that his scientific elders would ridicule any attempt to connect viruses and cancer. In writing up his experiments, he therefore settled on the euphemism "tumor agents" to designate whatever in the brew was contagious. Nevertheless, his work was paid little heed. A disgusted Rous himself at length gave up on the experiments, calling cancer research "one of the last strongholds of metaphysics." (He was vindicated at last in 1965, when he received a Nobel Prize a mere fifty-five years after his discovery, giving new meaning to the phrase "ahead of his time.")

As decades passed, other researchers demonstrated that viruses did indeed sometimes beget cancer in rabbits, mice, and frogs. But the vast majority of animal tumors seemed not to fit this mold. Dissection of the tumors failed to turn up a smoking gun in the form of any virus. Further, in spite of an intense search, no viruses could be implicated in any human malignancies. It began to look very much as if viruses, while able, in certain instances, to trigger cancer, could not account for the lion's share of the disease.

Researchers recognized that they would have to delve deeper for a fundamental cause. The most obvious feature of cancer is that it involves an aberration in the way cells divide. Cancer cells reproduce at a madly accelerated clip, with each new cancer cell inheriting the same manic behavior. It seemed plausible that the cancer process must begin with some inheritable disturbance in the mechanism of cell division.

The tumor virologist A. Harry Rubin, of Caltech, helped advance this theory in the late 1950s when he inserted a single cancer virus into a normal chicken cell. (The virus, not so incidentally, was the same kind studied by Peyton Rous, which by then had been isolated and designated the Rous sarcoma virus.) Within two days, the cell began multiplying uncontrollably, and all its offspring did the same. Well aware of the recent discovery that viruses work by harnessing the genetic material of their hosts, Rubin proposed that genes from the Rous virus had barged into the nucleus of the chicken cell and caused changes in the DNA that unleashed wild cell division.

An intriguing notion. However, for the theory to explain cancer in anything more than a narrow sense, it would have to account for the many malignancies caused by nonviral agents. For years, it had been known that ionizing radiation could induce tumors. By the 1960s and early 1970s, it appeared likely that many chemical substances—for example, benzene and cigarette smoke—could trigger cancer as well. Did these, like Rubin's virus, also work by disrupting DNA? The biochemist Bruce Ames, of the University of California at Berkeley, tested the hypothesis by spraying a veritable pharmacopoeia of suspected carcinogens onto *Salmonella* bacteria. Almost immediately, the bacteria started making amino acids no self-respecting salmonella had ever made before, a sure sign that their genes had become mutated by the chemicals. To Ames, the case seemed strong. Mutation was tantamount to carcinogenesis.

But monkeying around with just any DNA clearly does not beget tumors. There are at least five thousand diseases linked to genetic mutations, and few involve malignancy. In 1969, Robert Huebner and George Todaro of the National Cancer Institute argued that there must be certain critical genes in the cell nucleus whose mutation unleashes cancer. In the 1970s, outfitted with the new tools of genetic engineering, cancer biologists began questing for these "oncogenes" (from the Greek *onkos*, meaning "mass" or "tumor"). The hunt at first focused on the Rous sarcoma virus, which by its very nature was assumed to contain an oncogene. The Rous virus, it turns out, comes in two varieties—one harmless, the other deadly to chickens. When the two viral strains were juxtaposed, researchers were surprised to note that one was longer than the other. The tumor-causing version of Rous had four genes to the other's three. An extra gene on its tail—christened *src* (pronounced "sark") because of its link to sarcomas—gives the virus the potential to make cells run amok.

But now a second conundrum loomed. "It wasn't clear why a virus should have such a gene, because *src* doesn't do anything for the virus itself," says J. Michael Bishop, of the University of California at San Francisco, one of the early investigators of *src*. "Viral genes are generally there for one purpose, the production of new virus." Researchers suspected that, originally, *src* was not a viral gene at all, that the Rous virus, whose customary home is chicken cells, had pirated the gene from chicken DNA sometime in the evolutionary past, probably millions of years ago (cancer, after all, has been a part of the earthly scene for eons; there are even telltale signs of it in dinosaur remains). If this theory is true, then *src* genes must also be part of the normal chicken. Bishop and his coresearcher Harold Varmus began ransacking the cells of healthy chickens in search of *src*. Sure enough, in 1975 they turned up a gene in the cells of normal fowl that was a virtual carbon copy of *src*. On further investigation, they made a still more profound discovery. *Src* is present in the cells of virtually all species—fish, ducks, turkeys, rabbits, even human beings. From an evolutionary point of view, this means that the gene has been around for at least as long as fish, about half a billion years. A gene conserved like that throughout most of evolution probably serves a crucial function common to all species. The best guess was that in its normal state, *src* had something to do with development and growth, in other words, with cell division.

Soon, more oncogenes were discovered lurking in other animal viruses. But the field did not move into overdrive until 1980–81, when Robert Weinberg's group at MIT and rival laboratories at Cold Spring Harbor and the National Cancer Institute isolated the first human oncogene. To everyone's surprise, the gene, which was recovered from human bladder cancer cells taken from an unfortunate fellow who had died of his disease, proved identical to an animal oncogene called *ras*, which had been harvested a few years earlier from a virus that infests rats. The discovery that the same genes could unleash cancer in both human beings and animals came as a shock. When scientists got down to sequencing the specific mutation in *ras* that made it oncogenic, they received another jolt. It was an opportunity of sublime moment: the chance to pinpoint for the first time ever a genetic defect responsible for human cancer. But instead of the catastrophic mangle of dozens of base pairs that everyone expected to underlie so profound a disease as cancer, the defect was a simple point mutation—a change in just a single base. The alteration of only 1 of 6,600 nucleotides in a normal *ras* gene, resulting in the substitution of the amino acid valine for the customary glycine in the protein chain, was sufficient to convert it into a deadly killer.

To date, more than sixty oncogenes have been isolated from a variety of sources, including lymphomas and leukemias and tumors of the colon, lung, breast, and prostate. Each has been given a quirky name such as *ras*, *sis*, *myc*, *fes*, or *erb*. By studying the normal versions of these rogue genes, molecular biologists have, over the past decade, achieved one of the great feats in all of medical history—the demystification of cancer. No longer is the disease viewed as an inscrutable, alien plague. Instead, it is logical, explicable, and very much an internal affair. As it happens, all of us carry the seeds of our own destruction in the form of an elite cadre of powerful master genes, genes that are essential to life but that, under certain dark circumstances, are the deadly precursors of oncogenes. Should these miscreants get the upper hand in just one cell, one lone cell, the rest of the body may be doomed, for these wayward genes are capable of transforming normal cells into cancer cells with awesome speed and power.

Oncogenes start out life as law-abiding citizens whose job is to produce proteins associated with such vital tasks as cell duplication, differentiation, and growth. Bishop coined the term "proto-oncogenes" to describe them in their normal state. At present, there are four known classes of proto-oncogenes, though there are quite possibly more. Best understood are the ones that produce growth factors, a broad family of hormones that wield command over the number and kind of cells that our bodies produce. For example, the first growth factor identified was platelet-derived growth factor, which, upon release from blood cells after an injury, speeds the production of new skin cells to promote the healing of wounds. Other growth factors have been found to dictate the creation of new cells in such locations as the lungs, gut, liver, and connective tissue. Although growth factors broadcast messages to cells, telling them to divide and, when necessary, differentiate into specific kinds of cells—hair cells, bone cells, pancreas cells—the message must penetrate the cell membrane, or it will go ignored, the way a radio signal is useless unless it is pulled in by some sort of receiving device, be it Walkman, boom box, or car stereo. There is thus a second category of proto-oncogene that outfits the cell surfaces with receptors corresponding to specific growth factors. These receptors allow the growth hormones to deliver their message to the intended cell.

Once the message has been received, a third category of proto-oncogene swings into action, manufacturing enzymes that relay the signal from receptors to the cell's command center in the nucleus. These signaling enzymes are aligned in a kind of biochemical pathway, each one alerting its neighbor to the news by passing along phosphate molecules, in a manner reminiscent of athletes handing off an Olympic torch.

Finally, there is a class of proto-oncogene that operates at the end point of the process, that is, at the stage where the message has already been received by the nucleus. These proto-oncogenes are called transcriptional factors, and their job is to activate the target genes that carry out the actual mechanics of cell division and differentiation. They appear to work by regulating domains adjacent to the target genes called "enhancers" and "promoters," which turn the genes on and off and tell them how much of their protein to make.

Growth factors and receptors, as one might imagine, are particularly active in the developing fetus. Scientists at the Scripps Clinic, in La Jolla, California, have measured the levels of growth proteins in pregnant women and recorded some interesting data. The proteins switched on and off systematically, as if someone were punching directions into a computer. One protein decreased fifteenfold in a week's time, while several others were in the process of coming on-line. Three stayed constant during the last month of pregnancy, another was detected only in the last two weeks, and one was seen only in the final week before delivery. Clearly, growth factor genes are on the job before birth, doing something vital to help make a living creature.

But as life proceeds, these genes are supposed to switch off permanently or operate under strict control. Should they become damaged by a carcinogen, viral or otherwise, they can be transformed from respectable Jekylls into dangerous Hydes.

In a perversion of normal physiological function, the gene will start churning out too much of its protein at the wrong time, releasing powerful stimuli that, if accompanied by other alterations within a cell, can cause that cell to turn renegade. This scenario illuminates the way viruses, such as Rous sarcoma viruses, cause cancer in animals. They subject the genes they long ago shanghaied from their hosts to unfamiliar and powerful viral control signals, unleashing cancerous potential the genes would not have in their natural surroundings.

The flip side occurs when the genes for growth factor receptors go bad. Receptors may be produced that lack the ability to distinguish their matching growth factor from the rest. Such faulty receptors are like all-night restaurants; their doors are open all the time, to any growth factor that comes along. In fact, no hormone need be present at all. The receptor may send bogus signals to the cell nucleus, making the cell think it is under continuous instructions from a growth factor to divide, when the stimulus is, in fact, a figment. But trouble can also start farther along the line, with the signaling enzymes that, like an oldtime bucket brigade, transport messages in a cascade to the nucleus. There are several different kinds of such enzymes, including protein tyrosine kinases, and

GTPases, and any of them can become perverted, garbling the message as surely as if the growth factor or receptor themselves were faulty. The familiar *src* proto-oncogene, for example, calls forth a protein that lies just inside the cell membrane near the growth factor receptors. When *src* goes awry, its protein transmits a signal to divide even when the receptor is shut off. Farther along the pathway, meanwhile, lie proteins manufactured by the *ras* family of proto-oncogenes. *Ras* proteins are thought to play a pivotal role in getting the signal to the nucleus. This has been inferred from the fact that *ras* figures in a fourth of all human cancers, including tumors of the brain, breast, pancreas, lung, stomach, and colon.

Just when you thought it was safe, after the message has finally reached the nucleus, things can still go wrong. Since the genes that implement cell division must be triggered by transcriptional factors, a glitch in a relevant transcriptional factor can directly disturb the production of new cells.

Consider a proto-oncogene called *c-jun* (the *c* stands for normal cellular version). *C-jun* appears to stimulate enhancers adjoining a still-unknown gene that helps initiate the cell cycle, as biologists call cell division. Should a bastard version of *c-jun* come along and overstimulate this cell cycle gene, the result could be calamitous. (Presumably, *c-jun* could be altered in a way that might make it underestimate its target gene, or even fail to stimulate it at all, but these developments would be unlikely to cause cancer.)

Clearly, then, cancer can gain a foothold at any point in the long chain of signals that leads to cell division. But proto-oncogenes are merely part of the story, as cancer researchers have only just learned.

The finding was so surprising that, at first, Bert Vogelstein didn't believe it. The colonies of cancer cells had stopped growing, right in their tracks.

Vogelstein, a goateed man with dark, penetrating eyes, is not one to run off half cocked. He is dedicated to the scientific method, that pitiless leveler of high hopes that says one must rerun an apparently successful experiment several times before buying drinks for the house. "In ninety-nine cases out of one hundred, when a truly startling result is obtained in the laboratory," he notes, "it turns out to have been a mistake."

So Vogelstein and his lab crew at Johns Hopkins took the gene known as p53, the gene that less than a year earlier he had tied to colon cancer, and again inserted it into glass dishes teeming with bowel cancer cells. Once more the cell colonies, created twenty years ago from the tumors of patients with names long since forgotten, stopped dividing like magic.

"It completely suppressed the growth," Vogelstein remarks about the 1990 experiments. "There was no growth at all. We had suspected it might slow the formation of tumors, but the idea that putting in a single normal gene would have such a dramatic effect was totally unexpected."

P53 is a member—the most important member found to date—of a newly discovered class of genes called tumor suppressor genes, whose job is to hold cell growth in rigid check. When such genes become inactivated one way or another, the result is almost inevitably a cancer. And what Bert Vogelstein had just discovered is that if you airlift a healthy version of a tumor suppressor gene into a crowd of hostile cancer cells, the gene will shut those cells down cold.

For Bert Vogelstein, a pediatrician whose anguish at losing young patients to malignancies caused him to switch to the field of cancer research, it was a moment of keen, if tempered, satisfaction. Years of driving himself—of arriving at his laboratory in a converted Baltimore shopping mall at five and working in the predawn darkness before the rest of the building came to life, of staying till seven o'clock at night, six days a week—had yielded a precious nugget of knowledge.

Put simply, tumor suppressor genes act as a genetic brake on the cell cycle. Also called anti-oncogenes, they appear to work by producing proteins that intercept the message to divide before it gets to the nucleus. They thus counterbalance proto-oncogenes. Common sense had long suggested that there must be some yin-and-yang mechanism by which the body controls the proliferation of cells. An arm grows only so long and then it stops. Normal cells growing in separate colonies on the bottom of a petri dish cease to multiply when they touch one another. Clearly, some in-house memo is disseminated canceling the signal for cell division. And the signal is thought to come from tumor suppressors.

If a tumor suppressor gene should become disabled, or somehow expunged from the genome completely, then the message to make more cells is left unchallenged. Under such conditions, oncogenes are free to keep sending frenetic messages for growth proteins, which gush out in quantities sufficient to bring about tumors.

The first tumor suppressor gene came to light in the mid-1980s during studies of retinoblastoma, a rare cancer of the retina that strikes 1 child in 20,000 between birth and the age of four. (It never strikes older youngsters, apparently because retinal cells stop dividing at that time.) Retinoblastoma was uniformly fatal until the middle of the nineteenth century, when the invention of the opthalmoscope made it possible to peer inside the eye and detect the tumor before it spread to brain tissue. Surgery to remove the tumors soon followed, and for the first time retinoblastoma patients began surviving until adulthood. This led to a curious phenomenon. When patients married and had children, about half of their offspring developed this otherwise very rare tumor.

For the better part of this century, researchers puzzled over how an uncommon malady like retinoblastoma could have such a high inherited incidence among select individuals. Then, in 1971, Alfred Knudson, Jr., of M. D. Anderson Hospital and Tumor Institute, suggested an answer. Knudson, a onetime student of the pioneering geneticist Thomas Hunt Morgan, began by making an educated assumption: the affliction, he conjectured, is recessive; that is, it is caused by mutations in both genes for some unknown gene product. Knudson went on to speculate that in the familial form of retinoblastoma, children might be born with, in effect, one strike against them. In other words, they could inherit a mutation in one of these critical genes, so that it is present in every cell of their body, including the cells of the retina. At some point during early childhood, then, the infant might be subjected to some mutational event that damages the remaining healthy gene in a single retinal cell. Both genes would then become nonfunctional, the aberrant cell would begin to propagate, and a tumor would result. By contrast, in the non-familial version, both genes would start out normally but by tragic coincidence become mutated or lost in retinal cells in early infancy. The unlikelihood of such a coincidence, noted Knudson, would account for the rarity of the disease.

The first clue that Knudson was right came from Jorge Yunis, of the University of Minnesota Medical School, who found that retinoblastoma cells frequently had a noticeable deletion at a specific site on the long arm of chromosome 13. What was more, in children with familial retinoblastoma, the segment was missing not only in the tumor cells but also in normal cells throughout the children's body—and in the body cells of one of their parents. In nonfamilial cases, however, the deletion was present only in tumor cells, as it would be if it were not inherited.

Soon other researchers found that in retinoblastoma patients, a mutation was present at the same site on the other copy of chromosome 13. Webster Cavenee and Ray White, at the University of Utah, reported in 1983 that a certain piece of marker DNA, presumed to flank the retinoblastoma gene, was invariably missing on both thirteenth chromosomes in children who developed familial retinoblastoma. It was now possible to validate Knudson's theory in the light of demonstrable data. Retinoblastoma requires two hits, each inactivating one of the two copies of a person's retinoblastoma gene. Some children, as a result of family inheritance or prenatal mutation, are born with one intact and one defective copy of the gene. As long as the intact copy remains functional, they are protected; but if they should lose the intact copy—as happens 80 to 90 percent of the time—cancer develops. In the noninherited case, children are born with two good copies of the gene, but through a chain of circumstances lose both copies in a retinal cell while very young, again leading to cancer. Remember, all it takes for a cancer to begin is the corruption of a single cell. Multiplication takes care of the rest.

Not much time was required for scientists to reason that if deactivation of a gene such as that involved in retinoblastoma leads to cancer, the gene in its healthy state must do something to control cell growth. The discovery in 1986 of the actual retinoblastoma gene through a joint effort of the laboratories of Thaddeus Dryja, of the Massachusetts Eye and Ear Infirmary, and Stephen Friend and Robert Weinberg, of the Whitehead Institute, led to the revolutionary conclusion that there exists an entire category of gene whose function is to suppress the action of proto-oncogenes. (The terms "proto-oncogene" and "anti-oncogene" are, incidentally, disliked by researchers, because they imply that the primary function of these genes is to cause or prevent cancer. In fact, nature did not spend hundreds of millions of years evolving such potent genes with any such idea in mind. Their jobs are to regulate functions of cells.)

Further research has revealed that a breakdown of tumor suppressor genes may be a far more common cause of cancer than the derangement of proto-oncogenes is. This appears true even though the loss of suppressor genes is a recessive event, that is, it must occur in two genes, whereas only one proto-oncogene need go wrong to help trigger a cancer. Transformed proto-oncogenes have been found so far in only 20 percent of all human tumors, whereas examination of the chromosomes in various tumors reveals frequent deletions of chromosomal material, as one would expect to find where a gene has been lost.

A hint that researchers were onto something big with the discovery of tumor suppressor genes came when the retinoblastoma gene was linked not only to eye tumors but to a totally different form of malignancy. In pinpointing the gene, Dryja, Friend, and Weinberg noted that the gene also predisposes those who have defective copies to osteosarcomas, cancers of the bone. This explains why adolescent and adult retinoblastoma survivors are several hundred times more likely to develop such bone cancers than the average person.

Interestingly, many of these tumors occur in the orbital bone of the eye, previously radiated to kill the retinoblastoma. Almost certainly, the radiation mutates the remaining healthy gene in the orbital tissue. But the retinoblastoma gene's involvement in cancer apparently does not stop there. Its deactivation has also been implicated in breast cancers, small-cell lung cancers, and malignancies of the bladder. Hence it appears that the same tumor supressor gene can be at fault in a wide variety of tumors.

In 1988, a paradigm for how tumor suppressor genes work in concert with other factors to cause malignancy emerged in the form of an explanation for colorectal cancer, the second leading cause of cancer death in the United States. Bowel and rectal cancers, and the benign, mushroom-shaped polyps of the intestinal wall that often give rise to them, have long been known to run in certain families. This genetic predisposition was quantified by a group led by Mark Skolnick, of the University of Utah, who found that polyps appeared to be linked to a gene so common that up to a third of the population carries it. But it remained for another team, headed by Vogelstein, to clarify the actual genetic mechanism by which the polyps become cancerous.

The Vogelstein group found that a series of from four to six genetic changes, occurring in rough sequence over the course of a lifetime, are necessary to generate cancer of the colon or rectum. These changes are the same both in the inherited form of colorectal cancer, which accounts for about 10 percent of the disease, and in the far more common non-familial form. Very early in the development of the cancer, there is a loss of paired genes on chromosome 5. (Consistent with the Knudson theory, people with the familial form of the disease are born with one of these genes already flawed.) These losses permit the formation of intestinal polyps. Later in this slow dance of death, *ras* oncogenes on chromosome 12 frequently become activated, causing the polyps to grow, but the polyps are not yet cancerous. Next, there must occur the deactivation or deletion of a gene on chromosome 18, a gene identified by Vogelstein and colleagues in a paper published in early 1990. This gene, which they christened DCC (for "deleted in colon cancer"), is clearly a tumor suppressor. Vogelstein found it to be missing in some 70 percent of the bowel cancers he examined. DCC appears to belong to a clan of genes whose function is to make cells adhere to one another, instead of sliding apart, as they would do when a tumor begins to spread. Finally, according to Vogelstein, the all-important p53 gene on chromosome 17 must be lost. P53 appears to be a critical tumor suppressor gene whose loss administers the coup de grace that allows the polyp, which has by now bloomed into a large benign tumor, to turn into a cancer.

"A number of genetic changes are probably required for the development of these cancers," muses Vogelstein. "It is the accumulation of these alterations, rather than the order of their occurrence, that appears to be important in causing cancer growth. The fact that it may take years for these changes to accumulate explains why colon cancer occurs mainly in people over forty years of age."

In May of 1993, Vogelstein's group found evidence of still another gene that seems to predispose people to the familial forms of cancer—the gene seems to have the unique capacity to trigger all sorts of malignancies. Reporting in *Science,* Vogelstein and colleagues estimated that 1 in 200 people carries the predisposing gene, and 95 percent of them eventually will develop the disease. Of these, 60 percent will have colon cancer, while 40 percent will develop tumors of the uterus, stomach, pancreas, or urinary tract. Although the gene has yet to be found, the marker should quickly lead to better screening tests, especially for colon cancer.

Vogelstein's work has unmasked, for the first time, the workings of a major cancer. But even as the team was unlocking the secrets of a disease that produces nearly 150,000 new cases each year in the United States and kills some 60,000 people, another group headed by John Minna, of the University of Texas Southwestern Medical Center, was shedding light on another, even more deadly malignancy—lung cancer. It is Minna's belief that as many as ten to twenty genetic abnormalities may be involved in the development of a lung cancer. By examining the battered chromosomes of lung tumor cells, he has located at least six putative abnormalities. They include the alteration of *ras,* the overexpression of the proto-oncogene *myc,* and distinctive abnormalities in the p53 gene. Minna has also found segments of DNA on chromosomes 3 and 13 that are deleted, indicating that unknown tumor suppressor genes may be absent at those sites.

Discovery of p53's footprints in two of humanity's most feared cancers has prompted laboratories throughout the world to comb through a wide assortment of tumors to see whether the p53 genes are likewise abnormal in these cells. To date, missing or altered p53 genes have been found in an astounding 60 percent of tumors, involving malignancies not only of the colon and lungs but of the breast, brain, ovaries, esophagus, bone, bladder, liver, cervix, and adrenal cortex. The message is not that p53 is the sole culprit in any of these malignancies. In fact, all the tumors were marbled with deletions on various chromosomes. What seems increasingly likely is that cancer is the climax of a series of incremental mistakes in a single cell over the course of a lifetime, each involving vital genes scattered widely throughout the genome. But some of these genes, particularly certain tumor suppressor genes, such as p53 and the retinoblastoma gene, may be more crucial than others and involved in more than one kind of cancer.

The current thinking is that the formation of a tumor cell requires the interaction of a number of oncogenes and the failure of one or more suppressor genes. This phenomenon has come to be known as "cooperation" among cancer genes. Experiments indicate, for example, that the activation of two or more independent oncogenes, along parallel messenger pathways, is required for a tumor to arise, since it has been found that one oncogene acting alone seldom does the trick. Why is the system so exquisitely complex?

Normal cells have evolved multiple independent systems to regulate growth and differentiation—probably in order to make the process failsafe—and only after several untoward events can these controls be overridden.

Just as there are families with inherited changes in one retinoblastoma gene, it now appears that other kinships exist in which many, if not most, members are born with a missing or altered p53 gene in each of their cells. Among the most dreadful conditions in the medical annals is the so-called Li-Fraumeni syndrome, first identified more than twenty years ago. Li-Fraumeni families, while very rare—only one hundred of them are known worldwide—have an improbably high rate of developing multiple cancers in sites as diverse as the breast, muscles, brain, bones, marrow, adrenal cortex, pancreas, gonads, lungs, pancreas, prostate, and skin. These cancers strike early. A family member has a 50 percent risk of coming down with cancer by the age of thirty, a time of life when only 1 percent of the general population has developed the disease. By the age of seventy, 90 percent of those with the Li-Fraumeni gene defect will have fallen victim.

The nature of this defect puzzled researchers for years. Then, in November of 1990, Stephen Friend's laboratory, in conjunction with Frederick Li and Joseph Fraumeni, the National Cancer Institute epidemiologists who first identified the syndrome, announced a dramatic discovery. Examining DNA taken from five Li-Fraumeni families, the researchers found that p53 mutations were present in all five. The defect is almost certainly the predisposing factor in the syndrome. The next question to be answered is how many people walking around in the general population may themselves have been born with alterations in their p53 genes? "The suspicion," Frederick Li told Science, "is that the germ line mutation might be present in people who don't have these spectacular family histories." Adds Robert Weinberg, "Li-Fraumeni syndrome could be the tip of the iceberg."

Though Li-Fraumeni victims contract many kinds of cancer, the one with the highest incidence among them is, by far, breast cancer. The evidence seems strong, therefore, that the shutdown of p53 plays an important role in cancer of the breast, which is the third-leading cause of cancer deaths in the general population. Each year, about 180,000 women in the United States will be diagnosed with breast cancer, and 46,000 of them are doomed to die of it. Curiously, p53 is not the only gene on chromosome 17 that appears pivotal in breast tumors. The cancer researcher Mary-Claire King and her colleagues at the University of California at Berkeley have linked a form of early-onset hereditary breast and ovarian cancer to an unidentified gene elsewhere on the chromosome. Their study focused on 329 individuals from twenty-three extended families in which breast cancer is common enough to be called a family curse. Nearly half of the family members studied, or 146, had experienced breast cancer. A clue to the severity of the disease in these families can be gauged from the fact that it strikes as early as age twenty-three, that many of the cases affect both breasts, and that, in some cases, it has even affected men and can be passed on by them. Although the inherited form of breast cancer so far is believed to account for only about 10 percent of all cases of the disease, the King group's findings nonetheless may have much greater applicability. The same genes are very likely involved in the "sporadic" form of the disease in the general population. Understanding the molecular genetics of the hereditary kind is thus critical. Furthermore, such studies should shed light on ovarian cancer, which is believed to have a number of genetic flaws in common with breast cancer.

A fierce hunt is now on for further tumor suppressor genes. The specific way these genes work is still being sorted out. But researchers at the University of Washington, in Seattle, have linked them to certain cell receptors whose job is to *remove* phosphate groups from proteins. In other words, tumor suppressor genes may function like a good linebacker, stripping away the phosphates that carry from protein to protein the command to divide. If, as it now appears, both the proto-oncogene and the anti-oncogene work through the same communication network, the one deactivating the other, then it is likely that there are as many anti-oncogenes as proto-oncogenes. It is also thought possible, even likely, that anti-oncogenes provide the long-sought key to the mystery of differentiation—the process by which cells become tissue-specific by shutting off genes that are not needed by that tissue. If the cell does not receive such a message, because of inactivation of the growth suppressor gene, it may cease to differentiate—and become the gross, undifferentiated mass that is a tumor.

A clue to how tumor suppressors work was provided by researchers investigating the causes of neurofibromatosis, a disfiguring, sometimes fatal disease marked by an abundance of mostly benign tumors throughout the body. After a long search, the gene for this grotesque disease was isolated in 1990; its discovery soon led to the identity of the protein manufactured by the gene. Intriguingly, this protein, NF-1, was found to downregulate the *ras* proto-oncogene. Thus the neurofibromatosis gene almost certainly is a tumor suppressor gene, and its inactivation allows *ras* "to run amok stuck in the 'on' position," according to the University of Michigan's Francis Collins, who found the gene. The result is the proliferation of tumors. While neurofibromatosis tumors tend to be benign, Collins notes, cells in those tumors are far more susceptible to further mutations that could render them malignant. Most important, findings about NF-1 have provided the strongest evidence to date that tumor suppressor genes act by directly shutting down the activity of proto-oncogenes.

While both copies of a tumor suppressor gene usually have to be deactivated before a cancer can develop, neurofibromatosis and the Li-Fraumeni syndrome alike are inherited in a dominant fashion; that is, acquiring a defective gene from only one parent is enough to transmit the disease. Why this should be so remains a mystery. But some scientists believe that certain NF and p53 mutations may cause so much bogus protein to be made that it *overrides* the competent performance of the alternative good gene. Wheels within wheels!

From the evidence, the NF gene is supposed to snuff out the cell cycle signal on its way from growth factor receptor to nucleus. But there appear to be a number of tumor suppressor genes that work at ground zero, within the nucleus itself. Two transcriptional factor genes, *c-erbA* and *c-rel*, appear to activate genes that directly turn *off* the cell cycle. A hitch in *c-erbA* or *c-rel* would prevent these "damper" genes from coming on-line, tantamount to letting the cell cycle genes they normally control run wild. And a series of important experiments indicates that p53 itself may be a transcriptional factor. Its protein is found in the cell nucleus, and binds to DNA the way transcriptional factor proteins do. More important, it seems to switch on any gene that is placed next to it. Conversely, when a mutant variety of p53 is added, it diminishes the healthy p53 activity proportionately. That is, if equal amounts of normal and mutant p53 proteins are added to the mix, the expression of the gene next to p53 declines by 50 percent. More mutant protein reduces it as much as 95 percent. Vogelstein's laboratory has found that this is because the mutant p53 prevents the normal version from binding to the gene's promoter. The best guess at the moment is that p53 activates some crucial gene that ordinarily inhibits cell growth. When it is blocked from performing its role, cell growth proceeds unimpeded.

Switching off cell division, incidentally, is thought to be necessary to allow the cell to differentiate. For the most part, growth and differentiation are mutually exclusive, and differentiation is a critical counterweight to the tumor process. When a cell differentiates, it is highly unlikely to become cancerous. It is a hallmark of cancer cells that they remain undifferentiated and generic, just raw, insatiable protoplasm. By blocking differentiation, oncogenes may not directly cause a cancer, but they allow a mutant cell to proliferate until a second, more critical alteration ensues.

What factors make proto-oncogenes and tumor suppressor genes go astray? Essentially, there are four. One of these crucial genes may undergo a change in one or several nucleotides, an event known as a point mutation. Or it can be expressed in excessive amounts because of either stepped-up transcription or the bizarre circumstance in which a gene is duplicated many times, so it is present in a cell in as many as fifty copies. This process is known as amplification. In some cases, a gene is absent altogether, an event known as deletion. And in many other cases, the damage involves whole chromosomes, which in breaking apart and forming new connections, called translocations, activate oncogenes located at the break sites.

In fact, such a translocation phenomenon provided one of the first clues to the cancer process in 1960 with the startling observation that many people suffering from a form of leukemia called chronic myelogenous leukemia (CML) seemed to be missing part of chromosome 22. Because the original discovery was made by two Philadelphia researchers, the chromosome was dubbed the Philadelphia chromosome. But where did the missing material go? For thirteen years, investigators searched in vain for the answer. Then, in 1973, the University of Chicago's Janet Rowley announced that she had solved the mystery. Using sophisticated new techniques for staining chromosomes that enabled scientists to identify and track parts of chromosomes, Rowley found that a piece of chromosome 9 had broken off and switched places with a piece of chromosome 22. A study of 1,129 CML patients revealed that 92 percent of them had the specific translocation discovered by Rowley. Subsequently, it was learned than an oncogene called the Abelson, or *Abl*, oncogene, is right next to the breakpoint on chromosome 9. The exposed gene journeys to chromosome 22 as part of the migrating DNA, where it forms a hybrid with another gene, whose resulting protein appears to be a tyrosine kinase with ferocious activity. When this lethal mix-and-match occurs in a bone marrow stem cell, all of the cell's offspring will carry the translocation, and the result will be a fulminating leukemia.

In 1982, a translocation was implicated in the lymphatic cancer called Burkitt's lymphoma, which is common in central Africa. Harvard's Philip Leder, and Carlo Croce, of the Wistar Institute, in Philadelphia, found that 90 percent of Burkitt's patients had a swap involving chromosomes 8, 14, and sometimes 2 and 22. A gene that codes for human antibodies shuffles around in tandem with *c-myc*. In 1986, Janet Rowley followed up her earlier work by finding a translocation between chromosomes 9 and 11 in acute monocytic leukemia (AML). The schism splits up the alpha and beta interferon genes, which are located on chromosome 9, shipping the beta interferon gene off to chromosome 11. In return, a proto-oncogene called *c-ets-1* settles down on chromosome 9 next to the alpha interferon gene.

But why should a chromosomal game of musical chairs cause cancer? It is believed that as long as proto-oncogenes are in their normal surroundings, they remain under strict control. But when they move to a new neighborhood, they fall under the spell of gene control regions meant to apply to other genes. Signals from these alien promoters may switch the oncogene on when it is supposed to stay turned off. This, together with the loss of a protective tumor suppressor gene, could trigger the uncontrolled proliferation of cells we call cancer.

"It's as if I tore off a page of the *Chicago Tribune* and fused it to the *New York Times*. It wouldn't make any sense," says Leder. "By the same token, you destroy the sense of logic by which the gene was expressed when you break it off from its original position and put it in another."

Hence, in AML, *c-ets-1* apparently becomes activated by control signals from an interferon promoter after migrating to chromosome 9.

This presumably sets off a chain of circumstances directly related to the central problem in AML, which is a superabundance of monocytes, a type of white blood cell. One of interferon's jobs is to halt the proliferation of monocytes whenever it gets a signal that too many of the cells are being made. But if a translocation has occurred, the newly activated *c-ets-1* gene may stimulate overproduction of monocytes, while, by its presence, it may simultaneously block the feedback signal. Or even worse, in a meltdown kind of scenario, the feedback signal may get through—but instead of responding with the release of interferon, which would shut off runaway monocyte production, the interferon promoter triggers its new "go for broke" neighbor c-ets-1, which causes still more monocytes to be made.

Similarly, in Burkitt's lymphoma, the c-myc gene gets deposited next to the gene for immunoglobulin, the antibody protein. Since immunoglobulin is needed in vast amounts to fight infections, it must be controlled by a very powerful promoter. Allowing such a promoter to switch on an already aggressive gene like c-myc can be a recipe for disaster. Says Rowley, "Myc is one of the very early signals within a cell that tells the cell that it's going to undergo division. If there is some kind of derangement whereby myc is taken away from its normal regulator, and put next to a regulator that keeps it permanently switched on, it would forever be sending a signal to the cell, 'Get ready to divide.' So cells would keep dividing all the time, and the result would be cancer." (It is almost always cells that are dividing that are in danger of becoming cancerous. Cells that do not divide, such as neurons, are not prone to cancer as a rule. That is why nearly all of the more than one hundred forms of cancer fall into two categories: carcinomas and sarcomas. Carcinomas are cancers of the tissues that make up skin and the lining of organs, while sarcomas involve connective tissue such as bone and cartilage. In both kinds of tissue, the cells are constantly replenishing their numbers through division.)

Although some researchers, notably Jorge Yunis, believe that all cancers will ultimately be shown to involve chromosomal defects, not all of these defects are necessarily due to translocations. For example, the c-myc gene has been implicated in a variety of cancers at various sites in the body. However, rather than move to the wrong place, the proto-oncogene stays put; but because some force has caused it to mutate, it multiplies wildly within the cell. There may be anywhere from thirty to fifty copies of myc in a single cell, urging the cell to divide. Obviously, a greatly amplified protein will crank out enormous and inappropriate amounts of growth protein. Very likely, this perilous situation is the result of a failure of a tumor suppressor gene to hold an oncogene in check.

While cancer has been a continuing source of bafflement, the way in which chromosomes sustain damage that leads to tumors is no mystery. Legions of scientists have shown over several decades that our cells are prey to the mutational effects of a broad class of irritants known as carcinogens.

To be sure, genes can go wrong by falling into the clutches of a virus, as happens in animals. In taking a proto-oncogene captive, the virus does something to keep the gene in a permanently aberrant state. Thus, as we have seen, simple infection with a virus containing an oncogene is sufficient to cause a cancer in animals, and the probability that cancer will develop in an exposed creature is in fact very high.

In human beings, however, the role of viral infection in causing cancer appears to be exceedingly small. The likelihood is that we have evolved some way of shutting off viral oncogenes. That is not to say that viruses have nothing to do with human cancer. They do. In the early 1980s, after half a century of trying, science finally tied several human viruses to malignancy. Two rare diseases, T-cell leukemias and lymphomas, were linked to a retrovirus called HTLV-1. A connection was firmly established between the hepatitis B virus and late-onset liver cancer. An association was made between the Epstein-Barr virus and Burkitt's lymphoma. And the AIDS virus was linked with Kaposi's sarcoma, a once rare form of cancer now seen frequently in AIDS. Investigators have also associated the human papilloma virus and herpesvirus with certain genital cancers. But with the exception of Kaposi's sarcoma, which seems to be caused directly by a gene belonging to the AIDS virus, these afflictions do not appear to be brought on by viral oncogenes. Rather, the virus, by its intrusive presence in the cell, makes a proto-oncogene or anti-oncogene go bad, either by crashlanding on top of it and mutating it or by coming to rest next to it, allowing the viral control region to assume command.

Nonviral carcinogens are far and away the chief cause of human cancer. One need look no farther for these substances than our dinner table, our backyard, and our petty addictions, for these insidious substances include such familiar items as red meat, sunlight, and the nicotine in cigarettes—which Leder calls a "mutagen par excellence." More than one thousand carcinogenic materials and agents have been cataloged, including charcoal, pesticides, saccharine, vinyl chloride, nitrosamines, nitrites, and asbestos, with cosmic radiation being perhaps the most pervasive mutagen of all.

Cancer is not normally thought of as hereditary. But, in fact, it appears that certain people inherit a vulnerability to chromosome breaks or deletions at particular locations. If these locations encompass or abut an oncogene or tumor suppressor gene, such individuals will have a higher risk of developing a malignancy. It is surely no accident, for example, that four members of former president Jimmy Carter's family,

including his two sisters, his brother, and his father, all died of pancreatic cancer (his mother, Lillian Carter, died of breast cancer that metastasized to her pancreas).

Why isn't cancer even more common than it is? To be sure, one out
of five of us will eventually die of cancer. But given that we are under
daily bombardment by mutagens, constantly causing random breaks in
our genes and chromosomes, one might think that we would be developing cancer all the time.

Certain statistical factors protect us, of course. There are no more
than 100 or so proto-oncogenes in our cells, and possibly a like number
of anti-oncogenes. Yet we have an aggregate of 100,000 genes, and
enough total DNA for a million genes. Therefore, the chances that 1 of
these 200 key genes will be struck by a cosmic ray or a saccharine molecule are rather slight. Then, too, damage can occur in a proto-oncogene
or tumor suppressor gene and still not cause cancer, because a tumor
forms only when oncogenes switch on or suppressor genes fail in a cell
that is susceptible to a specific kind of malignancy. For example, if a
proto-oncogene translocates and comes to rest alongside an interferon
gene in a skin cell, it will most likely not induce cancer. Interferon genes
are supposed to be switched off in skin cells. But in a marrow cell, where
interferon is made, genes are operable and ready to interact if an oncogene happens by.

That is why cigarette smoke figures so strongly in lung cancer and
oral cancer. It clearly affects proto-oncogenes and tumor suppressors in
lung tissue, as well as similar genes in the mucosa of the mouth and
throat. Papilloma viruses, meanwhile, are sexually transmitted and thus
have access to the cells of the internal genitalia, which is why they contribute to cancers of those organs. Sunlight triggers changes in skin cells
that can become the relatively inoffensive basal cell carcinoma or the
highly virulent melanoma.

Another factor holding down the prevalence of cancer is our own
ability to repair DNA. As we have seen, cancer requires more than one
"hit" from a mutagen and, in adult cancers, many hits acquired over a
lifetime. In part, this is because a substantial number of genes must go
awry before a cell turns outlaw. But it is also because cell nuclei always
contain enzymes whose job is to repair DNA quickly. This "fail-safe"
system has evolved over the ages because of DNA's propensity both to
become injured and to make mistakes during cell division. Every time a
cell divides, it must reproduce three billion letters of the nucleotide al-
phabet exactly. No system is error free, and glitches constantly occur.
That is where the repair enzymes come in. "But sometimes these en-

zymes fail, and a mutation slips through," notes Phil Leder. "And that mutation, if it's the first one, is going to wait around until another one, and still another one, take place."

Frank Raucher, Jr., a senior research consultant for the American Cancer Society, said in an interview with the *Chicago Tribune*'s Ron Kotulak, "People very likely get cancer a couple of hundred times in their lives but our immune surveillance systems are so good that we get rid of those aberrant cells." This scenario probably explains why cancer is more common in the elderly. The older you get, the greater are your odds of sustaining two or more hits from a mutagen—and the less efficient is your immune system at repairing DNA.

The fruits of these conceptual breakthroughs may not be long in coming. One bright prospect is the early detection of cancer before it spreads to the rest of the body. As the proteins manufactured by oncogenes are identified, chemical tests for mutant versions of these proteins, or their presence in abnormal concentrations, can be devised. Cancer diagnosis will then consist of testing people for the presence of the proteins, perhaps by something as simple as a urine or blood test.

Scientists from the Scripps Institute, in La Jolla, and from Hammersmith Hospital, in London, have already succeeded in spotting oncogenic proteins in the urine of patients with cancer of the breast, prostate, bladder, lymph glands, or lungs. The researchers first injected oncoproteins into mice, which responded by making antibodies to the proteins. The antibodies were then set loose in the urine of cancer victims. As antibodies do, they quickly zeroed in on the molecules they had been bred to attack, in this case the oncoproteins. Not only did the test reveal the presence of cancer, but it discriminated as to the kind of cancer—for example, the protein pattern of lung cancer was different from that of breast cancer.

In April 1992, Bert Vogelstein's group announced that colorectal cancer could be diagnosed by screening people's stool for the presence of altered *ras* genes. The researchers first analyzed the intestinal or rectal tumors of twenty-four patients from whom stool samples had been obtained before surgery. Nine patients were shown to have *ras* mutations in their tumors. Stool from these patients was then examined to see whether the same mutations could be detected in cells shed from the intestinal wall. This was by no means a sure thing. Stool is packed with enzymes that can break down DNA. But using the PCR technique to amplify the genes present, the group was able to tease out *ras* mutations in the stool of eight of the nine patients. This is a much better batting average than that claimed by the current diagnostic test for colon cancer, which checks stool for the presence of occult blood and frequently yields false results. Moreover, the Vogelstein test was able to detect mutations in two patients whose tumors had not yet become cancerous. This is encouraging because 90 percent of colorectal cancers can be cured if still localized to the intestinal lining. The test is not ready for the big time, however. *Ras* mutations are seen in only 50 percent of colorectal cancers. For the method to have broad screening applicability, it must be able to pick up all the possible genetic mutations in colorectal cancer. Even then, the cost of PCR may make it impractical to screen everybody who comes to a doctor's office. It may be more useful for people who have a genetic predisposition to colon cancer or have had a previous tumor.

Cancer diagnosis is also likely to be improved by the budding ability of molecular researchers to tell, from the number of multiple copies of oncogenes in a malignant cell, which patients need particularly intensified forms of cancer treatment. This triage tool was first uncovered by Dennis Slamon, of UCLA, whose group showed the connection between the number of oncogenes present and the prognosis of breast and ovarian cancer patients. Tumors carrying multiple copies of an oncogene called *HER-2/neu* may signal faster-growing tumors. It was found that patients with tumors containing extra copies of these genes were more likely to suffer a relapse. The significance is that physicians may soon use gene probes to evaluate prognosis and give more-aggressive therapy to those with multiple copies of the gene.

Eventually, of course, scientists hope to go beyond diagnosis and begin taming cancer with the powerful arsenal that will soon be offered by gene therapy. Some laboratories, like those of Steven Rosenberg and Gary Nabel, are already attempting to reverse melanoma and certain other cancers by using gene therapy–enhanced immunotherapy. But other, even bolder treatment modalities are on deck.

Among the most dramatic and potentially far-reaching is Jack Roth's proposal for tricking oncogenes into putting themselves out of business. Roth bases his idea on the assumption that one need not tackle the full array of genetic changes in a tumor cell to halt cancer. "It may only be necessary to reverse one or two of these genetic abnormalities to break the chain," he says.

The specific alteration he wishes to attack in lung cancer cells is the mutant version of a *ras* gene called *K-ras*. In lung cancer, *K-ras* is almost always mutated in a specific place, the twelfth codon, a spot very near the beginning of the gene. It is a mutation, by the way, that correlates highly with cigarette smoking. Not surprisingly, it is also seen frequently in pancreatic cancer, another malignancy that predominates among smokers.

In early 1992, Roth learned that he could greatly reduce the development of tumors in nude mice by inoculating them with retroviral vectors containing short stretches of genetic material whose base sequence near the twelfth codon exactly mirrored the sequence of the mutated *K-ras* gene. When the tumor cells take up these complementary fragments—which they do with great abandon, Roth discovered—the fragments produce strips of messenger RNA that bond tightly to the single strands of messenger RNA produced by the *K-ras* gene. These *K-ras* mRNAs, now double-stranded in places, are like ruptured ducks; they stagger along, unable to make their way to the cell's factories, where they would ordinarily direct production of the mutant *K-ras* protein. The oncogene has thus been tricked into switching itself off. By drastically downregulating the production of *K-ras* protein, Roth was able to generate a fourfold decline in the number of lung tumors seen in his mice, not to mention a significant reduction in the size of the tumors that did develop.

Roth's 80 percent "cure" rate stands in startling contrast to the human survival rate in lung cancer, which, in spite of everything medicine can throw at it, remains an abysmal 14 percent.

The idea of crippling a defective gene by saddling its mRNA with an albatross is not new. First proposed in the early 1980s, the strategy goes by the euphonious name "antisense" because it seeks to disrupt the informational, or "sense," strand—mRNA. But while the concept has some mileage on it, Roth's group is the first to attempt an actual antisense experiment in human beings. More important, the experiment marks the first time anyone has tried to interfere directly with the action of an oncogene.

One would think Roth had enough on his plate. But at the same moment he is breaking ground with antisense technology, he is trying to pull off another coup. He is the first to attempt to replace the missing or mutated p53 gene in cancer patients lacking the crucial tumor suppressor. Again the experiment was preceded by research in rats. Insertion of the p53 gene in the rodents' tumor cells resulted in a dramatic, nearly 300 percent reduction in the development of malignancies.

As approved by the RAC, Roth's experimental design calls for using both approaches on a group of fourteen lung cancer patients, all of whom have failed the more traditional treatment methods of surgery, chemotherapy, and radiation. Each of these end-stage patients is experiencing difficulty breathing because their tumors are obstructing their lungs. "Whether we put in an antisense construct or a P53 gene depends on the mutation that is present in the patient's tumor," says Roth. "Some patients have *K-ras* mutations. Others have p53 mutations. Only a relatively small percentage of tumors have both. We're thus customizing the treatment to the defect."

Gene delivery is straightforward. Roth injects his viral vectors into the patients' airways directly, using a bronchoscope to visualize the tumor. The treatment is repeated for up to five days in a row, since there appears to be very little toxic side effect, which is decidedly not true of the TNF and other immunotherapy experiments being conducted elsewhere. Although Roth does not expect to succeed in infecting more than 30 percent of the tumor cells, he hopes to have an impact on many more cells thanks to the "bystander" effect, which was clearly at work in his experimental mice. The bystander effect makes the uninfected cells somehow begin to behave as if they, too, had received an influx of new genes.

The value of Roth's therapy does not apparently lie in reversing advanced cancers. "We will call it a victory if we just see some shrinkage of the tumor," he says.

"My goal," he notes, "is to use this treatment for *prevention*. It is applicable to the very early stages of cancer, and may even make it possible to treat people before a cancer has developed." As Roth envisions it, people with high-risk factors, such as smokers or chemical workers, could be screened for the telltale signs of premalignancy. Lung tissue shows characteristic changes before a tumor develops. So does the lining of the esophagus, which becomes similar to that of the stomach, causing victims to experience severe heartburn. "We have the screening capabilities right now," notes Roth. "Using PCR, we can sample very small amounts of tissue and amplify the genes in question. Then we can move in and stop the cancer process in its tracks."

All of which would be very gratifying to Roth, who, while head of the NIH's thoracic oncology section from 1980 to 1986, worked closely with Steven Rosenberg. Over the years, Roth has seen more than his share of human misery. Lung cancer kills a staggering 143,000 people annually. "For years," Roth laments, "we've been faced with very difficult, or almost impossible, clinical problems. It is tragic to have people come to us with cancer and be unable to do anything about it. Now we are finally developing the tools to treat this problem, or prevent it from ever happening in the first place. It is tremendously exciting."

In principle, using antisense constructs to "switch off" malfunctioning oncogenes, or replacing missing or deactivated tumor suppressor genes, would be the ideal way to deal with cancer. That such a technique is feasible was shown as far back as the 1970s, when researchers learned to their surprise that if normal cells are added to tumor cells, the tumor cells often revert to normal. Later, when the existence of tumor suppressor genes came to light, a number of laboratories showed they could reassert control over cancer cells by introducing tumor suppressor genes.

In one experiment, California researchers "cured" a severe childhood kidney cancer, known as Wilms' tumor, by putting a normal human version of chromosome 11 into a culture of Wilms' tumor cells, whereupon the cells lost their ability to produce tumors in mice. The disease is marked by the loss of genes on chromosome 11. Using similar methods, other researchers have suppressed cells from tumors of the prostate, marrow, and brain. And, as we saw, Vogelstein dramatically halted the growth of colon cancer cells with healthy copies of p53.

But there is a hitch. The gene delivery problems are immense. Imagine trying to get a normal copy of a tumor suppressor gene into every malignant cell in a cancer victim's body. Yet that is the challenge—no more, no less—for should one lone tumor cell manage to evade the net, the cancer process will start all over again. Time alone will tell whether science can figure out a way around this impasse. Some believe the answer may lie in monoclonal antibodies, those cellular hybrids that have some of the attributes of magic bullets, tracking down cancer cells wherever they are in the body, even if they have spread far from the initial tumor site. It remains to be seen.

As Roth labored with his bold new approaches, Ken Culver continued to refine his attempt to use gene therapy to cure brain tumors. Of the first six patients to receive herpes simplex genes and then ganciclovir, two showed some shrinkage of their cancers. But both suffered from tumors that had originated elsewhere in the body and spread to the brain. The other four had tumors that began in the brain, and they showed no improvement whatsoever.

Culver explained the failure in several ways. For one thing, the brain cancer patients in his experiment had been heavily dosed with radiation, the standard treatment for such tumors. The problem is, radiation creates large masses of scar tissue. "With all that scar tissue," said Culver, "there is no way for the bystander effect to spread throughout the tumor. It gets stymied." The bystander effect is so pronounced in rats, he noted, "because nobody gives rats radiation." Another impediment to success, Culver said, was the fact that it was very difficult to target the herpes genes into the actively growing parts of the tumor, a necessity if the treatment is to work. "You're injecting randomly into the tumor, and with a large tumor, you could make five injections and easily miss the active area of growth."

Showing the resourcefulness that has characterized his career, Culver came up with some new ways to deal with these problems. The first idea he planned to try at the Iowa Methodist Medical Center, in Des Moines, which he joined in July of 1993 after leaving NIH. The idea was to remove the center of the brain tumor surgically, leaving only the margins, where the actively growing regions of a tumor tend to congregate. "Then we can concentrate our injections of herpes genes into those tumor walls." Culver hoped to leave a shunt in the patients' skulls to allow further injections as needed. He was also considering using human cells instead of mouse cells to introduce the herpes genes into patients' brains, since the human cells would migrate within the tumor and the mouse cells would not. "Being mobile, they would reach more of the tumor cells to deposit the genes," Culver explained.

The other planned experiment was to be performed in conjunction with Los Angeles Children's Hospital. It would concentrate exclusively on small children with primary brain tumors. The reason was simple. Children under the age of six do not receive radiation, because it tends to retard the growth of the brain. "So this way we will test the idea that this treatment works better when there has been no radiation to create scar tissue." If the treatment works well, then Culver will have the ammunition to persuade regulatory committees to allow him to use his treatment in adults in lieu of radiation. "Right now," he said, "we don't have enough information to justify withholding the standard treatment from desperately ill patients. But I think we will have that proof in a very short time.

Indeed by mid-1994 a total of fifteen patients had undergone the procedure. Of these, six had responded positively and were still alive from six to eighteen months after gene therapy.

Culver was, in any case, realistic about the value of his quixotic experiment. "I don't think," he said, "that this will be the complete answer for brain tumors, but it is an important step along the way."

17

TRIUMPH OR TRAGEDY?

Trouble in gene therapy first showed up in the early 1990s as a sullen dispute that threatened to tear apart the brave and formerly united little world of neuromuscular diseases. It centered on a desperate treatment, long dreamed about, for a deadly genetic disorder, Duchenne's muscular dystrophy. Duchenne's is one of the most familiar of all inherited illnesses, thanks to the power of television. Its ravages have been portrayed widely, sometimes to the point of lugubriousness, by the annual Jerry Lewis telethons for the Tucson-based Muscular Dystrophy Association. But although a generation of such glitzy fund-raisers has provided torrents of tears and hundreds of millions of dollars to help strip away the mysteries surrounding Duchenne's, the clinical payback until 1990 was virtually nil.

Then, suddenly, whipping like a wind sock in a hurricane, the respected MDA, the very epitome of a savvy, humane charity, was to find itself being depicted as cruel and uncaring because of its refusal to back a man it had previously supported enthusiastically. Peter Law, a maverick researcher from Memphis, Tennessee, was claiming spectacular results with a novel form of gene therapy—the injection of billions of immature muscle cells called myoblasts directly into the legs of Duchenne's boys—that seemed to suggest that 390,000 otherwise doomed Duchenne's patients worldwide might be spared their tragic fates. Almost overnight, Duchenne's families had something to cling to: if Law was right, dead muscles somehow could be resurrected and Duchenne's boys soon would be able to throw away their wheelchairs and walk. Hundreds of parents implored Law to accept their sons in his on-going clinical trials.

But if Law was wrong, if no long-lasting benefit was showing up in the bodies of the boys, the saga of this heartbreaking disease, so long fraught with quack cures and dashed hopes, had again turned dismal.

Duchenne's is the most common and severe form of muscular dystrophy, a group of inherited disorders in which the muscles of the body progressively degenerate. Duchenne's strikes mostly boys at a rate of 1 in every 3,500 male births. About a third of all cases consist of new mutations; that is, there is no long-standing family inheritance of the gene. Duchenne's generally does not tip its hand until victims are at least three years old. As infants, they seem perfectly normal, with one exception: they tend to have unusually large calves, because of the replacement of muscle by fat and connective tissue. Ironically, parents often take this to be a sign that their son is going to be heroically muscular. Between the ages of three and five, the boy suddenly starts to stumble. He experiences difficulty climbing stairs or rising from the floor and is easily fatigued. He may walk on his toes or display a characteristic waddling gait. Gradually, the condition worsens, with muscle deterioration sweeping upward from legs to hips to chest. Eventually, the weakening respiratory and heart muscles become affected, leading to pulmonary and cardiac problems. By the age of twelve, the boy needs a wheelchair. By his early twenties, in all but the rarest cases, he will die. The immediate cause of death, as in so many other neuromuscular diseases, is respiratory or cardiac failure as a result of muscle weakness.

Despite the Peter Law controversy, for the first time since the ailment was described more than a century ago, there is a strong sense in the scientific community that a treatment and possible cure are nearly at hand. This expectation rests on two crucial breakthroughs: discovery of the gene believed responsible for Duchenne's, and identification of the muscle protein made by the gene that is absent or defective in its victims.

By starting to attack the disease at its biochemical basis, molecular scientists may have accomplished what generations of earlier researchers could not: the unearthing of the root cause of Duchenne's muscular dystrophy and, by implication, of dozens of similar disorders of musculature.

Both research milestones burst in a span of less than two years from the same cramped, little laboratory in Boston, tucked away in a nondescript research building adjoining that international mecca of pediatrics—Boston Children's Hospital. The proprietor of this minimalist enterprise was a Harvard geneticist, now in his early forties, named Louis Kunkel, whose disarmingly undergraduate manner and appearance—he has the compact build and boyish face of a college wrestling team manager—belie what is regarded as one of the most creative minds in all of DNA research. The resourceful strategies Kunkel perfected for ferreting out the Duchenne's gene possess a deductive elegance worthy of Hercule Poirot.

Like so many in the world of recombinant DNA, Kunkel backed into what was destined to become his life's work. There is something quintessentially American about molecular biology. Anyone with the requisite brains and interest is welcomed, regardless of age or background. People come to it in offbeat and roundabout ways, whereas in more mature fields, such as physics or mathematics, the track is usually joined early in life and late entrants are discouraged. For instance, Helen Donis-Keller, of Washington University, was a professional photographer and graphic designer when, with her husband, she went to Thunder Bay, Ontario, in the late 1960s because of the Vietnam War. While there, she took courses at the local college and grew interested in science, to which she had never been exposed before. Shooting pictures for a scientific publication, she became intrigued by DNA structures and resolved to go back to school to learn molecular biology. On a dare, she applied to Harvard and MIT, was accepted by both, and opted for Harvard. Today she is a major figure in human genome research.

Another late starter but one who rose to the stratosphere of scientific research, the distinguished theorist Susumu Ohno, of the City of Hope, in Duarte, California, was a veterinarian—"a horse doctor," he gleefully cackles—and later a diplomat for the Japanese government before he turned to molecular biology and became one of the world's foremost authorities in the evolution of genes.

In the same fashion, Lou Kunkel did not start out to be a molecular biologist. By his own admission an indifferent student, his grades never rose above a B average at his alma mater, Gettysburg College, a small liberal arts school in southern Pennsylvania. An aspiring botanist, he might never have been bitten by the molecular biology bug had he not spent two summers working at Cornell University's genetics research laboratory in his native New York City. At the suggestion of his bosses, who were impressed with him, he applied to the graduate school at Johns Hopkins University, where he studied human genetics. In 1978, he won a postdoctoral fellowship at the University of California at San Francisco, where his mentors in gene cloning were the inventors of the art form—Howard Goodman, Bill Rutter, and Herb Boyer. The boost that working under such luminaries gave Kunkel's career, in terms of both skill and résumé points, was considerable. Two years later, Kunkel was back on the East Coast, at Harvard, on a Muscular Dystrophy Association fellowship. Casting about for his own research niche, he had found himself drawn to the female X chromosome, which led him to develop a more than passing interest in Duchenne's muscular dystrophy.

That Duchenne's is caused by a defective gene on the X chromosome has been known since the early part of this century. The site could be inferred from the fact that the ailment is almost exclusively limited to males. Diseases show no sexual bias, but a different pattern prevails when the gene responsible is located on the X chromosome. The reason appears to be simple. Girls have two X chromosomes and hence healthy backup copies of all X-linked genes, which can pick up the slack in making essential proteins. But boys, with just one X chromosome, inherited from their mother, are out of luck. Unfortunately, knowing that the muscular dystrophy gene was on the X chromosome was not enough. Lacking even the most primitive technology for exploring chromosomes, researchers were no better off than they had been in the 1860s, when Guillaume Benjamin Amand Duchenne first brought his eccentric and brilliant intellect to bear on the disease that bears his name. Duchenne was a peculiar man, born into a family of sea captains in the town of Boulogne, on the French coast. Spurning a maritime career for the study of medicine, he set up practice in Boulogne in 1831, but the death of his wife during childbirth left him wild with grief and drove him to quit the profession for nearly ten years. When he finally resumed his career in Paris at midcentury, Duchenne had become a maverick of single-minded intensity, devoting himself to a little-known branch of medicine, the study of nerve and muscle disease. With no official standing at any of the Parisian hospitals or institutes, Duchenne became a researcher without portfolio, combing the city's hospital wards for interesting cases of neuromuscular disease, often following patients at his own expense. In what must have seemed bizarre behavior, he began experimenting with applying electrical currents to the human body, using electrodes to stimulate muscles in order to observe their function. In this way, he succeeded in mapping out the function of every muscle in the body. In his pioneering analysis of muscle disorders, he provided medicine with its first detailed description not only of Duchenne's muscular dystrophy, which he called pseudohypertrophic muscular dystrophy, but of polio and spinal muscular atrophy. But Duchenne was an unworldly and naive individual, rather inarticulate and absentminded. Had it not been for the urging of his friends and students, Duchenne most likely would never have published much of his work. As it was, his colleagues greeted his findings on muscular dystrophy with little enthusiasm. "I thought humanity to be inflicted with enough evils already," one physician is reputed to have grumbled. "I do not congratulate you, sir, upon the new gift you have made it."

Despite Duchenne's illuminating findings, a closer understanding of the illness eluded researchers for decades. Not until the 1950s did scientists launch a major offensive on the disease and begin to zero in on its biochemistry. By comparing Duchenne's muscle tissue with healthy tissue, they detected a wide range of abnormalities. They found defects in the outer membrane of Duchenne's muscle cells, in the calcium levels within those cells, and in the ability of the muscle cells to contract, the sine qua non of proper muscle function. Duchenne's patients were also shown to have unusually high blood levels of an enzyme called creatine kinase. But each of these abnormalities turned out to be a consequence of the disease, not the underlying cause. Something else was pulling the strings, and it gradually became clear that this something else was a pivotal protein, probably a muscle protein. Not until the emergence of recombinant DNA technology two decades later, though, were scientists able to go after their quarry—the fundamental protein and the gene that encodes it. Even with that technology, the hunt was difficult. The X chromosome is one of the larger chromosomes, encompassing 150 million base pairs, or 5 percent of the total human genome. That is a huge pool of DNA to go fishing in. Geneticists desperately needed at least an inkling of where in all that DNA the gene might lie.

The first clue emerged in 1977 when Oxford University researchers examining an X chromosome from one of the dozen or so girls in the world who suffer from Duchenne's made an intriguing discovery. A portion of the girl's chromosome had broken off and traded places with a piece from chromosome 21. Interest grew into excitement when the remaining girls were found to have a translocation at precisely the same spot. The trail was getting warm. The Duchenne's gene almost certainly had to be near the site of the break, because whatever had protected them by virtue of their sex was no longer there. The Duchenne's girls were like Duchenne's boys.

Confirmation came in 1983 from the laboratory of Kay Davies, another Oxford scientist. Davies had been attacking the Duchenne's problem from a different direction—reverse genetics—linking a disease-causing gene to a particular spot on a chromosome by statistical analysis. Davies and her labmates managed to deduce that the Duchenne's gene must lie in the area where the break had been found in the girls' X chromosomes, in what is known as the short arm of the chromosome, and this narrowed the search somewhat. Still, the area constituted an immense stretch of terra incognita, spanning ten million base pairs. The search could theoretically take years. At this critical juncture, the Muscular Dystrophy Association—borrowing pages from the March of Dimes' successful quest for a polio vaccine in the 1950s and the National Aeronautics and Space Administration's 1960s effort to put a man on the moon—decided to undertake a crash program to find the Duchenne's gene. The fruits of this decision might well serve as a model for all genetic research.

The MDA, founded in 1950, is a highly aggressive voluntary organization with a long history of funding Duchenne's research. Toward the end of 1982, the leaders at the association had come to recognize that the traditional "investigator initiated" approach was not working very well. Says Donald Wood, former director of research, "An agency like MDA normally operates by amassing a pile of money and waiting for those who want to do research to devise and submit proposals. The proposals are then competitively evaluated. That's how the National Institutes of Health work; that's how almost everyone works.

"But we began wondering how we might reorder things to really make a difference. In essence, we put our pile of money together and then went out and found the people to spend it on." A blue-ribbon task force was organized in early 1983. The task force, composed of all the experts Wood could drum up, was asked one overriding question: How do you locate a gene in a big hurry? Among the cognoscenti was the patriarch of DNA science himself, James Watson, who is said to have the keenest eye for research talent in all of molecular biology. Watson minced no words. "It's going to take you five years and cost you $5 million!" And that would be the easy part. The trick, he said, would be to find superior people willing to dedicate themselves for so long to little else but Duchenne's. The idea of concentrating on a single project runs against the grain of molecular researchers. They thrive on having many irons in the fire at once. This has little to do with their restless minds; primarily, it is insurance. The chances of being left at the starting gate scientifically are reduced as the number of projects one is involved in is increased.

Wood was undaunted by Watson's caveat. Working from a list of names provided by the panel, he began sending out invitations. The bargain he offered was a tempting one: the researchers' time and knowledge in return for access to the agency's fat bankbook and a chance to make history. The talent safari yielded a number of respondents, who were brought together in August of 1983 for a symposium at Watson's Long Island bastion in Cold Spring Harbor. Several more researchers were added a few weeks later.

Shrewdly, although it followed the time-honored precept of merely paying the bills and standing back and letting the researchers do their stuff, the association maximized its chances of finding the gene by ensuring that each laboratory pursued a different strategy. For example, Ron Worton, of Toronto's Hospital for Sick Children, was to devote himself to studying the translocations that had been discovered some seven years before. Kay Davies, of Oxford's John Radcliff Hospital, would continue to perform analysis, using DNA samples provided by the association from some one thousand muscular dystrophy families. And sure enough, by late 1985, she had found markers that were more than 96 percent accurate in prenatal diagnosis—but no gene. Nevertheless, she remained the odds-on favorite to find the gene because of her success with the probes. At Baylor, Thomas Caskey was following a third tack, using, of all things, a mouse. This special strain of mice was stumbled upon in 1984 by a Scottish veterinarian named Graham Bulfield, who found that the mice had a form of X-linked muscular dystrophy. Caskey believed that the little rodents, dubbed mdx mice, had a mouse version of Duchenne's, even though their symptoms were far milder, and that the gene's location on the mouse X chromosome, if it could be found, would approximate the location on the human X, given the tendency of important genes to be conserved among species. Caskey's plan was to do linkage analysis on the mouse in hopes the work would yield the precise location of its muscular dystrophy gene.

And then there was young Lou Kunkel. Kunkel did not belong to the original group recruited by the association. A postdoctoral fellow working unobtrusively under a small MDA grant, he had been suggested by his Harvard mentor Samuel Latt, who himself had declined the association's blandishments. But Kunkel is known to his colleagues as a risk taker. Very quickly, he decided to branch off onto a path more venturesome than that taken by the others. If seeking a gene among millions of anonymous base pairs is a task akin to finding the proverbial needle in a haystack, then Kunkel's strategy was to begin looking where the haystack wasn't. This required a leap of intuition that Kunkel was in a unique position to make. Early in 1985, he learned of a patient whose case had been reported by Yale's Uta Francke. This patient, a young man from Seattle named Bruce Bryer, had the misfortune of suffering from three X-linked diseases at once: retinitis pigmentosa, a form of severe tunnel vision leading to blindness; chronic granulomatous disease, a form of immune deficiency; and Duchenne's muscular dystrophy. Although the young man had died in 1983, scientists had created from his tissue an "immortal" cell line, making it possible to examine his DNA posthumously—until the end of time, if one wished. What caught Kunkel's attention about the youth's DNA was that, according to Francke, there was a tiny deletion, or gap, in the short arm of his X chromosome—in the same spot where others had shown the Duchenne's gene had to be.

The word "deletion" fell like celestial music upon Kunkel's ears, for he suddenly realized he had a way to find the gene. It was clear that the Duchenne's gene—and very likely the genes for chronic granulomatous disease and retinitis pigmentosa as well—were within the deletion. A number of laboratories, including Francke's, had already deduced the same thing. But exploiting such an insight to find the gene seemed impossible to other labs. There were certain acknowledged ways to find genes. You could, for example, work backward from a gene's protein product, using messenger RNA and reverse transcription to make, in effect, photo negatives or cDNAs that would lead to the proper DNA sequence. Or you could do linkage analysis, obtaining markers flanking the gene in question and then "walk" along the chromosome from either side, sequencing as you go. In the first instance, however, you had to know the gene product beforehand, which did not apply to Duchenne's muscular dystrophy. And to do linkage analysis, you had to have an intact chromosome to work with. The method is useless when the bridge is out, as it is with a deletion. Here Kunkel's ingenuity came in. Years before, while working on his fellowship in San Francisco, Kunkel had experimented with a novel way of attacking the male Y sex chromosome. Now he saw that he might try the same sleight of hand with the female X. Kunkel reasoned that if he took fragments from the short arm of the Seattle youth's X chromosome and let them react with homologous fragments from a normal X chromosome, he would get a musical-chairs result. All the DNAs on the dead youth's chromosome would stick like refrigerator magnets to the corresponding DNAs on the normal X chromosome, because of the way nucleotides bond with their complements. But some fragments from the normal chromosome would not be "grabbed," because there would be no DNA within the deletion to do the grabbing. These DNA fragments would correspond to the missing DNA in the deletion and would give Kunkel tangible pieces of chromosome to work with in the lab.

This improvisational technique was called subtraction hybridization. But although the method had previously been thought feasible, no one prior to Kunkel had put it to the test. The first step was to get a sample of the Seattle youth's cell line, which Francke, after some negotiations, supplied to Kunkel. Kunkel next let the youth's DNA interact with normal DNA taken from a cell line that contained four X chromosomes—Kunkel's way of bettering his odds. These normal X chromosomes contained the same stretch of three million base pairs of DNA that was missing from the Seattle youth. From this union of DNAs, Kunkel came up with ten tiny probes of normal DNA that failed to link up to points within that missing stretch. It should have been possible but exceedingly time-consuming to derive probes covering all three million of Bruce's missing base pairs. Fortunately, it was unnecessary. Kunkel needed only to "blanket" the area at a number of different and widely separated points, hoping for a handful of probes that would have sequences in common with most, if not all, of the relatively small number of genes within the deletion.

Now the problem was to find which, if any, of those probes might be responsible for Duchenne's. In the spring of 1985, Kunkel took eight of his probes and let them interact with X chromosome samples from fifty-seven boys with Duchenne's. In five of those boys, or about 9 percent, Kunkel found deletions that corresponded exactly to one of the probes, which Kunkel had given the name pERT87. Nine percent may not seem like much at first. "But," says Kunkel, "it's almost dead-on what happens with other X-linked diseases. You have to remember that not all cases of the disease will be due to a deletion at that site. The rest will be due to mutations or other deletions. In Lesch-Nyhan disease, only 9 percent or so are caused by deletions. So we were right in the ballpark." (In fact, it turns out that more than half of all Duchenne's cases are caused by deletions; but these deletions are located elsewhere within the same gene, not at the site occupied by pERT87.)

Kunkel had made a momentous find. PERT87 must be right in the middle of the gene responsible for Duchenne's. But he was like a downed pilot who knows which forest he has crash-landed into, but not how far he has to go to get out. In this case, the forest consisted of nucleotides, hundreds of thousands of them, and he had no real way to tell where one gene left off and another began. To make matters worse, as Kunkel and his tiny band, which included the graduate student Tony Monaco and the postdoctoral fellow Eric Hoffman, began to sequence outward in both directions, it was becoming clear that the muscular dystrophy gene was going to be enormous. Whereas the garden variety gene runs somewhere between 40,000 and 50,000 base pairs in length, and whereas the largest gene then known, that responsible for hemophilia, contained 230,000 base pairs of DNA, the Duchenne's gene was shaping up as more than two million base pairs long. This is a third the size of *E. coli*'s entire genome. The sheer magnitude of the gene probably accounts for the high rate of new mutations in Duchenne's. It is not unknown for mothers to have several normal children, then one child with Duchenne's, and later several more who are normal. The freak occurrence of Duchenne's is due to one defective egg among thousands in the mother's ovaries. Such a random hit usually involving a single nucleotide out of place in the chain is called a spot mutation, and it seems to happen ten times more often in Duchenne's than in other inherited diseases. Presumably, the explanation is that a large gene contains that much more DNA that can be damaged by exposure to radiation, toxic chemicals, and other mutagens. The pinpointing of the Duchenne's gene—which, as hoped, had the salutary side effects of leading to the isolation of the genes responsible for chronic granulomatous disease and retinitis pigmentosa—brought immediate acclaim to Kunkel when it was announced in September of 1986. He, in turn, was unusually gracious, sharing bylined credit with all the other MD researchers. For gene hunters, this was arguably the biggest game yet ensnared, viewed in terms of the heartbreak it has caused throughout the centuries. But much more remained to be done. In order to identify the underlying fault in Duchenne's, and map a strategy for finally conquering the disease, it was necessary to identify the protein product of the gene.

The task was not an easy one. Isolating and sequencing the entire gene, a precondition of working out the amino acid chain of the protein, would take months. Researchers began looking for shortcuts to the protein. Their rush caused them to implicate a false suspect, a mysterious muscle protein dubbed nebulin, which had been discovered only six years earlier. Nebulin represented a shot in the dark. It came under suspicion primarily because it was big. Suspicion intensified when biochemists at Columbia University discovered that it was either missing or barely detectable in the muscle of Duchenne's victims, whereas there seemed much more of it in victims of other forms of muscular dystrophy and in healthy individuals. So much for the plus side. On the down side, nebulin failed the antibody test miserably. Antibodies raised to nebulin attacked not only nebulin molecules but molecules of other muscle proteins as well, indicating that nebulin's structure mimicked that of other proteins and thus was unlikely to cause a disease like muscular dystrophy, whose cause appeared to be the total absence of a protein. In the acid test, cDNA clones of nebulin were matched with fragments of Kunkel's gene. If nebulin had been the gene's product, the two should have hybridized or entwined like nestling worms. But they didn't. Nebulin was not the gene.

Kunkel, meanwhile, had sensed that nebulin was an imposter. Being a hunch player, he pursued another tack, exploiting antibody technology, but in a different way. In his laboratory, the gene was yielding itself grudgingly, in molecular bits and pieces. Each time Kunkel's team sequenced a new stretch of the gene, it would splice the segment into bacteria. The bacteria, incorporating the new DNA, would then manufacture the corresponding segment of the gene's protein. These protein segments were thereupon injected into rabbits, which obediently produced antibodies to the interloper. As Eric Hoffman put it in a magazine interview, the hope was that "the antibodies would lead us to the protein much like hunting dogs might lead hunters to their prey." Each of these antibodies, as they came off the rabbit assembly line, was exposed to an assortment of muscle tissues. The tissues were derived from Duchenne's patients, from healthy individuals, from mdx mice, and from normal mice. In the fall of 1987, the hoped-for breakthrough came. A specific antibody detected an unknown protein in samples taken from normal people and normal mice. Significantly, it did not detect the protein in Duchenne's patients or mdx mice. Here, at last, was what the researchers had been looking for: a protein that was missing only in individuals with Duchenne's muscular dystrophy. Further proof that the protein was the culprit came a short time later when, with the entire deleted gene finally delineated, the amino acid sequence of the protein was found to correspond to the code embedded in the gene.

Kunkel and his colleagues christened the new protein "dystrophin" and announced its discovery with great fanfare just before Christmas in 1987. As they had assumed, dystrophin proved to be a protein molecule of great size, explaining why scientists had managed to miss it despite decades of studying the makeup of muscle tissue. Because of its unusual length, the molecule broke into minute, unrecognizable fragments when handled during anatomical studies. Not only that, the huge molecule was present in almost unbelievably small quantities in muscle, making it all the more easy to overlook. Follow-up experiments showed that dystrophin constitutes only .002 percent of the total protein found in muscle cells.

Here was a quirk of nature indeed, that a substance so rare should be the hinge of fate for so many youngsters. Soon a further finding about dystrophin added to the wonder surrounding that crucial .002 percent. There is another degenerative muscle disease of childhood called Becker muscular dystrophy. Much milder than Duchenne's, with a slower progression, it is seldom terminal. For most of this century, it was assumed that Becker and Duchenne's were entirely different forms of dystrophy caused by entirely different genes, but, as Kunkel and other researchers learned in early 1988, the dystrophin gene seems to be the villain in both diseases. The distinction is that in Duchenne's the protein is totally absent—deletions and mutations are sufficient to halt its production completely—whereas in Becker the protein is present, but in insufficient quantity and inappropriate form. "The bottom line," says Donald Wood, "is that any dystrophin is better than no dystrophin, even if that dystrophin tends to be abnormal." The Duchenne's researchers were in the odd position of detectives who with a single arrest had been able to clear up a number of crimes. But there was no time to exult. It remained to find out what dystrophin's job in muscle normally is. The key to this, as later to the cystic fibrosis protein, was to discover the protein's functional position within the cell.

At this point, Kunkel faltered for the first time. He asserted in the journals *Cell* and *Nature* that dystrophin appeared to be located in structures in the muscle cell called triad junctions. Too bad for Kunkel; it was not so. In Toronto, Ron Worton had been skeptical about Kunkel's findings from the outset. In a speech he gave in Milwaukee in February 1988, nearly two months after dystrophin had been publicly named as the guilty protein by the Muscular Dystrophy Association, Worton was still treating it as an unproven hypothesis. A few weeks later, presented with unassailable data, Worton was forced to accept dystrophin as the Duchenne's protein. But, he grumbled, that didn't mean Kunkel was right about the triad junctions. Using sensitive assays, Worton and his colleagues found that antibodies to dystrophin congregated not at the triad junctions but in clusters on the inner surface of the cell membrane. Worton hypothesized that it was the function of dystrophin to help anchor the internal working structure of the muscle cell to the membrane.

In the months that followed, various laboratories confirmed and refined this conclusion. Dystrophin is now known to lash tightly the long cylindrical bundles called muscle fibrils, which contract on command. These fibrils allow muscles to work, and they are embraced by a delicate latticework of rings that fasten to the membrane in a series of bands that resemble the straps on a subway car. Without dystrophin, the current thinking goes, these rings fail and the fibrils begin to tear away from the membrane. Over time, for lack of secure anchor, these critical muscle fibers start to shear and tear and eventually lose their contractile ability altogether.

While dystrophin is critical to muscle cells, it is found in still smaller quantities elsewhere in the body—notably in spleen, kidney, lung, and brain tissue—and its absence in the brains of certain Duchenne's patients is considered a possible explanation for the occasional learning disabilities and mental retardation seen in these patients. The discoveries pouring out of dozens of Duchenne's labs around the world make it a strong likelihood that, within the next several years, patients will at last be in line for lifesaving treatment. In fact, the potential beneficiaries may extend well beyond Duchenne's patients. In the spring of 1989, Kay Davies and her colleagues detected a gene on chromosome 6 with a nucleotide pattern startlingly similar to the dystrophin gene. This meant there may be a family of dystrophin-like genes on various chromosomes that may perform related structural functions in muscle. Indeed, research in recent years suggests that dystrophin acts as an anchor for a complex of proteins called dystrophin-associated proteins, that serve to link muscle cells to connective tissue. Lacking dystrophin, the associated proteins float off and leave muscles to wither and harden (sclerose). These genes and proteins may account for some of the eleven other forms of muscular dystrophy, including a number of severe adult forms. Now, mdx mice already have figured out a way to elude such deadly tinkering by their keepers; they use a related protein, utrophin, to step in and save their muscles, especially their hearts. Research presented in *Nature* in December of 1992 by Kevin Campbell, a Howard Hughes investigator and physiology professor at the University of Iowa, showed that human utrophin closely matches dystrophin (it is even referred to as "dystrophin-like protein"), although it is usually confined to a small area in muscle, the neuromuscular junction, where a muscle meets the nerve that nourishes and fires it. Perhaps if the muscles of Duchenne's patients could be brought back to life by supercharged utrophin genes to imbue the muscle with substitute protein, the door might open to a new treatment. As Donald Wood says, "Now all the real excitement starts."

Under ordinary circumstances, one would expect protein replacement therapy to be the initial card played. It would be wonderful if it were possible to synthesize dystrophin and inject it directly into a Duchenne's patient. Sadly, this approach almost certainly will not work. Merely to drip dystrophin into the body, according to Eric Hoffman, "is like trying to solve a plumbing problem by dumping a load of pipes on the doorstep of a building." "Dystrophin appears to be a structural protein, rather than an enzymatic protein," explains Donald Wood. "Enzymes tend to be floating loose in cells, and all you need to do is simply get it into the cell and you're done. But with a structural protein, you not only have to get it into the cell, but insert it into the right spot in that cell. Then we have problems with the huge size of the molecule. How do you take something so huge and get it past the other structures in the cell? It's like trying to manuever a twenty-foot-wide pole through a doorway without having the option of turning it end on."

With protein replacement of dubious merit, it was gene therapy, the supposedly more difficult technology, that came to the rescue. In the early days, shortly after his isolation of the dystrophin gene, Kunkel was pessimistic about the prospects for gene therapy in Duchenne's muscular dystrophy. The problem of getting dystrophin into cells, he told us, seemed to pale before the difficulty of getting a normal gene into all the major muscle groups of the body. One might be able to remove portions of a muscle, through biopsy, but once these extracted muscle cells had been outfitted with the new gene, how could you put them back? You could not count on them to reintegrate into the muscle from which they came, and even if they did, they could not be expected to travel to neighboring tissues, as blood or marrow cells do, rendering it impossible to sow the new genes elsewhere. Hence, you would need to repeat the procedure for the entire muscle group, and then all the muscle groups of the body. "It's like this," Kunkel said. "You can take a piece of muscle out, but it's not going to regenerate and transfer the gene to all the rest of the muscle cells. So the rest are all defective. So maybe you've corrected this little area of muscle, but not the whole thing. If I was to predict, I'd say Duchenne's will not be readily amenable to gene therapy."

But as time passed, an alternative form of gene therapy seemed to beckon. This was another one of Kunkel's hunches. It was known from previous research that immature muscle cells, called myoblasts, fuse with muscle fibers and begin to express certain genes; but it still seemed a long shot that anyone could exploit the phenomenon to reverse a particular genetic defect.

The mdx mouse, which makes no dystrophin of its own, was a convenient vehicle in which to try the experiment. Scientists face an eternal problem in testing medical innovations such as new vaccines. No one wants the testing of unknown and possibly dangerous remedies to begin in human beings. The preferred approach is to start the testing in animals. The trouble is, for many human diseases, there are no animal models—that is, no animals suffer from conditions similar enough to people's that they could serve as a dry run for a human experiment. The situation is beginning to improve as molecular biologists such as Rudolf Jaenisch at the Whitehead Institute, Mario Capecchi at the University of Utah, and Michael McCune, formerly at Stanford and now in private industry, perfect techniques for reproducing human physiological defects in virtually any animal. But thanks to Graham Bulfield, the Scottish vet, there seemed on hand a suitable animal model for Duchenne's muscular dystrophy. Myoblasts fuse with muscle fibers because it is their job to help repair fibers that have torn as a result of injury or exercise. Although myoblasts are only embryonic versions of adult muscle cells, they contain a number of switched-on muscle genes. Kunkel's group hoped that among these would be the gene for dystrophin. Normal mouse myoblasts were injected by the millions into the shin muscles of a population of young mdx mice. After a number of weeks, the mice were killed and the shin muscles removed. Of the seventy mice that were analyzed, thirty-nine, or more than half, showed unmistakable signs that the injected cells had fused with the host muscle fibers. More significantly, antibody testing revealed that the cells had begun to make dystrophin, as much as 30 to 40 percent the normal amount. Furthermore, up to 40 percent of the time, the dystrophin had found its normal cellular location in the membrane of the muscle fiber cells.

Similar results, meanwhile, were being achieved by Canadian George Karpati and his colleagues of the neuromuscular group of the Montreal Neurological Institute. Karpati's group injected human, rather than mouse, myoblasts into the thigh muscles of two dozen mice. Some dystrophin production was detected in every mouse, and those segments of muscle fiber that began making dystrophin resisted subsequent deterioration, while those segments with no dystrophin broke down. Although there were no guarantees that mouse research would work in people, it seemed time to try a human experiment.

In early 1990, four North American research centers, including Karpati's and Peter Law's at the University of Tennessee at

Memphis, prepared to test myoblast transfer in young boys. Law's experiment called for injecting eight to ten million myoblasts obtained from fathers and brothers into their son's muscle controlling the big toe. For the sake of comparison, a sham injection of salt water was administered to the boy's other big toe. By summer, Law was reporting striking results in three of the eleven youths he had tested. In one of them, eleven-year-old Jeremy Brown, the big-toe muscle of the right foot reportedly had become 80 percent stronger than that of the left. A long-range "toe" watch now began. The value of the treatment would not be proven, Law said, unless the increase in muscle strength could be sustained for up to a year. "We are very optimistic and very excited," said Jeremy's mother, Judy, whose older son, Jeff, had donated the myoblasts to his brother to minimize the chances of tissue rejection. "But I want the medical profession to move as rapidly as they can because we are running out of time. The disease is still progressing."

In Montreal, meanwhile, George Karpati announced even more startling news. He had performed myoblast transfer not in a small muscle like that of the toe but in a major muscle—namely, the bicep. Karpati is a professor of neurology and associate professor of pediatrics at McGill University, where he holds the Isaac Walton Killiam chair of neurology, and also serves as a senior neurologist at Royal Victoria Hospital and the Montreal Neurological Institute. In a controlled double-blind study that ran from April 30, 1990, through February 12, 1992, Karpati made fifty-five injections into one bicep muscle of eight young boys, giving them a total of fifty-five million myoblasts provided by their fathers. The other bicep received comparable injections, but without myoblasts, as a control. Neither patient nor personnel were aware of which side was which. All the youngsters received the drug cyclophosphamide for immunosuppression for six or twelve months. No serious complications were observed, demonstrating the safety of the procedure; however, the overall therapeutic efficiency was "poor," according to Karpati, as determined by a wide range of quantitative and biochemical tests.

Reporting his results in the *Annals of Neurology,* Karpati measured the maximum power generated by the boys, and went on from there. He studied the dystrophin content of their muscle by repeated biopsies. He examined cross sections by magnetic resonance imaging. He analyzed the injected tissue for signs of paternal genes—donor-derived DNA and messenger RNA—in short, he did everything he could think of to evaluate the efficacy of the procedure.

His conclusion, reiterated in a November 1993 interview with the authors: "Myoblast transfer in the present form is not sufficiently effective. I believe this even though at least two boys showed a statistically significant increase in force generation in the injected muscle.

"One boy remained the same and the rest deteriorated to a significant extent. But even in the two who showed improvement, it was so modest that it could not have made any significant difference in daily living, which is the standard I use to define functional improvement. In fact, we couldn't even ascribe with confidence that the improvement was due to the injected cells. We couldn't find donor-derived DNA or messenger RNA."

Once myoblasts are injected into muscles, Karpati suspects, they run into formidable, perhaps impossible, hurdles: the cells have a tough time migrating through connective tissue barriers; they often fail to fuse with muscle fibers; they seem unable to replicate. Although Karpati administered immunosuppressive therapy to his patients, he believes it is likely that many of the injected myoblasts were wiped out by killer T cells on routine patrol.

Although disappointed by his results, Karpati laid to rest a major question: Could enough myoblasts be bred in the laboratory from a tissue sample to supply the needs of patients for gigantic numbers of cells? Karpati found they could, but he's not sure that it matters. In fact, he fears that laboratory manipulation of the cells may contribute to their inefficiency after injection. "I don't have any clever ideas how to improve efficiency," he says, "but I've started to think that when we culture myoblasts in vitro, the unnatural environment may reduce their power tremendously. So we're testing the idea of *not* culturing them.

"We want to take out the precursor cells from the donor and, without giving them a chance to go astray in the dish, inject them right away. Obviously, we will get only thousands, not millions or billions, this way; but if they do the job when larger volumes can't, that may be a way to go."

Even as the early data had come in, Kunkel also was frowning about myoblast treatment. Publicly he called it "promising," but privately he worried about two difficulties. There was the practical problem that has always dogged Duchenne's researchers: how to treat all the major muscle groups. Myoblasts presumably would not migrate more than a few millimeters, meaning that injections would probably need to be spaced no more than two-fifths of an inch apart. It might require hundreds, perhaps thousands, of shots in order to make a difference throughout a youngster's body. But, noted Kunkel, even if one could undertake to give a patient so many shots, dealing with the heart muscle in this fashion would be impossible. The cardiac muscle is unusual in that it will not fuse with myoblasts. In other words, the patient might be "cured" in all other respects, but retain the severe cardiac abnormalities that would, in any event, kill him in his early twenties—unless, of course, he was able to undergo a heart transplant, and with the scarce organs going to otherwise healthy people, that avenue did not seem very promising.

Content to let better-equipped laboratories cope with these and other problems, Kunkel, the hunch player, moved on to yet another promising avenue of therapy, a kind of protein modification. Kay Davies's hypothesis of a family of dystrophin-like proteins intrigued him almost immediately. "She found one, but there are more than that," he said. "We think there are a number of proteins related to dystrophin that let the mdx mouse and young human infants escape disease symptoms, and we are pursuing them. Once we find a protein that functions like dystrophin but doesn't quite do the job, we'll figure out why it can't and learn how to make it do the job. And we can do that pharmacologically. Therein lies the potential for real treatment."

Another possibility might draw on the gene therapy technology envisioned by James Wilson at the University of Pennsylvania. The gene, with a coding sequence of 140,000 base pairs, is far too big to be placed in retroviruses that would then be used as vectors, and retroviruses would not be of any use, since muscle cells do not divide. But a way might be found to outfit shell-like carrier cells with receptors specific to muscle cells. Inside the carrier cells would be plasmids containing the dystrophin gene, which would enter the muscle cells. The carrier cells might then simply be injected into the bloodstream, and they would find their way to individual muscle cells the way Wilson is trying to do with liver cells in patients with severely elevated cholesterol (right now, he is using a catheter and dumping in the therapeutic cells directly). In such a fashion, the entire voluntary muscle system might be covered without an inordinately burdensome series of shots. A second strategy is being pursued by C. Thomas Caskey in Houston. Caskey envisions using a gene "gun," called the Bioblaster, which has been developed by the Du Pont Chemical Corporation, to shoot dystrophin genes into muscle groups. The gun, originally designed for modifying the genes of plants, fires a broad pattern of genes into cells.

"We have already had some success using the gene gun to shoot genes into liver cells," says Caskey. Now he is ready to try the approach in a mouse model of Duchenne's muscular dystrophy that his laboratory created by putting a defective human dystrophin gene into mouse embryos. If the dystrophin gene, fired by the Bioblaster, cures the mice,

Caskey will consider trying the same approach in a human patient. "This is really exciting," he says.

Other delivery systems were also starting to appear. On March 19, 1990, a pediatrician-geneticist named Jon Wolff, director of biochemical genetics at the University of Wisconsin at Madison, found that mouse muscle cells had the unique ability to "suck" up injected genes and express them. "If we're right, it means you don't need a retrovirus. The DNA doesn't get incorporated in the chromosome—it's swimming free inside the nucleus like another bit of chromosome, yet it expresses its protein," explained Wolff. Tinkering with several "reporter" genes that make enzymes that tell scientists where the gene product is being made in the cell—including luciferase, which makes fireflies glow, beta galactosidase, which makes cells turn blue, and growth hormone—Wolff was able to track his effect rather easily. The effect was transient and gene expression meager—"We'll have to increase it tenfold," he said. But he believed the startlingly simple concept had worked. "To treat Duchenne, you'd have to give between five and ten shots in a muscle group," he predicted. And what if the solution to Duchenne's ultimately proved to be that simple? "It's mind-boggling," Wolff said. "It's like God said, 'I'm going to create this horrible disease that kills these beautiful children, but then I'm also going to create this back door, if you're smart enough to find it.'"

When James Watson said it would take five years and five million dollars to find the gene and the protein responsible for Duchenne's, he was right on the money. The significance of those five years and what it will mean for patients has yet to be seen. But fifteen thousand families across America are watching Memphis very closely. It was there that the Duchenne's story turned weird.

Early in 1991, frustrated that no cheering throngs had greeted his proclamation that cell therapy could cure the underlying biochemical defect in Duchenne's, Peter Law suddenly bolted. He quit his tenured professorship in neurology at the University of Tennessee, obtained private backing from investors, and on January 15 founded his own institute—the Cell Therapy Research Foundation—and announced plans to begin injecting massive dosages of myoblasts into the legs of thirty-two young boys from the United States, Canada, Poland, and Italy and to give them cyclosporine to keep their bodies from rejecting the foreign cells. Why Law left remains open to debate. He said he faced unnecessary restrictions, including a special institutional board set up solely to review his work. Yet, if true, the establishment of such a body does not seem out of line; after all, he was attempting an unproven treatment on children. The university declined comment. Law, for his part, declined to name his backers. But on December 1, 1992, the *Washington Post* reported he had received $2.5 million in individual donations and $800,000 in support from unspecified foundations (a total that had increased to $4 million by the time of a 1993 interview with the authors).

At any rate, Law's announcement of his determination to plunge ahead, no matter what, did draw a crowd. By the following May, trying to find out what was going on, the MDA declared it was "very nervous" and worried that desperate families might support drastic measures. "The step is very brave, but we fear that it's premature and potentially dangerous," said the neurologist Larry Stern, director of the neuromuscular clinic at the University of Arizona at Tucson and the MDA's director of research. The MDA argued that strengthening a single toe didn't justify massive myoblast transplants into the legs and thighs of disabled boys. The effect of antirejection drugs was another big unknown, said the organization, contending that other experiments might soon be proposed that were safer. "If it turns out that Dr. Law's a genius and we're all scaredy-cats, we'd obviously be ecstatic," said Stern. "But right now, we're very worried about what he wants to do."

Not to worry, reassured Law. "Myoblasts fuse with muscle fibers and fix them after they are torn due to injury or exercise," he said. "I believe they can deliver many active muscle genes and proteins, not just dystrophin." Law then predicted that the experiment would have a 70 percent chance of success and said he hoped to see some results in six to nine months after myoblast transfer.

Accordingly, on March 5, 1992, Law began reporting spectacular news, though not in *Nature*, or *Annals of Neurology*, despite those eminent journals' known interest in a possible cure for muscular dystrophy. Rather, his study of twenty-one Duchenne's muscular dystrophy boys aged six to fourteen, who received only forty-eight injections under general anesthesia into twenty-two major muscles, appeared in the second issue of *Cell Transplantation*, the journal of the Cell Transplant Society. A new journal for a new field, owned by a respected scientific publisher, Pergamon Press, it is edited by Paul R. Sanberg, a researcher in neural transplantation at the University of South Florida. Law is listed as a member of its editorial board. Similar to French Anderson's gene therapy journal, *Cell Transplantation* is designed as a forum for the work of researchers in all aspects of the fledgling field of cell therapy, whether it involves insulin-producing islets of Langerhans cells for diabetes, fetal tissue transplants for Parkinson's and Huntington's, or other procedures. Sanberg says that he viewed Law's study as "preliminary . . . reporting early results . . . by no means conclusive." It was submitted for peer review and received the approval of other "cell transplanters," although Sanberg will not confirm that they were myoblast researchers, a field that struck him as politicized and often nasty, despite its small numbers.

Of the first thirteen boys cited in the study three months after transplant—which involved sixty-nine muscle groups, including knee extensors, knee flexors, and plantar flexors—81 percent showed "functional improvement," Law declared in a press release and in his paper. He also showed that more than 5 billion myoblasts can be cultured from a gram of normal muscle biopsy, "providing unprecedented numbers of cells for MTT," and that "myoblasts frozen over a year retain the ability to proliferate from 10 million to 5 billion and to form normal myofibers." Moreover, Law found that injections of 5 billion myoblasts "have not provoked any immunological rejection symptoms in the Phase II subjects, 11 of whom received 8 million myoblasts in the Phase I [trial] a year ago."

Did the children really get stronger? All claims of efficacy must boil down to that single question. But how can you tell?

It is not generally known that this question has caused immeasurable frustration in muscular dystrophy research. More than a decade ago, neurologists supported by the MDA set out methodically to trace the natural history of the disease: the progression of Duchenne's muscular dystrophy muscle by muscle. The reason for doing this, as one of the investigators, the Canadian Michael Brooke, said was that if you put one child in a room with a hundred neurologists, you would get a hundred different opinions about the status of that child.

In its meticulous study, the team discovered that one muscle could deteriorate at a rate slightly different from that of another muscle, and that during a growth period of a child—for instance, between the ages of five and ten—some muscle groups showed no deterioration and others even a slight strengthening. But the general trend was ever downward, and all boys with Duchenne's dystrophy would be in wheelchairs no later than at age twelve.

The study results explained why the field has been plagued by conflicting efficacy data regarding drugs and treatments: depending on the individual child and the muscle group being studied, a researcher can easily be confounded.

The study also suggested that, in looking at the therapeutic efficacy against Duchenne's, one must chart the progression of the disease, muscle by muscle, before administering the drug; determine any improvement while on the drug; and then document what happens when it is removed.

The research demonstrated, above all, that objectivity is hard to come by in tests of muscle strength. Donald Wood explains, "If you're a physician in a room with a kid, you can change the course of the exam, even if you have hooked the boy up to monitors. Pat him on the back and encourage him. Perform the test in the morning, not the evening. Do it before a meal. All these factors can influence results. Muscle tests must be performed under certain conditions.

"So then Peter Law puts in his paper that the kids' muscles showed 'functional improvement' 81 percent of the time—and that's it. How much force is 81 percent? How much is that in kilograms per square centimeter?"

Law's impressive results were accompanied by a slickly professional videotape sent to the nation's science writers to document apparent progress. Duchenne's boys were shown pushing the pedals of bicycles, standing and walking in water, climbing stairs, and performing other activities they had supposedly been unable to do before.

Although the tape, which featured testimonials from parents, was very poignant, it raised more questions than it answered. Use of anecdotal evidence, the effects of cyclosporine, the lack of a control group or DNA analysis, the possibility of the placebo effect, even the postures of some of the seated children that might have let them increase their muscle power temporarily—not to mention Law's use of promotional videotapes and publication of his findings outside the scientific mainstream—all led one to wonder what he was really up to.

Over the years, other researchers had claimed to have found the cure for muscular dystrophy, and nearly a hundred clinical trials have been held since the 1950s. More than eighty chemicals were tried and abandoned. In the 1960s came claims of efficacy with steroids, but investigators tried to repeat the results and were unable to document more than a transient boost in muscle strength. Later, an Indiana physician claimed that a drug called adenylosuccinic acid improved the muscle strength of youngsters. The doctor soon had a following, and many families believed that their children got better. But nothing ever came of it. As recently as 1989, the steroid prednisone was documented as temporarily improving muscle strength in DMD boys. Again the explanation is unclear, but some findings suggest that an immunosuppressive drug such as prednisone may reduce the number of cytotoxic T cells and macrophages, part of the inflammatory process that may play a role in muscle fiber degeneration. Thus immunosupressives may interfere with the production of interleukin, various cytokines, and gamma interferon by T cells, perhaps either preventing muscle fiber degeneration or allowing damaged fibers to recover. However, therapeutic doses of prednisone, Imuran, and other antirejection drugs administered to dystrophin-deficient mdx mice failed to change the course and prevalence of muscle degeneration.

Was myoblast transfer therapy safe? Law claimed that his study had been "critically evaluated and approved" by two institutional review boards that were in accordance with FDA regulations about human experimentation, but he refused to identify them further, because they allegedly wanted to remain anonymous to review the controversial work without interference. The FDA said it did indeed allow such IRBs to be formed—in effect, for hire—to evaluate human research that is not affiliated with a medical institution or university. But here's the catch: the FDA also said that it did not regulate cell therapy, such as myoblast transfer. "Cell therapy is unregulated; it's brand-new," says Wood. "The academic community, with the exception of Peter Law, has not gone public yet. It remains in the province of high-risk experimentation done under rigorous controlled conditions."

At any rate, responding to complaints, the FDA in May sent an investigator to Law's lab, although the man, David C. Benstein, was in an odd position. The FDA did not regulate cell therapy, so what was he supposed to look for? Nonetheless, Benstein filed a report that listed, in meticulous bureaucratese, nineteen areas of deficiency. Most involved sloppy record keeping:

- "Reported data does not include variations in the isometric exercise series according to the quantity of repetitions and the rest periods between repetitions."
- "Force data generated and recorded from each isometric muscle contraction was not always consistantly [sic] reported in the data summaries."
- "Planar angles for resting the various muscle groups in the isometric contractile series were not always consistantly [sic] reported."

The investigator also wrote that Law had failed to follow his own protocol in several technical aspects, including the numbers of patients treated, points that Law and his attorney later disputed.

In earlier experiments, Law's team had injected myoblasts in one muscle group and sham injections in a counterpart, in accordance with accepted scientific practice. Neither the children nor the researcher had known which muscles had received new cells, so the possible benefits of the placebo effect had been prevented. The Memphis doctors had earlier also given the children cyclosporine to prevent rejection, and had noted no muscle enhancement from the powerful drug.

This time, though, Law skipped sham injections and had new cells injected into both legs of his youngsters. He did so, he said, because his earlier sham injections had proven unnecessary and only the muscles receiving new cells had shown improvement. Moreover, he had destroyed critical muscle tissue in children who needed all they had. He wouldn't do that again, he vowed. Besides, he had already shown that myoblast transfer worked. If others didn't believe him, too bad.

However, without controls to rule out the placebo effect, other scientists quickly dismissed Law's study. They pointed out that because the dystrophic muscles of the children contain mostly fat and connective tissue and were unlikely to be harmed by solutions of saline, Law's humanitarian explanation for a lack of controls did not hold water. Interestingly, none of his critics would take him on directly. "I'm not saying he's wrong," explained Harvard's Robert Brown, echoing the sentiments of colleagues. "I'm saying that I cannot interpret his data."

By this time, Peter Law declared that he had grown very mistrustful of the MDA, as well as of his competitors in the race to cure muscular dystrophy. It was *he*, Law insisted, who had first conceived of the idea of myoblast transfer therapy, in 1975, and with more than twenty years of animal work behind him, he was not going to be held back by a bunch of latecomers and nervous Nellies.

"The MDA loves their own people," he said. "They've sunk a lot of money into them." He had been a good scout until the first human experiments, Law claimed. "Then, I proposed to go ahead and move into a Phase II [efficacy] study in which we would be injecting five billion, not million, myoblasts in twenty-two major muscle groups as a potential treatment to allow the Duchenne boys to discard their wheelchairs. But the MDA reacted with shock. They said it was much too early. That we should give it more time."

From then on, Law said, he was forced off the beaten scientific path because the MDA controlled everything having to do with the disease. For instance, each time he tried to publish a paper in a major scientific journal, such as *Nature*, "the first thing [the editors did] was to call the

MDA and ask if this is a significant breakthrough in science and technology?" To Law's mind, he was being blackballed. And for what purpose?

"I've opened up a nonprofit agency that is in direct competition with the MDA. I think that is why we haven't received favorable reviews from the journals. Our paper always would be turned down because the reviewer has not been fair to us."

Stunned, the MDA's medical advisory committee hotly denied Law's charges. "We wrote to him," states its chairman, the Cornell University neurologist Leon Charash, "and asked if he would allow a world-class group of pediatric researchers to come and look at his program. We proposed that he reject any of the people we chose, if he felt they lacked objectivity, or if he didn't respect their work in muscle research. Peter wrote back indicating his dismay at my criticism of him. He said there was no need for anyone to come and see his program."

Charash, who has found himself in the uncomfortable role of Peter Law's nemesis, insists that, no matter what the public has been led to believe, no true unanimity exists among researchers about the safety and efficacy of myoblast transfer.

On March 31, 1992, Stanford University rushed in where other institutions feared to tread. Stanford announced that it had gone ahead with its own version of myoblast transfer therapy—a solidly established effort with a commercial tinge, supported by the MDA, NIH, the March of Dimes Birth Defects Foundation, and the pharmaceutical firms of SmithKline Beecham and Sandoz. The results were published in the April 2 issue of *Nature*.

The Stanford researchers, led by Dr. Helen Blau, in affiliation with physicians at the California Pacific Medical Center, had worked with eight boys. Under anesthesia, each youngster received eighty to one hundred closely spaced injections of 50 million myoblasts in a single shin muscle. The same number of injections of a placebo, a salt solution, were administered to the same muscle of the opposite leg. Neither researchers nor patients knew which leg received the myoblasts. The sensitive lab test, PCR, showed that in three of the eight boys the donated cells had fused with muscle fibers and—as long as six months past transplant in one of the three boys—the new genes were working and producing dystrophin, even though no improvement in muscle strength was noted.

That was a crucial point: *No improvement of muscle strength had been noted.*

On the other hand, officially sanctioned human myoblast transfer had moved from the big toe, up to the biceps, and then back to a single shin muscle.

Law termed the Stanford experiment "brutal" and said it was crazy to reinvent the wheel. "Think of it!" he exclaimed. "One hundred injections in a small area of muscle. You don't have to do that many. It's a very brutal operation."

Law went so far as to claim that his own experiment—he avoided any mention of French Anderson's and Michael Blaese's—represented the actual birth of human gene therapy: "Myoblast transfer therapy is the first genetic engineering procedure to have produced any functional significance in the treatment of mammalian genetic disorders."

This was highly debatable. But in his enthusiasm to herald the Stanford experiment, the MDA's Leon Charash with patrician disdain pretended that Peter Law did not even exist. Charash called the Stanford finding "seminal" and claimed, "It is the largest trial yet to prove that cell transplants can cause dystrophin to be expressed."

The California researchers also said they were preparing to publish a study showing that the antirejection drug cyclosporine did indeed seem to have a transient effect on strengthening muscles in Duchenne's boys (which seemed confusing, because the Stanford team in its own paper was already saying that no increase in muscle strength had occurred). At any rate, this was an unexpected finding about cyclosporine that had never been reported either in animals or in humans, Law angrily pointed out. And he had good reason to be furious: if true, the mysterious effect echoed that of other steroids and immunosuppressives and could account for the seemingly miraculous improvement that he was reporting in Memphis, in which case his entire program was imperiled.

As letters to the editor flashed back and forth between Law and his detractors, finding a neutral party willing to discuss these matters was becoming impossible. One who believed he qualified, because he worked only with dystrophic mice and not with people, was willing to give it a try: Terence Partridge, whose title is "reader" in experimental pathology (a reader ranks somewhere between an assistant and full professor) at Charing Cross and Westminster Medical School, London.

Partridge was a heavyweight. In 1988, working with Lou Kunkel and Eric Hoffman, he had shown that normal muscle cells, when injected into mdx mice, would make enough dystrophin to cure the mice. That was a major discovery, and his work was frequently cited by Law.

Now, though, Partridge did not know what Law was up to. "I'm not convinced that he has positive results," he says. "I could be convinced if I saw the data in a better form."

"It gets very difficult. Nobody knows what to make of Peter," says Partridge in summing up the situation. "My main gripe against his current work is that he's not putting the data out in a form that you can evaluate. If there was a genuine effect [on the children's muscles], all of us should be able to study his data and determine it. I think it's very mischievous to put out data that is likely to excite people's interest in a particular patient in a form that can't be assessed."

On July 14, 1992, Law announced that he had received a patent (number 5130141) for myoblast transfer therapy, citing it as "the first successful demonstration of a human genetic treatment for Duchenne muscular dystrophy." In an interview, he said he "felt sorry" for Jonas Salk, inventor of the polio vaccine, because Salk had not personally profited from his discovery. "A scientist is just another human being who is trying to make a living through talent and hard work," Law said. "I don't see anything wrong with that."

Spokesmen for the MDA asserted that the answer to the disease would come from gene therapy to replace missing muscle proteins, not from massive transplants of myoblasts into the legs of crippled children. But Law sarcastically predicted, "We'll all be retired by the time gene therapy comes along."

He promised that he would eventually be vindicated like Albert Einstein and the pioneering organ transplanter Thomas Starzl: "They had the same problem: They were ahead of their time. I formed my organization to speed up the work." The MDA was furious when he did that, Law alleged in the interview.

In the March 1993 issue of *Neurology*, however, Robert Miller and colleagues at California Pacific Medical Center, in San Francisco, reported their findings. The headline was blunt and cold: "Cyclosporine Increases Muscular Force Generation in Duchenne Muscular Dystrophy."

Miller investigated the effect of cyclosporine in fifteen boys with DMD by taking monthly measurements of the isometric force generated by both shin muscles. Seven boys were wheelchair bound; six had ankle braces. The experiment gathered as many data—charts, graphs, measurements—as any scientist could have wished; it fit the model of the classic protocol for DMD, with patients comfortably strapped in place, connected to electromyograph machines, which recorded voluntary and involuntary muscle activity (when electrically stimulated). The boys were studied for six months. During four months without medication, both the force and the maximum voluntary contraction (MVC) of their leg muscles declined. However, during eight weeks of treatment with five milligrams of cyclosporine per kilogram of body weight (taken orally), their force rose as much as 25 percent, and the MVC rose 13.6 percent. When the drug was withdrawn, the strength vanished. Despite some gastrointestinal side effects, the drug was tolerated well by all the boys. The Miller team called for further controlled studies over a longer period to confirm the observations and shed light on the mechanism of increased generation of force.

With the experiment, the neurology establishment provided a scientific explanation for the miraculous effects seen on Peter Law's videotape. The ball was now in his court. It was up to him to prove that the transplanted myoblasts, and not the drug designed to prevent them from being rejected, were helping his boys.

In November of 1993, Law again attempted to answer his critics with a major new study bristling with data—charts, graphs, and statistics—published in *Cell Transplantation*. By then, the FDA had gotten around to regulating cell therapy, or "somatic cell therapy," as it was called, and had granted Law approval for the first phase (safety testing) of a new biological drug product, which was how it viewed myoblast transfer therapy. This time around, Law named his institutional review board in the paper (the Essex Institutional Review Board Inc. of Lebanon, New Jersey) and reported that the protocol had also been approved by the Patient Participation Committee of the Baptist Memorial Hospital Medical Center in Memphis.

Law described the results of myoblast transfer in thirty-two boys with Duchenne's, a continuation of his previous three years of controversial human trials. The data seemed encouraging: Law reported that 88 percent of the ankle plantar flexors, 49 percent of the knee flexors, and 45 percent of the knee extensors in his patients "showed either increase in strength or did not show continuous loss of strength" in measurements of isometric contractile forces nine months after myoblast transfer. Ankle flexors, in which the greatest success had occurred, are the calf muscles known as the gastrocnemius and soleus attached to the Achilles tendon, which allow us to push down our foot and walk on tiptoe.

Law's patients, aged six to fourteen years, received five billion myoblasts in twenty-two major muscles in both lower limbs from a series of forty-eight injections while placed under general anesthesia for ten minutes. The youngsters were given cyclosporine for six months after the injections, then were weaned off the drug.

Unlike Karpati—who had injected millions of cells, saw scant benefit, and no longer wanted to culture them before transplanting them—Law believed it would take no fewer than five billion myoblasts to strengthen both lower limbs of a DMD boy.

Also unlike Karpati and others who perform DNA analysis on transplanted muscle to look for dystrophin production from the precursor cells, Law believed that the role of dystrophin had yet to be proved in Duchenne's and that by injecting myoblasts he was giving the boys all the genes already found in healthy muscle. Hence he didn't care which genes did what—just so long as they worked.

Another important distinction—perhaps the most crucial of all—that separated Law from his colleagues was their insistence on sham injections to eliminate the possibility of the power of suggestion on beneficial results. Law believed, as he noted in the paper, that such injections "would undoubtedly cause damage because dystrophic muscles exhibit abortive regeneration. Accordingly, the myoblast-injected lower limb would be stronger than the control, and would result in imbalance of the subject, possibly increasing his risk of falling."

Here was a novel idea, to say the least. But if one followed this line of thought, Law argued, "the most feasible control with least risk to the subjects is for each subject to serve as his own control. Muscle function is assessed before MTT vs. after MTT. The study thus would involve the same person, same muscles, same progression of dystrophy, same physical condition, and same activity level."

In an interview with the authors on the eve of publication of the paper, Law seemed upbeat. "The children showed more functional improvement—by two or three times—than has ever been reported in any other therapy," he declared. "It is the most significant finding of functional improvement in the whole history of muscular dystrophy research."

"But," he cautioned, "we have not cured anybody yet. What is happening is the gradual development of a potential treatment. We are making the right moves in the right direction, moving at a pace much more rapid than any other group. To be honest, there is no cure or treatment, but we are the closest to it. I don't see gene therapy even in sight for the next decade."

Law estimated that it will require between twenty and twenty-five billion myoblasts administered to the major muscle groups of the whole body of a DMD boy to show what he terms "dramatic behavioral improvement and possible life prolongation," adding, "Fewer myoblasts will be needed by the younger patients who demonstrate less degeneration and weakening. Perhaps the ideal age of MTT treatment for DMD boys is between three to five years, when muscle growth is rapid."

Despite Law's conclusions and the impressive statistics he provided to back them up, other researchers didn't rally to his cause.

"There is no proof that a single cell he injected survived," asserted George Karpati, admittedly a competitor. "Survival can't be determined without a marking system. There are accepted ways of looking for the new cells and evaluating whether they had contributed to any change in the force generation that was measured. He didn't do that."

Karpati furthermore contended that when researchers measure the force generated by muscles, it is important to express the data in absolute terms rather than in relative ones. "To say you have improvement of 50 percent sounds impressive, but it might be negligible for activities of daily living. For example, if I increase my personal wealth by 400 percent, I sound like I'm rich. But if I only had a dime to start with, I've now got 40 cents."

The bottom line for Karpati: "I don't know if we are any further ahead in understanding how these injections into muscles can improve function at all."

After his own experiments, Karpati had not felt that he could in good conscience keep raising what he viewed as false hopes. "It wasn't fair to the families and children to go on with that expectation. . . . We all agree that the procedure is safe and feasible, but the efficacy situation is still suboptimal—grossly suboptimal—and I don't see anything in Dr. Law's paper that would offer a credible way of improving this procedure so that it really becomes a therapeutically efficient modality for Duchenne's."

After a study of Law's data, further reservations were expressed by Michael Brooke, professor of medicine and director of the Division of Neurology at the University of Alberta in Edmonton. Brooke, an expert in organizing and conducting clinical trials for neuromuscular diseases, serves on the editorial boards of the two prestigious journals *Muscle & Nerve* and *Neurology*.

"I have to say that any attempt to develop treatment for this illness is laudable," Brooke said in an interview. Politeness accomplished, he got down to business: "This time, the problem I have in interpreting his results is that the muscle for which he claims success is one that doesn't get weak."

Brooke meant that, for reasons unknown, the ankle flexors, which Law reported had an overall 88 percent success rate, do not deteriorate in Duchenne's like other muscles. "That particular group of muscles remains strong throughout the course of the illness," Brooke said. "As the boys grow, their ankle flexors normally increase in strength to support their weight. They walk on their toes—that's a characteristic of these patients—and they do it for a long time, until they finally stop walking."

Brooke suspected that even though Law removed cyclosporine for his patients, the reported improvement could still be attributable to the drug. "He could have used prednisone. He could have used everything." Moreover, Brooke could not accept Law's avoidance of routine controls. "Unless you do sham injections and then compare the two muscles, you can have a tremendous psychological effect. In our studies, we found that children responded quite well when they were given *nothing*. So you have to be very careful about the power of suggestion."

Another myoblast researcher, Jerry R. Mendell, professor of neurology and pathology at Ohio State University College of Medicine in Columbus, called Law's study "interesting and provocative." He elaborated, "As I interpret the data, the only muscles that seem to have improved are the ankle flexors. The others really didn't change, or got worse."

Mendell then echoed Brooke. "If you examine a boy who is wheelchair bound and try to manually overcome his ankle muscles—if you have him push down as hard as he can—you wouldn't be able to do it. The muscles are that strong."

Mendell insisted he had no bone to pick with Law. "If there's a difference between us, it's this: Peter Law has accepted myoblast transfer as a form of treatment for muscular dystrophy. Others, like myself, consider it an experimental treatment. That's a basic, fundamental difference. And it's precisely why we need to have controlled studies."

Mendell and other scientists had problems with the statistical methods used by Law. "There are differences of opinion that get highly technical," Mendell said. "Our statisticians [at Ohio State] don't agree with certain things he did. But based on his data, I'd say that the next thing we need is a controlled, randomized trial to see if we can repeat his results."

Mendell, however, did express disappointment that the results Law reported were not more spectacular. "It's too bad after injecting five billion cells into the legs of Duchenne's children, we didn't achieve any real functional improvement, at least as the rest of us measure it.

"That worries me. But I'm willing to say that even if Dr. Law found improvement only in the best group of muscles, that should provide us with some hope and serve as a stimulus to look carefully at double-blind controlled trials."

Law had by then already moved on. With FDA approval, starting in August of 1993, he had begun a clinical trial involving multiple injections of 25 billion myoblasts into the lower and upper bodies of six boys, five of whom had Duchenne's and one of whom had infantile facio-scapulohumeral dystrophy. On August 20, he gave a fifteen-year-old named Chris Furlong 12.5 billion myoblasts via 188 injections into twenty-eight muscles of his shoulders, arms, and back while under general anesthesia. The procedure took forty-five minutes. Furlong was to return in a few months to receive an additional 12.5 billion myoblasts in twenty-six muscles of his lower body.

As the wrangling continued, other researchers quietly took to latching onto myoblasts and tinkering with them in very interesting ways. For example, in November of 1992, the gene therapy pioneer Inder Verma and his team at the Salk Institute, in La Jolla, along with Mark Roman of the University of California at San Diego, removed myoblasts from mice and inserted a gene that made the immature muscle cells manufacture and secrete factor IX. This blood-clotting protein is deficient in about 15 percent of all patients with hemophilia; the remaining 85 percent have deficiencies in factor VIII, which could presumably be supplied to the body by the same technique. Muscle cells do not normally produce blood-clotting proteins, but these cells did, with gusto: when injected into other mice, the myoblasts churned out factor IX for more than six months—and the injected mice showed no ill effects! Verma called the experiment a major step toward the treatment of hemophilia, but its implications extended to such disorders as diabetes and a genetic form of emphysema, which also might be treated by genetically modified cells in the bloodstream.

Two months later, the scientific mainstream weighed in yet again when a Harvard team headed by Robert Brown published in *Neurology* a fascinating discovery: when myoblasts are injected into a major artery, the embryonic cells can reach and fuse with existing muscles and usually bypass other tissues. Brown successfully delivered new muscle cells to existing muscles in rats via injections into the aorta. Could it be that simple? Could Duchenne's be treated in this manner?

Brown was quick to urge caution, noting that the type of myoblast used for the experiment was a "souped-up," laboratory-modified cell that cannot be used in humans, because of its potential for causing a tumor. He added that the method for determining how many cells survived the transplant is subject to error. But, at the time of the experiment, the researchers noted, it was known that bone marrow cells implant themselves in bone marrow after being injected into the bloodstream, and other teams were trying to get infused cells to go to the pancreas and liver. Injecting them into an artery had not been studied very much. Perhaps the myoblasts were able to home in on the rats' muscles through some lock-and-key receptor interaction that had yet to be probed.

There may have been no agreement on the efficacy of myoblast transfer, but the method was under active investigation. By December of 1993, Brown was injecting natural myoblasts into dogs, gathering evidence that "appears promising."

By that time, nearly a hundred Duchenne's boys had received myoblast injections in the United States and Canada. Efficacy, outside of Law's claims, had not been proven. But one thing had become very clear: "It's safe," said Brown. "That means we can seriously think about myoblasts to deliver drugs and other genes. It opens this up as an approach to cell-based gene delivery."

No one was talking to Peter Law, but George Karpati was talking about how it might make sense purposely to injure a muscle before inserting donated myoblasts, because the microscopic cells enjoy fixing strained muscles, as any power lifter can attest. Then a French team headed by Thierry Ragot, at the Institute Gustave Roussy, in Paris, described how their attempts to whittle down the huge dystrophin gene had resulted in a serviceable minigene that they could cram into a retrovirus. Human cells and mouse muscle precursor cells infected with the recombinant virus expressed the dystrophin protein.

Furthermore, the gene stuck around. Injection of the virus into the muscles of mice that lacked a functional dystrophin gene resulted in the presence of the minidystrophin protein in as much as 50 percent of the muscle fibers. The virally produced protein, which continued to be made for at least thirteen weeks, was found to be situated in a place in the muscle where it might help Duchenne's children, making systemic delivery appear more feasible. And that might open the door to solving the stubborn problem of how to deliver dystrophin to the heart and respiratory muscles, which are not accessible to local injection but which ultimately kill all patients with Duchenne's muscular dystrophy.

Despite setbacks and controversy that had divided Duchenne's researchers since the discovery of the gene, advances were being made to answer the crucial question: Will anything be able to correct the defect in dystrophin-deficient muscles? In August of 1993, University of Michigan researchers led by Jeffrey Chamberlain showed they could thwart the disease process in specially created transgenic mdx mice that received a dystrophin minigene when they were embryos. That lab feat marked the first time the disease had actually been cured, although the therapy had occurred at the embryonic stage. But a few months later a French team showed that the disease could be prevented for six months in newborn mice that received human dystrophin minigenes in their skeletal muscles. The researcher Michel Perricaudet, of the Institut Gustave Roussy in Villejuif, France, reported his findings in the November issue of *Nature Genetics*. Suddenly, gene therapy, which had gotten off to a slow start, was overtaking myoblast transfer as the best hope for children with Duchenne's. In fact, as of this writing Robert Brown was predicting that the first MDA-approved gene therapy trials for the disease were no more than a year or so away.

18

THE THIEF
OF BREATH

One exhausting day in August 1989, the researchers Francis Collins and Lap-Chee Tsui and a handful of their scientific colleagues piled into two small propeller-driven planes that had been hastily rented to fly them to Toronto and Washington, D.C., for back-to-back press conferences. The purpose of the media blitz was to announce a secret that was already out of the bag: Collins, Tsui, and colleagues had just pulled off one of modern biology's greatest coups. They had identified the gene and protein defect responsible for the deadly disease cystic fibrosis.

"This is a great day for us," Robert Dresing, president of the National Cystic Fibrosis Foundation, confided as he prepared to go before reporters at the National Press Club in Washington. "We are terribly excited. We're flying." Even as he spoke, parents of cystic fibrosis victims, alerted by news of the discovery that had already leaked out to the press, were besieging the Maryland offices of the foundation with telephone calls. "They all start out the same way, asking, 'Is it true?'" noted Robert Beall, the foundation's executive vice president for medical affairs. "And when you say, 'Yes, we've found this gene,' there isn't a single one of them who doesn't become choked with emotion. This has us all running on adrenaline."

It was easy to understand. The achievement did more than signal an end to half a century's futile search for the underlying cause of cystic fibrosis, the most common lethal inherited disease in the Western world. It opened the door at last to drug treatments aimed at the basic fault in the disease rather than at mere symptomatic relief. Overnight it catapulted cystic fibrosis to the top of the list of diseases believed suitable for gene therapy.

Time has only enhanced that perception. Racing the clock, researchers have brought the gene from the lab dish to the bedside with unprecedented speed. By the summer of 1994, no fewer than five gene transfer experiments had been launched against cystic fibrosis, and the ingenuity behind the strategies seemed destined to set the pace for the entire field of gene therapy for years to come.

"It looks too promising to hold back," said Ronald Crystal, who set the ball in motion by filing the first application to treat cystic fibrosis with new genes. He was echoed by Collins, who, together with James Wilson, was pursuing a rival gene therapy protocol aimed at the disease. "Gene therapy is the *big* push right now," observed Collins. "Cystic fibrosis offers a wonderful opportunity to test the technique."

This energy was focused on a pitiless disorder that strikes once in every two thousand births. Cystic fibrosis is the most common genetic disease among Caucasians, affecting an estimated 35,000 young people in the United States and Canada. Although the life expectancy of victims has lengthened in recent years, thanks to improvements in nursing care, most still die by their early thirties. Each year, the disease claims the lives of some 1,500 young adults. It respects no one. The former television personality Jimmy ("the Greek") Snyder has lost three children to cystic fibrosis; the former *Sports Illustrated* writer and newspaper publisher Frank Deford has lost one.

Cystic fibrosis affects a number of body systems at once. Its hallmark is the production of abnormal mucus, the vital substance that cleanses dust and germs from the lungs and breathing passages. Mucus is supposed to be thin and slippery, but in cystic fibrosis patients it is thick and sludgelike, obstructing the lungs and airway and providing a medium that invites bacteria to grow. Repeated lung infections begin in early childhood, and the incessant fight for breath gradually robs a patient's lungs of elasticity. Meanwhile, the same gummy mucus clogs the pancreatic ducts, preventing digestive enzymes from reaching the small intestine. Deprived of these enzymes, the body cannot break down proteins, fats, and carbohydrates, and food passes through the system unabsorbed. Most cystic fibrosis victims are slight in build because they cannot get enough nutrients. They suffer from muscle weakness and ravenous hunger, as well as cramps and bowel obstructions because of their impaired digestion.

The roots of the disease are ancient. Experts think it was born roughly forty thousand years ago when one of our Stone Age ancestors was silently speared by a cosmic ray somewhere in the Middle East. The radiation burned a point mutation into a germ cell, whereupon it was passed along to a son or daughter. Soon thereafter, the gene defect began to circulate. In small tribes, where intermarriage is the norm, mutations can spread throughout a population in just a few generations, a phenomenon known as the founder effect.

Medical detectives are able to surmise when and where cystic fibrosis began, because the disease is utterly unknown in Africa, the cradle of human evolution. Logically, the mutation must have originated after *Homo sapiens* began migrating out of Africa, some forty millennia ago, through the natural funnel formed by the Arabian peninsula. On the other hand, it could not have happened much later, because after that the wave of migration split in two, one group heading northward and one traveling east. Cystic fibrosis is found in either direction, in places as widely scattered as Europe, Turkey, and Pakistan, and its statistical frequency is strikingly similar. What is more, the polymorphisms, which, as we shall see, are signature stretches of DNA that are inherited along with specific forms of a gene, are identical in disease victims in each of these far-flung locales, indicating a common ancestor.

Cystic fibrosis has caused generations of heartbreak for couples such as John and Mary Kaye Bottorff of Evanston, Illinois. At the age of three months, their son, Eric, was in the thrall of a pneumonia that refused to go away. Moreover, he exhibited severe growth problems, weighing only a few ounces more than his birth weight, when he should have nearly doubled in size. Doctors administered the standard clinical test for cystic fibrosis, which consists of measuring the sodium chloride content of a child's sweat. Children with cystic fibrosis have uncommonly high sodium chloride levels in their perspiration, giving rise to an old wives' tale that if a baby's skin tastes salty, the child will not live very long. The results of the sweat test in Eric's case were positive. When the doctor conveyed the bad news to the Bottorffs, they looked at each other in consternation. Neither had ever heard of the disease. Their first instinct was to shrug. "It didn't sink in how serious it was," recalls Mary Kaye, "until the doctor used the word 'fatal.' And then it hit us like a ton of bricks."

The Bottorffs resigned themselves to the possibility of losing Eric at a tender age. "It is hard," says John. "You can't watch movies where people die. It tears you up. When you go to funerals, you end up crying for yourselves." Loath to throw in the towel, however, they have tirelessly battled Eric's illness. Twice a day they give Eric, thirteen, breathing

therapy, which involves back-pounding and rigorous exercise, all to promote the drainage of mucus. Mary Kaye cooks high-protein meals and
administers massive doses of vitamins in an effort to pour calories into
his rail-thin body. Recently, the Bottorffs started putting Eric in the hospital several times a year for the more sophisticated regimen that cystic
fibrosis patients refer to as tune-ups. Eric receives megadoses of antibiotics, huge infusions of enzymes to replace lost nutrition, and postural
drainage, a process in which the patient is tipped at a thirty-five-degree
angle on a machine while his chest is vigorously banged.

More than a decade ago, the desire to conquer cystic fibrosis and ease
the burden of people like Eric Bottorff spawned an impassioned race to
identify the gene at fault. The quest was marked by high drama and high
dudgeon as some of the most fulminant egos and original minds in the
gene business vied for the prize.

Events began unfolding in 1981, when Lap-Chee Tsui was a postdoctoral investigator working in Manuel Buchwald's laboratory at the Hospital for Sick Children in Toronto. The young refugee from mainland
China was interested in isolating the cystic fibrosis gene, but he had no
inkling how to proceed. Then one day he learned of a new method of
mapping genes called "riflip" analysis. Riflips—short for restriction
fragment length polymorphisms—are bits of DNA that occur in the
same places on everyone's chromosomes, but whose nucleotide sequences vary slightly from person to person. These variations can be
uncovered and compared by means of restriction enzymes, hence the
name. The value of riflips lies in their polymorphic nature; their differences can be tracked from generation to generation and thus used as
markers. Imagine that someone's great-great-uncle had a genetic disease, and that he also possessed a distinctive riflip marker adjacent to
the disease gene. Since neighboring genes are almost always inherited
together, the same combination would occur only in descendants who
themselves had the disease. Finding the defective gene would therefore
be a matter of identifying markers exclusive to family members who
suffer from the disease and then, knowing that the disease gene must be
near, mounting an intense search of the chromosomal neighborhood for
the gene.

All Tsui needed, he realized, was some good markers with real polymorphic potential and a number of large families with cystic fibrosis on
which to test them. Taking advantage of the Toronto hospital's world-
renowned cystic fibrosis clinic, Tsui and his mentor, Buchwald, had
soon put together a collection of fifty families, each of which had two or
more living children with the disease. Then they set about probing the

DNA of each family member with a series of about fifty riflip markers, in the hope that one of the probes would show a linkage to cystic fibrosis and could serve as a marker.

It is a tedious process, and as Tsui struggled, other researchers were hot on the trail. Across the Atlantic, there was Robert Williamson, of St. Mary's Hospital Medical School, in London, who had turned his attention to cystic fibrosis after a long involvement with Duchenne's muscular dystrophy. In Salt Lake City, Ray White, with his access to the vast genealogies of the Mormon church, was also in the chase. So was a relatively young biotechnology firm called Collaborative Research Inc., of Lexington, Massachusetts.

Collaborative's motivation had little to do with curing cystic fibrosis as such. Smelling profits, the firm had its eye on creating the first linkage map of the entire human genome. With such a map, teeming with riflip probes located at regular intervals along the chromosomes, Collaborative could detect the presence of a variety of disease genes present in people's DNA, and thus corner the potentially lucrative market for prenatal diagnosis and carrier testing. This would fulfill the dream of Collaborative's founder, the academician turned entrepreneur Orrie Friedman. In Friedman's vision, physicians from all over North America would send blood samples to Collaborative, and the company would screen them against their disease probes, returning results in a matter of days. Since cystic fibrosis is far and away the most prevalent inherited disease in white America, the demand for diagnostics, he calculated, would be quite large. To that end, more than a dozen Collaborative scientists, working under the direction of Helen Donis-Keller, had spent two years cloning random segments of DNA and then painstakingly testing them to see whether they were capable of the variation among individuals that would qualify them as usable riflips. Out of 1,700 clones tested, Collaborative had by 1985 succeeded in stockpiling some 500 polymorphic probes, several hundred more than their nearest competitor, Ray White. Surely, one of these probes was near enough to the cystic fibrosis gene, wherever it might be in the genome, to be a suitable diagnostic marker.

But by 1985 morale among those seeking the cystic fibrosis gene was low. The disease was not lending itself easily to linkage analysis. In this downbeat environment, Collaborative came to Tsui and Buchwald with a proposition. Would the Toronto team make available to Collaborative its large collection of informative families if in exchange the company gave the hospital access to its library of probes? It was an offer that the Toronto group, which was running out of probes, could ill afford to refuse. By July, the new partnership of Collaborative Research and Tsui/ Buchwald was in full bloom.

That August brought big news. While Tsui was in Helsinki for a conference, his laboratory obtained genetic linkage with cystic fibrosis, using one of the first probes lent by Collaborative. The linkage was not close—the probe was judged to be approximately fifteen million base pairs away from the cystic fibrosis gene, judging by the frequency with which it was, or was not, coinherited with the disease. (The farther away a riflip marker is from a gene, the more often the two are separated from each other during the shuffling of genes that accompanies the manufacture of sperm and ova.) Nevertheless, the probe *did* appear to be a marker. When Tsui returned home, he quickly informed Collaborative. "It was wonderful," recalls Donis-Keller, who has since left the firm for Washington University, in St. Louis. "We hadn't expected to get results for a year or more. We thought we would have to go through our whole collection. It was just a piece of blind luck."

Now the story got sticky. It remained for Collaborative to map which chromosome the probe was on. Its researchers had found it more efficient to use random probes without knowing their chromosomal home, figuring they could easily map the probe later if it proved to show linkage. Donis-Keller told Tsui the firm would map the marker immediately; but would he please not try to map it himself? The mild-mannered Tsui reluctantly agreed.

Several weeks followed with no word. Tsui began to get edgy. The results should have taken only a week. When he finally contacted Donis-Keller, she pacified him by saying that preliminary work showed that the probe was on chromosome 7. But she said more work was needed to confirm it. Time continued to pass. Tsui grew angry, feeling that perhaps he was being cut out of a major discovery. Finally, he ran the experiment himself. Sure enough, the probe mapped to chromosome 7.

Now more trouble ensued. Tsui told Donis-Keller he wanted to announce the finding at the American Society of Human Genetics meeting that October, in Salt Lake City. Donis-Keller reportedly balked. She would agree only to announce that a marker had been found, not what chromosome it was on. Tsui had no choice but to go along, though he felt that the company was trying to sit on the gene's location for commercial reasons. "I think it is very clear the company wanted to keep it a secret," he later told an interviewer from *Science*, noting his belief that the firm, afraid competitors might beat it to the punch, wished to keep the location quiet until it could find probes that were a good deal closer.

According to Donis-Keller, however, Collaborative had not verified the gene's location to its satisfaction by the time of the October meeting. "We just found it at the end of August. It took us a month to find the chromosome, and we needed another month to confirm it," she said in an interview, asserting that the firm was wary of losing its credibility if there was a mistake. Part of the problem, she went on, was that the company had made a previous agreement to let a French group do the actual mapping, and these investigators "took a very long time." When the French group finally sent its results, "they were ambiguous." Donis-Keller maintained, "Our hands were tied by this agreement and our loyalty to these outside collaborators. But these guys let us down."

She cited other reasons for the firm's reticence. For one thing, another prominent woman researcher was claiming that the gene was on a different chromosome. "I felt if our company was shown to be wrong about such an important finding," Donis-Keller said, "we'd never hear the end of it. So maybe I was too conservative."

In any case, Donis-Keller vehemently denies that the company kept quiet about the chromosome's identity for proprietary reasons. "There was very good scientific reason," she says, "not to assert that it was on chromosome 7 at that meeting. We really weren't certain at that point. Things still had to be sorted out. In those days you had to confirm findings by two different methods. But remember, it was only a matter of weeks before we announced the location. It's not like we were withholding data for a year."

The rumor mill being what it is, however, by the time of the Utah meeting, everyone had heard the news that Collaborative had mapped the probe to chromosome 7. Thus poor Tsui, who had to deliver the paper announcing the marker, was in the awkward position of claiming that the marker's location was unknown when the entire audience knew better.

The scene now shifted to the National Cancer Institute's Frederick Cancer Research facility, in Frederick, Maryland, where in 1984 a vicious little oncogene called *met* was isolated by the facility's director, Dr. George Vande Woude, and one his fellows, Michael Dean. The oncogene mapped to the seventh human chromosome, which sent Dean's blood pressure soaring, for that chromosome is associated with several human leukemias. Had the two of them happened on a gene for bone marrow cancer? Dean sent a copy of *met* to the laboratory of Janet Rowley, of the University of Chicago, a leading authority on the genetics of leukemia. Rowley's lab was able to pinpoint the gene further, to a part of chromosome 7 known to be significant in leukemias. The trail seemed to be growing hot.

Wearing out a path to the post office, Dean now sent the gene to Ray White's lab in Salt Lake City, where there are DNA libraries from Mormon families with a high incidence of cancer. Dean asked White whether he would employ *met* as a probe to see whether it clung to any of the DNA samples of the cancer victims. Unfortunately, it did not, and Dean's hopes came crashing down. "We ran the probe through all of the cancer families," says White, "and the answer was negative." With a sigh, White filed the probe away and turned to other matters. That was in May of 1985. That October, White picked up on the rumor concerning Collaborative's cystic fibrosis marker on chromosome 7. He was intrigued by the possibility of finding a marker that was even closer.

He began rummaging through all of his probes for any that were on chromosome 7. Bingo. There was the *met* probe that had been gathering dust for six months. White and his associates dropped the *met* probe onto some single-sheeted nitrocellulose filters containing the DNA of cystic fibrosis victims. They crossed their fingers. They muttered some prayers. And lo, it was immediately apparent that there was a tight link between *met* and the gene responsible for cystic fibrosis. White's team repeated the experiment on forty more cystic fibrosis families. Forty more times, the *met* probe came back positive. Forty out of forty seemed such a perfect correlation that, for a time, he entertained the idea that *met* was the cystic fibrosis gene itself. Eventually, however, White encountered a negative, indicating the two genes had "crossed over"—in other words, had been shuffled—and he satisfied himself that *met* was not the gene. But the linkage was clearly very tight—no more than a million base pairs away—much closer than that of Collaborative's probe. Hurriedly, he put together a paper on his discovery and sent it off to *Nature* on November 1.

Meanwhile, in England, Williamson was also busy. Just three days after White submitted his paper, Williamson turned in to *Nature* an article of his own, reporting that he had found a probe he called J3.11, which also was within an estimated million or so base pairs of the still-unknown cystic fibrosis gene. For his part, Williamson insists that the Collaborative rumors had nothing to do with steering him to chromosome 7; by the time of the October conference, he maintains, he already had mapped the gene.

The team of Tsui and Collaborative Research was now in danger of being scooped on its own discovery. Donis-Keller's paper regarding

Tsui's finding was not due to be published in *Science* until December, and *Nature*'s schedule called for publication in late November. Fortunately for Tsui, a German colleague tipped him off that both Williamson and White had papers in press in *Nature*. Enraged that two other scientists should reap laurels for what he, Tsui, had achieved, he put through a call to Donis-Keller demanding that she make things right. Donis-Keller, now chastened, immediately got on the phone to *Nature* and, backtracking over events of the previous months, presented her case that it was Tsui and Collaborative investigators who had originally discovered linkage to chromosome 7. The others, she said, were Johnny-come-latelies who had pirated her findings. She threatened dire consequences if *Nature* did not publish a paper by her team. Donis-Keller proceeded to write the paper in two days. *Nature* accepted it a day later.

All three articles appeared simultaneously in *Nature* on November 28. An accompanying editorial tried to right matters by giving the historical background. It noted that White and Williamson had been "spurred on by the rumors" that Collaborative's marker was on chromosome 7.

Later, *Science* reconstructed the affair in a two-part series in which everyone's ethics but Tsui's were questioned. The series treated Collaborative roughly for purported foot-dragging, then left the implication that White and Williamson had improperly tried to make hay from Tsui's and Collaborative's findings.

Donis-Keller is philosophical about the actions of Williamson and White, noting, "It is very common for people to find out something by some route and then try and beat you into print." As for Collaborative, she defends its role and attributes *Science*'s criticisms to "anticompany bias." But in a sense Collaborative had, in the eyes of many, been hoisted by its own petard. Had the firm been more open with its findings, others would not have been able to steal its thunder. Moreover, its behavior, while consistent with customary business practice, was considered bad form by scientific and academic standards. Biotechnology is not like other industries. Corporate researchers more often than not share data with public-sector scientists and thus profit from discoveries financed by tax dollars. Many critics contend that they have a corresponding obligation to be less proprietary than businesses in other fields. With its allegedly outlaw behavior, Collaborative made itself some enemies, a situation not helped by statements attributed to Orrie Friedman, who, with his purring voice and high-pressure manner, at times has an unfortunate tendency to come across like a snake-oil salesman. Friedman allegedly once boasted to a science writer that his company "owns chromosome seven." Some joked that he intended to patent it.

In December 1986, a workshop was convened under the auspices of the Cystic Fibrosis Foundation. Held in Toronto, it brought together representatives from virtually every major group working on cystic fibrosis. At the meeting, a very important question was answered. It was known that the *met* and J3.11 markers had to be very close to the cystic fibrosis gene. But if they were both on one side of the gene, it meant that the researchers would have no idea in which direction to "walk" to find the gene. They had a 50 percent chance of going the wrong way. After the labs agreed to pool all their data on the frequency of coinheritance, however, the realization emerged that, providentially, *met* and J3.11 must sandwich the cystic fibrosis gene between them. The scientists now knew exactly where to look for the elusive piece of DNA. It had to be somewhere on the 1.7 million base pair span separating the *met* and J3.11 markers.

"I can't emphasize too much how lucky we were that they were on opposite sides," says Williamson, "because it meant when you started walking, you had somewhere to go. Now we knew we could just start walking from *met* on our way to J3.11 and be assured we'd find cystic fibrosis somewhere along the way." Williamson soon hosted a follow-up meeting in London, to which Collaborative scientists were pointedly not invited. There seems little doubt that the snub owes much to Williamson's distaste for the presence of private industry in scientific inquiry. Williamson, the son of a Cleveland, Ohio, political activist who moved with his family to England very early in Williamson's life, is a confirmed socialist who is incensed that a firm like Collaborative would profit from tax-paid work on cystic fibrosis and refuse to share its findings. For all his rectitude, though, Williamson himself was about to falter.

Williamson, like everyone in the cystic fibrosis field, had pursued a number of false trails since joining the hunt in 1979. One of his most amusing miscues came early on, when he decided to see whether cystic fibrosis was a disease in which the gene was deleted from the genome rather than simply mutated. To test the idea, Williamson compared genes expressed in normal people against genes expressed—or rather, not expressed—in cystic fibrosis patients. He was ecstatic to find a DNA sequence among patients that contrasted strongly with an analogous sequence from normal people on his own lab staff. Then he learned the embarrassing truth. The DNA sequence was different because the cystic fibrosis patients were all males and his laboratory staff was coed. "The Cystic Fibrosis Trust," says Williamson, with his dry sense of humor, "wanted to know why we spent £250,000 to get a gene probe that distinguished men from women, which most people manage without extensive genetic analysis."

In trying to narrow the ground between *met* and J3.11, Williamson's competitors, notably Tsui, had decided on a conservative strategy—namely, saturation mapping, a brute-force method by which one isolates dozens of random markers along the chromosome. One then maps these markers, by noting how often they are inherited along with cystic fibrosis, to determine how far away they might be from the gene. Williamson resolved instead to try and pull the gene out all at once, a high-risk tack. He took advantage of the fact that *met* was an oncogene, now known to be associated with the uncontrolled growth of cells that is seen in human sarcoma. Placing fragments of chromosome 7, some of them presumably containing *met*, into mouse cells, he waited for the oncogene to do its dirty work and transform the cells into precancerous ones. The cells that were transformed he knew must contain *met*, and those which also contained J3.11 he knew must contain the cystic fibrosis gene. The trouble was, an oncogene like *met* has unpredictable effects on its surroundings. It might be that the oncogene would "scramble" the location of its neighbors, leaving the cystic fibrosis gene out of the picture entirely. Other labs considered it risky. But Williamson persevered.

Using a new technique, whimsically called Mach One, which enables researchers to recognize at once the starting sequences characteristic of a coding gene, Williamson's laboratory in early 1987 stumbled on a gene halfway between *met* and J3.11 that appeared very, very promising. Williamson's daredeviltry seemed to be paying off. The gene showed no crossovers in families. It satisfied a test of authenticity known as disequilibrium, which refers to a phenomenon that allows you, when on top of a gene, to match the surrounding terrain of genes against a set of known landmarks, the way one can identify bombing targets by superimposing on the radar screen a matching profile of buildings and hills around the target. Finally, it was expressed in considerable quantity in lung tissue, just as one would expect of a cystic fibrosis gene—as well as in pancreas, intestines, kidney, and placenta.

Williamson was soaring. In April 1987, he and his colleagues published a paper in *Nature* announcing to the world that they had found a "candidate gene" for cystic fibrosis. The gene, which makes a substance related to a mouse protein called *int*, was thus named *int*-related protein, or IRP for short, as in Wyatt Earp.

When Williamson's competitors learned of his achievement, they felt like fools. Had they only had more brass, they too could have been on top. Williamson was scrupulously careful in public to call the gene merely a "candidate." But his body language told everyone he was sure. "We never issued anything saying it was the gene," he says in retrospect. "But privately, to be honest, we were sure we had it, we really were. We are professional enough to be careful what we say to the media. But we had convinced ourselves."

The upshot of Williamson's report was that the search for the cystic fibrosis gene all but ground to a halt. Some labs, notably Ray White's, totally suspended work. Others, among them those of Tsui and and his new collaborator, Francis Collins, found their NIH funding suddenly threatened. The NIH study section that reviews grant proposals reasoned, Why bother to spend more on finding the cystic fibrosis gene when the gene has already been found? When, in late July of 1987, Tsui and Collins got their "pink sheets"—the evaluation forms the study section sends out to applicants telling them how well their proposals had fared—"We all did terrible," recalls Collins. "We got completely unfundable scores."

Meanwhile, as the months passed, an ominous silence settled on Williamson's London laboratory. Tongues began to wag. Where was Williamson? Where were more data? The supportive evidence in his *Nature* article had been sketchy to begin with. At the time, everyone assumed he was simply withholding information to keep others from catching up with him. But now, as he continued to play his cards close to the vest, people began to wonder whether he had possibly been wrong.

Williamson was in fact experiencing the exquisite agony of finding that he did not have the gene after all. The first chink came when his lab workers completely sequenced two versions of the candidate gene—one from a normal person and one from a cystic fibrosis victim. The two sequences should have differed in some way if a mutation in the gene was responsible for the disease. Instead, they were the same in every way. "It was an awful thing to face," Williamson says—so awful that, at first, he did not face it. He decided he must have made a sequencing mistake.

But then other doubts appeared. The disequilibrium data wasn't exactly right. There was some blurriness when the chromosome terrain was superimposed. More damning still, there began to appear evidence of crossovers—cases in which mutant IRP was inherited by people without cystic fibrosis. The numbers were small—only two cases were verified—but if the two genes were identical, there should be no crossovers at all. "But what really told us, what cinched it," says Williamson, was that the candidate gene was not expressed in all the tissues it should have been if it caused cystic fibrosis. It was not expressed in the sweat glands, for example, or in the lining of the nose, two places invariably affected by the disease. By October, Williamson had to bow to the inevitable. IRP was not the gene.

Fortunately for Tsui and Collins, NIH had already recognized the obvious and had plowed ahead and funded their grants. But enmity toward Williamson died hard. Tsui resented having nearly lost his financial support. Others scored Williamson for slowing down research when thousands of children's lives were at stake. Had he been more forthcoming with his data in the original *Nature* piece, they claimed, the IRP gene could have been quickly discredited and work would have continued. Williamson himself is properly contrite, and blames zeal for his self-deception. He also points out that IRP was still a major breakthrough, a fact admitted by his competitors. That there were only two crossovers in twenty thousand cases studied meant that IRP was very close to the actual gene, and that narrowed the search.

In October of 1987, as Williamson and his team were, as he puts it, "picking ourselves up off the floor and dusting ourselves off," a fateful collaboration was in the works. Tsui was attending the annual meeting of the American Society for Human Genetics, in San Diego, when he fell into discussion with Francis Collins, who had entered the cystic fibrosis race only the year before. As it turned out, each had what the other wanted. Tsui had probes. After screening 250 markers with his saturation-mapping method, he had finally isolated two that were not only between *met* and J3.11 but much closer to the cystic fibrosis gene than either one. On the other hand, Collins had a skill that made him molecular biology's equivalent of Michael Jordan. He could jump.

Walking along a chromosome in search of a gene is extraordinarily tedious work. Starting from a given position along the chromosome, you must take a cloned segment of DNA no more than a few thousand base pairs long and sequence it until you get to the end. Then you pore through your library of clones from that same chromosome and find one that overlaps the previous clone. When you come to the end of the second clone, you return to your library for a third clone that overlaps *that* one. Each step, even if you are fairly good at it, takes almost two months. In that time, you will cover, if you are lucky, about 20,000 base pairs, meaning that if the distance you have to travel is about half a million base pairs, as it was in the hunt for the cystic fibrosis gene, it could take considerable time.

To complicate matters, every 100,000 to 200,000 base pairs the molecular traveler encounters DNA segments that cannot be cloned for study. That is, bacteria reject them and will not let them be spliced into their chromosomes for making copies. Generally, these pariah segments take the form of very long, repetitive sequences, which suddenly start running in reverse, forming anagrams, and in general acting in an unstable manner. As a practical matter, one cannot go any further. Unclonable DNA stretches are basically walls that cannot be climbed.

But in 1983 Collins, then a postdoctoral fellow at Yale, helped invent chromosome "jumping," a technique that enables researchers to cover huge distances of DNA in a single bound, leaping over walls as if they were not there. Here, in brief, is how jumping works. One prepares very large strips of DNA, say, 100,000 base pairs long and, having used chemicals to make their ends "sticky"—that is, eager to bond with other DNA—one tacks on a tiny piece of "flag" DNA at the very tip. Then the researcher fools the DNA segment into thinking that the most attractive thing its sticky end can glom onto is its own tail. So each piece greedily curls up and forms a circle.

What has happened is that one end of the DNA segment has hooked up with the other end, so the place the researchers started at and the place they want to get to are right next to each other, instead of being 100,000 base pairs apart. One simply finds the flag and snips out the DNA surrounding it. The result is a short segment that contains a beginning and an end and bypasses all those boring sequences in the middle. How much time does it save? One can do in a week and a half what used to take two months.

Jumping raises difficulties of its own, however. "Every time you take a jump," says Collins, a lanky, bespectacled man with a ministerial look and bearing, "you have to ask the question, Am I there yet? Or even more so: Did I go too far?" This is the bugbear of all gene finding: figuring out when you have arrived at the gene. No fireworks go off to mark the occasion. You must stop periodically to take stock.

"That's where you need family analysis to help you," explains Collins. Whenever one has established a new base camp, 100,000 base pairs farther along the DNA trail, one must test the new DNA against cystic fibrosis families to see whether one has stumbled on the gene. When Collins and Tsui started jumping in 1987 from one of Tsui's clones, they encountered crossovers between the probe and the disease. After the first two jumps, they were still getting crossovers. But by the third jump, crossovers disappeared. The absence of crossovers did not mean that the DNA segment they had found was part of the cystic fibrosis gene. But it did mean that the new probe was so close to the gene that it never failed to be inherited along with the disease.

Williamson's clone had helped. Collins had begun jumping from both *met* and J3.11, working toward the middle. When it became apparent that Williamson's gene was not responsible for cystic fibrosis but was indeed within 200,000 base pairs of the gene, as determined by statistical analysis, Collins and Tsui were able to narrow their field considerably.

By January of 1989, they had hit part of the gene but didn't know it yet. Their hopscotching had landed them on a DNA segment that was rich in cytosine and guanine—a sign that they had hit a so-called CpG island. Since CpG islands generally occur at the beginning of genes, Tsui and Collins had a clue that a gene was in the vicinity. But whether it was the one they were looking for could be answered only by a battery of tests.

The first order of business was a zoo blot. This involves comparing the new piece of DNA with DNA from other species in order to see whether the sequence has been conserved throughout evolution, a tipoff that the segment is part of a gene with an important function. Immediately, bells began to ring. The sequence was strongly conserved—that is, it was virtually the same—in mice, cattle, and dogs.

But they had obtained such results twice before, and the genes had proven to be unrelated to cystic fibrosis. Now it was essential to learn whether this new gene made a product and, if so, whether it did so in cells known to be affected by cystic fibrosis. At this point, enter John Riordan, director of the Cystic Fibrosis Research Development Program at the Hospital for Sick Children. Riordan had constructed a library of active genes taken from sweat gland cells, one of the cell types involved in the disease. A test of the new DNA sequence against the DNAs in Riordan's library resulted in a match with one of the sweat gland genes. The gene was demonstrably expressed in sweat glands, at the very least.

At this point, Tsui and Collins were intrigued enough to use the starting sequence as a probe to clone the entire gene—a task that took until late April. Still, there was no proof that this rather large chunk of DNA—the gene consisted of nearly 250,000 base pairs—was the cystic fibrosis gene. Collins notes a little regretfully that the case for the gene was built so gradually that it almost robbed the researchers of the exhilarating "aha!" experience.

"I always figured," he says, "that one day, all of a sudden, wham, we were going to know that we had the CF gene, and we would pop the champagne corks and have a wonderful celebration. But that didn't happen, because with this kind of technique, the hardest part is deciding when you really have the right gene."

More tests followed. The gene was found to express in the rest of the body tissues involved in cystic fibrosis—notably, the pancreas and the lungs—but not in other tissues. Moreover, analysis of the gene's sequence showed that it probably makes a cell membrane protein, and years of research had brought scientists to the conclusion that the site of the disease is the cell membrane.

But the discovery that told Collins and Tsui that they had the gene, at last, required the services once again of Jack Riordan's sweat gland library. Riordan had actually made two libraries. One contained DNAs from normal sweat glands; the other, DNAs from the sweat glands of cystic fibrosis victims. In May, Tsui withdrew from one library the normal version of their newfound gene and compared it with its counterpart from the cystic fibrosis library. Tsui's eyes glittered as he beheld the results. Along the cystic fibrosis version was a place where three base pairs were missing!

Within weeks, they determined that the mutation, never found on normal chromosomes, was present on 70 percent of cystic fibrosis chromosomes. "At that point," says Collins, "it became a little hard to imagine that this was all one big coincidence." In other words, it was time to pop the corks.

In the next two months, the excited researchers learned that the gene codes for a protein that contains 1,480 amino acids and that the omission of the three base pairs results in the loss of one amino acid—phenylalanine—at the 508th position. Although the other 1,479 amino acids are perfectly normal, the absence of one phenylalanine somehow dooms a child.

Meanwhile, Collins's penchant for helping people, the same dogged humanitarianism that makes him chase after disease genes, was getting him into trouble. An elder in his church, Collins found time that summer to work with his wife, Margaret, in an isolated mission hospital in Nigeria. The two of them came home with malaria.

Still shaky and fatigued, Collins returned in time to participate in the announcement of the gene's discovery that August. For the occasion, the research team named the protein cystic fibrosis transmembrane conductance regulator (CFTR) and then set out to identify the mutations responsible for the remaining 30 percent of all cases of the disease. It was estimated there would be four to six such mutations. Once these genetic errors were all known, one could screen prenatally for the disease with

100 percent accuracy, using the disease genes as probes. That would obviate the pain faced by thousands of parents who have had one child with cystic fibrosis and desperately fear having another—parents like the Bottorffs, who agonized over whether to bring a second child into the world, then went ahead and did it and spent the first two weeks of her life licking her skin, wondering whether it was too salty. Beyond prenatal diagnosis, completely reliable carrier screening would also become available and enable anyone to learn whether he or she is a carrier of the gene. Even mass screening of entire populations would be feasible. The fault is common enough to warrant it, perhaps. In the United States, one in every twenty persons carries the gene for cystic fibrosis, or as many as twelve million people. Nevertheless, the search for more mutations would prove far more difficult than anyone suspected.

Other researchers were swiftly capitalizing on the new information to zero in on the nature of the basic fault in cystic fibrosis. For years, the most baffling thing about cystic fibrosis had been how one genetic flaw could explain symptoms as diverse as the sludgelike airway mucus, digestive problems, and salty sweat. The saltiness of the perspiration in cystic fibrosis patients implied that the chemical components of salt—chlorine and sodium—were somehow involved, but no one was certain in what manner. Scientists also surmised, from the viscosity of cystic fibrosis mucus, that it did not contain enough water. But over the past several years the pieces of the puzzle have fallen into place.

Various researchers reported they had discovered a curious fact about the cells that line the respiratory tract in cystic fibrosis patients. The cells cannot transport chloride ions across the cell membrane to the outside. Any chloride ions they contain stay trapped inside the cells. The ions, in turn, trap their customary traveling companions, sodium ions and water. In a sense, water is retained in the cell in the same manner that moisture in the air is captured by common table salt through deliquescence. What does this have to do with cystic fibrosis? In the simplest terms, water dilutes mucus the way turpentine thins paint. If water cannot get out to the mucus, the mucus will stay as thick as molasses.

Normally, chloride ions exit through a tiny gate in the cell membrane called a channel. Researchers considered two possibilities: either the chloride channel was somehow defective in cystic fibrosis, like a door that is warped and unable to open and close, or else the channel was all right, but there was a problem with the way the gate was regulated; that is, no one was around to open the door at the proper time. It now appears from the structure of cystic fibrosis transmembrane conductance regulator that the protein *is* the chloride channel. The structure resembles that of a family of proteins known to act as channels, or tiny pumps.

Why wouldn't the channel work? The answer lies in the site of the phenylalanine deletion. The region contains a binding site for adenosine triphosphate (ATP), a substance that gives cells the energy to perform various jobs. If ATP is prevented from binding to the cystic fibrosis protein, the omission may sap the protein to the point that it cannot perform. The region is also next to a target area for certain compounds called kinases, which initiate protein reactions. The absence of the phenylalanine appears to make it impossible for the kinases to rouse the channel into action.

The initial fruits of these developments were the first drug treatments that actually aimed to remedy the inherited defect in cystic fibrosis. One plan called for trying to administer ATP directly to the lungs in aerosol form, for it had been observed that extra ATP activates an alternative, backdoor pathway that causes cells to expel chloride ions. This approach was to complement a drug called DNase, whose function is to destroy DNA. Prodded by continual bacterial infections, the mucus of cystic fibrosis patients contains excessive numbers of germ-fighting white cells, whose deaths release DNA, increasing the viscosity of the mucus. DNase digests the surplus DNA and speeds up mucus flow.

Neither of these pharmacological strategies, however, had the high drama of gene therapy. More important, no drug, no matter how sophisticated, offered the same promise of a permanent cure. It is significant that, in the months following isolation of the gene, the Cystic Fibrosis Foundation became the first prominent medical research organization to throw its full weight behind the pursuit of gene therapy. Its decision gave instant credibility to the technology, even as it made the financial resources of an aggressive fund-raising agency available to those who would bring the idea to life.

"Gene therapy has been in the forefront of our plans from the start," affirms the foundation's Robert Beall. "The cost is no object. We spent much more than a million dollars just to find the gene. Our commitment is total."

Cystic fibrosis presents novel opportunities for gene therapy. Absent is the problem of finding stem cells in which to insert the gene. There is no bone marrow to fuss with, no tissue to be excised and laboriously coaxed into accepting new genes. Instead, researchers have the luxury of millions of target cells whose very function is to absorb, to soak things up like a sponge. For these are the cells that line the interior of the lungs, legions of easily accessible cells that connect to the outer world via the trachea and bronchial tubes.

And how might the virus reach all those waiting, willing cells? "Ultimately, through aerosols," Jim Wilson forecasts. "Patients breathe antibiotics through nebulized air. Why can't they breathe vectors that contain a normal gene?" The straightforward economy of the idea is compelling. Cystic fibrosis patients breathing live virus, taking in lifegiving genes as they struggle to sip the very air.

No sooner had the gene been located than researchers came roaring out of the chute, seeking to satisfy themselves that gene therapy would work on the disease. In Francis Collins's laboratory, the virologist Wilson began putting the cystic fibrosis gene into retroviral vectors within weeks of its isolation. At 250,000 base pairs, it is a large gene, far too big to fit into a virus, but much of it is meaningless piffle. Like a soufflé that is 90 percent air, the gene is riddled with spacer DNA, noncoding regions of no known function. The business part of the gene consists of only 7,500 base pairs. That is small enough to stuff into a retrovirus, although it is a tight squeeze.

Size having been shown to be no issue, the next step was to take viruses loaded with normal CFTR genes and let them loose in a lab culture of cystic fibrosis cells. A number of institutions tried it and were elated to find that, upon acquiring the proper gene, the cells began to spew out chloride ions.

In early 1992, Ron Crystal of NIH announced that he had taken the quest to a yet higher level by putting the human gene into live animals, specifically into a strain of rodent called the cotton rat. Two days after the gene was inserted, Crystal reported, human CFTR was already being produced in the lining of the rats' lungs. Signs of the protein were still detectable in their lungs six weeks later. As an encore, Crystal introduced the CFTR gene into rhesus monkeys, with similar results. The experiments amounted to a stunning triumph, not merely because they were the necessary prelude to a human trial but also because they marked a historic departure for gene therapy. Crystal had become the first researcher to abandon the retrovirus as his vector. He had turned instead to a DNA virus familiar to anyone who had ever experienced the miseries of the common cold—the adenovirus.

As vectors, adenoviruses offer a number of advantages in dealing with the lung. Unlike retroviruses, which insert themselves only into cells that are dividing, adenoviruses will infect cells regardless of what phase they are in. The distinction is important because in the airway the number of cells undergoing division at any given time is very low, only

2 percent of the total lung cells. Altering such a small fraction of cells is probably not enough to effect a cure. Hence, the use of retroviruses as a vector requires doing something to boost their infectiousness. One might, for example, have the patient inhale ozone, a mild abrasive agent that would damage the lung cells slightly. The trauma would be enough to prod them into dividing. But this is not desirable from the point of view of safety. In contrast, adenoviruses love the lung. They are a primary cause of colds, pneumonias, and other respiratory infections, and they customarily operate in the airway. As a bonus, adenoviruses are spacious enough to accommodate the CFTR gene and its promoters more comfortably than retroviruses do.

After his success with rats and monkeys, Crystal was champing at the bit to go into humans. "We can only get a limited amount of information from animals," he said. "There is no good animal model for cystic fibrosis. And the adenoviruses we use don't infect primates. So if we are going to push this field along, we have to get into man as soon as we can."

Francis Collins was not so sure. "I'm a little less optimistic than some people that we have this figured out," he said in the summer of 1992. "There are still a number of hurdles." The usual gene therapy questions loomed, of course. How many cells must one correct? (Studies showed that 6 percent might be enough to cure cystic fibrosis.) Is overexpression of the CFTR protein dangerous? (It hadn't been in rodents.) How long could expression be expected to last? (It depends on the kind of vector.) But cystic fibrosis presented some unique issues of its own.

First of all, there was the question of which cells one must hit with the new genes in order to cure the illness. Early on, researchers burned to know where in the respiratory tract the CFTR protein is primarily made. The way one determines such things is to make antibodies against the protein and then unleash them. Like bloodhounds, the antibodies will lead you straight to the protein's lair. But when Jim Wilson turned his antibodies loose, they revealed something disconcerting. In the windpipe and large bronchi at the top of the lungs, more CFTR is produced at the level of the mucus glands beneath the airway lining than in the lining itself. The finding was both surprising and deflating. If the gene had to get into these glandular cells, which are much harder to reach with aerosol sprays than are the more accessible surface cells, then the prospects of gene therapy dimmed.

But as time passed and Wilson marched farther down the bronchial tree with his antibodies, the picture began to brighten. In the smaller bronchioles that make up the lower portion of the lungs, a site where there are no glands, CFTR was highly expressed in the surface cells. This meant that in the lower lung, at least, gene delivery would be less of a problem. "In the final analysis," said Wilson in early 1993, "it may still be necessary to correct the defect in both places. CF is active in both the upper and lower lungs. But at least we should be able to correct the fault in the lower lung."

Crystal put an even more positive spin on the situation. "The fact that more CFTR is made in the glands does not mean that is the seat of the disease," he said. "Everyone agrees that the disease starts in the small airways, and there are no glands down there." Even if the glands turn out to be crucial, he added, researchers could still solve the problem by injecting the genes into the pulmonary artery, which in a series of experiments in sheep transferred the gene to the glands quite nicely.

Equally, if not more, important was another issue. What was the likelihood that the immune system would reject the new gene? Rejection is always a possibility when gene transfer occurs in the living body instead of in a lab dish. Put new genes into cells before the cells are returned to the patient, and the immune system will be none the wiser. The cells look the same as ever from the outside. But let the gene-carrying vectors loose into the bloodstream or the airway, and the immune system might spot them, sound the alarm, and try to destroy them before they can reach sanctuary in the target cell. Adenoviruses induce a vigorous immune response. Their use virtually assures that the immune command center will pull out all the stops. This may not be an immediate problem, because the virus may very well make it safely into the cell the first time. Unfortunately, adenoviruses do not integrate directly into the host's DNA the way retroviruses do. Being DNA viruses, they work outside the chromosomes, hijacking the cell's reproductive machinery by brute force. Their action eventually kills the cell, making it necessary to go in and alter more cells. Thus, the treatment of cystic fibrosis with adenoviruses would probably require repeating the treatments periodically, probably once every three months, an approach that has been likened to using genes as "drugs." And here is where the immunity problem may get sticky. The body may easily recognize the adenovirus from previous treatments and zero in before infection can occur.

However, the immune system must still catch the virus before it gets into the lung cells. The virus may very well win the race. "So while the immune network may prevent repetitive dosing, there is no guarantee that it *will*," notes Crystal.

In a similar vein, Wilson points out that most studies of the immune system have concentrated on how that system operates to fight disease.

"Few studies have looked at it in the context of gene therapy, where the system is being exposed to millions of virus as opposed to someone coughing on you and exposing you to perhaps ten particles," he notes. In gene therapy, the immune system may very well be outmanned.

Beyond these uncertainties were a number of safety concerns. If the gene therapy patient were to catch a cold, would the disabled adenovirus used to make the vector combine with the cold-causing adenovirus and regain its full powers? Would the vector then reproduce and overrun the system? Could the viral protein hurt the patient? Could aerosols contaminate the environment with the virus? Could the virus get into the germ cells? Crystal thinks most of these questions have been answered satisfactorily by animal research, though some details still need to be worked out. "I think we can never minimize the risks to zero," he says, "but it's a question of whether the potential benefits outweigh the risks. This is a fatal disease, the most common inherited one in the West. Either we're going to deal with it or we're not. And we'll have to take some risks."

Crystal's initial human experiments, which were to be conducted at NIH even though in early 1993 he accepted a new appointment at Cornell University Medical College, in New York City, were to involve ten cystic fibrosis patients. All would be over the age of twenty-one, to eliminate problems with informed consent. They would have only a mild to moderate degree of the disease, because in the early experiments, aerosols would not be used. Instead, the genes would be put into the body via a bronchoscope, in a surgical procedure that might cause complications in someone with severe cystic fibrosis. The first two patients would receive only a low dose of the new genes, and only half of one lung would be targeted. If all went well, the next two would get a higher dose, and so on. "We expect we will start to hit an effective dose around the middle of the trial," predicted Crystal. The dose did not need to be very high, it appeared. No more than 6 to 7 percent of the cells would probably need to be corrected, because lung cells seem to communicate chemically somehow. "If you look at the cilia, the fingerlike projections in the airway that help clear the lung," said Crystal, "they move in unison like seaweed. It's like a Jacques Cousteau movie. So the cells have to be communicating." Studies have already shown that if one injects dye into some lung cells, it quickly spreads to other cells. If one corrects one cell in ten with new genes, all ten cells will be found later to have been corrected. "It's grounds for optimism," Crystal noted.

Wilson's experiment was to be similar to Crystal's, with one important difference. Only 10 to 15 percent of one lung was to be infected, the rationale for a smaller-scale treatment being that the adenovirus might well cause inflammation of the lung. It did so in Wilson's monkey experiments. Lung inflammation could be catastrophic for a CF patient, even one with a moderate case. "We don't want to compromise anyone," said Wilson. Twelve patients would be treated, with the first three getting a low dose and the next set of three getting higher doses.

Just as mindful of safety were Michael Welsh and his colleagues at the University of Iowa College of Medicine. Their protocol for cystic fibrosis, approved by the RAC at the same time as Crystal's and Wilson's, called for putting "very small" amounts of gene-laden adenovirus into the nasal passages of three cystic fibrosis patients. They would test the effectiveness of the treatment by using a small electrode to measure any improvement in the tiny voltage that normally exists across the lining of the nose. In patients with cystic fibrosis, the voltage is abnormal. Welsh was clearly going slow for fear that adenoviruses might backfire as vectors.

It was Wilson's view that the cystic fibrosis experiments were to be a proving ground for adenoviral vectors in general. "A huge amount of investment has been going into these new vectors," he says. "If they cause inflammation or have other flaws, the field needs to know now before we get in even deeper."

By the middle of 1994, experiments were well under way in all three labs. Welsh was the first to report success in getting some gene expression, although transient, into the nasal tissue of his CF patients. Crystal, meanwhile, hit a temporary roadblock when his third patient, responding to a boost in gene dosage, developed an inflammation in his airway. Lowering the dosage in the next patient negated the probem. As of this writing, Crystal had treated eight patients, noting, paradoxically, that lower doses brought the best results. "The patient in whom we used the lowest dose produced the most positive expression, reaching 5 percent of cells, which we think we will need to bring about a cure," said Crystal.

Wilson, meanwhile, had stopped his experiment in midstream to change vectors. His first adenovirus had provoked an immune response, threatening follow-up treatments. He proposed to resume the experiment with a new adenoviral vector that had been further disabled so as to make it less likely to ruffle the immune system. Even so, Wilson predicted that gene therapy for cystic fibrosis would eventually need to be given at yearly intervals "in concert with a weeklong regimen of immunosupressant drugs."

Will gene therapy cure the disease? "I can't answer that," says Crystal. "My own clinical experience tells me the effective treatment is going to end up as something we don't envision now. It may be a combination of this vector and something else. But unless we move soon, we're never going to get there."

If they did nothing else, the cystic fibrosis experiments showed the power of the new technology. Bob Williamson, who is still in the midst of cystic fibrosis research, relates an amusing tale from his youth. In 1959, as he was about to begin his Ph.D. at London's University College, his adviser, Ernest Baldwin, asked him on what he planned to do his dissertation. Williamson, a biochemistry major, replied that he wanted to investigate the ribosomes and RNA, which were still mysteries in those days. Baldwin, a crusty biologist from the old school, trained to study whole plants and animals, not molecules, shook his head sadly. He was convinced his young charge was squandering precious time on tommyrot. "Mark my words, Williamson," he admonished. "No good will ever come of that nucleic acid stuff."

19

IT TAKES A
WORRIED MAN

It became one of the most frustrating rituals in science. For ten years, an international band of gene hunters would huddle every few months and try to feel their way out of a molecular jungle. Gathering at agreeable locales, ranging from Los Angeles and Cape Cod to the sleepy Florida Keys community of Islamorada, the fifty-eight scientists representing six American and British teams would camp out in private homes and take stock. But until mid-March of 1993 one of the deadliest of all mutant human genes thumbed its nose at their collective pride and experience. The quest for the gene responsible for a terrifying neurological disorder, Huntington's disease, became a debacle.

The human brain weighs but three pounds, no more than 2 percent of the normal adult body weight. Yet its care and feeding occupy the efforts of a grossly disproportionate share of genes. Approximately half our genome is thought to be involved in running the brain and its fiefdom, the central nervous system. That leaves the door open for a lot of mutational mischief, and, not surprisingly, a staggering number of genetic diseases display, as their key feature, disruption of intelligence, mental stability, or motor coordination. The hunt for the genes that control neurological and mental function is a top priority among molecular biologists.

Bagging these genes was a goal often cited in support of the Human Genome Project because it would provide a long-sought explanation for Alzheimer's disease, Down's syndrome, Huntington's disease, multiple sclerosis, amyotrophic lateral sclerosis, autism, and many other baffling neurobiological conditions affecting hundreds of millions of people worldwide. But a successful effort would do more than that. Presumably, it would also uncover the suspected organic roots of mental illness. Confirmation of a link between genes and psychiatric disease would lift the stigma from an unfortunate population habitually denied the empathy accorded those with ailments of obvious physiological origin. Fueling the search is the hope that an understanding of brain-based illnesses could lead to improved drug therapies and even a plan for treating these frequently incurable maladies by means of gene replacement. However, so far in the early days of gene therapy, researchers have been dogged by a series of false starts, illusory leads, ephemeral data that looks good one minute and turns to dust the next, and outright error.

Neurobiologists, in all fairness, are like divers performing dives of a high degree of difficulty. It is very likely that many or all of the diseases they are tackling involve a subtle pas de deux between genes and life's experiences, or the breakdown of more than one gene on more than one chromosome. This complicates the picture and makes the task of identifying individual genes that much harder. Setbacks and premature bow taking are nothing new to science. But never before have the spotlights of the media been so bright or the expectations of the public so high.

In 1983, the spotlight fell on one of genetic medicine's greatest triumphs. James Gusella, a shambling, bespectacled, and virtually unknown thirty-year-old molecular geneticist at the Massachusetts General Hospital, in Boston, published a historic paper, "A Polymorphic DNA Marker Genetically Linked to Huntington's Disease," which appeared in the November 17 issue of *Nature* and made front-page headlines worldwide. Wrote Gusella's team, "We have now identified an anonymous DNA fragment from human chromosome 4 that detects two different RFLPs in a HindIII digest of human genomic DNA. This polymorphic DNA marker shows close genetic linkage to the Huntington's disease gene in two different families. . . . We infer that the Huntington's disease locus resides on chromosome 4."

In plain English, what Gusella had found was a 17,000-nucleotide swatch of DNA that could indicate the presence of the gene causing Huntington's disease in someone, even though the gene itself had not yet been found. The discovery was momentous because the researchers had had no clue beforehand where the gene was hiding out, and the finding was later hailed by James Wyngaarden, then director of the National Institutes of Health, as the most important discovery in all of science in 1983.

The idea of using linked genetic markers in genetic counseling to predict the presence of disease-causing genes had been discussed as early as 1947, but it was Gusella who came up with a workable mapping strategy. It could not have happened at a more propitious time. Like many other young Harvard researchers rushing to make discoveries and not lose their funding, Gusella was running out of time—federal cutbacks meant his lab was about to close. He had bet his future on the new DNA techniques to see whether he could link the Huntington's gene to the genetic markers called riflips, or restriction fragment length polymorphisms (RFLPs) that dot everyone's chromosomes and indicate points where the genetic code harmlessly varies slightly among individuals. Studying DNA from Huntington's families in Venezuela and the United States and looking for differences, Gusella took an educated shot in the dark. Unlike his competitors working in other diseases, he quickly drew blood: he found a stretch of DNA on the short arm of chromosome 4 that seems to lie so close to the Huntington's gene (only about 1.5 million base pairs away, as it would turn out) that the two travel together and are nearly always coinherited. The restriction enzyme called HindIII will cut that piece of DNA differently, resulting in DNA fragments of different lengths, depending on whether someone has the marker and hence the gene.

The technique Gusella used is called in situ hybridization, in which cloned radioactive DNA probes are mixed with preparations of chromosomes fixed on a slide. The probes will hybridize, or bond, with their complementary DNA sequence whenever they find it on a particular chromosome and create a discernible blot on a piece of x-ray film. Thus is revealed the chromosomal location of the probe, and that is sufficient to map the marker gene to its home within a few million base pairs. Gusella's random probe, known as G8, hybridized both to the short arm of chromosome 4 and to the DNA fragment bearing the gene linked to the restriction enzyme site that is different in families with Huntington's disease. That meant the gene for the disease mapped to the short arm of chromosome 4, even though the gene's precise location had not yet been determined.

Gusella's lucky hit ensured that his funding would continue; in fact, his allotted lab space in Mass General increased dramatically as visitors dropped in from around the world. Yet even as he accepted congratulations for the first major triumph to emerge from the revolution in gene mapping, Gusella kept urging caution and pointing out that he now faced a horrendous problem: he knew the gene was somewhere on chromosome 4 but had no idea in which direction it was situated relative to the marker. He was like a traveler standing at a crossroads staring at arrows pointing both ways. And while all journeys begin with the first step, Gusella knew better than anyone that an awful lot of steps lay ahead of him: the fourth human chromosome consists of about 250 million base pairs of DNA, and the Huntington's gene could have been anywhere.

Moreover, even at those early stages of the journey, Gusella abruptly presented science with a moral bombshell. As its discoverer, a young Long Island physician named George Sumner Huntington, noted in 1872, Huntington's disease usually is a late-onset disorder; symptoms do not, as a rule, begin to show up until midlife. Huntington had grown up watching his father and his grandfather, both also physicians, attempt to treat four generations of neighbors afflicted by a progressive inherited dementia. Young Huntington first encountered the disease as he made the rounds with his father and met an affected mother and daughter, "both tall, thin, almost cadaverous, both bowing, twisting, grimacing."

The disease, Huntington later noted, "is peculiar in itself and seems to obey certain fixed laws. [It] is confined to certain and fortunately a few families, and has been transmitted to them, an heirloom from generations away back in the dim past." Once it starts, though, the disease progresses inexorably, Huntington reported, "often occupying years in its development, until the hapless sufferer is but a quivering wreck of his former self."

The natural history of Huntington's disease haunted Jim Gusella. What had he wrought, anyway? The statement "What we did by finding the marker was to create a diagnostic tool in a fatal disease where there's no therapy" reflects the researcher who always seems both fatigued and preoccupied. "A healthy person, thus, could learn that in fifteen years they're going to look like their sick parent and there's no hope. That's a *hell* of a quandary, let me tell you! It put tremendous pressure on all of us to find the gene because if we know what the defect is, we have at least the possibility of developing some kind of rational therapy."

Gusella's luck did not hold. The quest turned into the world's most persistent gene hunt and often verged on grotesquerie. It became an odyssey haunted by misleading data, false trails, embarrassment, and backbreaking work. For the next decade, Gusella and his fellow lab chiefs—including his mentor David Housman at MIT, John Wasmuth of the University of California at Irvine, Francis Collins at Michigan, Peter Harper of the University of Cardiff in Wales, and Hans Lehrach and Anna Marie Frischauf in England—trekked after the gene, sometimes smoothly, sometimes fractiously, but basically together. They gathered for family reunions every few months and kept glancing over their shoulders at other competitors in the race—Stanford's David Cox and Richard Myers and the Canadian researcher Michael Hayden at the University of British Columbia in Vancouver.

The person who kept them at their benches, Nancy Wexler, has since become almost a conscience for the entire field. One of the few leading figures in genetics who is not a geneticist herself, Wexler is, in fact, not a natural scientist at all. She represents the Other Side—the social sciences that deal with human feelings, attitudes, behavior. A Ph.D. clinical psychologist in the department of neurology and psychiatry at Columbia University, Wexler is a blond, vivacious, and extremely thoughtful scientist whose family foundation—the California-based Hereditary Disease Foundation—in 1984 formed the Huntington's Disease Collaborative Research Group. She brought the research teams together, helped fund them, carefully tended to their morale, divided their labors so that they would not be competing, and scoured the world to find genetic material for them to work with. Besides being merely a "cheerleader," as she whimsically terms herself, Wexler had a vested interest—she has a 50 percent risk of carrying the Huntington's gene and developing the incurable disorder that killed her mother and many of her relatives. In her forties now, well into the decade when Huntington's symptoms usually start to appear, Nancy and her older (by three years) sister, Alice, have become lightning rods for genetics research: millions of people have seen the Wexler sisters on TV shows like "60 Minutes" or have read about them. They symbolize the urgency that pervades gene medicine, and their presence has served to bond a diverse band of virtuoso gene scientists whose gut instinct is to compete rather than to collaborate. Francis Collins, world class gene finder, has noted, "Knowing Nancy's status, you can't look her in the eye and say, 'I can't work with so and so.' "

Yet Wexler comes across as anything but a martyr. From the outset, long before other genes were found by riflip mapping, she maintained that the Huntington's quest could serve as a model for finding and perhaps understanding all the defective genes that cause so much human misery. Wexler's name appeared second, after Gusella's, on the *Nature* paper that proclaimed the initial marker. She, more than anyone else, managed to keep the research in motion over the next decade, even though this disease, which afflicts only one person in ten thousand, threatened to make bloody fools of some of the biggest stars in recombinant DNA research. It took all her skills as a psychologist and as a shrewd observer of human behavior to keep her collaboration together, but her people skills are so formidable that they led James Watson, who is quirkier than hell, to choose her to chair what should prove to be a real hot seat: the crucial committee that will weigh the ethical questions of the Human Genome Project as they occur over the next dozen or so years.

Each year without fail since 1981, Wexler has led medical expeditions to the torrid jungles of Venezuela where sit three isolated villages on the garbage-strewn shores of oily Lake Maracaibo. The villages, called San Luis, Barranquitas, and Laguneta, have the world's highest rate of Huntington's disease. By empathizing with victims (assisted constantly by Fidela Gómez, a nurse whom Wexler calls her right arm), by begging for blood samples from patients and their relatives, and by painstakingly compiling family lineages, Wexler has been able to trace the inherited disease as far back as the early 1800s, to a woman aptly named María Concepción. In fact, Wexler and her colleagues have managed to piece together a staggeringly complex, yet meticulously documented, pedigree of María Concepción's clan that now numbers 12,000, living and dead—among them, 371 with Huntington's disease. Currently, 9,000 members of the family are alive and under the age of forty—a young population, indeed, but one that may be accursed like no other generation in history: Wexler estimates that no fewer than 800 children carry the lethal gene, and, unless something is done by science, they are doomed.

Familiar to many as the disease that killed the American folksinger Woody Guthrie, Huntington's is a total mental and physical catastrophe, much like having cancer, muscular dystrophy, and Alzheimer's at the same time. It destroys the mind as well as the body and turns human beings into grotesquely dancing puppets. The disease kills the nerve cells in the basal ganglia in the upper portion of the brain that coordinates movement; it also devastates cells in the cortex that control cognitive functions.

Normally, these neurons make two brain chemicals, the neurotransmitters acetylcholine and gamma aminobutyric acid (GABA), and have receptors that bind to another neurotransmitter, glutamic acid, to cause normal nerve excitation. The cause of neuronal death is a mystery, but it seems to be produced by a single autosomal dominant gene and is accompanied by dramatic loss in acetylcholine and GABA, and hence in the ability of the nerves to communicate with one another. The spiny neurons, instead of sprouting vigorously to make connections, bend back on themselves, useless, like power lines downed by a storm. At the same time, another neurotransmitter, dopamine, builds up in the basal ganglia, causing alternating spells of excitement and depression, which may be misdiagnosed as schizophrenia.

Unlike Parkinson's disease, in which only one population of cells dies—the dopamine-producing neurons in the brain—Huntington's disease involves the destruction of several populations of neurons that control different transmitters. As a consequence, instead of the Parkinsonian symptoms—tremor, rigidity, difficulty in standing, and slowness of movement—the more than 30,000 Americans who suffer from Huntington's disease are slowly losing control of their reason and their bodies. Moreover, an estimated 150,000 other family members are at risk of developing the dominant disease, because from the moment of conception, if it is passed on, the gene is pitiless.

No one knows why, but when the gene is inherited from the male parent, symptoms generally begin in adolescence. When it comes from the mother, symptoms do not usually start until middle age. (In this process of "genetic imprinting," genes inherited from a male seem to be more active.) But sooner or later, symptoms will begin to appear. By then, the unsuspecting victim may already have had children, even grandchildren, all of whom are in mortal peril. In affected families, the Huntington's disease gene hangs like a sword of Damocles that can destroy them emotionally and financially.

Ordinary people think little of dropping a water glass, stumbling when they walk, or forgetting a name or phone number. But when such trivialities happen to Nancy Wexler, she shudders. "You just keep wondering if it's finally starting," she says.

A Huntington's patient, as he battles the first symptoms, grows irritated, depressed, tearful. He may go from doctor to doctor, few of whom can accurately diagnose the disease. Inevitably, the victim's body slowly, over months or years, begins to work as if by remote control.

It starts to move.

Normal movement may slow down and twist, forcing the legs to shuffle instead of walk and the arms to writhe. Over time, the limbs start to flail, faster and faster, involuntarily jerking, fighting off all attempts at control. As the disease destroys the brain, speech grows indistinct, and a repeated grimace or tic may progress to a frenzy of facial movements: the eyes roll, the tongue darts in and out, the eyebrows glide up and down. The entire body may gradually become a horror of contorted, uncontrollable movements. *Chorea,* this is called, from the Greek root for "choreography." As the symptoms escalate over the years, the wild movements finally confine the patient to a wheelchair or bed. Sometimes, he may stiffen up like a board and lose the ability to swallow. Mental functions also deteriorate, personality changes are common, and eventually sanity disappears.

Huntington's can take twenty years to reach full cry. Drugs such as Haldol or Prolixin may help control movement in some patients; Valium and antidepressants can modify depression. Yet no treatment devised so far has been shown to slow the course, let alone halt it. Every afflicted person and his family must engage in lonely battles against the inevitable, struggling day after day to maintain hope and keep fighting. The Huntington's patient can look forward to little but years of gradually decreasing capacity, followed by total disability and certain death. The suicide rate among these patients is more than seven times above normal. Most must eventually be institutionalized. Infection often ends the story, although many victims actually die of malnutrition because their movements are so fierce that they cannot be fed. Others die of aspiration pneumonia after they accidentally inhale food. Grieving families, by the bedside, often say they are amazed to see their loved ones finally stilled. The devilish dancing—in the Middle Ages the afflicted would offer prayers to Saint Vitus, who they believed could cure them— has finally ceased. Ironically, victims look as they did before the long suffering began.

El mal de san Vito is the name the people of San Luis have given to George Huntington's disease. *El mal*—an apt description for a brain as it slowly dies. Shunned by the rest of rural peasant society, the Venezuelans are forced into remote ghettos isolated by water and reachable only by boat. Like modern-day lepers, they seek solace in one another. Their Roman Catholic faith instructs them to marry and prohibits birth control. Abortion is illegal in their country. Generation upon generation has been locked in the grip of the ghastly legacy. As the villagers work as fishermen and loggers, raising their young and caring for their dying, they patiently wait for the implacable enemy to strike again. Who will it be? Who will next start to show the first, subtle signs? " 'The eyes have it'—that's what they say," notes Nancy Wexler. "You can always tell it in the eyes. Each year when I arrive, they always look deep into mine. And they're not very subtle about it."

La catira (the blonde), as she is known, keeps her own fears to herself. "I try to be as counterphobic as possible," she says of her own risk. "I'm trying to do something about my possible fate, rather than just being passive. Once you're part of a family with Huntington's, you share blood with people around the world. You're not a private family any more."

Nancy Wexler's own world came crashing down in 1968 when she was twenty-two and she and Alice were carefree college students. Their father, Milton Wexler, a prominent Los Angeles psychoanalyst, summoned them home for a family meeting to tell them that their mother and his ex-wife, Leonore Sabin Wexler, had been diagnosed as having Huntington's disease. "We were utterly stunned," Nancy recalls. "She'd never mentioned that she might carry the gene. She had believed, as it turned out, that only men got Huntington's. Her father had died from it, and her brother later was diagnosed. That's when she learned that she could be a carrier. By then, my sister and I had been born. So she was hopeful that she had escaped."

A classic victim of the disorder who once was stopped by police for "drunken behavior" the same way Woody Guthrie was, Leonore Wexler suffered for more than a decade. A biologist and teacher of biology, she was very aware of her fate, and tried to commit suicide by taking an overdose of sleeping pills. When her daughters found her, she had propped their photographs next to her on her bed.

Near the end, the family was forced to pad her hospital bed with lamb's wool because her thrashing left her black and blue. And when she died, Nancy says, "she looked like an inmate at Dachau."

When the disease struck his family in 1968, Milton Wexler decided to start his own research foundation and fund it directly. The family motto of the Hereditary Disease Foundation, Nancy says, is "To get this disease before it gets us." The idea was to place their trust in recombinant DNA research as a means of attacking not just Huntington's but a whole range of other genetic diseases. A scientist himself, Milton Wexler sought out some of the leading biologists, including the legendary *Drosophila* geneticist Seymour Benzer and the biochemist William Dreyer at nearby Caltech, in Pasadena, and asked what his foundation ought to do. Benzer and Dreyer advised against building some sort of formal institute. Instead, they told Wexler to scout the nation's up-and-coming young biologists and tempt them with the possibility of solving such an ancient and horrible brain malady as Saint Vitus' dance, caused by a single gene. Wexler recruited a Benzer graduate student, Ronald Konopka, to go after the hottest postdocs he could find and lure them to the first of what have since become annual interdisciplinary workshops sponsored by the foundation at various institutions around the country. A major find was molecular biologist Allan Tobin, then at Harvard, now at UCLA, who was so inspired by a workshop that he became the foundation's scientific director in 1978 and has been recruiting other molecular biologists to the cause ever since.

When the Wexler sisters realized they might be carrying a genetic time bomb, they faced a question common to other Huntington's families: Should they get married and have children? Nancy and Alice decided not to have children. Many other people do, knowing that if it turns out that they carry the gene, those they love the most may face the same fate. Yet if they forgo marriage and parenthood, and the gene never shows up, they have worried and sacrificed in vain. As a psychologist, Nancy reacted to the Huntington's threat with what she calls "massive implosion therapy." She couldn't just sit still and wait. For several years "married," as she says, "without grace of law" to Herbert Pardes, dean of medicine at Columbia University and one of the world's leading psychiatrists, she decided to dedicate herself to the family cause.

To begin the effort, Tobin sought out David Housman at MIT, with whom he happened to share a baby-sitter in Boston. As luck would have it, Housman already had Huntington's in mind because Joseph Martin, the neurology chief at Mass General, was setting up a Huntington's center. Housman and Tobin organized a workshop at NIH in October 1979 that brought together many of the big guns in genetics. "It was complete pandemonium, total chaos," Nancy Wexler recalls. "Everyone was yelling and screaming. David Botstein [now at Stanford but then at MIT] would go to the board and scribble furiously," and then somebody else would, too. Botstein was arguing about the new riflip markers that he, Ray White, and others had recently developed. Why not look for riflips in the DNA of Huntington's families and see whether any of them were consistently inherited with the gene? That might narrow the location of the gene down to a specific chromosome.

"Basically, the idea is the following," explains David Botstein. "When you inherit your chromosomes from your parents, you don't get them exactly. Pop's chromosomes are mosaics, a mixture of all his parents' and ancestors'. So are Mom's. And in the process of giving them to you, the chromosomal deck shuffles again. What we call a 'crossing over' happens.

"Riflip mapping is very simply the idea that we all have DNA markers that are different in each of us, and by following them, we can see segments of parental chromosomes being inherited. So I can tell whether it's your grandma's chromosome, or your grandpa's chromosome at point X on, let's say, chromosome 4.

"If there's a disease there, too, the odds are that the disease will go along with the markers that are nearest to it. So this type of mapping is merely using the marker to follow the disease-causing gene. This way, I can predict in a newborn child whether or not he's going to have Huntington's disease fifty years later. That's basically all there is to it."

Nevertheless, in 1979 the idea seemed farfetched—except to Housman, who took it to Gusella, then his grad student, and did a selling job. The studious young scientist assumed command of the experiment, first at MIT and then in Martin's new unit at Mass General, the Huntington's Disease Center without Walls.

It was after the 1979 workshop that Nancy Wexler learned of the world's largest known Huntington's family, down in Venezuela. Soon she became executive director of a congressional commission on Huntington's disease and had by 1981 drummed up enough funding for an expedition to Maracaibo. Nowhere is the Huntington's heirloom more evident. In the twenty-five-square-block barrio of San Luis, for instance, the Huntington's rate is 700 times higher than normal. In the isolated village of Laguneta, in a remote lagoon eighty miles south of Maracaibo, some two hundred people live in twenty-one tin shacks built on stilts over the water and accessible only by boat or helicopter. Many of the residents are descended from one married couple who were both Huntington's victims. Up to 25 percent of the villagers are doomed.

"The first time we went down there," recalls Wexler, "we drove from Maracaibo to another town and took a little outboard motor boat for a couple of hours to Laguneta. It was very hot and muggy, and along the shores were swamps and jungles. I spoke no Spanish at all. I'd never seen a stilt village before. I had no idea what to expect.

"What was so odd to me was to pull up to an alien-looking shack and see a woman sitting cross-legged on the porch. When she saw us, she stood up, and the movements started coming out, sort of flourishing—when people are at rest, the movements will subside, but when they start doing anything, they get a lot more intense. So, suddenly this woman flowered into movement. It was totally bizarre. A different world. A different culture. A different language. . . . And she was acting just like my mother."

The villages have proved to be fertile ground for a deadly gene because big, inbred families are the rule—twelve to fourteen children are common—and everyone seems related to everyone else. No new blood has a chance to widen the gene pool, because outsiders are afraid to venture within. There is nothing to stop the spread of the gene, nothing to stop its victims from dancing to a merciless tune. The residents of the closed communities seldom leave. The men earn their living as fishermen, going out in the lake each night in their long, thin wooden boats powered by large outboard motors. When they get sick, the men start falling off their boats. They lapse into depression and, frequently, attempt suicide. They drink gasoline. They attack each other with knives. When the mysterious disease strikes a wife, her husband may leave her. The disease, though, can claim a victim of any age. Children as young as two have come down with it, as have adults as old as eighty-two. Shunned by the rest of rural society, the villagers have lived with their executioner for so long that they tend to be fatalistic about it. It seems that every other household has been struck by *el mal.* The ostracized clans seek solace in one another, refusing to isolate their suffering and dying family members, but integrating them into daily life.

Visitors to San Luis are shocked by the struggling residents in evidence everywhere, trying to make their twisted bodies walk, their minds think, their mouths speak. Wexler was introduced to them by Americo Negrette, a former Venezuelan public health physician who identified Huntington's in the unique cluster he had first mistaken for drunkards, in 1952, and who, with his former student Ramón Avila Giron, a psychiatrist from the University of Zulia, had refused to abandon them.

As the thirteen-member Wexler team set to work, doctors and nurses drew blood and interviewed families. Psychologists tested for early signs of impairment. Neurologists checked for physical deterioration. The scientists laboriously charted family pedigrees and collected thousands of blood samples for DNA analysis. "It was always hot and humid—about one hundred degrees. And, of course, there was no air-conditioning," Wexler remembers. "Many people had never given blood before and were afraid to lose their fluids. Men were more recalcitrant than women. Some men believed they couldn't drink if they gave blood, and they didn't want their drinking interrupted. So we had to resort on occasion to drawing blood from one arm and giving them a beer in the other. The hardest people to collect from were the old and healthy. They had escaped the illness. They were scared because we wanted their blood, too. What did that mean?"

Once someone gets sick, there's little that can be done, but these patients are active and out in the community all the time.

"The problem in the United States with any disease like Huntington's," says Wexler, "is that the family will park the patient in the back room and turn on the TV. Patients just sort of die of sensory deprivation. They can't go outside, because they look funny, they're laughed at. They can't go to restaurants, because they make a mess. The family's embarrassed, so they become like hermits. Other relatives don't visit."

But in Venezuela there is no back room. "There may be eighteen people watching TV, and the patient is in the middle of it. Often it's the kids, in fact, who take care of the older people—little five-year-olds. So, in a way, by accident, people get much better care there. I've never seen a bedsore.

"People live with diseases who would be dead in a week in nursing homes in the States. It's incredibly dramatic to find a Venezuelan woman lying in her deathbed, and her family is all around her taking care of her. They know absolutely that some of them will get this disease, and, indeed, some already are starting to show signs. So they're looking at their future mirror image, while they're giving her water and changing her."

Most heartbreaking of all are the children. "There's one little girl who's absolutely stunning. When we first went down there, she looked perfectly fine. Suddenly, now, as they say in Venezuela, 'she's lost.' On the next trip, we looked at her, and oh, my God, something *was* going on. She just clung to my neck and started sobbing and saying, 'Don't go' and 'Take me with you.' That just rips you apart.

"You're trying desperately to find something that's going to stop this thing. You're trying to turn back history. It's like trying to stop a tidal wave."

As Gusella and fellow molecular biologists studied materials obtained by Wexler, they found themselves grappling with a gene that seemed to delight in double-crossing them. Everything had to be checked and double-checked. In the meantime, P. Michael Conneally at Indiana University had identified a huge multigenerational midwestern family and was supplying Gusella with blood to check against that of the Venezuelans. Conneally also ran thousands of paternity tests on the Venezuelan blood samples to verify that fathers listed on the family tree were actually the biological fathers. On paper, the strategy looked unbeatable—the pedigree was huge, which gave the researchers a huge boost over other genetics searches—but the team desperately needed a flanking marker located on the other side of the gene. Then it would have been a simple matter to squeeze in from both sides until there was nothing left but the gene. But the Huntington's team was never to be favored by a flanking marker; indeed, its members looked on with longing as their colleagues in the race for the cystic fibrosis gene found first linkage and then a flanking marker that put them within 1.5 million bases of that gene.

As spectacular triumphs occurred in work on other diseases, thanks to the riflip technique Gusella had pioneered, the Huntington's group had no choice but to slog along chromosome 4, as sixteen more markers were found that seemed to lie closer and closer to the Huntington's gene: each new family test put them farther out in the terra incognita at the end of their chromosome.

By 1988, the search was getting ridiculous. Pretty soon, they would run out of chromosome, but there was still no gene.

As the scientists kept hacking their way through the underbrush, the Huntington's gene seemed to skip and dance out to the farthest tip of the chromosome, called the telomere, like a phantom. It appeared to be hiding out there, somewhere, amid a hundred thousand base pairs in a region dotted by endless random, nonsensical DNA repeats and odd bits of gene detritus deposited eons ago by foreign viruses—"the twilight zone of gene mapping," one researcher has called it.

Finally, in January of 1989, one of the collaborators, Gillian Bates, a postdoc in Hans Lehrach's laboratory at the Imperial Cancer Research Fund, in London, managed to clone the end of the chromosome—no mean technical feat. At long last, the teams had an end point. Surely, they could simultaneously move forward along the telomere and backward from it, and trap the gene in between. Surely, it was nearly at hand.

But then, in 1990, without warning, as if it had bounded gazelle-like over two million base pairs, the gene's probable location was shoved backward to the middle of the final chromosomal band, meaning that the hunters had been following the wrong tracks for years.

Disaster struck with all the subtlety of the opening salvo of Huntington's disease itself. In June, a sole errant chromosome from a Venezuelan man suffering from Huntington's threw "a monkey wrench" (Gusella's words) in the works. The man had the requisite markers, but only to a point about halfway up the chromosome, not far from the site of Gusella's original 1983 marker. The patient had no markers after that, yet he obviously had the disease. This was devastating news to the researchers, because it indicated that the gene could not be at the end of the chromosome but had to be located several million base pairs in.

Gusella, assuming a mistake had been made, rechecked his work several times. Not only did the data stand up; Gusella found another patient with the same marker pattern. A research group outside the collaboration then found a third. The possible explanations of how this could happen are very complex and involve genetic events, such as double mutations or double recombinations of the chromosome, which are rare, and Gusella and the other collaborators were still inclined to believe that the gene was actually situated out at the tip of chromosome 4.

But then Peter Harper at Cardiff and Michael Hayden in British Columbia found linkage disequilibrium several million base pairs from the tip. To gene hunters, linkage disequilibrium usually means they are in the right neighborhood of their quarry, although in this case the matter was complicated further because Harper and Hayden found not one but two places that showed tantalizing hints of the presence of the gene, and those places were a million base pairs apart.

As the quest drifted ever closer to the shoals of fiasco, all the researchers started having what had long been a recurring Gusella nightmare: instead of being a single gene that controls production of a protein, as everyone had hoped, the Huntington's gene might actually be some alien DNA chunk transported from another chromosome, or it might itself signal wandering genetic material to land at the wrong sites, setting off a series of mishaps that resulted in the disease. In that case, all their efforts were for naught and they would never figure out the disease in anything like the near future.

By 1993, the team, past frustration by now, was rummaging around in a region about a million base pairs north of the original marker. Using the technique not so whimsically called "brute force," the investigators were probing, sequencing, and studying about a hundred candidate genes, one at a time.

One bleak Boston afternoon in late February, Marcy MacDonald, Gusella's senior postdoc at Mass General, was rather proudly examining the sequencing gel of a gene that was inside a 500,000-base-pair region of promising DNA. She had had a hell of a time cloning this gene segment. IT-15, it was dubbed. Just one of many. Interesting transcript-15.

Gusella's lab had discovered the gene a year earlier, but could not seem to clone its beginning—the initiation site that geneticists call the 5-prime (5') end. Something weird was going on there—something that confounded the probes MacDonald had. It was a big gene, about 200,000 base pairs long, with a coding region of about 10,000 bases.

With plenty else to do, the team had not been concentrating on the stubborn gene. "It was very difficult to clone," Gusella explains, "and we'd also been working on so many other genes at the same time. It wasn't getting special attention."

But MacDonald persisted. This late in the game, you don't get sloppy. You bear down. Finally, following the time-honored technique of sequencing the gene by hand, MacDonald ended up with a 14-by-17-inch gel with ladderlike black smudges indicating the elusive nucleotide sequence of the gene's start. There it was. MacDonald had done it. Staring at the gel, she could see why she had had a problem. The gene began with the same three base pairs, C-A-G, repeated over and over and over again.

As her eyes raced up and down the gel, she counted forty-eight repeats of the trinucleotide sequence. Quickly, MacDonald compared the site to a gel characterizing the beginning of the same gene from a normal person. She counted some repeats—more than twenty C-A-Gs—but nowhere near the number in the Huntington's patient's gene. MacDonald felt something click in her mind.

The Huntington's gel revealed a recently discovered phenomenon in molecular genetics called a "polymorphic trinucleotide repeat," the molecular equivalent of a stutter. MacDonald's heart was racing. She knew that the same type of stutter defect—which seems to cause segments of a gene to expand unnaturally—had within the preceding two years been found in the genes responsible for other disorders, including an inherited form of mental retardation (fragile X syndrome) and two neurological diseases, myotonic (muscular) dystrophy and spinobulbar muscular atrophy. The number of repeats can multiply enormously; in fragile X, for instance, they can stretch into the thousands.

Suddenly the dawning.

MacDonald rechecked her work. It was correct. Time to tell Gusella, who was working at home on his computer. She didn't try to phone him. Instead, she exitedly typed an electronic-mail message, which instantly flashed on his monitor:

"Bingo!"

This time it was Gusella's heart that missed a beat. "I knew what she was talking about," he says now. He didn't jump up and down. He didn't run to the lab. He didn't even yell, "Yes!"—"I didn't get excited. It was clear what she had to do next. It could have been a chance polymorphism," he speculated. Maybe only that particular gene in that particular patient had so many repeats. "Marcy had to do the next experiment. I knew she already was doing it."

MacDonald concentrated on that DNA segment. She probed the DNA of other Huntington's patients, amplified the segment using PCR, and compared it with that of normals.

This went on for a week. Soon, Gusella joined in, and they worked full tilt around the clock, performing the humdrum but necessary chemical tests. Their spirits soared as Huntington's patient after Huntington's patient revealed the stutter. Imagine the deliciousness of the moment when Gusella realized he had the ancient monster in his grasp at last! The greatest thrill in science comes when a researcher finally knows something *that nobody else knows*. A secret of nature has been unmasked. The discovery; the final experiment that nails it down.

No matter where Gusella, MacDonald, and their colleagues probed, the finding was the same. They probed the DNA of the gigantic Venezuelan family. Yes. And the DNA from seventy-three other Huntington's families from around the world—Yemenites, Japanese, Finns, Europeans from every nation, even a native tribe from New Guinea. Yes. Yes. And yes again.

"We kept finding the same thing," Gusella remembers. "The overwhelming feeling was relief. We'd been under tremendous pressure to get this done. External pressure, as well as internal."

In all seventy-five Huntington's patients tested, the gene always began with a long repeat—at least forty-five triplet C-A-Gs in a row, with the number sometimes rising into the eighties.

When he systematically examined the same gene in seventy-five healthy people, Gusella found some repeats, *but no more than forty.* That was the legal limit, then. The body could deal with that. But just five more C-A-Gs spelled doom.

Gusella had convinced himself a year earlier that it was not going to be a trinucleotide repeat. Technical reasons, he says. Now he instantly reversed himself.

A copying error? A simple stutter? Was that all there was to it? Was that Huntington's? Why not? More and more killers and cripplers—among them cystic fibrosis, the Lesch-Nyhan syndrome, ADA deficiency, sickle-cell anemia, the thalassemias—are turning out to be caused by the most minuscule miscues in the world: *a point mutation,* a single nucleotide out of place among hundreds of thousands. Genetic diseases often hinge on gross errors of nature—additional copies of chromosomes, translocations, deletions—yet catastrophe can indeed be due to something as mundane as a stutter, a lethal repetition, a simple copying error repeated millions of times in cells of the brain. Gusella left nothing to chance. Not after so many years, so many detours, dead ends. But one can't do better than seventy-five out of seventy-five. Bingo.

The repeats he found were wildly unstable; even brothers and sisters with Huntington's had different numbers. But they were all abnormal.

"And when we typed early-onset patients, we found something chilling: one boy was only three years old when he got sick. He had the longest number of repeats. He had eighty-six copies."

Another door unlocked. The greater the number of abnormal repeats, the earlier the onset of the disease. That finding *did* give Gusella pause. First, he had given society a dubious gift, a marker test that could free some people but condemn others. Now he could tell them not only whether they carried the gene that would kill them but even approximately when the nightmare would start.

Michigan's Francis Collins was off skiing in the Upper Peninsula with his daughter and her boyfriend when Gusella finally reached him by telephone at a ski resort.

"What's up?" Collins asked.

"Oh, everything's okay," said Gusella, laconic as ever. "Just wanted to tell you that I'm trying to draft a paper here."

"So, what's new?" joked Collins.

"This one will be different . . ."

"Oh, really? What have you got?"

"Well, we've got the gene."

You could have scooped Collins off the floor.

"All of us had been thinking of these fragile X and myotonic dystrophy repeats for a year and a half. We'd all been searching this interval of the fourth chromosome looking for a trinucleotide repeat. We all missed it. But Marcy found it."

As Collins impatiently stood stomping his feet around the reception desk at the ski resort, Gusella faxed the whole file of data to him a page at a time. "I tied up their fax for hours," Collins says, not the least embarrassed. "But the paper was elegant. And it was absolutely convincing."

Nancy Wexler was about to leave on her annual March trek to Venezuela when Gusella called her. Unlike Gusella, she jumped up and down. In fact, she couldn't stop. And she couldn't stop crying. "We've got it! We've got it, thank God!" she screamed to an interviewer. "It's been a long day's journey into night!"

Wexler couldn't stress enough what the finding meant. "We've jumped across *continents* with this discovery. There are a lot of floating pieces left to be picked up, but I know this will bring treatments and the cure."

Within days, she would be in Venezuela to tell the villagers the news: *Mal!* They had found the cause of *el mal*. On *"cromosoma cuatro,"* as Wexler always has called it in fractured Spanish.

Gusella's immediate concern was security. "I wanted to publish as quickly as possible," he says. "We knew we had it. We had to keep quiet. Results like this have a strong tendency to leak everywhere. Families invariably get the wrong rumor. We wanted to make sure the news got out accurately." Even after ten years, the group found itself in a race to skirt an embarrassment: the competing Canadian researcher Michael Hayden was about to announce in *Nature* that *he* had found a gene that caused Huntington's. He hadn't, the collaborators knew, but *Nature* had already gone to press and had no choice but to ballyhoo its gene. Gusella submitted the paper to the journal *Cell*, for both its prestige and its quick turnaround time. The journal deviated from its customary practice for the March 26 edition in order to scoop *Nature*; it let the collaborators hold a hastily called press conference in Boston to announce the discovery.

Interestingly, no first author was cited on the historic paper. It bore an unusual collective byline—The Huntington's Disease Collaborative Research Group—a byline that had been decided on years earlier.

Scientists around the world breathed sighs of relief. It seemed fitting that the final victory had been won in Gusella's lab. And now, there was the terrible gene. Captured, finally. Everyone agreed. Stored safely in a bottle up in Boston.

The most sought-after gene in the history of molecular genetics is a 200,000-base-pair stretch of DNA situated about 3.5 million base pairs from the tip of the short arm of chromosome 4. The triplet codon C-A-G produces the amino acid glutamine. But no other protein known to science contains as much glutamine as that produced by the Huntington's gene. The next task facing Gusella and colleagues will be to determine what the protein does normally. Then it will be possible to determine what too many repeats mean physiologically.

In the most simplistic scenario, the brains of patients are apparently producing a toxic protein that by midlife has destroyed massive numbers of neurons in the basal ganglia and cortex, causing dementia and the involuntary jerking of arms, legs, torso, and facial muscles that characterize Huntington's.

But how did the gene come to exist in the first place? Again, no one knows. When genes duplicate themselves as cells divide, the double helix splits apart and free-floating nucleotides pair up and bond and form another strand. A total of about three billion base pairs of DNA must be faithfully copied, and although nature shows great skill in getting this job done, errors do occur. And an error in the wrong place can bring catastrophe. The same kind of error can strike before conception, in the making of sperm cells and egg cells. The gene IT-15 was big, about the size of the CF gene, with lots of nucleotides to copy, and, in his mind's eye, Francis Collins could see the disaster happen.

Up to forty C-A-G repeats, forty glutamines in an amino acid chain—

perhaps the cell could handle that as it set out to duplicate the 200,000-base-pair gene. But as Marcy MacDonald had found out, if there was something chemically strange about the beginning of the gene that made it a particularly tough stretch to sequence, the copying mechanism, speeding along, might, in effect, brake too slowly, faltering, stuttering, adding just a few triplet codons too many. And when a certain, crucial threshold was crossed, an entire family line would be cursed forever.

"This gene," muses Nancy Wexler, "is so sinister, so sly. It's like there's an orchestra playing, and there are these foul notes. At first, they're very soft, but they keep getting louder and louder. Finally, they're all there is."

The Huntington's gene is so powerful that gene therapy may not be easy. Huntington's is a true dominant disease, meaning that the mutant gene is sufficiently toxic that it always expresses itself over the normal copy. Merely adding another normal version of the gene to the brain would not help, because the abnormal one will still be there. A possible strategy would be to introduce an antisense RNA that slowed down the mutant allele or turned off the abnormal gene altogether.

The cancer researcher Jack Roth, at M. D. Anderson, in Houston, is already trying such a strategy for lung cancer, and many AIDS researchers are experimenting with complementary DNAs to turn off the transcription mechanism of HIV.

"Or maybe drug therapy for Huntington's might make a lot more sense," Collins surmises. "But that's just conjecture when we don't know know what the protein actually does. Could a drug turn it off? We'll have to see."

After twenty-five years of wondering whether she carries the gene, Nancy Wexler still will not say what she personally will do now. A simple blood test. That's all it would take. Gusella could tell her within a few hours. But like all people at risk, she would love to know that she doesn't have it, but doesn't want to know that she does. And she would never want to be an "escapee" and learn that her sister wasn't. "It's absolutely the hardest choice there is," she says, fully aware of the irony that it is a choice she brought on herself by her determination to find the gene.

At any rate, Wexler is now a giant step closer to her dream of boarding a jet for Venezuela and bringing her huge extended family the cure for *el mal*. "They've given everything they have to us," she says. "Without their help, we never would have found the gene. Imagine the incredible joyousness when we finally bring them the cure, with everybody hugging each other, and everybody crying for joy. To be able to stop this thing in its tracks. I can't imagine anything in the whole world that would be better."

Throughout the long siege, no matter what skeptics said, she never doubted that the Huntington's gene would eventually be found. Wexler had too much faith in her friends to waver. They wouldn't let her down. They wouldn't let the families down. Most of all, they wouldn't give up. That's why she picked them. "I always knew the wonderful day would come," she says. "I just wanted to be around to see it."

20

DELIVERING
THE GOODS

The brain nests in its bony hideaway, baronial and aloof, the most intimidating, yet arguably the most important, target for gene therapists. What Woody Allen once called his second-favorite organ is scarcely more accessible than the sheer, icy face of an alpine peak. It is one thing to have in hand the gene for Huntington's disease or for any of the several thousand neurological or psychiatric disorders that plague humankind. It is quite another to get restorative versions of those genes into the brain's convoluted mass.

Established forms of gene therapy will not work. Brain tissue cannot be removed for exposure to virus, nor can the organ be approached easily via the bloodstream, because of that anatomical bundling board—the blood–brain barrier. And since brain cells do not divide beyond early childhood, they are, for all intents and purposes, impervious to retroviral vectors, which eschew nondividing cells.

Like hundreds of molecular researchers worldwide, Beverly Davidson and Blake Roessler, of the University of Michigan, dreamed of cracking the problem. Their line of attack was to pursue a number of potential gene delivery strategies at once. They wrapped genes in fat globules to help slip the genes through the membrane of neural cells. They sidestepped the retrovirus problem by letting the finicky viruses put genes into fast-dividing *fetal* brain cells before transplanting the cells into rodent brains. They auditioned herpes and other viruses that favor central nervous system tissue to see whether any of them would make workable vectors.

In the end, though, it was a virus that hangs its hat in a much different part of the body that stepped into the limelight and, for the time being, seems to have won the role. This unlikely ingenue was the humble adenovirus, already being tested as a vector for the cure of cystic fibrosis.

For Davidson and Roessler, who, along with two French groups working independently and a group at NIH headed by Ronald Crystal, uncovered the adenovirus's latent talent, the breakthrough has brought instant fame. For the rest of us, the discovery means that there may soon be an avenue for curing some of humanity's most horrible and intractable illnesses, those that strike the intellect, the emotions, and the nervous system.

It was a common interest in one of these diseases, Lesch-Nyhan syndrome, that early candidate for gene therapy, that first brought Davidson and Roessler together. A biochemist and molecular geneticist, Davidson earned her Ph.D. by identifying some forty new mutations responsible for the palsylike ailment whose signature is retardation and self-mutilating behavior. Roessler, a physician trained in rheumatology and infectious diseases, came at Lesch-Nyhan from the clinical perspective, having treated a number of young men with the syndrome. Under the aegis of their former mentor at Michigan, William Kelley, who helped discover the HGPRT gene responsible for the disease during his own student days, they began working jointly on ways to deliver genes to the brain to alleviate Lesch-Nyhan's catastrophic neurological symptoms.

"Lesch-Nyhan patients touched me personally," says Davidson. "We can treat some of the systemic ramifications of their disease with drugs that control uric acid, but the nervous system manifestations go unchecked."

"These patients live a low quality of life," adds Roessler.

Working under Kelley, the two colleagues helped make some profoundly important discoveries about the disease. For a long time, it had been conjectured that the primary seat of the disease was not necessarily the brain and that the psychomotor deficits could be treated or prevented by making corrections outside the brain. But in recent years it has been confirmed that the missing HGPRT gene is most active in brain cells and that it is critical for the conversion of certain cerebral metabolites needed to maintain proper communication among neurons. Even minimal production of HGPRT in the brain—that is, anything above 4 percent of normal—is sufficient to keep a person healthy. A percent production will cause a person mild problems. However, if the level falls below 1 percent, the Lesch-Nyhan syndrome is the inevitable result. That the central venue for HGPRT is the brain itself probably accounts for why the attempt to alleviate the disease with a bone marrow transplant—preparatory to marrow-based gene therapy—failed. Feeding HGPRT into the system via the marrow, and hence the blood, will not supply enough of the essential enzyme to the brain to do the patient lasting good. The distance is too great and the blood–brain barrier too strong.

Davidson and Roessler's study of the Lesch-Nyhan syndrome at length led them to scrutinize a second bane of the central nervous system, Parkinson's disease. "It turns out," says Roessler, "that many of the neurochemical abnormalities in Lesch-Nyhan are akin to those in Parkinson's. Lesch-Nyhan is thought to be due to irregularities within the pathways that produce dopamine. The pathways are altered in ways that are analogous to alterations seen in Parkinson's."

"You see the same kind of spasticity—," says Davidson.

"—and choreoathetoid movements—," interjects Roessler.

"—in both diseases," concludes Davidson.

Davidson and Roessler share a rare professional chemistry. They resonate together so perfectly, are so mutually attuned, that not infrequently they finish each other's sentences. The habit is a source of amusement among their colleagues at Michigan's Howard Hughes Medical Institute. But nobody jokes about their work.

In the spring of 1992, the idea came to them to try adenoviruses as vectors for forwarding genes to the brain. Their Michigan colleague Jim Wilson was touting the irksome cold viruses for treating not only cystic fibrosis but other diseases in which the target cells do not, as a rule, divide. But it seemed farfetched to think that the pesky things would work on brain. "There is no a priori reason why adenovirus should infect the brain," notes Davidson. "Ordinarily, the virus seeks out the surface of the lung. But adenoviruses were hot, so we just figured, hey, let's try it. It was a spontaneous idea."

Stripping the adenovirus of its reproductive genes, Davidson and Roessler filled the gap with genes for the enzyme beta galactosidase, whose function is to break down a form of milk sugar. The enzyme is put out by E. coli bacteria and is the main reason why mammals harbor E. coli in their digestive tracts. Why did the two researchers choose such an odd gene to shoot into the brain? Because it provides a splendid way to tell whether the gene has made it into target cells, thanks to its ability to turn cells a characteristic blue on exposure to a staining agent.

The final step was to outfit the construct with a sequence from the SV40 (simian) virus that tells the virus to lodge itself in the cell nucleus.

This is necessary because adenoviruses do not normally infect nuclei. When the handiwork was complete, the researchers injected the customized adenoviruses directly into the right hemispheres of the brains of mice, using a device similar to the one Ken Culver employed to put herpes genes into the cerebral tissue of brain tumor patients. The focal point of the injection was a region of the brain called the caudate putamen, which is affected in both Parkinson's disease and Lesch-Nyhan.

If Hippocrates ever envisioned an era in which medicine would be inserting genes from a bacterium found in the gut into the brains of mice using human cold viruses equipped with control signals from a monkey virus it is not recorded.

Over the next eight weeks, the mice were dispatched at regular intervals and their brains analyzed for the presence of beta galactosidase. When stained, the brain cells promptly glowed bright blue. The adenoviral vectors had placed the test gene into two different kinds of brain cells—namely, neurons, the primary cells that convey thought and motor impulses, and glia, the network of supportive brain cells. Moreover, the vectors had carried out the task with great efficiency. "We had very high expression, extremely high," says Davidson. So high that the team immediately dropped its experiments with fat globules and fetal cell transplants.

If the quantity of expression was encouraging, so was the fact that it could be detected for at least eight weeks. Two months in mouse time is equivalent to several years in a human being's life, so it seemed clear that adenoviruses could bring about long-term therapeutic effect in brain cells. "And eight weeks was just the arbitrary time frame we placed on the experiment," says Davidson. "We chose to cut it off then. We have no reason to think expression wouldn't go on longer."

Finally, there was the issue of safety. In previous experiments, vectors made from herpes simplex virus had managed to place genes into brain tissue, too, although at a lower rate of efficiency. But they had a dismaying tendency to cause disease. By contrast, the modified adenoviruses caused no perceptible harm to the mice. In fact, a set of control mice that received "wild type" adenoviruses—adenoviruses as they are found in nature—showed no ill effects either.

Davidson and Roessler have since gone on to duplicate their experiment in rhesus macaque monkeys. Again they reported "excellent" gene transfer with long-term expression.

Within weeks of Davidson and Roessler's initial experiments, two French teams, unaware of what was going on in Ann Arbor, performed virtually the same trials, using rats as their test animals. In one experiment, the genes were inserted into the hippocampus, a region of the brain associated with memory, and the substantia nigra, which figures in motor disease. In the other, the genes were placed in the thalamus, striatum, and substantia nigra. In each case, the number of cells infected and the amount of expression were high.

Almost at the same time, a group at NIH led by Ronald Crystal used adenoviruses to put both genes for beta galactosidase and an enzyme normally made in the human liver into the brain of rats. One experiment called for injecting the beta gal gene directly into a brain structure called the globus pallidus. As the other research groups found, the gene showed pervasive expression at the injection site and in regions surrounding it.

But Crystal's group added a new twist. It also put the beta gal gene, and a second gene, for the liver enzyme alpha-1 antitrypsin, into the ventricles of the rats' brains. Ventricles are lagoonlike cavities inside the brain that are filled with spinal fluid. The ventricles reach extensive areas of the brain as they go about their job of nourishing the organ and keeping pressure uniform inside the central nervous system. The Crystal group found that the lining of the ventricles took up the new genes eagerly, as might be expected, since adenoviruses favor the surface areas of organs such as the lungs. Almost immediately, the cells began making beta galactosidase and alpha-1 antitrypsin in substantial amounts, easily detectable because neither enzyme is normally found within a foot of the brain. The value of such an approach is that a single injection of genes into the ventricles could reach a much larger area of the brain, at least in theory.

All of the groups were delighted with the findings. One of the French teams, directed by Marc Peschanski, of Paris's Faculté de Médecine, wrote in the journal *Nature Genetics*, "Adenovirus vectors represent clearly a promising means to transfer foreign genes into the brain with a therapeutic goal. Relatively [small amounts] of adenovirus can transfer foreign genes into a significant number of brain cells without triggering pathological effects. . . . If efficiency and safety of foreign gene transfer into brain cells by adenovirus is confirmed, it will open new avenues in the treatment of many genetic and acquired neurological diseases for which this method may work as an alternative to drug treatment or brain transplantation of fetal tissues."

The use of adenoviral vectors to transfer genes to the brain has several potential applications. It will, for example, enable neuroscientists to study the effect of specific genes on the function of the brain. It lets them *add* a gene to the mix to see what happens. For example, one theory suggests that the overexpression of cytokines starts a cascade of biochemical events leading eventually to Alzheimer's disease. Up to now, it has been impossible to model such a situation in a living animal. But it will soon be possible to insert genes for cytokines into the brain of, say, a rat and watch what develops. Scientists at MIT and other institutions have already learned to knock specific genes *out* of an animal's genome to observe the effect, but the new technique is the obverse of such "knockout" technology.

Nevertheless, the most exciting applications remain the therapeutic. How useful the new vectors might be in treating brain diseases remains to be seen.

A major stumbling block is the limited scope of the vectors. They do not seem to infect tissue very far from the site of injection. That would greatly impede those trying to reverse diseases affecting large areas of the brain, such as Alzheimer's and the Lesch-Nyhan syndrome. Obviously, one cannot wantonly perforate the brain with too many widely scattered injections.

There are, however, several ways around the problem. First of all, the "bystander effect," as hypothesized by Kenneth Culver and Michael Blaese, may double or even triple the area of the brain affected by a single injection, if the theory pans out. Then, too, axons, the threadlike projections that nerve cells use to send messages to each other, may ship some of the corrective DNA to other cells. There is evidence that such transport sometimes occurs in nature.

Most promising, however, is the idea of injecting the adenovirus into the ventricles of the brain. A single dose of vector delivered to a ventricle could ultimately bathe a huge area of the brain. You would get a lot more bang for the buck that way.

Davidson and Roessler have begun trying to get genes into the ventricles of monkeys. At the same time, they have a large number of associated experiments on the drawing board. In one trial, they will inject the vectors into mouse spinal columns to see whether adenoviruses can transfer genes into other kinds of central nervous system tissue. If so, it will facilitate treating motor diseases such as amyotrophic lateral sclerosis that are not confined to the brain.

In another study, in collaboration with Tsien Sen Li, at the Harvard Eye and Ear Infirmary, the Michigan team members plan to investigate the technique's ability to deliver genes to the eye to repair retinal degeneration, a common cause of blindness in the elderly. With Fred Lederhoff, of Columbia University, they will determine whether the vectors can carry nerve growth factor into nerve tissue to speed the healing of brain, and possibly spinal cord, injuries. Finally, in conjunction with laboratories in San Diego and at NIH, they intend to study the use of adenoviruses to correct the symptoms of Parkinson's disease. In one experiment, undertaken with the NIH researcher Edward Ginns, the gene for tyrosine hydroxylase is being placed into rodents that have a form of Parkinson's. Tyrosine hydroxylase converts the substance tyrosine into L-dopa, which is in turn transformed into the neurotransmitter dopamine. It is dopamine that is lacking in Parkinson's disease. (Recent experiments at the University of Wisconsin, under Jon Wolff, have shown that long-term expression of tyrosine hydroxylase and, thus, correction of Parkinson's symptoms are possible in rats. Wolff's method of gene transfer, however, is indirect. He genetically engineers immature muscle cells to produce tyrosine hydroxylase and then transplants them into the rat brains. The Ginns experiment would directly alter the brain cells.)

All of these experiments will be conducted on animals. The crucial question is when the time will be ripe to move into human beings. Davidson and Roessler are willing to be seers. Projects Davidson, "Assuming the animal studies continue to be successful—"

"—in other words, safety and efficacy are high—," says Roessler.

"Then I'd say we can go into people in, say—"

"—five years."

"Yes, five years," says Davidson.

Five years, it so happens nine times out of ten, is the life expectancy following diagnosis of a neurological disaster, amyotrophic lateral sclerosis (ALS). Since March of 1993, when the first gene mutation to be associated with the disorder was pinpointed, ALS has become a prime target for gene therapists. Before then, nothing positive had happened in ALS research in 124 years.

Whereas Parkinson's and Huntington's are marked by quivering and violent movements, ALS involves a slow and agonized *quieting*. The uniformly fatal disease is characterized by a progressive paralysis, a gradual but profound physical silence of the body that drags on for years, while the alert mind trapped within can do nothing but grieve, celebrate the temporary plateaus, and catalog the growing loss of sensation.

As patients are forced to rely first on canes, then on crutches, wheelchairs, and beds, and, finally, on ventilators to keep them breathing, their brains remain as sharp and bright as ever. They can *feel things—*

hunger, thirst, pain, itches, cramps, full bladders, sexual desire—but cannot respond. The motor pathways to their voluntary muscles have shut down until, in the end, they cannot speak, cannot move. This grave and total quieting makes ALS one of the most terrifying of all human diseases—the mere diagnosis has led patients to commit suicide.

All human movement depends on a kind of large nerve cell called a motor neuron. Great masses of motor neurons form embedded cables that run up and down the spinal cord, transmitting impulses from the brain at speeds of up to 200 miles an hour, scattering along networks of nerve trunks to all parts of the body, flashing orders to muscles, and nurturing them. Unlike sensory neurons, which receive sensory information and communicate it up to the brain, motor neurons form the critical return-trip link between mind and body. In degenerative nervous system diseases, cells mysteriously wither and disappear; hence the motor neurons of the cerebral cortex, brainstem, and spinal cord of ALS patients harden (sclerose); whereupon muscle fibers served by the nerves atrophy (become amyotrophic), making the patient grow increasingly paralyzed.

Best known as the disease that killed the legendary Yankee slugger Lou Gehrig in 1941, ALS usually strikes people in their fifties and sixties, although Gehrig died just two years after having been diagnosed on his thirty-fifth birthday. Twice as many men as women fall victim, with about 4,600 new cases in the United States each year (about the same number as those of multiple sclerosis) and a total of 100,000 worldwide. As in Huntington's, the earliest signs of ALS sneak in so subtly—a lazy vocal cord, weakness in a finger, a twitching muscle in a forearm—that the diagnosis is often missed. Relentlessly, though, all the skeletal muscles innervated by motor neurons break down. Destruction is unpredictable and heartbreaking hiatuses of hope may intervene, but ultimately only the muscles of the eyes may be spared. Patients usually die from suffocation or respiratory failure as a result of muscle incapacity.

Identified in 1869 by the French neurologist Jean Martin Charcot, ALS has mystified scientists ever since. The cause has been attributed to some unknown virus, to a breakdown of the immune system, even to heavy-metals poisoning. Eleanor Gehrig once described what had happened to her husband, the mighty "Iron Horse," who set a record by playing in 2,130 consecutive American League games. "At first," she told the U.S. Senate, in hopes of gaining funding for ALS research, "he simply couldn't play baseball with the skill that won him a place in the Hall of Fame. Then he couldn't play well enough to stay in the Yankee

lineup. Finally, he couldn't play baseball at all. As the disease progressed, he couldn't dress himself, he couldn't feed himself, he couldn't
walk."

Curiously, despite the specificity of the neural defect, ALS
has three different rates of progression. Nine out of ten ALS patients die
within five years. But death may also occur in a matter of months or
even weeks. On other hand, the paralysis may progress at an extremely
slow rate, as it has in the great British astrophysicist Steven Hawking,
who for nearly thirty years has kept unraveling the origins of the cosmos
although almost totally paralyzed and confined to a wheelchair.

Most cases of ALS seem to arise spontaneously, but a variant, familial ALS (FALS), which may account for 5 to 18 percent of all cases, is
passed from generation to generation in an autosomal dominant fashion.
However, the clinical symptoms of sporadic and familial ALS are virtually identical, giving rise to the belief that they may be the same
disease.

Neurology is not the most cheerful of medical specialties. Clinicians
often can do little for patients except provide accurate diagnoses and
supportive treatment. Fed up, Teepu Siddique in 1984 decided that the
answer to the riddle of ALS would be found in molecular genetics. He
won a five-year teacher-investigator development award from NIH to
learn cloning techniques under the mentorship of Allen Roses at Duke
University, one of the few neurologists who was investigating genes.
Siddique transferred to Northwestern University in 1991, having developed a fruitful collaboration with the Harvard neurologist Robert H.
Brown, Jr., as the two spearheaded a nine-year international ALS gene
hunt involving thirty-two researchers at thirteen institutions.

In 1988, the Les Turner ALS Foundation, in Chicago, formed the Familial Amyotrophic Lateral Sclerosis Collaborative, a tight cooperative
effort that enlisted researchers smart enough to realize that only by
pooling their resources (multigenerations of DNA represented by their
families) could they ever find the FALS gene. "ALS is so lethal," explains
Brown, "that you don't have a lot of people surviving in the same family.
We had to share information to prove genetic linkages." Most members
of this unusually large group were clinicians dismayed by their inability
to do anything for their ALS patients except buy time by treating the
symptoms aggressively until death finally crept in. Genetic material
from 150 families was collected and pooled, while ten large multigenerational families were selected for genetic linkage analysis. The idea was
to attempt reverse genetics, running several hundred probes over the

human genome, looking for markers that might map the FALS gene to a specific chromosome.

On May 16, 1991, the powerful technology paid off. In a landmark paper in the *New England Journal of Medicine,* with Siddique as senior author, the team announced it had linked the FALS gene to the long arm of the smallest human chromosome, chromosome 21. Suddenly, a gene for ALS looked possible. The neurologists began to close in, pooling data constantly and nailing down flanking markers on both side of the still-unknown candidate FALS gene.

At the time, Siddique and Brown could not help noticing a very large gene that loomed just to the south, rather like Mount Rainier, about four million base pairs (4 centiMorgans) in from the outermost marker, numbered D21S58, which is just a hop, skip, and jump, in terms of genetic distance.

No secret, this gene. Its protein product has inspired about ten thousand scientific papers in the fields of enzymology and the biology of aging. As a matter of fact, the gene was a hometown favorite at Duke, where Siddique had trained. It makes the enzyme superoxide dismutase (SOD), found in every cell of the body. SOD is a powerful antioxidant, a leading member of the family of chemicals whose job is to quench loose breakdown products of oxygen known as free radicals. Oxygen free radicals are unstable and dangerous atomic particles created by the routine metabolism of oxygen molecules and by the effects of radiation, toxic chemicals, and other environmental agents. As cells burn oxygen for energy, they release free radicals that career around the system, triggering other crucial chemical reactions in the process, but also looking for trouble. To counter this, the body manufactures SOD and other scavenger antioxidants such as glutathione, beta carotene, and vitamins A, E, and C to mop them up. But if this delicate balance ever gets tipped in favor of the radicals, watch out. The superoxide radical, for instance, reacts with naturally occurring hydrogen peroxide in the system, creating the toxic breakdown product—hydroxyl—which has been shown to shatter genes and demolish cells. SOD controls the superoxide radical by holding it in check and breaking down superoxide into water through various steps. The radical and its controller were discovered by teams led by the Duke biochemist Irwin Fridovich, whose work starting in the late 1960s provided the first evidence that the human body made both free radicals and antioxidants against their onslaughts. Free radicals, in fact, have been implicated in many common diseases—Alzheimer's, schizophrenia, heart disease, Parkinson's, Down's syndrome, cancer, stroke, paralysis, and arthritis. They are believed to contribute strongly to the cellular wear and tear that makes all creatures age, and they may influence human longevity.

Thus, when Siddique and Brown found that their linkage had brought them within range of one of the three known SOD genes, an unexpected possibility confronted them. Among SOD's many cellular duties is the protection of nerves from free radical damage.

"But so far as anyone knew, SOD had nothing to do with ALS," says Siddique. He was no SOD expert; nor was any neurologist. Still, when he wrote the paper announcing the linkage of the FALS gene to chromosome 21, Siddique thought it would be helpful to other researchers to point out the proximity of Mount Rainier. In the original draft, a key sentence reads, "Free radical damage has been implicated in neurodegenerative disorders; therefore we will examine the gene for Cu-Zn superoxide dismutase as it lies only 4 cM from D21s58."

Oddly, though, that sentence never appeared in the final paper. Siddique says that "another scientist" on the team urged him to remove the reference to SOD. Was the gene too farfetched for ALS? Or was the other scientist worried about tipping off the competition? Siddique didn't think a lot about it. "We didn't have any data about SOD, so I deleted it. It was merely a strong hypothesis."

What occurred two years later, however, tells a lot about the politics of awarding credit in modern science. Immediately after establishing linkage, the ALS collaborators swarmed over the SOD gene, throwing probes at it like mad, studying DNA of ALS patients from families in which the disease was linked either to the SOD locus or to neighboring markers on chromosome 21. Finally, in October of 1992, a member of Brown's laboratory at Massachusetts General, Daniel R. Rosen, found the first mutation of the gene in an ALS patient. By January, sharing genetic material, the team had documented SOD mutations in thirteen ALS families (seven of which Siddique had supplied to colleagues). When they sequenced the DNA fragments, the collaborators detected a difference between affected and nonaffected family members, showing that all the ALS patients had point mutations that would alter the code for an amino acid resulting in a defective enzyme. What might that mean? One possibility is that in FALS patients SOD's activity is reduced from birth, leading over decades to a buildup of the poisonous superoxide radical and to the destruction of motor nerves. Alternatively, if the mutation bollixes up a gene promoter, excessive SOD may be produced, leading to damaging levels of hydrogen peroxide and the rampaging hydroxyl radical, which can be formed when hydrogen peroxide reacts with a metal such as iron.

By late winter, the group had enough material to publish. *Nature* accepted the paper posthaste and scheduled it for the March 4, 1993, issue. This was big news—the first crack in the impenetrable armor of Lou Gehrig's disease. By prearrangement, Rosen was listed as first author (Siddique came second) and Brown as senior author. That seemed fair: Siddique had found the marker; Brown got credit for the gene mutation. Then another player entered the picture. Brown had been collaborating with Robert Horvitz, a Howard Hughes Medical Institute investigator at MIT, who is a world expert on how genetic defects make cells die. In addition to his expertise, Horvitz had contributed financially to Brown's quest for the gene that had killed Horvitz's father. The friendship between the two scientists resulted in the *Nature* paper's being dedicated to the memory of Oscar Horvitz. Again unusual but touching.

Later, however, in several publications, including the *New York Times*, Horvitz was elevated to the status of team leader with Brown. Massachusetts General announced that its researchers had discovered "the long-sought gene for Lou Gehrig's disease," which was simply not true. The SOD gene had been implicated only in FALS, not in the far more common sporadic ALS, and then in only half the FALS families studied. "An equal measure of skill and serendipidity led to the breakthrough," crowed the PR apparatus at Mass General, explaining how the Brown team had found the first mutation in the faulty SOD gene a few months earlier. "We didn't know whether this discovery was significant," Brown is quoted as saying, which seemed a bit disingenuous in light of old Mount Rainier. As other institutions spit out their own press releases claiming an equal share of the discovery, matters got so confusing that some TV networks interviewed Brown, others Siddique, as the discoverer of the gene. The *New York Times* went with Horvitz. *Time* magazine ran Siddique's picture. The national wires ran a photo of Siddique and Brown, with the latter misidentified as Horvitz.

The discovery means that researchers must search for another locale for a related gene, or cascades of genes, or controlling elements that may be responsible for ALS. And these are situated on other chromosomes. The ALS riddle is far from solved. "What matters," notes Siddique, "is that we may be able to link ALS to free-radical damage, and without reverse genetics that might have taken us another ten years."

Brown nevertheless predicted that the first ALS therapy was imminent. "Megadoses of antioxidant vitamin regimens can be tried immediately. Vitamin A, C, E, beta carotene, will be administered on a systematic basis. And SOD, sure. Lots of SOD."

Siddique, too, is hopeful. "Now we can design therapy. We can investigate efficient scavengers of free radicals. We can think of how to give SOD to susceptible people. And in families where individuals are at risk, we can follow them closely."

The next step, when such a gene is found, is for specialists to "make a mouse," and that's what Siddique did in 1994. "We created a transgenic mouse, to see if by mutating this gene we could get ALS symptoms."

In fact, such a mouse already existed. Within weeks of the gene announcement, teams at Johns Hopkins and at McGill University, in Montreal, published studies in *Cell* reporting the degeneration of motor neurons in genetically engineered mice. The animals express excessive amounts of proteins that form skeletonlike, supportive structures in nerve cells. This mouse model of motor nerve degeneration could be used to test ways to reduce the levels of these proteins and prolong survival of motor nerve cells, according to a Hopkins professor of biochemistry, Don W. Cleveland. That might lead to a reversing or slowing of the course of ALS.

Cleveland notes that when excessive levels of the damaging proteins, called neurofilaments, accumulate, they seem to block the passage of vital nutrients and other factors from nerve cells into axons, the extended arms of nerves that stretch out to muscles. Nutritionally deprived, the axons wither and fail to stimulate the muscles, which, in turn, atrophy and waste away. "The axons get choked to death by the neurofilaments," says Cleveland. He speculates that excessive amounts of SOD from a defective gene might contribute by choking off axons—which can be three feet long—that run from the motor nerve cell body in the brain or spinal cord to muscles throughout the body.

Besides causing great excitement in ALS research for the first time, such findings may be even more momentous. Scientists funded by the National Institute on Aging have studied the SOD gene for years and come up with mounting evidence that SOD protects against oxidative damage caused by free radicals. According to Gene Cohen, acting director of the NIA, "This link with ALS adds further to the notion that we may be able to slow the rate of normal aging and age-related loss of function by preventing oxidative damage. We're very excited by this finding. It has extremely far-reaching implications."

What if the total physical helplessness seen in ALS as a result of the annihilation of motor neurons had an intellectual counterpart? What if one were to lose the precious neurons that enable one to think, remember, and cope? Alzheimer's disease is the mysterious scourge of later life that is the principal cause of dementia in human beings and America's fourth-leading killer. It begins with a gradual memory failure and intellectual decline that progresses inexorably to near-infantile helplessness and death. These events parallel an inexplicable and catastrophic loss of brain cells.

For a long time, the number of Alzheimer's sufferers in the United States was thought to be about two and a half million. But researchers at the Harvard Medical School have found that its prevalence far exceeds that estimate. Fully four million Americans are now thought to suffer from Alzheimer's, and the number could climb to as high as fourteen million by the middle of the next century, as the population continues to age.

The disease was originally described in 1906 by the German neurologist Alois Alzheimer, who, in conducting an autopsy on a woman whom he had treated for what was then called presenile dementia, found two curious brain abnormalities. The first was plaques, thick growths on the ends of the nerve cells of the woman's cortex. The plaques blunted the tips of the branchlike spines, known as dendrites and axons, through which nerve cells talk to each other by means of chemicals called neurotransmitters. Today these plaques are known to consist of dead or dying nerve tissue surrounding a core of tough, fibrous protein known as beta amyloid. The second aberration Alzheimer found was called tangles— twisted filaments *within* the nerve cells whose frizzled appearance resembles burned-out wiring. Plaques and tangles are seen in the brains of virtually all deceased Alzheimer's patients, and are the disease's hallmark. They congregate chiefly in the cells of the cortex and hippocampus, the two areas thought to be the seat of memory, speech, and reasoning. But it has never been determined whether plaques and tangles are the cause of the ailment or merely an effect.

In recent years, it has also been found that Alzheimer's patients produce markedly less of the neurotransmitter acetylcholine. Acetylcholine, made in a brain structure called the basal nucleus and distributed to the cortex and hippocampus, has been linked in many studies to memory function. The degree to which acetylcholine is lacking corresponds directly to the severity of memory loss in Alzheimer's. More recently, shortages of other neurotransmitters—specifically somatostatin, noradrenaline, and serotonin—have been found in Alzheimer's brains. Since each type of nerve cell makes and uses its own signature neurotransmitter to converse with its own kind, something global is clearly going on in Alzheimer's disease. But how these biochemical shortages fit into the larger puzzle of Alzheimer's, and what triggers the deficiencies in the first place, remains unknown. For the moment, it appears that neurotransmitter loss is a consequence rather than a cause of Alzheimer's. In other words, something else is making brain cells die by the millions, and when they die, they no longer make normal quantities of the neurotransmitters. The result is dementia.

Many theories have been advanced as to the source of Alzheimer's disease. Some scientists think a slow-acting virus is at fault. Others have suggested a toxic accumulation of aluminum in the brain, since high concentrations of aluminum salts have been found in the brain tissues of Alzheimer's victims. This theory, however, has lost ground in recent years. Still others lean toward a possible genetic explanation. But is Alzheimer's genetic? The issue is not clear.

It has been known for some time that the early-onset form of the disease, in which victims begin to show symptoms in their midforties, runs in certain families and follows a pattern of dominant inheritance. Children and siblings of victims have a 50 percent chance of getting the disorder. However, the early-onset strain of familial Alzheimer's (FAD) seems to represent only a small fraction, perhaps 10 to 15 percent, of the total incidence of the disease. The question intriguing some investigators was whether a hereditary pattern might also be present in the seemingly more random, late-onset cases. Such a pattern, they reasoned, could be masked by the fact that people tend to die of other causes before they reach the age at which an inherited late-onset Alzheimer's gene might kick in. Instead of concentrating on a few unusual families, researchers began looking at large populations of individual Alzheimer's patients, such as might be found in nursing homes. As each patient was identified, the fate of his or her parents, siblings, or children was ascertained. The size of the patient pool provided enough relatives who survived into their eighties and nineties for researchers to make some inferences about late-onset Alzheimer's. The findings suggest that even when the emergence of the disease is delayed, close relatives of victims share an increased risk of themselves getting the disease. These studies have convinced some authorities that far more Alzheimer's than originally thought is genetically determined. At the very least, a susceptibility to some environmental factor might be inherited. As James Gusella, who devotes a large share of his time to Alzheimer's research, says, "Who knows? If people routinely lived to 120, we might all develop Alzheimer's."

But if a genetic process is involved, how would one begin tracking it down? Molecular biologists had one important clue. For unknown reasons, people with the birth defect Down's syndrome almost always develop the plaques and tangles of Alzheimer's if they survive to the age of forty. Moreover, Down's syndrome is unusually prevalent in the families of Alzheimer's victims. But what in Down's syndrome might tie it to Alzheimer's? To researchers, the answer lies on chromosome 21. Down's syndrome, the single most common cause of mental retardation, is the result of a biological accident in which the victim inherits all or part of an extra copy of chromosome 21. If a gene or genes predisposing to Alzheimer's disease were located on that chromosome, then Down's victims, with a triple copy, would show an accelerated tendency to get the ailment. Investigators began ransacking chromosome 21 for possible genes.

The initial findings seemed too good to be true. In early 1987, a research team from Harvard Medical School, led by Gusella and featuring his assistants Rudolph Tanzi and Peter St. George-Hyslop, reported that, by doing linkage analysis on four large FAD families, they had mapped the gene responsible for the clans' woes to a particular region of chromosome 21. That very week, the same team and three others announced that, working backward from the beta amyloid protein that clogs the brains of Alzheimer's victims, they had identified a parent protein called the amyloid precursor protein (APP), whose gene lay on chromosome 21. The amazing thing was that the FAD gene and the APP gene mapped so close to each other on the chromosome that researchers dared to think they were one and the same. When investigators at Boston University School of Medicine found an extra copy of the region containing the amyloid precursor gene in each of fifteen Alzheimer's patients, but not in twelve normal controls, visions of Stockholm began to dance in everyone's head. It seemed to pull together into one neat theory that an inherited tendency to get Alzheimer's disease is due to the inordinate production of amyloid, owing to duplication of the amyloid precursor gene or to a mutation in the gene.

Alas, further analysis showed otherwise. Cases began to crop up in which familial Alzheimer's was not inherited along with characteristic forms of the APP gene, as it invariably would be if the two genes were identical. Such crossovers shoot candidate genes down like ducks on the wing. Moreover, studies by both Gusella's group and a European group headed by J. A. Hardy, of St. Mary's Hospital Medical School in London, failed to turn up any more cases of Alzheimer's in which the amyloid gene was present in excessive copies. The all-purpose theory had to be discarded. "It would have been tremendous," sighed St.

George-Hyslop. "You would have sort of had a hole in one. But unfortunately it didn't turn out that way."

Not to throw out the baby with the bath water, however. Hardy in London continued to check for linkage of the APP gene to his Alzheimer's families and found two clans in which there was a tight and unshakable association. Further study revealed that in these families there is a characteristic mutation in the APP gene at the 2,149th base pair, causing a C to be changed to a T, with a consequent substitution of the amino acid isoleucine for the normal valine in the resulting protein. The mistake seems to make the protein unstable in some way, causing it to spit out the amyloid that figures in Alzheimer's disease.

Geographically, the amyloid precursor protein is a membrane protein, that is, its home is on the surface of cells. It is 695 amino acids long and doubtless has a normal function in the body, perhaps as some sort of receptor, though its purpose has yet to be pinned down. Not all of the protein contributes to the production of amyloid. Just a small, 42-amino-acid segment does the deed. "For some reason," says Gusella, "as the larger protein is broken down, this amyloid-making chunk gets snipped out and combines with other pieces of the same kind, precipitating and forming this insoluble substance, beta amyloid. Then you've got all this garbage hanging around in your brain that you can't get rid of." Nearly everyone will develop some of these amyloid plaques if he or she lives long enough. But Alzheimer's patients undergo an exaggerated buildup.

To understand the way the process begins, imagine a splinter in your finger. Like a splinter, the protein is embedded in the membrane of cells. Most of it protrudes outside the membrane, but a bit resides inside the membrane and below. The external segment is normally sheared off by proteases, natural enzymes that act as the body's sanitation crew, breaking down used proteins. But the beta amyloid segment is not part of this projecting piece. It is primarily housed within the membrane. It, too, should degrade, but somehow it escapes this fate. Huntington Potter, of Harvard Medical School, has suggested one possible explanation. The brain tissue of some Alzheimer's victims contains abnormally long versions of the amyloid precursor protein. These variants have up to 770 amino acids, instead of the customary 695. Among their extra baggage is a protease inhibitor that binds precisely to the 42-amino-acid beta segment. This inhibitor may shield the segment from destruction like a bulletproof vest. Another possibility is the failure of an allied gene, as yet unknown. Such a gene may make an enzyme whose job is to regulate the production or breakdown of the amyloid precursor protein. "If there is a dysfunction in the manufacture of that enzyme, then it follows that there would result an excess of the amyloid protein," says Gusella.

Nevertheless, as long as the beta amyloid splinter stays put within the membrane, it should cause no harm. Only when the crucial piece is dislodged, perhaps by some sort of damage to the membrane or to the cell itself, is it free to accumulate as plaque. What form that damage could take is a mystery. It is here that a potential environmental agent such as a virus, a toxin, even a blow to the head, might come into play. These agents might injure or kill nerve cells, causing them to relax their grip and release the beta amyloid.

In the case of the mutation explored by Hardy's group, a defect in the amyloid precursor protein itself appears to be enough to cause the beta amyloid segment to be expelled. If such mutations in the APP gene were common, they would provide an easily understood mechanism accounting for most of Alzheimer's disease. Unfortunately, the mutations are extremely rare. It is estimated that errors in the APP gene account for substantially less than 1 percent of all Alzheimer's.

It was clear that investigators would have to look elsewhere for the major genetic causes underlying Alzheimer's disease. Before long, a group at Duke University led by Allen Roses found that the disorder is linked strongly to chromosome 19, specifically to a gene for a protein called apolipoprotein E, which has also been implicated in heart disease. People with a certain form of the gene, called apo E4, have a greatly increased risk of getting Alzheimer's.

Every cell in the human body contains two copies of the apo E gene. About 79 percent of the population has at least one copy of an apo E type called E3. Another 7 percent has a variant called E2, which appears to offer some protection against Alzheimer's. But some 14 percent of all people have at least one copy of apo E4, which renders them five times more likely to get the disease than those who have no E4 genes. And the Duke research, which has now been replicated by other labs, found that people unfortunate enough to have two copies of the E4 gene have an up to seventeen times greater risk of developing Alzheimer's. According to the researchers, 90 percent of people with two E4 genes will get the disorder by the age of eighty; they also are more liable to acquire the disease at a younger age.

It appears that the apo E4 gene is not directly responsible for Alzheimer's. Rather, it may be the lack of apo E3 or apo E2 that renders the person vulnerable. The investigators contend that both E2 and E3 seem to confer some protective effects. In any case, drug treatments are now being envisioned that imitate the natural protective action of E2. Tests for the E4 variant may one day also assist in diagnosing the disease. Research is continuing.

Meanwhile, the lab of Rudolph Tanzi and James Gusella has linked two other genes to Alzheimer's. The genes, whose products have been christened amyloid precursor–like proteins, or APLPs, are on chromosomes 19 and 11, respectively. APLPs seem to belong to a family of cell surface molecules that act like receptors and help carry signals to the nerve cells. "One possibility," says Tanzi, "is that mutations in these genes alter the transduction of signals to the neuron and an improper signal leads to cell death. Then the death of the cell releases large amounts of amyloid—up to 1 percent of the RNA in a neuron consists of APP—and forms plaques."

Tanzi's group is exploring the question why beta amyloid, which is normally water soluble, should precipitate out of solution to form the hard plaques. In laboratory experiments, the researchers have found that high levels of zinc, as are sometimes found in the brain, can cause amyloid to fall out of solution. Conversely, high concentrations of copper make it soluble once again. "We hypothesize that, for whatever reason, if zinc levels get too high or copper levels too low, it allows the formation of amyloid," says Tanzi.

Possibly the most promising avenue of research lies on chromosome 14, where more than a dozen flanking markers have been isolated for a gene that appears to be closely linked to up to 90 percent of all early-onset Alzheimer's. The fruit of these labors is a piece of DNA a million base pairs long that almost certainly contains the gene. Tanzi and Gusella's lab, as well as laboratories in Montreal, Florida, and Belgium, are attempting to pull the gene out of this trackless stretch of DNA.

"I think once we get this gene on chromosome 14, we are going to understand the bulk of the Alzheimer's disease process," says Tanzi, "because here is a gene that leads to the earliest onset of the disease. It is directly causative, while genes on the nineteenth and eleventh chromosomes appear to be primarily *susceptibility* genes—that is, they cannot by themselves cause the disease, they only make one vulnerable to an environmental trigger. This gene will tell us what goes wrong at the most basic level."

Of course, how one thinks the question will be resolved depends on where one stands on the chicken-or-egg question of the role of amyloid. "People who believe amyloid causes the disease think the gene will be found to affect how APP is processed," says Tanzi. "And those who think the death of neurons precedes and causes amyloid buildup believe the gene has to do with neuron function, specifically the signal pathway.

But it's all speculation. We can hypothesize over beers all we want. What we really have to do is get the gene. Then we'll know."

At first blush, it seems strange that so many different genes could be responsible for the same end result, Alzheimer's disease. But Gusella suggests that one think of it as a long chain of biochemical events. "Any disruption of the pathway at any point could cause a breakdown," he says.

In any case, as Tanzi says, the denouement of the mystery seems nearly at hand. "We are on the verge," he says.

In early 1990, guarded optimism arose over the possibility of treating Alzheimer's disease with nerve growth factor (NGF). NGF is a naturally occuring protein vital for the growth and health of nerve cells. The gene has been cloned by two biotechnology companies, Genentech, of South San Francisco, and the Syntex Corporation, of Palo Alto, both of which plan to mass produce the protein by recombinant methods. Studies in rats and monkeys show that injecting NGF into the brain can stop nerve cell destruction in the acetylcholine pathway of the basal nucleus and hippocampus. One study, by Fred Gage's laboratory at the University of California at San Diego, even showed that NGF could improve the learning behavior of elderly rats that had lost their ability to learn to swim through a water maze. The protein stopped the animals' nerve cells from shrinking and promoted new connections between these and other brain cells. Similar results have been achieved in aged monkeys at Johns Hopkins by Donald Price and his colleagues.

Franz Hefti, of the University of Southern California, an NGF pioneer, calls the drug "the most promising approach for the first effective treatment for Alzheimer's disease." "Instead of trying to replace missing neurotransmitters, which is the purpose of current pharmacology," he told the Chicago Tribune, "we are trying to use NGF to save brain cells that would otherwise disappear."

Sure enough, in 1991 a pioneering Swedish researcher, Lars Olson, showed the beauty of the approach. He installed a NGF drip directly in the right lateral ventricle of a seventy-year-old patient, identified only as Mary, and made medical history. Under her right breast, Olson implanted a refillable pump/reservoir powered by a battery that would last for two years. A thin cathether was run up to her neck and into her brain via a hole drilled in her skull. Mary was unaware of any of this; her mind had been virtually destroyed by her disease. But when Olson primed the pump with a syringe through the skin and NGF started dripping into the woman's brain, the halls of the Karolinska Institute began to glow with wonder. Like a rusty machine creaking back to life, Mary's brain began to remember again. After a few days, her memory came flooding back. Her husband's eyes filled with tears of joy. He had his wife back, at least for a while. When the infusions were stopped as part of the experiment, Mary's memory began to fade again. She was to be scheduled for a second round of rejuvenation, and the dangers of long-term exposure to the hormone remain to be determined. But Olson had made his point in about as dramatic a way possible: he had temporarily cured Alzheimer's with NGF.

Enthusiasm for the treatment is not universal, however. Both Carl Cotman, of the University of California at Irvine, and Dennis Selkoe, of Harvard Medical School, have found disturbing evidence that runaway nerve cell growth may be the very thing that is wrong in Alzheimer's and that NGF might make things worse. Specifically, they found that beta amyloid can stimulate the sprouting of new dendrite branches in nerve cells cultured in the laboratory. This caused Cotman to conjure up a vicious cycle: damaged cells deposit beta amyloid, promoting new outgrowths of dendrites in neighboring nerve cells, which in turn strangles more nerve cells, leading to the accumulation of still more beta amyloid. For this reason, everyone is treating the application of NGF to Alzheimer's disease with caution.

On the other hand, animal testing has shown no such dangers to date. If human trials are given a go-ahead, Gage is waiting in the wings with a form of gene therapy. His group, which includes Theodore Friedmann, has already introduced an NGF gene into fibroblasts, which were then implanted into the brains of rats. The NGF-producing cells prevented deterioration of the rats' acetylcholine network. Gage hopes his technique will get around a basic problem with NGF therapy—how to get the protein across the blood–brain barrier.

Recently, another encouraging development in Alzheimer's research was reported. Selkoe's group announced that it had, for the first time, uncovered beta amyloid in tissues outside the brain of Alzheimer's patients. These tissues included spinal fluid, skin, blood vessels, and the gut. The finding could lead to the first practical diagnostic test for Alzheimer's disease. At present, physicians must be content with a working diagnosis, which they arrive at by ruling out all other disorders known to cause dementia. A more definitive test for Alzheimer's requires slicing up the brain in a search for plaques and tangles, obviously not an option in a living patient.

Selkoe's finding came on the heels of a report from several laboratories that abnormal changes resembling plaques and tangles had been found in the nasal lining of people with Alzheimer's. The finding may be of critical importance for two reasons. First, the nerve cells of the nasal passages can be extracted from living Alzheimer's patients; second, they are the only neurons in the adult human body capable of reproducing. In other words, they can be grown in the laboratory; that enables scientists to study directly the abnormal molecular changes leading to Alzheimer's disease.

The ability to look at living Alzheimer's tissue affords an opportunity to follow up on a controversial discovery made at Yale University in 1988 that seemed to indicate that a virus is at fault in the disease. Previous attempts to demonstrate viral transmission by injecting tissue from deceased Alzheimer's victims into animals had always failed. But some researchers speculate that the virus may leave the body long before the time of death, dooming any experiment that uses posthumous tissue.

In an effort to overcome this, the Yale scientists took white cells from eleven close relatives of Alzheimer's victims and injected them into the brains of hamsters, reasoning that the relatives might be in the initial stages of the disease. Sure enough, five of the eleven samples triggered brain degeneration in the animals after a year or so of incubation, whereupon brain tissue from the ailing hamsters was able to spread the symptoms to other hamsters. The researchers tentatively concluded that the five volunteers had in their blood an infectious agent of some kind. Though no one has ever shown that Alzheimer's disease can be passed from one human to another, two other progressive diseases of the brain are known to be caused by viruses. One of these, kuru, strikes only the natives of the New Guinea highlands and stems from their ritual practice of eating each other's brains. The other, Creutzfeldt-Jakob disease, can affect people years after exposure to contaminated needles, surgical instruments, even electrodes once used to treat patients with the disease.

So much is going on in Alzheimer's research that an effective treatment of some kind will very likely be introduced soon. Creighton Phelps, vice president of medical and scientific affairs for the Alzheimer's Association, recently declared, "Within the next five years I think that there will be discoveries that will allow us to diagnose Alzheimer's patients before the symptoms set in, and new treatments to keep these people functioning at a higher level longer."

Even as the molecular machinations of Alzheimer's disease are being worked out, thanks to clues provided by Down's syndrome, other scientists are making an ambitious attempt to decipher the genetics of Down's itself. At first glance, this seems a daunting task. Just knowing that this relatively common genetic disorder, which strikes 1 out of 800 babies, is caused by an extra copy of chromosome 21, is of little help. There are more than a thousand genes on the chromosome, most of them anonymous, and no one knows why having a third copy of some of them should produce the syndrome's characteristic traits. These include retardation, short stature, an enlarged tongue, a single crease across the palm, a distinctive Oriental facial appearance that gave rise to an objectionable term for the condition ("mongoloidism"), immune system problems, a greatly heightened risk of leukemia, premature aging, and, frequently, congenital heart defects. "We know you get 50 percent more protein product from these extra genes, but we don't know how that causes harm," says the researcher Charles Epstein, of the University of California at San Francisco. A few years ago, investigators concluded, to their relief, that they may not have to comb through the entire chromosome in pursuit of the genes that underlie Down's syndrome. It appears that most of the genes responsible for the disorder are located at the bottom of the chromosome in a region that has been given the scientific name of 21q22. Scientists were led to this assumption by the fact that a handful of people with Down's syndrome do not have a complete third copy of chromosome 21. The only thing they have in triplicate is the 21q22 fragment. It is a rather small region, with only perhaps one hundred genes. How many figure in the genesis of Down's syndrome is unknown, but Epstein conjectures there may be fifteen or more. Unfortunately, he points out, other regions on the chromosome seem to contain genes that also produce manifestations of Down's syndrome, so one cannot concentrate solely on 21q22. It may be that those Down's patients with the extra fragment constitute partial or milder cases. Says Epstein, "Probably the syndrome is caused by many genes scattered along the chromosome, some with a large effect, some with a small, and we're going to have to tease them out."

A veteran researcher who played the leading role in the RAC's approval of the first gene therapy experiments, Epstein is after the genes responsible for the condition's most disabling features—namely, mental retardation and heart problems. As new genes are isolated on the chromosome that appear to be expressed in the brain or the heart, he inserts them into mice to see whether an extra copy causes symptoms similar to Down's syndrome. In an earlier approach, his laboratory actually created a disease in mice that mimics the syndrome. Exploiting the fact that some genes on human chromosome 21 have counterparts on the mouse's chromosome 16, Epstein began breeding mice with a third chromosome 16. Interestingly, the rodents displayed congenital heart defects, brain abnormalities, and immune system disorders remarkably similar to those seen in Down's syndrome. But the animal model was of limited value. There was no way to examine the effect of individual genes, nor would it have been useful to do so, since chromosome 16 of mice contains a number of genes that do not tally with those on chromosome 21 of man. Do the Down's-like symptoms in the mice stem from genes they share with us or from genes unique to their species? It is impossible to tell.

On the other hand, the newer strategy of putting individual human genes into mice has already begun to bear fruit. Recently, in collaboration with Yoram Groner, of Israel's Weizmann Institute, Epstein outfitted mice with extra genes for the powerful antioxidant superoxide dismutase (SOD). Life cannot exist without SOD, as we have seen, and the gene is located on human chromosome 21. But is 50 percent more SOD—as Down's patients have by virtue of three copies of 21—too much of a good thing?

Epstein put the question to the test, using as part of his experiment a substance well known to pot smokers: paraquat. Paraquat is the herbicide the Mexican government formerly sprayed on marijuana to make it undesirable to smoke. The stuff hurts plants (and people) by producing oxygen free radicals. When Epstein took skin cells from his transgenic mice, which were brimming with an excess of SOD, and exposed them to paraquat, the cells, as expected, proved more resistant to the effects of the herbicide. But brain cells were another story. They seemed actually to become *more* vulnerable to damage, especially when Epstein exposed them to other treatments—namely, high-oxygen and low-oxygen environments that stimulate the production of free radicals, particularly that bugaboo hydroxyl. "Then you're in trouble," says Epstein, "because hydroxyl is the most toxic of oxygen radicals. It's a DNA breaker, and is bad for cell membranes."

So much for cells in a test tube. But does this lead to retardation in living people? It's quite possible, considering that the things we eat and drink, the very air we breathe, are constantly driving free-radical production up and down in our brains. "We are trying to get a handle on whether too much SOD causes the death of brain cells," says Epstein. "We plan to take our SOD mice and put them in an environment where we can challenge them. Radiate them a bit, generate free radicals, and then see what their brains look like a year down the road."

Working on his own, meanwhile, Yoram Groner has made a series of striking discoveries about the SOD gene. SOD mice, he found, have perfectly normal hearts, lungs, and other internal organs. But they do have one structural abnormality in a very significant place: the tongue. Actually, the defect is in the junctions where the nerves link up with the tongue muscles. This flaw causes the mouse tongues to be weak and limp, which is of more than passing interest, because a common feature of Down's syndrome is a large and flaccid tongue. When Groner and the pathologist Rena Yaron, of Jerusalem's Hadassah Hospital, examined tongue muscle removed from Down's syndrome patients who had had cosmetic surgery on their tongues, they uncovered something very intriguing. The Down's patients had the same neuromuscular defect as the mice. A SOD glut clearly affects the central nervous system.

In 1989, Groner went that discovery one better. For years, it had been observed that Down's victims have low levels of the neurotransmitter serotonin, but no one knew what causes it. Then Groner found that SOD mice also have diminished blood levels of serotonin, whose presence is essential to mental activity as well as a number of vascular and intestinal functions. Groner was able to identify the problem as a lowered acid level in the membrane of cells responsible for absorbing and storing serotonin—apparently brought on by a surplus of SOD. Since serotonin helps brain cells communicate, Groner's work suggests that a SOD imbalance contributes to the mental retardation of Down's syndrome. And it raises the possibility that the SOD gene or its mates on chromosome 21 are responsible for the drop-off of other neurotransmitters, such as noradrenaline and acetylcholine, which has also been observed in the disorder.

What are the therapeutic goals of all this effort? Does anyone seriously think it is going to be possible to intervene against Down's syndrome? The answer is emphatically yes. Once the offending proteins are known, a treatment may be envisioned in which the protein levels are brought into something approaching balance. The treatment would not cure all problems associated with Down's syndrome, because many genes affecting multiple body systems are at fault. But it might stop the most devastating consequences, those involving the brain. "I would use PKU as an example," says Stanford's David Cox. "There the brain starts out normal, but because a child cannot metabolize a certain amino acid, phenylalanine, retardation eventually sets in. We deal with that by putting the child on a special diet devoid of phenylalanine.

"There is some evidence that much of the brain degeneration in Down's children happens after birth. They definitely seem to go downhill. So if it is due to a metabolic thing, we may be able to ameliorate it through metabolic therapy."

On the other hand, it is clear that by the time a child with Down's syndrome comes into the world, substantial brain damage has already occurred. The number of neurons in certain regions of the brain is noticeably smaller in Down's infants than in the normal newborn population. Even so, the picture may not be bleak. "Whether the function has been irretrievably damaged, we don't know," says Epstein. "The central nervous system is very plastic. It can respond to all kinds of terrible trauma, relearning and forming new circuits. If the problem turns out to be with the hardwiring, we probably can't fix it. But if it's a software problem, maybe we can redo it a little bit.

"If you take sensory neurons from a fetus with Down's syndrome and culture them, you find they have abnormal electrical properties that affect neurotransmission, and conceivably learning and memory. One might find ways of modulating that. You could change certain neurotransmitters, raise the levels of some, lower the levels of others. One might even do it in the developing fetus, if you knew from prenatal diagnosis that the child had Down's syndrome. At ten weeks, the nervous system is just beginning to get cranked up. If you knew that a half dozen extra genes were involved in perturbing the nervous system, conceptually one might just be able to go in and shut them down."

These might seem idle ruminations if they came from someone else. But Epstein is clearly not a man to think small: "If you were to ask me where we stand in Down's syndrome research, I would answer that I think we are finally starting for real."

Multiple sclerosis is another mysterious disease that is being illuminated by molecular biology. The disorder, which generally begins in young adulthood and affects some 250,000 Americans, is marked by patchy and inflammatory destruction of the myelin sheaths that insulate the nerve fibers of the brain and spinal cord. The loss of this protective coating halts the transmission of messages between nerve cells much the way frayed electrical wires short out and stop carrying current. Visual and speech disturbances, labile emotions, and a progressive loss of muscle control are the results.

It has been known for some time that susceptibility to multiple sclerosis is powerfully influenced by heredity. Children and siblings of patients have a thirty to fifty times greater risk of getting the disease than members of the general population. Studies have shown that identical twins, who by definition share the same genes, have a significantly higher likelihood of each contracting the disease. That the illness is not, however, entirely genetically determined is highlighted by the fact that the rate at which the disease strikes each twin is about 26 percent—considerably less than 100 percent. Therefore, most scientists believe that both a genetic factor and an environmental agent, possibly a virus, are involved.

Specifically, they theorize that these factors combine to produce an autoimmune reaction. That is, the patient's body attacks its own myelin tissue as if it were a foreign invader. But how can this come about? Molecular biology is providing the answer.

Science now knows that all of us carry a unique logotype on the surface of our cells. This logo is called our HLA (human leukocyte antigen) type, and it is put there for a very good reason. It enables the leukocytes, or white cells, of our immune system to distinguish between the good guys and the bad, between our own tissues and those of an alien antigen such as a virus or bacterium. One's HLA type consists of a series of marker proteins dictated by a critical cluster of genes on chromosome 6 called the major histocompatability complex, or MHC. There are quite a few genes and variable gene segments in the MHC cluster, and they permit a huge number of combinations, giving each of us a self-tag nearly as personalized as a set of fingerprints.

Over the past few years, it has been found that improbably large numbers of multiple sclerosis patients share the same HLA subtypes. Some 50 to 60 percent of those in the group at highest risk for the disease, people of northern European descent, carry what is known as the DR2 antigen. An even higher percentage have the so-called HLA-DQwl or DPw4 antigens. But none of these HLA markers occurs in all patients, and thus none is sufficient by itself for development of the disease. Moreover, it is not uncommon for perfectly healthy people to have these antigens with no ill effect. It therefore seemed that an additional genetic factor must be at work. In early 1989, a consortium of researchers from NIH, Caltech, and UCLA reported finding what appeared to be that second factor. In studying the immune system's T-cell receptor genes, the investigators noted that certain combinations of these genes occurred more frequently in multiple sclerosis patients than in healthy controls. And in people who have both the unusual T-cell receptor gene pattern and high levels of DR2 antigen, the chances of developing multiple sclerosis are more than three times as great.

"What this suggests," said Dale McFarlin, a member of the consortium from NIH, "is that there is a genetic component in MS, but there is more than one gene."

The helper T cell is the key disease-fighting cell of the immune system. When alerted that a specific interloper has penetrated the body's perimeter defenses, it calls out the guard in the form of macrophages, killer T cells, and antibodies. Normally, the T-cell receptors give a pass to the body's own tissue type antigen, such as DR2. But if McFarlin and others are right, something causes a multiple sclerosis victim's T cells to mistake the HLA markers on the patient's myelin sheath tissue for those of a trespasser.

This could occur if a tiny genetic defect in one's HLA markers or T-cell receptors fouled up the intricate self-recognition system. But it could also occur by infection with a virus. Suppose a virus contains protein sequences that mimic those of the tissue it is laying siege to—for example, myelin or the cells that produce myelin, known as oligodendrocytes. Or suppose the viral protein is very close in sequence to the HLA markers themselves. Or what if the virus lifts sequences out of the host's myelin tissue and incorporates them into its own outer shell? In any of these scenarios, the immune system could be tricked into thinking its own body tissue is part of the virus and go on the attack. A number of viruses have been suggested as possible villains, including those that cause mumps, measles, and parainfluenza. The prime suspect, though, is the HTLV-1 retrovirus, which is already known to cause leukemia, lymphoma, and a paralyzing disease called tropical spastic paresis. Sequence similarities have been found between HTLV-1 and the nerve tissues of multiple sclerosis patients. But the findings are currently in dispute, and the issue is far from clear.

As the battle over a possible viral link to the disease rages, other researchers are concentrating on an animal model for multiple sclerosis that they have created in the laboratory. In this way, they are working out a basic understanding of the genes that code for myelin and its parent cells, the oligodendrocytes, in the event that the disease proves to be a disruption of the normal function of these genes. Much of this work emanated from Leroy Hood's lab at Caltech, where he had long been working with a strain of mice that lack the gene for myelin basic protein, a major component of myelin. These mice, aptly called shiverer mice, lead a dreadful life marked by tremors, convulsions, and early death. But in 1987 Hood reported that he had been able to introduce the gene for myelin basic protein into shiverer mice embryos via microinjection. The result? The adult mice no longer shivered or died prematurely. The shiverer mouse's problem is not directly analogous to multiple sclerosis, nor is Hood's therapeutic mode applicable to humans. But his ability to cure a lethal neurological disorder by gene manipulation startled the neurobiology community.

Another shock came in March of 1993 when other new data caused a medical advisory panel to urge the FDA quickly to approve the first drug for the treatment of MS. It is the antiviral agent beta interferon, shown to reduce nerve damage and severe attacks of MS by as much as 50 percent. The drug works by reawakening helper T cells and stopping the immune system's overzealous response to viral invaders, thus preserving myelin. The drug's success seems to strengthen the argument that MS is an autoimmune disorder caused by a viral agent.

A mood of cautious optimism has suddenly entered the multiple sclerosis picture. "If you look at the genetic work, it's all come within the last year or two," observes Steven Rheingold, national scientific director of the Multiple Sclerosis Society. "I can't give any estimation of how close we are to a solution, but we are at a level of cumulative knowledge in molecular genetics and immunology that is making a significant difference in the way we think about the disease process. It's a very exciting time now."

21

SOMETHING IN THE BLOOD

In the late 1980s, brisk winds of excitement swept through the psychiatric world. A research group headed by the veteran investigator Janice Egeland announced in *Nature* that it had tracked one of the most disabling of mental disorders, manic depression, to a gene on chromosome 11.

The finding represented more than another tour de force in the contemporary art of mapping genes. If valid, it marked the first time in medical history that a genetic cause had been definitively established for a psychiatric disease.

To those who wished that psychiatry could tie mental illness to a physiological, and therefore tangible, source, a door seemed to have suddenly opened. But more surprises lay ahead. Within three weeks of Egeland's report, another research team spread the news that it, too, had located a gene for manic depression. This time, the offending DNA was traced to the X chromosome. To the uninitiated, it might seem that the finding contradicted Egeland's. But the idea that the same clinical result can be caused by two different genes was not at all strange to geneticists, who had seen it happen before. Rather, the mapping of *two* genes responsible for manic depression only strengthened the growing conviction that defective genes lay at the heart of many psychiatric illnesses.

The dust had scarcely settled when yet another discovery bolstered the genetic hypothesis. A Canadian group reported that it had uncovered a piece of DNA believed associated with the ruinous disease schizophrenia. By now, the psychiatric community was abuzz. The gene hunters had extended their magic to two of humanity's most ancient psychic curses. Had diseases of the mind at last been convincingly linked to inborn errors of brain chemistry or structure? If so, would they one day be treatable, not by talk therapies that required years or by systemic drugs with terrible side effects, but in ways that zoomed right in on the underlying biological cause? Could the brain chemistry of patients be brought into line with drugs that worked with laserlike precision, or might their genes themselves be corrected once and for all through gene therapy?

The view that inherited biochemical faults make one vulnerable to mental illness has been kicked around in psychiatry for generations. Freud himself suggested as much, but lacked the technology to pursue the idea. In the absence of anything better to go on, Freud's heirs of the 1930s and 1940s focused on the dogma of the day, that early emotional scars were the root cause of psychiatric disorders. The age-old debate over "nature versus nurture" seemed to have been resolved by default in favor of nurture.

But cracks in this picture began to appear. Patients with the most intractable conditions did not get better by "talking things out." Drugs emerged in the 1950s that brought marked improvement to certain patients, a tip-off that something biological was going on. And eye-opening data came along showing that parents, siblings, and children of mental patients were far more likely to develop psychiatric illness than the kin of normal people. For example, close relatives of manic-depressives are twenty-four times as likely to suffer from manic depression; with schizophrenia, the risk is eighteen times as great. Identical twins, who are carbon copies genetically, are 300 percent more prone to both develop schizophrenia if one has it than are fraternal twins, who have fewer genes in common.

Of course, there were those who dismissed such findings. They pointed out that immediate family members share the same households and social backgrounds. Environmental factors that affect one might easily affect another. But when researchers examined children who were separated at birth from their natural parents and raised in adoptive homes, they found a fascinating thing. Such children tend to get the same mental disorders as their biological parents, even though they have never spent a day under the same roof.

The evidence was piling up that heredity was a more decisive factor than had previously been thought. When Janice Egeland and her colleagues announced their discovery in November of 1987, it seemed to provide the clincher.

The human laboratory for the team's research was a population of

Old Order Amish living in Lancaster County, Pennsylvania, about sixty miles west of Philadelphia. For nearly three centuries, this pious, agrarian community has adhered to a simple lifestyle that is at odds with the modern world. Its members are forbidden to have electricity or telephones in their homes. They eschew machines, tilling their fields by hand and getting about by horse and buggy. Violence is taboo. So is any deviation from the sect's strict standards of plain dress and behavior.

All the more reason why certain flamboyant conduct stood out in bold relief and left the Amish scratching their heads. There was the young Amish man who on sudden impulse called up an auto dealer to order a sporty new car. And one who, in the middle of spring planting, packed a bag and announced he was going on a "vacation." And an Amish woman whose moods alternated between black despair and periods of religious euphoria during which she could not sleep for days. The strange behavior typified the wild mood swings and grandiose thinking of manic depression, which affects some two million Americans. For years, the Amish had puzzled over such happenings and over the reality that they seemed to be concentrated in certain families. Of twenty-six suicides in the community over the past century, nineteen had occurred in the same four families. One family alone accounted for seven such deaths and another for five. To the Amish, it was apparent that, like certain physical ailments, mental problems run in families. *"Sis im bloot,"* the elders said. "It is in the blood."

To Egeland, professor of psychiatry at the University of Miami School of Medicine, and her colleagues at Yale, MIT, and Washington University, the Amish community presented an ideal opportunity to probe for genetic linkage to manic depression. This was so not because the mood disorder is more prevalent among the sect—it is not—but because by centuries of intermarriage the clannish Amish have kept their gene pool virtually free of contamination by outsiders, making it easy to track the inheritance of individual traits back through the generations. All fifteen thousand members are descended from the same thirty couples who emigrated to Pennsylvania from southern Germany and Switzerland in the early 1700s.

The Amish are attractive laboratory subjects for other reasons. Their families tend to be large and rooted in one place, so DNA samples from up to four generations can readily be extracted and compared in a hunt for riflip markers. Investigators can pursue the trail even farther back, to ancestors who died in the nineteenth century and before, thanks to the community's lively tradition of family history. A network of kindred "scribes" is charged with keeping a record of genealogical matters, including the traits, mental status, and cause of death of forebears who died long ago. As if all of this were not enough to make the Amish "a geneticist's dream," as Egeland puts it, there is their ascetic lifestyle. Anyone tracing genes for psychiatric disease must be able to tell conclusively which family members suffer from the disease in question. The sect's prohibition on the use of alcohol and drugs, agents that often mask depression, makes diagnosis significantly easier.

A conscientious woman with a warm, toothy grin and large glasses, Egeland has spent nearly thirty-five years winning the confidence of the reclusive Amish. Her kinship with them dates back to her days as a student at Johns Hopkins in the 1950s, studying with the grand seigneur of American genetics, Victor McKusick, who himself has studied the Amish extensively. Egeland's involvement went far beyond the call of the traditional genetics investigator. True to her early training as an anthropologist, she literally moved in with the Amish in the 1960s as a graduate student at Yale, breaking the ice at first with a few families, who let her live in their homes for increasingly long periods of time, then gradually gaining favor with the entire community—so much so that when her father became ill with leukemia, the Amish donated large quantities of blood. There was a mutual attraction between the quiet, folksy academic and the devout rustics whose way of life she came to admire so much. But there was more. The Amish consider mental illness to be the worst ailment afflicting mankind because, according to Egeland, "it interferes with the function of the mind and the wholeness of the spirit." Hence, for ten years, thousands of Amish volunteers cooperated with Egeland's group, helping in the tedious task of ascertaining who suffered from manic depression and later allowing the researchers to draw blood and perform detailed chromosomal studies in a search for riflip markers associated with the disease.

Between 1976 and 1980, working out of her headquarters in the town of Hershey, Egeland uncovered 112 people in thirty-two families who were classified as mentally ill. Of them, thirty-two were diagnosed as manic-depressives. Egeland discovered that all had a long family history of the disease in a pattern suggesting that a single, dominant gene was responsible. One family in particular, which she decided to make a target of her hunt for markers, produced 9 of the 32 manic-depressives. Going back four generations, Egeland learned that all had descended from a couple in which the husband, who died in 1898, possessed an obsessive-compulsive personality. Of his 8 children, 6 suffered from psychiatric illness, and of his 35 grandchildren, 8 were diagnosed as mentally ill. By the fourth generation, 25 of 191 great-grandchildren were found to have clinical psychiatric disease. Old family records, in the meantime, turned up a key ancestor, born in 1763, who suffered recurrent depression and whose depressed brother committed suicide.

That suicide clustered in a handful of families was a striking finding in itself. Taking one's own life is rare among the Amish, who consider it a great sin. That made the details of each occurrence notorious and simple for Egeland to ferret out. She learned, for example, that twenty of the twenty-six had hung themselves, primarily in barns, four had shot themselves in the head, and the two others had drowned themselves. One of these, a man, left a note, lashed himself to a boulder, and tied his Amish hat to his head before jumping into the deep water of a local quarry. When his body was discovered months later, identification was made from the initials in his hat. The other was a woman who was found floating in a washhouse cistern at home, having been so intent on suicide that she climbed up and over a boiler to enter the cistern. Twelve of these luckless people, Egeland determined, suffered from manic depression.

Like many gene researchers, Egeland was floored to learn in 1983 that the Huntington's disease gene had been localized to chromosome 4 by the use of riflip markers derived from a large colony of extended families in Venezuela. She resolved to try the same technique on her Amish. Thus it was that she began systematically assembling blood samples from one large set of related, extended families she called Pedigree 110.

Using supercomputers to sort the families' DNA, and testing the DNA initially against some markers on chromosome 11, she and her collaborators struck gold almost immediately. They found what was described as a "strong predisposition" to manic depression in subjects who possessed particular markers near the tip of the short arm of the chromosome—namely, telltale versions of the Harvey *ras* oncogene and the gene for insulin. People who had inherited characteristic polymorphisms of these genes also developed the mental disorder 63 percent of the time, a strikingly high correlation, indicating that a gene for manic depression was in the vicinity. A further calculation by the team suggested that as many as 85 percent of those with the gene would eventually develop manic depression.

In a 1987 interview, Egeland was asked whether it wasn't a lucky shot that the markers were found on the first chromosome studied. Replied Egeland, "Absolutely, like penicillin flying in the window and landing in the petri dish."

Unfortunately, the assumption was premature. By the late summer of 1989, it began to look as if Egeland's group had been hasty in assigning the gene to chromosome 11. Not that the team's methodology was in question. In studies of eighty-one subjects, it had obtained such high scores in favor of linkage between the disease and the markers that anyone would have published the results. These scores, called lod scores, measure the odds that what appears to be linkage is legitimate, as opposed to the product of pure chance. A lod score of three, which signifies odds of a 1,000 to 1 against chance, is generally considered the threshhold necessary to prove linkage. But Egeland's lod scores had been nearly five, or almost 100,000 to 1 against coincidence.

Even so, as Egeland added blood samples from some forty new subjects to the study, the case for linkage began to come apart. First of all, two of the original eighty-one members of Pedigree 110 fell ill with depression for the first time in their lives. One of them, to the team's chagrin, possessed neither of the markers believed to be linked to manic depression. In the riflip game, a single exception to the rule can destroy your case. The second man *did* have the markers, but was considered an anomaly anyway—he has several siblings with identical markers who are not sick, making coexistence of the disease and the markers in the brother look like happenstance. Taken together, the two men lowered the lod scores substantially.

Still more disappointment awaited. Thirty-one of the forty new subjects belonged to an extension of the pedigree who were related to the central family by marriage. There was a pattern of psychiatric disease, notably manic depression, in this extension as well, so Egeland had high hopes that they would also show linkage to the markers on chromosome 11. But it was not to be. The gene for manic depression in this new group displayed absolutely no linkage to the chromosome 11 markers. The lod scores plunged to an abysmal one.

Egeland and others had long known that a complex and variable disease like manic depression is probably caused by more than one delinquent gene, with the gene varying from group to group. For example, ample evidence had accumulated over the years that in some people the disease is strongly linked to the X chromosome. But Egeland had felt sure that, within an isolated society like the Amish, one gene would probably account for all the cases. The hitch was that this new branch of the pedigree included a woman who four generations ago had married into the Amish and could have introduced an alien gene. By including the branch in the study, Egeland was taking a risk, but it was a risk demanded by scientific integrity. The thesis that there is a gene on chromosome 11 responsible for all manic depression in the Amish had to stand on its own merits. "In our heart of hearts," says Egeland in retrospect, "we were hoping there was just a single gene." But, as is often true in research, nature is not so willing to be pinned down.

What did the setback mean scientifically? In effect, the possibilities came down to two: either there is no gene on chromosome 11 responsible for manic depression among the Amish, or the sect has *two* such genes, and one of them may be on chromosome 11.

Bolstering the latter argument is the fact that one of the people who became newly ill during the study, the man with no markers, did not fit the profile of the manic-depressive. Not only was he not a manic-depressive; his illness was not the sort indicating a genetic cause. His depression's onset at the age of forty-two was due to overwork and excessive family responsibilities. Egeland compares him to a "horse that foundered" and notes that he would not have been included in the study except that dropping him would have required changing the rules in midstream and made it look as if she were trying to stack the deck to drive the lod score higher. As for the second anomalous subject, the man whose siblings had the markers but no disease, there remains the possibility that his siblings will develop manic depression down the road, which would send the lod score climbing again.

Egeland, however, is dubious that there is a major gene for manic depression on chromosome 11. "The highest probability," she says, "is that there is a single major gene, and it's not there. I'm so sure because the extension to thirty more subjects gave us a very significant rule-out."

Why then did they get such tight linkage at the outset? "It could have been a complete statistical fluke, even with that incredible lod score," she says. "You would have put money on it but lost. From now on, I won't get excited until we have a lod score of ten. Actually, I never broke out the champagne. The most I did was eat a Hershey bar. The lod score of five sounded fantastic, but something inside said this was too easy, too fast."

Chromosome 11 may contain no major gene, but Egeland still thinks something there is related to manic depression. After all, many authorities think psychiatric diseases may be the result of two or more genes acting in combination. "I'm not sure we can throw out what we found on the chromosome," she says, a trifle wistfully. "It may be a modifer gene of some sort. Manic depression is a cyclical disorder. In the early days before electroshock and pills, when people suffered through dog days and blues, as Churchill called it, they recovered spontaneously after nine months of depression. And on the manic side, before lithium came along, people were tied down and restrained until they recovered.

So this is not a gene that exists as a permanent defect of some kind. In biology, if the illness is waxing and waning, it must relate to other health, environmental, or medical genetic influences. Something turns the disease on and off. So that's why I have to wonder if we still don't have to keep an open mind about the prospect that what we found in the original pedigree is some modifying influence."

In any case, Egeland and her colleagues, who include David Pauls of Yale and Edward Ginns of NIH, are back to square one. "We've regrouped," she says, noting that a new effort has been mounted to find the gene or genes. Instead of concentrating on one chromosome, as they did before, they are now doing a saturation screen of the Amish DNA samples with probes covering the entire genome. The undertaking is being carried out in part by Collaborative Research Inc., which helped produce the world's first genetic map and has nearly a thousand probes to throw at the chromosomes in a search for linkage. To date, nearly three hundred polymorphic markers, spaced some twenty million base pairs apart on the genome, have been used, along with a newer kind of probe called short tandem repeats.

One thing Egeland has vowed: "You can bet we are going to be stricter as to who qualifies for the study from now on." Only those positively diagnosed as manic-depressives or people with recurrent depression will be included this time around. "No more false positives," she avers.

The pool of individuals in the study has been widened to 169, but while the number of affected persons rose to 37 under a loose definition of manic depression, a stricter definition lowered the size of the affected group to only 17.

The team's experience highlights the difficulties inherent in trying to locate genes linked to psychiatric disorders, as opposed to those at work on other parts of the anatomy. Not only is diagnosis more a matter of "feel," in the absence of purely objective clinical criteria—a schizophrenic cannot be identified by use of a patch test, for example—but very likely mental illnesses originate with defects in more than one gene. Moreover, those genes do not always show equal penetrance—that is, they do not always reveal themselves outwardly to the same degree. Thus, the same gene may cause severe, cyclical manic depression in one person but only recurrent depression in another, or infrequent episodes of barely noticeable disturbance in yet a third.

"That's the most sobering thing of all," says Egeland, "that there are difficulties at both levels, in diagnosis and in gene detection. Nevertheless, with the improvements molecular biology is making each year, we will get there. It's simply a matter of time. There is absolutely no reason to think we won't be able to make a major breakthrough in discovering the genes behind common psychiatric disorders."

Even as Egeland's findings were collapsing in ruins, those who tracked manic depression to the X chromosome seemed, for a time, to be enjoying better luck. Perhaps "luck" is the wrong word, since these researchers, unlike Egeland, enjoyed the luxury of knowing where to look for the gene. For years, scientists suspected a link between manic depression and the X chromosome. For one thing, the disease is seldom transmitted by fathers to sons—a quirk easily explained if the gene is on the X, because males never pass an X chromosome along to sons, only a Y. For another thing, the disorder not infrequently seems to be inherited together with color blindness and a form of anemia, conditions long ago mapped to the X.

Miron Baron, a blunt-spoken, Israeli-born psychiatrist and geneticist, was intrigued by this genetic clue when he joined the faculty of Columbia University's New York State Psychiatric Institute in the late 1970s. The tools did not yet exist for molecular analysis, so Baron kept the hypothesis on the rear burner, an intellectual puzzle to return to whenever he had time. But several years later, when riflip mapping techniques became available, Baron saw his chance. He resolved to go after the gene, using as his study population certain large Israeli families in which manic depression had been found to cluster.

"The X chromosome hypothesis had been around a long time, and it seemed logical to pursue something where we had an inkling of the location rather than try to find a completely unknown," says Baron. "We figured if we headed in that direction, there might be some light at the end of the tunnel."

Baron formed a collaboration with two Israeli doctors, Batsheva Mandel and Rahel Hamburger, and between 1982 and 1985 they set about gathering pedigrees at the Jerusalem Mental Health Center. Patients were sought who had been diagnosed with manic depression and had at least one immediate relative with the disease. Their suitability grew if family members also suffered from color blindness and G6PD deficiency, a kind of X-linked anemia. The two conditions had been tied to manic depression in a controversial 1969 study. To enhance the odds of finding families, the researchers also asked the Israeli army to pass along to them data on new recruits with color blindness and G6PD deficiency.

Since X-linked manic depression was the paramount interest, families in which there was male-to-male transmission of the illness were excluded.

As 1986 dawned, Baron and his colleagues found themselves with five extended families who fit the bill. These families contained 161 adults, 47 of whom were suffering from manic depression or related disorders. An extraordinary degree of linkage was found between the disease and the X-linked conditions, with the lod score exceeding nine. Although unable to determine the location of the defective gene, the team assumed that it was near the tip of the X chromosome, where the two markers, the genes for color blindness and G6PD deficiency, were known to be located.

For several years thereafter, Baron tried to repeat his findings, using additional individuals and more refined techniques. "We don't want to say prematurely that we have proof there is a manic depression gene on the X," Baron said guardedly, mindful of Egeland's misstep. "The finding is encouraging, but one needs to be very careful."

It was a good thing Baron hedged, because early in 1993 he had to backpedal. In his earlier study, he and his team had used the simple physical presence of the two medical conditions as their markers. This time, they tested the patients with actual DNA markers on the tip of the X, a method that permits much more precise analysis. In two of the families, the lod score dropped dramatically, showing what the team called "greatly diminished support" for linkage of manic depression to the X chromosome. In a third family, however, the lod scores remained positive, keeping the hypothesis alive for that clan, at least. Still, Baron's finding marked another stiff jolt for those who had rejoiced several years earlier at the thought that the secrets of psychiatric diseases were at last being revealed.

Even if the Baron team's finding had been confirmed, however, it would not by any stretch of the imagination have accounted for all manic depression. In fact, one of Baron's collaborators, Neil Risch, estimates that no more than 25 percent of all cases of the illness can be attributed to an X-linked gene. As if to bear Risch out, studies of other large families predisposed to manic depression in remote areas of Iceland and North America have turned up no genetic markers for the disease at all. The Amish and Jerusalem genes were specifically excluded as culprits. The conclusion to be drawn is that manic depression has multiple causes. Almost certainly, different genes are involved in different populations, all producing a disease entity called manic depression. Some of what is diagnosed may not be due to heredity at all, but may stem primarily from environmental causes. Nevertheless, the importance of the Amish and Jerusalem studies is that they have established beyond dispute that, in certain cases at least, a well-known psychiatric disorder has a genetic origin.

As Egeland later put it in accepting the Victor M. Cannon Award for research into manic depression, "This is giving hope to thousands of Americans who suffer from mood disorders. People can now conceive of this illness as a medical or biological problem and know that swings in mood and energy are not something that they can control simply by spiritual well-being or will power. That's a very positive message. My hope is that these findings will reduce or eliminate the worst thing about mental illness—its awful stigma."

As the mystery of manic depression appeared at first to be yielding to the efforts of molecular biologists, so for a time did the shroud seem to be lifting from yet another widespread and tragic mental disorder—schizophrenia.

Schizophrenia has been called "arguably the worst disease affecting mankind" by the world's premier science journal, *Nature.* Affecting 1 percent of the human population, the ailment tends to appear, often with terrifying swiftness, in a patient's late teens or early twenties. It is marked by grotesque and dehumanizing symptoms, including delusions, hallucinations, disorganized thought patterns, flattened or inappropriate emotions, withdrawn or bizarre behavior, and, in some cases, total inability to distinguish reality from fantasy. For years, it was assumed in the psychiatric community that schizophrenia somehow resulted from early childhood experiences. In the 1960s, the villain was assumed to be an overbearing mother. This, of course, mainly served to heap guilt on parents already stricken with sorrow.

Twin and adoption studies eventually established that genetics, not apron strings, played a pivotal role. But the molecular complexities of the disease seemed as daunting as the mysteries of the cosmos itself. Then, in May of 1987, researchers at the University of British Columbia reported that, by accident, they had hit upon a stretch of DNA they thought might contain a culprit responsible for at least some cases of schizophrenia.

The Canadian researchers happened on their discovery because of some unlikely physical resemblances between a Chinese man and his twenty-year-old nephew, both of whom suffered from clinically diagnosed schizophrenia. The younger man had been brought into the university hospital's emergency room by his parents, who in anguish told the psychiatric resident on call, a young woman named Ann Bassett, that for weeks he had holed up in his room, laughing insanely and talking to himself. Bassett began to treat the youth in subsequent weeks, but a chance comment by the mother opened her eyes to the wider implications of the case.

One day, the mother mentioned in passing to Bassett how odd it was that the boy did not resemble her or her husband, but looked very much like her brother, who himself had schizophrenia. Her interest piqued, Bassett examined the uncle and nephew together and received a shock. Each man had a protruding forehead, widely spaced eyes, and an abnormally short fourth toe on each foot. Still more intriguing, the nephew, a college student, had a congenitally defective left kidney, while the uncle had been born with no left kidney at all.

Bassett now ordered some chromosome studies performed on the two men. The results were mind-boggling. In both uncle and nephew, a segment of the fifth chromosome had been duplicated and affixed backward onto one of their copies of chromosome 1. In other words, each had three copies of this segment, whereas a normal person has only two. Subsequent tests on the student's mother revealed that she had the same translocation of a piece of chromosome 5, but the corresponding piece was absent in her other chromosome 5. Therefore, she only had two copies of the segment, and, while she could pass along the translocation to her offspring, she did not herself suffer from schizophrenia or any of the physical abnormalities. The researchers immediately theorized that the segment, which is located on the "long arm" of chromosome 5, contains a gene or genes responsible for the men's schizophrenia. The long arm is also home to hundreds of other genes, some of them presumably responsible for the uncanny physical similarities between the two men.

In November 1988, the Canadians' serendipitous findings were compellingly affirmed by a team led by the British scientists Hugh Gurling and Robin Sherrington. In *Nature,* they reported having studied genetic material from members of five Icelandic families and two clans from Britain. These families were rife with mental disease. Of the 104 total family members, 39 were schizophrenic and 15 had various other serious mental conditions. Taking their cue from Bassett's study of the Chinese uncle and nephew, Gurling and colleagues tracked the long-arm segment of chromosome 5 through the seven families and found a pattern. A distinctive form was inherited exclusively by those with schizophrenia and the other disorders. The odds against this pattern's being due to pure chance were 50 million to 1, a convincing lod score, to say the least. Moreover, the fact that all of the traditional types of schizophrenia were represented in the families, ranging from paranoia to near-infantile regression, meant that the entire range of this unfortunate disease could be caused by a single gene. Then, too, the connection between the chromosome 5 segment and disorders other than schizophrenia seemed to indicate that a schizophrenia-susceptibility gene may manifest itself in alternative ways. The importance of this, asserted the Gurland group in its paper, is that it may throw light on genetic or environmental factors that occasionally change or limit the expression of a schizophrenia gene.

But, alas, the same issue of *Nature* that presented Gurling's findings also exploded any hope that all schizophrenia could be blamed on this one gene. An accompanying article by Kenneth Kidd of Yale reported on a study of an extended family of Swedes living in relative isolation above the Arctic Circle. Like the families in the Gurling study, the Swedish clan has an unusually high prevalence of schizophrenia. Out of eighty-one subjects examined, thirty-one were schizophrenic. But when Kidd and his colleagues tested family DNA against markers on chromosome 5, they could find no genetic linkage. Clearly, genes on still other chromosomes must be able to produce the same clinical result, not to mention the possibility of environmental causes in many cases. Obviously, schizophrenia, like manic depression, is not going to yield its secrets easily.

Several factors complicate the hunt for psychiatric genes. One is that such genes may contribute only 10 to 20 percent to a person's susceptibility, as opposed to the 100 percent seen in single-gene disorders. Environmental influences or other genes may do the rest. Another problem is that a given gene may, for a variety of reasons, cause a number of different psychiatric problems.

One man has been working long and hard on the theory that a single defective gene can manifest itself in various forms of mental illness. He is David Comings, director of medical genetics at the City of Hope National Medical Center, in Duarte, Calif.

Comings's views set off an unbecoming scene at the 1988 meeting of the American Society of Human Genetics, in New Orleans, where, in a conference hall swarming with press, geneticists began shouting at one another passionately. On the receiving end of a dressing-down was Comings, the incoming president of the society. How had Comings raised the dander of so many peers? After all, in addition to presiding over the prestigious genetics society, he is a respected figure in the hunt for genes associated with psychiatric and neurological illness. And in his area of special interest, Tourette's syndrome, he is among the world's foremost authorities.

Tourette's syndrome is a bizarre familial disease marked by uncontrollable muscle tics and grotesque vocalizations, including compulsive barking, grunting, and yelping and involuntary outbursts of profanity. It is generally considered a rather rare disorder. But during his presidential address at the New Orleans meeting, Comings had the temerity to insist that the Tourette's gene is much more common than heretofore thought and, more shocking still, may be a kind of supergene responsible for many different and seemingly unrelated mental conditions. Specifically, said Comings, it may cause phobias, depression, manic-depressive mood swings, obsessive-compulsive disorder, panic attacks, schizoid behaviors, hyperactivity, dyslexia, stuttering, conduct disorders, compulsive overeating, and alcoholism both in Tourette's patients and in family members who show no outward signs of Tourette's. Comings went so far as to estimate that from 10 to 30 percent of some of these mental problems in the general community could be blamed on this single inherited gene.

Skeptics in the audience thought Comings was overreaching, basing his conclusions on anecdotal evidence gleaned from surveys of Tourette's patients and their families—evidence that many found scientifically shaky. "I disagree with you completely," snorted the venerable geneticist Arno Motulsky, of the University of Washington, a former mentor of Comings. For Motulsky to lead an attack on his onetime protégé was visibly painful to Comings. But Comings stuck to his guns, insisting that one protean brain gene can manifest itself in kaleidoscopic ways.

Tourette's is primarily a disease of childhood, although its most famous victim is an adult—the Philadelphia Phillies ballplayer Jim Eisenreich. Its onset is generally between the ages of two and fifteen. As recently as 1980, the disorder was thought to strike no more than 1 in 100,000 children. Later studies raised that frequency to 1 in 1,000 and added the recognition that besides its hallmark signs—muscle tics and vocal noises—Tourette's can also be associated with hyperactivity, learning disabilities, and unwanted behaviors such as exhibitionism and disciplinary problems. But flying in the face of most contemporary theory, Comings, and his wife, Brenda, a clinical psychologist, assert that Tourette's syndrome actually has a much higher incidence—1 in every 100 boys and 600 girls, which would make the number of its victims in America on the order of several hundred thousand. To get that kind of prevalence, they say, the gene must be carried by from 5 to 15 percent of the general population, making it a very common gene indeed. Their most radical contention is that Tourette's represents the extreme end of a spectrum of ailments that go well beyond learning difficulties and hyperactivity and that are all caused by the same putative gene. In other words, an unknown degree of psychiatric illness in people who show no classic symptoms of Tourette's may actually be caused by the Tourette's gene.

The evidence for their assertion comes from their study of Tourette's patients and their families over a period of years. They report that in addition to the disease's classic symptoms, its patients show an unusually high incidence of other behavioral abnormalities. For example, mania and severe anxiety were found to be twenty times more common in Tourette's patients than in controls. Obsessive-compulsive disorders, the maddening affliction that causes people to repeat certain ritual acts such as handwashing several hundred times a day, is five times as common. Chronic aggressive and violent behavior is seventeen times as common. Stuttering is five times as common. Dyslexia is six times as common. Tourette's patients are thirteen times more prone to panic attacks, eleven times more likely to develop episodes of depression, and three times as vulnerable to multiple phobias.

But the Comingses go beyond these startling observations, reporting that close relatives of Tourette's victims also show a heightened tendency toward many of these disorders. On the basis of extensive family histories, they assert that half of all carriers of the TS gene display some symptoms of the Tourette's spectrum, including panic attacks, obsessive-compulsive behaviors, depression, mood swings, and hyperaggression. Most recently, the Comingses added substance abuse and obesity to the list of problems that can plague Tourette's families.

Typically, says David Comings, parents will come in with their child, who is displaying tics and vocal noises. "Then you find out that the father had multiple tics as a child, and has a very short temper, and in fact he was divorced twice because of temper. And then you learn that there's a nephew who displays very violent, antisocial behavior and has a lot of trouble with the law. And the mother had severe agoraphobia and panic attacks at one time. And you keep finding these things all through the pedigree, a whole family with varying behaviorial symptoms, and it's all linked together by this single major neurological gene. We've seen it in our families for years, where people can carry this gene and have problems with other behaviors and not have the tics. We wouldn't call them Tourette patients. We diagnose whatever it is that they're having, be it phobias or whatever, and as a footnote say they are related to a TS patient. But it's all coming from the gene."

Comings's critics are not prepared to make such a large leap. Chief among them is David Pauls, who is also pursuing the Tourette's gene. Pauls concedes that obsessive-compulsive disorder is more common in Tourette's families, showing up in about 27 percent of patients' relatives, but in a presentation at the New Orleans conference he directly contradicted Comings: "Our data do not support the hypothesis that other psychiatric and behavioral abnormalities are associated with Tourette's syndrome." Pauls bases his findings on a significantly smaller sample than Comings's, but he contends that his methods, which involve direct interviewing, are more reliable than Comings's emphasis on questionnaires.

Many studies in recent years have linked various psychiatric disorders to heredity. One 1983 study, for example, found that the frequency of agoraphobia is twice as high in the relatives of agoraphobes than in the general population. Similar findings have been made with obsessive-compulsive disorder. The Comingses claim that a substantial portion of such afflictions may be due to the operation of an unsuspected Tourette's gene.

Comings thinks he may already know the identity of the gene. He has tested a candidate gene that directs the metabolism of tryptophan, an amino acid that is ingested as part of the diet and that is the biochemical precursor of serotonin. Comings was led to the gene because of laboratory findings that Tourette's patients tend to have low blood levels of tryptophan and serotonin.

But what might serotonin have to do with Tourette's? "That's the exciting part," responds Comings, who is also testing dopamine receptor genes for their relationship to Tourette's. "Serotonin is the inhibitory neurotransmitter in the brain. It is very rich in a part of the brain called the limbic system." The limbic system, a group of structures deep within the brain, is associated with emotions and feelings—notably, anger, fear, sexual arousal, pleasure, and sadness. It is Comings's theory that when serotonin levels are too low in the limbic system, this critical part of the brain becomes "disinhibited." Victims lose their self-control. That loss of restraint can take the simple form of tics, says Comings. Or it can manifest itself in the inability to suppress obsessions and compulsions, in rage, or in the urge to shout obscenities or to masturbate in public.

Comings has already cloned the tryptophan gene. Now his laboratory is trying to determine whether it is genetically linked to Tourette's syndrome. Once a Tourette's gene, or a marker, is in hand, it will be possible to verify or debunk the Comingses' hypothesis.

David Comings is very careful to stress that he is not implying that all, or even a majority, of cases of such illnesses as manic depression are due to the TS gene. "Other genes can cause these same manifestations and do. But I do believe that this one bad gene contributes to a lot of difficulties."

Unlike many of his peers, Comings believes that the time frame for solving the riddle of brain genetics is not going to be too long. "I've always felt that probably in the next ten or fifteen years we'll have specific genetic tests so that when a patient walks into a psychiatrist's office, maybe instead of sitting right down and talking with him, the first thing that might be done is to draw his blood and find out if he's got any genes for manic depression, schizophrenia, TS, autism, whatever. There might be ten or fifteen genes that account for most mental disorders. By being able to do that test ahead of time, it would allow one to very rapidly key in on the most specific and effective treatments. It could be drugs, or a combination of drugs and psychotherapy."

Among the first behavioral afflictions expected to be diagnosed genetically is alcoholism, a disease that distorts the lives of an estimated eighteen million people in the United States. A genetic predisposition to alcohol addiction has long been suspected. In the 1960s, Henri Begleiter of the State University of New York Health Science Center, in Brooklyn, began studying the brain waves of alcoholics and found that their evoked potentials (EPs)—that is, brain wave patterns that can be induced by exposing test subjects to various stimuli, such as flashing lights or a series of photographs—have a characteristic signature. The amplitude, or height, of the wave in alcoholics is reduced, compared with that in normal people. Wondering whether these patterns might be innate, rather than a consequence of alcohol abuse, Begleiter investigated the EPs of the offspring of alcoholics. He took two groups of boys aged six to eighteen, one group the sons of alcoholics and the other the sons of nonalcoholics, and found that as many as 35 percent of the alcoholics' sons had the wave pattern typical of alcoholism, whereas less than 1 percent of the control group did. For a certain subcategory called type 2 alcoholism, in which the victim's father is always an alcoholic, drinking problems surface early and are generally accompanied by antisocial behavior, 89 percent of the sons showed the characteristic EP pattern. This correlated with the findings of a Swedish study that a third of the children of type 2 alcoholics later become alcoholics themselves—even when they grow up in adoptive homes where the adoptive parents are teetotalers!

Research has turned up some other strong indicators of the heritability of alcoholism. At the University of California at San Diego, Marc Schuckit discovered that the sons of alcoholics are more tolerant of the effects of alcohol than the sons of nonalcoholics are and perform better on tests of hand-eye coordination after drinking the same amount. Other researchers have developed strains of rats that differ genetically in their preference for consuming alcohol in the presence of other food and drink.

But no one had linked an actual human gene to alcoholism until a pair of scientists caused a worldwide stir in early 1990 by announcing that they had finally accomplished the feat. The gene unearthed by Ernest Noble, of UCLA, and Kenneth Blum, of the University of Texas Health Science Center, was an atypical form of the receptor gene for dopamine, a neurotransmitter known to be associated with a sense of pleasure in the brain. Its correlation with alcoholism was so great as to seem almost too good to be true.

The method the two researchers used to find the gene was unorthodox. Blum and Noble studied the brains of seventy deceased individuals, half of whom were incurable alcoholics. The brains, which were obtained from a national brain bank at a Los Angeles veterans hospital, came from men and women from a variety of different ethnic groups.

After verifying from medical records and family interviews which brains were from alcoholics and which were not, Noble sent tissue samples to Blum's laboratory in Texas. The samples were identified only by code numbers, a procedure called "blinding," so that Blum could not tell ahead of time which tissue samples belonged to alcoholics. Blum then tested the samples against nine DNA probes, each of which had been proposed as a candidate alcoholism gene from previous research.

Of the nine probes, only the dopamine D2 receptor gene showed an affinity for the cadaver DNA. But what a strong affinity it was! The gene was present in 69 percent of the alcoholics but absent in 80 percent of the nonalcoholics, a correlation that the *Journal of the American Medical Association* called "striking."

The dopamine D2 receptor gene, located on chromosome 11, occurs in two different forms. A majority of all people have a form known as the A-2 allele, but some have a less common version called the A-1 allele. It was the A-1 allele that was linked so closely with alcoholism.

Blum and Noble have expressed the opinion that the atypical allele may direct the manufacture of defective dopamine receptors on the surface of brain cells. If this occurs, the cells may not be able to get enough dopamine, which is part of the brain's "reward pathway." This in turn could cause an intense craving for substances, such as alcohol and cocaine, that stimulate the production of more dopamine.

For now, the theory remains unproven. Even the study's results are at present in question, since the sample was so small. But if the findings are borne out by further investigation, there may soon follow a blood test to identify children with a weakness for alcohol and other drugs—in time, perhaps, to spare them a lifetime of misery by teaching them to avoid such substances before they develop a taste for them.

No one suggests that the dopamine D2 receptor will completely solve the riddle of alcoholism. Most scientists, including Blum and Noble, believe that, as with many other mental disorders, several genes are responsible for the disease and that cultural and social factors also play a large role. But the isolation of at least one gene associated with a disease that kills one hundred thousand individuals a year would represent a milestone in the understanding of the genetics of the brain.

Meanwhile, early 1993 brought news that a gene associated with attention deficit disorder had been identified at the National Institutes of Health. Attention deficit disorder affects about 4 percent of the school-age population and is responsible for serious learning and behavioral problems. Children who suffer from it are impulsive, restless, and easily distracted.

A group headed by Peter Hauser found that in certain people the disorder is linked to a defect in the gene for the thyroid hormone receptor. Such individuals seem to have a normal output of thyroid hormone. But the cells of the brain lack proper receptors for the hormone and, as a result, cannot process it. "Thyroid hormone is essential for normal brain development," noted Bruce Weintraub, who oversaw Hauser's work. "There are several ways that faulty thyroid hormone receptors could affect that development and result in behavioral abnormalities."

The link was discovered when NIH researchers compared forty-nine patients who had the faulty thyroid receptor gene with fifty-five relatives who did not. Half of the subjects were adults and half were children. What leaped out at the research team was that 70 percent of the children and 50 percent of the adults with the thyroid problem also had attention deficit disorder. By contrast, only 20 percent of the children and 7 percent of the adults without the thyroid defect had the disorder.

Clearly, there is a strong association between the two conditions—although the absence of 100 percent linkage shows that other factors must be at work as well.

The team expressed the opinion that the finding should promote a new perspective on attention deficit disorder. Said the study coauthor Alan Zametkin, "It's not bad parenting, overcrowded schools, or unmotivated kids. [It] is a neuropsychiatric problem based on brain physiology."

Discoveries like these are certain to have a profound effect on our concept of the brain and its relationship to thought and awareness. On the spot is the celebrated mind/body distinction made famous by the French philosopher René Descartes. As the functions of thought and memory are more and more reduced to their organic components, there may come a strong temptation to dismiss "mind" altogether as an outmoded concept, to be replaced with a mechanistic view of intellect as nothing more than the gene-choreographed ballet of proteins, neurons, and synapses. One hopes, though, that we will not come to speak of the brain as did one prominent neurologist who several years ago contended that the brain produces thoughts "the way a kidney produces urine."

Indeed, the more we find out about the physical workings of the brain, the more elusive seems an explanation for that incorporeal blend of personality and consciousness that we call mind. Will we ever be able to reconcile a biological understanding of the brain with the impudent and illogical existence of the self? Only time will tell.

Philosophical considerations aside, the pulling back of the curtain surrounding mental illness is of great importance. If psychiatric disorders are among the cruelest afflictions of mankind, they also dwarf all other categories of disease in terms of sheer numbers. It has been estimated that about 25 percent of the world's inhabitants will develop a neuropsychiatric disease at one point or another during their lifetime.

What a boon it would be if science were to isolate the brain-based genes that figure strongly in the development of personality, behavior, and psychopathology! We will probably never entirely settle the nature-versus-nurture debate. Even with our still-scanty knowledge of the genetics of the brain, we know that many people can carry genes that predispose them to such disorders as schizophrenia and depression and yet never develop the afflictions unless exposed to certain environmental stresses and influences. Not all identical twins raised apart fall prey to shared afflictions. Nevertheless, the role of heredity as a critical factor in the shaping of our psyches is becoming indisputable. And the message has not been lost on the psychiatric profession. Practitioners are increasingly aware that the future lies in treatments grounded at the molecular level as well as in the cushiony confines of the couch.

"Molecular biology will have a clear impact on the science and practice of psychiatry in the near future," writes Steven Hyman, of the departments of psychiatry and molecular biology at Massachusetts General Hospital. "It is not too early to urge psychiatric residents, and psychiatrists in general, to become literate in human genetics and the basic molecular concepts used in genetic analysis. Without developing sophistication in these areas, psychiatrists may mistakenly consider genetic research a threat to psychological theories rather than an opportunity for important psychological research aimed at understanding epigenesis and, potentially, preventive interventions."

These interventions, certain to include more-finely-honed drugs, may someday also include gene therapy. For example, faulty dopamine receptors have been implicated in schizophrenia by a number of researchers. The successful cloning of the genes for these receptors in 1989 and 1990 opened the possibility of somehow altering these genes in patients. Other possibilities will undoubtedly present themselves.

"Looking down the long road, I don't see why molecular genetic techniques used for alleviating any other disorder might not be applied to psychiatric diseases as well," says Herbert Pardes, dean of Columbia University's College of Physicians and Surgeons and a psychiatrist who has championed a molecular approach to mental illness. "It may well be that some of these diseases will eventually be treated by genetic manipulation."

Heretofore, the delivery of genes to the brain posed obvious hurdles. But the breakthroughs at the University of Michigan and in France in the use of adenoviruses to target brain cells may point the way toward eventual treatments. Pardes himself suggests some novel ways around the delivery problem, including putting normal genes into fetal tissue, which might then be implanted into the brain. Fetal brain cells multiply rapidly—though they lose this ability shortly after birth—offering a way to get enough corrective genes into the brain via retrovirus to perhaps reverse disease.

Pardes traces much of his optimism to the contagious enthusiasm of his friend Nancy Wexler, director of the Hereditary Disease Foundation and a prime mover in the effort to isolate the gene behind Huntington's disease. Many times, Pardes has traveled with Wexler to Venezuela and brought back for analysis blood samples from Huntington's victims. "I got good advice—some of it from Nancy—and took it. Also, I saw for myself what was going on with regard to the explosion in genetics. Psychiatric research has been groping for a handle for several decades, and now suddenly there is this hot set of possibilities. It is extremely attractive."

As a result, Pardes says, he has put his money where his mouth is. When he came to Columbia a few years ago as director of the New York State Psychiatric Institute, which is affiliated with the school, he organized a major molecular genetics research program, funded by the William Keck Foundation, of California. The institute has lured many of the top investigators in the field and is examining the molecular roots of psychiatric ills ranging from schizophrenia and manic depression to anxiety disorders. "Young people have spotted the excitement and are coming into the field in droves," Pardes says.

The most tantalizing area for near-term progress, he asserts, will be "transgenic" studies, in which genes believed to be related to psychiatric disease and human behavior will be spliced into mice and other animals to see what effect they have—the same method Charles Epstein is using with potential Down's syndrome genes.

"You have a beautiful sequence for first finding a gene, getting its product, then sticking it into a mouse and seeing what it does. It's really elegant, man," says Pardes. "If you can find out what systems a particular gene product disturbs, then you've got an entry point in terms of how to fix it. For so long, the brain has been inaccessible to us because it is protected by the skull. We've had to look at by-products secreted in blood and urine, which are such a great distance from the basic problem in the brain that it is difficult to come up with drug and other strategies. It's like trying to figure out what's wrong with a car without looking in the motor. Suddenly, we've got techniques for getting right to the basic mechanisms of these diseases. So ways of fixing them will occur to us in much greater numbers."

Treatment strategies would, of necessity, be more complicated than they might be for a neatly defined, single-gene disorder such as ADA deficiency or cystic fibrosis. Just diagnosing such variable diseases as schizophrenia is a problem. When is it schizophrenia, and when is it not? What kind is caused by genes and what kind not? Is one gene at work or more than one? There are only a few ways the human body can express disturbances of behavior: hallucinations, delusions, mood upheavals, agitation, and panic, to name the bulk of them. But there could be many different pathways leading to the same observable result.

Complicated it may be. But Pardes believes it may still be eminently feasible to interfere with the disease process. "You might have a number of genes interacting to produce the disease," he notes. "But if you knew what the critical genes were, maybe you could pull one of them out and nullify the disorder." He draws a parallel with the work being done in lung and colon cancer, which shows that a number of genes must go haywire to precipitate the cancer, conditioned by exposure to an environmental toxin. Treatment might involve repairing one gene along the chain. "Maybe it's the same way with schizophrenia and the other mental diseases," speculates Pardes. "If one of the abnormal genes could be made normal, then you could get hit with the environmental cause and it would be irrelevant. It would have no context without the gene. So that might offer a therapeutic strategy, removing or modifying one gene in the pathway.

"One thing is for sure. We're probably going to come out of this with a million different ideas."

22

GYPSY FORTUNE-TELLERS OF TECHNOLOGY

What profits wisdom
when there is nothing to be done?
—Sophocles, *Oedipus Rex*

It has been very consuming emotionally," the geneticist Francis Collins says with a headshake, "counseling women who have lived under a cloud all their lives." As he speaks, Collins reflects the bewilderment of the reluctant prophet who's being shoved into a new age before he's ready.

The cloud is breast cancer. Since 1991, Collins has been counseling women about BRCA1, the gene that predisposes 600,000 American women to breast and ovarian cancer. Collins has intensely researched a group of midwestern mothers, daughters, sisters, and cousins who represent the 5 to 10 percent of the women haunted by a form of breast cancer that can devastate whole families, striking female members when they are in their forties, thirties, even twenties. As of this writing, BRCA1 (for Breast Cancer-1) has not yet been found. Indeed, until the Berkeley geneticist Mary-Claire King mapped it to chromosome 17, in 1990, Collins had not even known there *was* a breast cancer gene. Since then, he and King have been collaborating in a race to isolate the gene. Competing teams have narrowed the search to a region of about 750,000 pairs close to the center of the chromosome—quite a long stretch, but one that may have been traversed by the time this book appears.

The researchers are not waiting around. Marker testing already is under way at three U.S. institutions—the University of California at Berkeley; Creighton University, in Omaha, Nebraska; and the University of Michigan, in Collins's lab.

Collins's pioneering venture into genetic screening for breast cancer began the day a thirty-three-year-old woman, known as Susan, marched into the office of his partner, Barbara Weber, director of the breast cancer clinic at the university, and announced that she had scheduled surgery to have both her breasts removed.

"I want 'em off," she declared. "I've got two kids I want to see grow up. I can't live with the fear any longer." That put Weber, director of the breast clinic, in an odd position. Susan's request for prophylactic mastectomies made sense; indeed, a surgeon had already scheduled an operation in a few days to remove her breasts. Susan's mother and her aunt had developed breast cancer in their forties within six months of each other. Susan's sister Fay died of the disease in 1988 at age thirty-six. That same year, her cousin Renee died of ovarian cancer. She was only twenty-nine. Six months later, another cousin, Audrey, died of breast cancer at thirty-two. In addition, a few months earlier, Susan's older sister, Janet, forty-one, had been diagnosed with cancer in her right breast.

Susan was thirty-two and scared to death. Weber made a snap decision. Just that week, the results of a revolutionary DNA test developed at Michigan convinced Weber that Susan would not get cancer. "I was sure," Weber recalled, "nearly totally—98 percent. What could I do? I had to tell her. I just *had* to."

Susan hadn't wanted to see Weber. "I was all up for the surgery. I didn't want her telling me I already had cancer or was doomed to get it." But Weber insisted, so Susan brought along Janet and their husbands for moral support. Nervous herself, Weber stammered a bit, then cut to the coda.

"You don't have the marker," she said. "I'm telling you to not have surgery."

Susan was dumbfounded. "My mouth fell open and I felt faint. For the first time since the nightmare began, there was hope in our family that maybe some of us would escape the curse. We're a very religious Catholic family. But for years I had wondered, 'God, are you listening?' "

Soon, Weber and Collins were able to tell Susan's forty-year-old cousin, Kathy, that she was a likely carrier of the breast cancer gene and persuade her to get a long-postponed mammogram. There it was: a 6-millimeter invasive cancer, small and innocent, but deadly. Such early cancers are highly curable, but Kathy sacrificed her breasts and ovaries. Again: she said she couldn't stand the fear any longer.

With that, a new era of genetic counseling was born. Another cousin, Trisha, learned that the breast amputations and reconstructions she had undergone five years earlier had been for naught. She was not a carrier.

Conversely, Janet learned that she had passed on the gene to her eigh-teen-year-old daughter and was overcome by guilt.

Francis Collins has a tiger in his Ann Arbor lab. "When Susan came to us in agony," he said, "we realized that we had genetic information that could dramatically change people's lives—right now, not some time in the future. We couldn't sit on this."

Collins and Weber went for broke. On a brisk autumn day, they in-vited forty-two members of this large Michigan family to the medical center and, one by one, told them how they had fared in the genetic lottery. The breast cancer susceptibility gene is dominantly inherited, meaning that one either gets it or is spared. Men learned, to their amaze-ment, that they could pass on breast cancer to their daughters. Women learned not only whether they had escaped but whether their children had, too.

Nine of nineteen men were found to have the gene, and eleven of twenty-three women. No family members were tested until they turned eighteen.

As a genetic counselor, Collins believes in passing out the facts, no matter what; it's a veritable credo. "Our hard and fast, unbreachable golden rule, our mission to the public," Collins says, with uncharacter-istic hyperbole, "is to give them information that enables them to make an informed decision about what is right for them. Consequently, with the breast cancer gene, we've had every situation here that you can imagine."

Compassionate, thoughtful, Francis Collins seems to move grace-fully through often fractious corridors of genetics. His charisma derives from his achievements. He is codiscoverer of the genes responsible for cystic fibrosis, neurofibromatosis, and Huntington's disease. He also is the inventor of chromosome jumping, the landmark lab strategy that helped make those discoveries possible. Furthermore, as James Watson's successor as director of the National Center for Human Genome Re-search, he will help lead the government's three-billion-dollar effort to track down and sequence all the genes that encode a human being. But Collins is a physician first, and at this juncture in his career, the restless forty-five-year-old gene hunter finds himself agonizing over BRCA1. The gene affects 1 woman in 150, he estimates. Its discovery would lead to the screening of all women to see whether they carry it.

"We've suddenly come to realize *that breast cancer may be the most common genetic disease there is,*" Collins says, with considerable sur-prise. Why should this be so? He speculates that if the gene is damaged over the decades, perhaps from exposure to environmental toxins, it can kick off a cascade of mishaps leading to breast cancer and to ovarian cancer, too.

Breast cancer is one of the most common of all cancers, striking 180,-000 American women each year, killing 46,000, and forcing the rest to join 1.5 million other worried cancer survivors. The causes of breast cancer are imperfectly understood, to say the least. Studies pointing to various risk factors often contradict each other, and up to 75 percent of all breast cancers seem to occur randomly. Changing lifestyles—an aging female population characterized by early menstruation (age ten or eleven), delayed childbearing, and late menopause (above age fifty), which mean constant exposure to the powerful cell-dividing hormone estrogen, plus lifelong exposure to high-fat diets and to environmental agents such as radiation and chemicals—may all be contributory.

In light of the confusion, a solid finding, such as that of the gene for the hereditary form of breast cancer, would be the first major breakthrough. In 1971, Henry Lynch, professor and chairman of preventive medicine at Creighton University, described what is known as the hereditary breast-ovarian cancer syndrome. Confusingly, the syndrome differs from so-called familial breast cancer, which accounts for up to 25 percent of all cases and basically means that the disease seems to run in a family yet cannot be tracked genetically and may be due to mischance.

Hereditary breast-ovarian cancer leaves no doubt. It is characterized by extraordinarily early onset, by involvement of both breasts, by tumors that are more highly proliferative than sporadic (random) cases, and by genetic linkage. Establishing linkage, however, does not hinge on a simple blood test. Family lineage data and DNA must be laboriously collected from as many generations as possible. Treatment usually includes prophylactic removal of the ovaries—there is no adequate early test for ovarian cancer, and these women stand a lifetime risk approaching 100 percent. Lynch also performs linkage analysis and has worked with more than fifty individuals so far, about half of whom turned out positive and half negative.

Significantly, both he and Collins have found that no matter the outcome of genetic screening, the members of these terrified families share the same paranoia: they do not want their insurance companies to find out they have been tested. The researchers say that as many as half the patients at risk do not even want their personal physicians to know of their family secret, lest the information be included in their medical records and they be cut off.

Such deception, although understandable, can be dangerous once the women return home for care: management strategies for such patients include physician's exams starting at age twenty, annual mammography starting at twenty-five, and ongoing education so that the women understand why they are being monitored. No women patients in medicine must be watched more closely all their lives.

For twenty years, Berkeley's Mary-Claire King has been one of the few scientists who believed that a single defective gene passed from parent to child could lead to breast cancer. More experienced scientists disagreed. Clusterings within families were far too random to be attributed to one gene (speculation about the inheritance of breast cancer goes back to the early Roman medical literature). And so it was thought that, yes, perhaps a predisposition to the disease does exist, but it is probably influenced by a number of mutated genes that have to be triggered by environmental agents before the cancer can arise.

But King, after studying all the known risk factors, hypothesized that breast cancer is a side effect of an affluent, industrialized, well-nourished Western society and, unlike lung cancer, cannot be prevented. Consequently, the only way to save lives is to get down to the molecules involved. The 1 woman in 150 who inherits the bad gene has an 80 to 90 percent chance of developing cancer. And 1 in 150 is an appallingly high rate of transmission for an inherited disorder.

As limited as it may be, the first genetic test for human breast and ovarian cancer is emblematic of both the problems and the benefits that genetic screening has for telling people more about themselves and their reproductive probabilities. As a potential boon, such advances have been long in coming, but they obviously are fraught with complications—including surgery, sterilization, abortion, family conflict, and the possibilities of stigmatization, discrimination, and creation of a biological underclass of medical untouchables and societal pariahs. In the United States, 164,000 applicants for medical insurance are now turned down for medical reasons each year, according to the investigative arm of Congress, the Office of Technology Assessment, which has warned that, "applicants for insurance plans are already being asked to provide information to prospective insurers related to genetic conditions like sickle cell anemia. Some experts fear that individual policies will become increasingly difficult to acquire as more genetic screening tests become available."

On the other hand, this year in the United States, more than a hundred thousand babies will be born with a genetic disease. The joy attending the births of some 5 percent of all newborns will be transformed into grief and uncertainty. There are at present no cures for most genetic diseases; half of them are fatal, and the vast majority are seriously disabling; and treatment of any kind is available for only about 15 percent. In the years to come, no matter how much love such children receive, no matter how much happiness they bring, no matter how much they deserve a chance at life, the hard truth is that unavoidable factors like parental guilt, financial burden, and sheer fatigue will cause many of their families to break up.

Already, though, there have been a few incidents in which insurance companies have threatened to not cover the medical expenses of such children, once mothers have been notified on the basis of prenatal tests that they are carrying an affected fetus. Such a threat may give a pregnant woman no choice about whether to have an abortion, if her only option is to give birth and face financial ruin. Several critics of genetic screening, including Paul Billings, chief of genetic medicine at Presbyterian Medical Center in San Francisco, have collected dozens of cases of apparent bias against innocent people, indicating that ignorance and misunderstanding about genetic screening pervade the issue. No group has become more alarmed than the nation's geneticists. "There is no more urgent issue before us," declared Stanford's David Cox, a Down's syndrome researcher, who was representing the American Society of Human Genetics before a blue-ribbon scientific advisory panel of NIH. The panel urged the government to launch a crash program to explore the implications of mass screening, beginning with cystic fibrosis.

In the meantime, genetic testing has become a growth industry in the United States. More than four hundred laboratories are now testing DNA for such traits as Down's, sickle-cell anemia, cystic fibrosis, Tay-Sachs, or spina bifida. Every baby born in a U.S. hospital is screened for genetic disease. Pregnant women in their thirties routinely seek prenatal diagnosis—in the early 1970s, when amniocentesis was a novelty, only about 200 were done each year; now more than 300,000 are performed annually.

In the years ahead, adults will increasingly be able to find out whether they are carriers of defective genes or predisposed to something dangerous, that in twenty years they may suffer a stroke, cancer, a heart attack, or a major depressive illness. The power of such prophecy is profound: almost everyone carries several serious gene mutations—one out of every twenty-five Caucasians is a cystic fibrosis carrier, for example; and one out of twenty-five Caucasians whose ancestors were eastern European Jews carries one copy of the Tay-Sachs gene. Because everyone has two copies of every gene, inherited from each parent, the good copy normally masks or "covers" the bad, protecting the body from disease. But not always. Not in a dominant disease.

"We all know we're going to die—yes. But most of us don't know how, or when," points out Kimberly Quaid, a slender, attractive Indiana University psychologist in her thirties, who since 1986 has been telling people whether they carry the gene for Huntington's disease. Formerly at Johns Hopkins, Quaid was first interviewed in her Baltimore office, which was warm and cozy with the muted earth tones that characterize genetic-counseling offices everywhere. She delivered a lot of good news there—and bad news, too. At the instant of revelation, she says, it doesn't matter whether the news is bad or good. People are in shock. They don't hear anything she says, which is why she always makes her clients bring along a friend who can explain and interpret later, once the tears of joy, or anguish, dry.

"Right now, those of us in this field are becoming the gypsy fortune-tellers of technology," complains Quaid. "What I see in my job is merely the tip of the iceberg, but it's very distressing. Frankly, I don't think we're ready for this."

Quaid has been a pioneer much longer than Francis Collins. Thanks to the new genetics, society since the mid-1980s has had a marker test that can tell today's healthy people that they carry a gene that will kill them sometime in the future, and that at the moment there's no hope. And with the discovery of the Huntington's gene itself, in March 1993, such people can now even be told the approximate age at which they will begin the long downward spiral.

"Many people who come to us don't want to know for themselves," Quaid says. "They want to be able to tell their children whether they need to worry or not. Others, though, say they have an overwhelming need to end the agony of not knowing."

In 1986, predictive testing of the DNA of healthy people at risk for Huntington's began at Hopkins and at Mass General. As other centers slowly came into focus around the nation, it grew apparent that the DNA test has for pioneering testers produced agonizing dilemmas, which will become increasingly commonplace in the years ahead as new specific genetic defects that predispose people to future physical and mental illnesses are discovered. Such tests are altogether novel to human culture. Do people really want that information? Not Huntington's disease families, apparently—only about 200 members of the 150,-000 U.S. families at risk of developing the disease have chosen to take the genetic test. On the other hand, dare society keep such knowledge from them? On the average, half of those haunted by Huntington's will be freed; they and their descendants need not worry about the deadly gene. People can get on with their lives. By 1993, the Johns Hopkins psychologist Jason Brandt, who pioneered the test with Quaid, calculated that at least six babies had been born to Huntington's "escapees," as those freed of the curse are called.

But what of the rights of other groups in society, such as insurance companies and employers? Should they have such information? What would they do with it? The presymptomatic test shoved medical science to the edge of an ethical abyss.

From its inception, the marker test has been viewed with ambivalence by Nancy Wexler, who offered it, then stopped offering it, then offered it again at Columbia Presbyterian Medical Center. She is no fan. Her attitude typifies that of other professionals: "It's not a good test if you can't offer people treatment."

Wexler has been struck by those who come in for testing and already are showing signs of the disease. "It really indicates how ambivalent people are. There's a tremendous need to deny what's going on. These people say they want to have this information. But they don't *really* want it, or they would deliver it to themselves. They don't need a fancy DNA test. If they just looked at their fingers and toes, they'd say, 'Well, that's it, all right.' " Denial, she notes, is a crucial coping strategy for human beings. How else can those living under a threat like Huntington's be expected to get through the day? All of us have health fears that we deny all the time. "But we're cracking people's healthiest defense by making them attend to the fact that they will actually get this disease," Wexler says. "We're opening deep wounds."

The tepid response to the Huntington's test suggests that the public may not welcome such bounty from science—at least, not so long as there are no cures.

Yet the dream of forecasting the future is as old as humanity itself, and the DNA oracle, as Kim Quaid has noted, is the most formidable fortune-teller the human species has ever had. Scientists have become so skilled that they can examine the genetic material in a single, microscopic sperm cell, as was accomplished a few years ago at the University of Southern California. The feat, which also killed the sperm, was viewed as a technical tour de force with no practical application—akin to the lighting of a match; once lit, it's destroyed. But the strategy extended gene analysis to sperm, something formerly considered impossible, and demonstrated the newfound capability of biologists to invent enzyme microassays and gene tests for single embryonic cells that allow them to be checked out in detail before being allowed to implant in a uterus and make a fetus. Nor is the technique impractical: for instance, an experimental one now under research at the University of Tennessee called uterine lavage—in which an early embryo consisting of a few hundred cells is flushed out of a woman's womb a few days after intercourse, checked for health, and returned—promises one day to make genetic testing of any potential pregnancy cheap and simple.

DNA testing is already settling paternity suits and identifying missing children. It could replace dental testing and fingerprinting in forensic medicine. Since 1985, the controversial new "DNA fingerprinting" tests (based on the same riflip technology that found the Huntington's marker and gene) have been used in hundreds of cases of rape and murder in nearly all fifty states and promised to figure prominently in the sensational O. J. Simpson murder case. At least fifteen foreign countries have embraced the technique, and it has been proclaimed fundamentally sound by the Office of Technology Assessment and by the National Academy of Sciences. DNA fortune-telling is creating strange bedfellows, with world-class academic geneticists advising police agencies, including the Federal Bureau of Investigation, about how best to apply the new science of genetic identification and establish national standards for accuracy.

But no aspect of the new technology is more ingenious, or more bizarre, than the movement that has quietly arisen to see what can be done about the genes of a child before it is even conceived.

The brainstorm struck the Chicago embryologist Yury Verlinsky at an international conference in Jerusalem on test-tube baby technology: a clever way to screen human eggs for defective genes before pregnancy and ensure that abnormal eggs never get fertilized. That discovery in the spring of 1989, which burst into a new science called preimplantation genetics, seemed farfetched, but gene splicers are bewitched by possibilities.

Basically, in these early days of gene therapy, two species of researchers represent the far ends of the spectrum—those who are exploring the very earliest days of life (embryologists) and those who are devoting themselves to extending its end or even reversing the process of aging itself (biogerontologists). Embryology and the biology of aging deal with the temporal extremes of human existence. For those of us alive now, it may turn out that time, not space, is the final frontier. Fanciful flights about its passage—how to slow down time, how to speed it up—have long been the fodder of science fiction prophets such as Jules Verne and H. G. Wells who inspired generations of dreamers and made them become scientists and engineers. But real science has surpassed fiction and promises to redefine not only what is usually meant by gene therapy but what it means to be a genetically unique human being.

As usual, the momentum of feverish discovery often leaves implications in the dust. There is something irrepressible about Verlinsky, a short, dapper, exuberant Soviet émigré in his early fifties, whose round, neatly goateed face flushes with excitement as he shares the secrets he is deciphering about the reproduction of life. Since coming to the United States more than a decade ago and finding work in Chicago—he is currently at the Illinois Masonic Medical Center, where he serves as director of cell genetics—Verlinsky has been helping physicians push back the time in pregnancy when they can determine the health of a budding embryo—the earlier the better, so far as he is concerned. "We want to prevent tragedy as early as we can," he says. The ultimate goal: "to prevent abortions, miscarriages, and hopelessly ill children by screening human eggs, sperms, and zygotes in preparation for pregnancy." Verlinsky, who learned his genetics at the University of Kharkov, in the Ukraine, personifies the younger generation of Soviet scientists who grew up professionally in the late 1960s, post-Lysenko climate of relative enlightenment. In Chicago, Verlinsky has become one of the first scientists to merge genetic engineering with in vitro fertilization (IVF), the "test-tube" baby technology in which a woman's egg is fertilized by her husband's sperm in a petri dish and then injected into her womb, in hopes that it will burrow into the uterine wall and result in a pregnancy. Originally designed as a last-resort therapy for couples unable to conceive a baby naturally, IVF has given genetics researchers a way to manipulate pregnancy at its earliest stages, an advance that brings with it important philosophical questions. Although Verlinsky lacks the big reputation and big institutional backing of a French Anderson, Francis Collins, or Richard Mulligan, his work is nonetheless at the cutting edge of clinical research. It is lawful work, bound by medical ethics—although, like most genetic screening in the United States, it escapes conventional government regulation by being privately funded through his hospital's Reproductive Genetics Institute. Still, no one knows better than Verlinsky, who grew up under communism, how such techniques might be perverted by tyrants and fools. Yet its potential for good seems to consume him.

Anything adverse was the farthest thing from Verlinsky's mind in April of 1989 when he went to Jerusalem to join thirteen hundred fertility experts from forty-one countries to discuss their mutual quest to perfect the controversial, decade-old IVF technique. Ethical issues surrounding conception outside the body have led the Vatican to condemn IVF and the United States to ban federal funding of such research, which may help explain why the technique remains inefficient—IVF costs several thousand dollars per attempt, and a baby will result only about 9 percent of the time (on the average) in the United States. Since 1978 and the birth of Louise Joy Brown in England, some twenty thousand test-tube babies have been created worldwide, but even in the best of laboratories a fertilized egg stands only a 10 to 15 percent chance of implanting in its mother. For this reason, most IVF clinics try to replace three or four eggs.

IVF boils down to a numbers game, explains Robert Edwards, the bold Cambridge University embryologist who invented IVF in the face of almost overwhelming opposition and who has "fathered" more than 3,500 babies at his Bourn Hall clinic. "With two fertilized eggs implanted in their womb, 25 percent of women become pregnant and 20 percent of those have twins," Edwards calculates. "With three fertilized eggs, up to 30 percent of women achieve pregnancy, a rate approaching that of nature in normal couples, even though every other type of infertility therapy has failed on these patients. These aren't bad numbers, although we wish we could do better."

Experts such as Edwards estimate that up to 15 percent of all couples will prove to be infertile, as a result equally of the wife's tubal disease, which prevents sperm and egg from meeting, or of the husband's defective sperm. So, despite being a long shot, privately funded IVF has gained widespread social and medical acceptance and become part of obstetrics services in many hospitals.

One afternoon in Jerusalem, Verlinsky and a colleague, Northwestern University's chief of reproductive genetics, Eugene Pergament, were winding down after the day's sessions by visiting a gallery of modern art in the King David Hotel. Verlinsky found himself contemplating an untitled 1935 painting by the great Spanish abstractionist Joan Miró. The painting fascinated him. In Miró's typically droll style, it featured two disks floating in space, one red and the other yellow. But just underneath the red disk was another round object, black and very tiny—a black dot that looked like an afterthought. Numbed by fatigue, Verlinsky kept staring at that painting for a long time. The more he examined Miró's colored disks, the more they looked to him like human eggs, which, after all, were very much on his mind. "Maybe the red disk became the yellow disk after it kicked out the black dot," he mused, wryly. A flash of insight jolted him. "I took a business card from my wallet," he says, "and scribbled down the idea so I wouldn't forget it." He eagerly shows you the card. It bears the cryptic words "polar body."

No sooner had Verlinsky returned home and begun experimenting with his idea than it was deemed so remarkable that he was summoned to deliver a plenary lecture before thousands of medical geneticists at the November 1989 meeting of the American Society of Human Genetics, in Baltimore. Verlinsky declared that he had found a way to spot mutations in human ova that were ready for in vitro fertilization, something that other scientists considered so improbable that only a few fellow pioneers were bothering to research it at all. By checking human eggs for abnormal genes before they are used to make test-tube babies, Verlinsky claimed, he could now assure couples from families plagued for generations by lethal genetic diseases that if they do get pregnant by the IVF technique, their baby will not be born with the disease. No longer would such people be bound by the crude rules of the old genetics— prenatal diagnosis followed by the agonizing decision about abortion if the fetus is abnormal. "Our patients are at high risk of passing on genetic diseases but don't want to have abortions," Verlinsky explains. "Usually, these people already have an affected child. They may have tried to have a normal baby, underwent prenatal diagnosis, and were forced to terminate a pregnancy. They found that experience so traumatic that they don't want to take that risk again. So they didn't have any more children, until they heard about this new technique."

Delicately, Verlinsky examines microscopic human ova that have been flushed from a woman's ovary in preparation for IVF. From each egg, he teases out the extra twenty-three-chromosome package—called the polar body—that an egg normally jettisons before it gets ready to be fertilized by a male's sperm. The black dot in the Miró painting, which Verlinsky mused had been squeezed out of the red disk had brought it to his mind. Once the polar bodies are removed from the eggs, specific genes therein undergo amplification for analysis through polymerase chain reaction by Verlinsky's partner, Charles Strom, director of medical genetics at Illinois Masonic. Eggs that prove to be healthy are reinserted into the woman's uterus after fertilization in a test tube, in the hope that at least one will implant and grow into a normal, full-term baby. Defective eggs are frozen and stored, Verlinsky says, awaiting the day that science can figure out a way to repair them. And so far, the idea of shelving unwanted eggs has bothered neither the donors nor the doctors.

Preimplantation genetic analysis by polar body biopsy is such a simple idea that it could be dreamed up only by a mind primed to leap from gallery art to human physiology. Every lunar month like clockwork, women release an egg—*omne vivum ex ovo*, all life comes from the egg, William Harvey taught physicians in the seventeenth century—but the genesis of that egg goes back to the time when the woman was herself a fetus. In the earliest stages of her mother's pregnancy, about two thousand amoebalike germ cells, or oogonia, migrated from the yolk sac to the ovaries of her own embryo. These cells multiplied, and at birth a baby girl's ovaries contain nearly 600,000 egg cells called oocytes, all of them descended from the oogonia. Each oocyte has the potential, if fertilized in the future, to develop into a complete human child, but only some 440 will ultimately ripen and make it to the big leagues—to be released during menses. Before she reaches menopause, the average fertile woman has the potential to produce about twenty children. The average man, by comparison, emits between 200 million and a billion sperm cells each time he ejaculates, and during his lifetime could conceive some 12 *trillion* citizens. Nature is profligate where reproduction is concerned. However, every single spermatozoon is descended from only one or two thousand spermatogonia that migrated into the male fetus's embryonic testicles during his mother's pregnancy.

When a girl reaches womanhood, her monthly ovarian cycle begins with a cascade of pituitary hormones that flood the ovaries, awakening about fifteen or twenty oocytes, which have been slumbering in their follicle for lo those many years. Just before the woman ovulates, each oocyte divides. Chromosomes are shared equally by the two new cells, but one cell steals most of the cytoplasm of the original cell and becomes what is known as a secondary oocyte. The other daughter cell, its days numbered, clings to the new oocyte on the way to becoming a polar body.

By grabbing the bulk of the cytoplasm, the secondary oocyte leaps into the race to become a woman's egg-of-the-month. To win, it must develop faster than its fifteen or twenty sisters so that it gets released from its ovarian follicle to embark on its odyssey, drifting toward the womb along one of the fallopian tubes, the slender ducts that point from the ovaries to the uterus. Because this cruise by the egg can take as long as ten days, only by snaring most of the available nutrients can the ovum be assured of enough food to nourish an embryo, should a sperm cell happen along and fertilize the ovum.

The polar body's fate is less dramatic. The tiny bundle of extra chromosomes drifts down to the outer rim of the maturing egg, as if excess baggage or, to Yury Verlinsky's mind, a black dot in a Miró painting.

But that dot, Verlinsky reasoned, actually represents one-half of the woman's genetic makeup. It holds twenty-three chromosomes arbitrarily cast off during meiosis, the winning twenty-three having been dispatched to the nucleus of the egg to constitute the maternal contribution to an embryo (the father's sperm cell provides the remaining twenty-three).

"When a mother is a silent carrier of a genetic disease, she is protected by her healthy copy of the defective gene," Verlinsky says. "But she may pass on either copy; the choice is random. Each of the eggs she is born with may contain either the normal gene or the abnormal, but the mature egg cannot contain both. It occurred to me that the polar body of an egg will contain the copy that won't get passed on."

To Verlinsky, that meant if he could figure out a way to examine the chromosomes in the polar body, he could deduce what copy of a specific gene was to be found in the nucleus of the egg itself, without subjecting the egg to possible lethal analysis. "If I know which gene is in the polar body," he says, "I can tell by elimination which gene remains in the egg cell.

"For example, if the woman is at risk of passing on cystic fibrosis, we look for the CF gene in the polar body. If it's there, we know the egg itself is good [because CF is a recessive disorder with each parent contributing a defective gene] and suitable for fertilization and implantation. But if we find the normal gene in the polar body, that means the egg carries the CF gene.

"The polar body is chaff," Verlinsky stresses. "It doesn't divide. It dissolves and disappears. There is no moral issue here. The only big technical issue was whether I could remove the polar body without hurting the egg." He knew that was possible because of the pioneering work of Jacques Cohen, of Cornell University, who had developed an important infertility therapy, known as partial zona dissection, used in IVF. A colorful character, outspoken and driven by nervous energy, Cohen specializes in spermatozoa. He is a magician at making the most of the few viable sperm that many infertile men are able to produce. The idea is to render the sperm's job easier by clearing a path into the egg, and Cohen has helped his male patients produce more than a dozen babies so far. "Sometimes," Verlinsky explains, "when we are dealing with men who don't have a lot of good sperms, we poke a tiny hole in the tough outer covering [the zona pellucida] of the egg, so that a sperm can swim inside more easily. When we do that, we aim the pipette to avoid harming the polar body. But sometimes we accidentally destroy it, yet everything is okay. We have been able to fertilize eggs that way and had pregnancies result, demonstrating that the egg isn't hurt by the procedure."

In Baltimore, Verlinsky and colleagues reported they had successfully analyzed five ova from a woman known to carry a recessive gene for a dangerous enzyme disorder, alpha-1 antitrypsin deficiency. Her husband also is a carrier of the disease gene. Both parents are healthy because their normal genes protect them. However, each child born of such a union stands one chance in four of inheriting both bad genes and the incurable chemical imbalance that leads to early emphysema and liver disease. AAT deficiency afflicts as many as forty thousand Americans who lack the normal-functioning gene that produces a liver enzyme crucial to the protection and maintenance of the air sacs in their lungs. By age forty, most patients have lost lung capacity. A lung transplant offers the only hope for some of them. By age sixty, only 16 percent are still alive.

After Verlinsky teased out the microscopic polar bodies from the patient's ova, the geneticist Strom analyzed the samples, looking for the AAT gene. Then the scientists inseminated all five eggs with the father's sperm. Of the three early-stage embryos that developed, only the two that came from eggs with healthy nuclei (hence defective polar bodies) were transferred into the female patient. At best, any resulting baby would have been born with two normal AAT genes. At worst, the baby would still be healthy, though a silent carrier like the parents, because it would carry the father's defective AAT gene. In this case, all the efforts were for naught: no pregnancy resulted. "This is common with IVF," Verlinsky says. "It can take a few tries before it succeeds. The key point is that we are not dealing with infertile couples. We think the success rate eventually will become much better, perhaps 50 percent or higher."

The initial triumph of preimplantation genetics did not involve polar body analysis; in fact, it was the work of an innovator who does not seem to show much interest in Verlinsky's deductive theory. He is the British pioneer Alan Handyside, of the Institute of Obstetrics and Gynaecology, Royal Postgraduate Medical School, at Hammersmith Hospital, in London, who in 1990 produced the world's first babies that had been screened genetically as embryos before they implanted and grew into fetuses. Remarkably, considering the usually disappointing IVF success rate, Handyside achieved five pregnancies in seven women, including two sets of twins, although one of the babies was stillborn, a tragedy that Handyside attributes to obstetrical complications unrelated to embryonic testing.

The self-effacing, scholarly Handyside, who looks a decade younger than his thirty-eight years, works in what he calls "a hut" nestled deep in the Dickensian bowels of Hammersmith, a huge institution that dates from the Victorian era. His surroundings are so bleak that the hallway outside his little lab was leased by a TV soap opera that is supposed to be set in a prison. "Many times, I've had to walk through a cell to go to work," Handyside says, smiling. "It's all rather foreboding." There are some Britons, though, who would like to see Handyside wearing prison stripes. Viewed as a hero to family-planning advocates and families at risk of passing on genetic diseases, he has been attacked by antiabortionists, who demanded that his work be banned. Handyside was a major protagonist of a parliamentary inquiry into the status of IVF and abortion in England in the spring of 1990. However, he was able to present not just pregnancies but real babies conceived after preimplantation analysis. Ecstatic families told their stories. Many of them had gone so far as to submit to tubal sterilizations to make sure they did not pass on any more diseases. Motions to ban IVF and embryo research in England were overwhelmingly defeated, although scientists are not permitted to experiment on embryos more than fourteen days old. "We can only keep them alive in vitro for seven days," notes Handyside, "So we're legal by a comfortable margin."

Life's earliest stages proceed with the stateliness and metronomical regularity of a baroque gavotte: as a fertilized egg divides, in thirty hours it has split into two cells; in two days, four cells; in another half day, eight cells. By three days, it has reached sixteen cells and under the microscope resembles a raspberry, a stage called the morula. By four days, although still no larger than the fertilized egg, the "pre-embryo" speck has become a blastocyst and has started to differentiate, with some cells striking off to make the placental membrane that will protect the fetus, while other cells prepare to start constructing the fetus itself.

As all this is happening, the fertilized ovum peacefully floats down its fallopian tube and by the blastocyst stage reaches the womb, where it drifts aimlessly for two or three days. However, by day seven or eight, it attaches to its mother's tissues, announces its presence chemically (to sensitive tests), and, no longer peaceful, starts to seize control of the pregnancy and dig into the uterine wall for the duration.

Because the clumpy, multicell blastocyst could give geneticists hundreds of cells to analyze instead of a scant one or two, English researchers have attempted to pioneer blastocyst biopsies. However, following IVF, only 40 to 60 percent of all normally fertilized embryos can be maintained in culture to reach the blastocyst stage, and the transfer of blastocysts has generally been unsuccessful. Consequently, Handyside tests embryos at the eight-cell stage, three days after insemination. At this early stage, the cells have not yet differentiated; all are genetically identical. Hence when Handyside removes one cell and analyzes it for genetic defects, he knows that the results will apply to the other cells, too.

Such work takes place in small, silent, sterile rooms, where the embryologists, masked and garbed like surgeons, peer through the lenses of expensive microscopes. As Handyside concentrates, his hands delicately move joysticks called micromanipulators, which reduce his movements many times and allow him to maneuver individual cells. Performing microsurgery with the deftness learned by work on mouse and hamster eggs, Handyside immobilizes the embryo with a hair-thin glass pipette while he drills a hole in the zona pellucida with a fine spray of acidified medium from another micropipette pressed up against the surface. He then pushes a larger pipette through the channel and sucks out one or two cells for genetic analysis. "Basically, we're sexing embryos," he says. "By using DNA amplification [PCR], we examine the cells for particular nonsense repeats of DNA sequences—they don't do anything, they're just junk DNA—that lie on the male Y chromosome. Such repeats occur hundreds of times and give us a big target to probe for.

"We then cull out the male embryos and freeze them and transfer only female embryos to the woman. We do this for sex-linked diseases— hemophilia, muscular dystrophy, various forms of X-linked mental retardation—and ensure that only female babies will result."

Handyside's live births of 1990 were considered to mark a milestone in reproductive biology, although he readily admits that the procedure is wasteful. All male embryos are frozen or, with the parents' permission, used for research. However, the rules of Mendel dictate that fully half of those male fetuses would be *normal*, unafflicted by sex-linked disease. Also, some of the female babies created by this method will be silent carriers, like their mothers. "When they grow up and start having babies, they will face the same problem," admitted Handyside at the time. "So we definitely needed better techniques. This was merely the first step in preimplantation genetics. We've shown that it can be done. Naturally, as more genes are found, we will be able to probe for them directly."

In March of 1992, Handyside was able to do exactly that. He achieved another milestone, and preimplantation genetics progressed to recessive disorders with the birth of Chloe O'Brien in Burnley, England. She is the daughter of Michelle and Paul, both cystic fibrosis carriers, and the sister of Martin O'Brien, who was born in 1988 with CF. Baby Chloe, a new British citizen, all seven pounds of her, arrived fourteen years after the first test-tube baby had been conceived.

Handyside, working with an international team, had the most common cystic fibrosis mutation to screen for—DeltaF508, which appears in 75 percent of all patients—and he found the gene in three couples who asked for his help.

IVF techniques were used to recover oocytes from each woman and fertilize them with her husband's sperm. Three days after fertilization, embryos in the cleavage stage (seven or eight cells) underwent biopsy and removal of one or two cells for DNA amplification. That gave the team multiple copies of the genetic material to analyze. In one woman, only two eggs had fertilized normally: DNA analysis of one of the embryos failed, and the other was diagnosed homozygous for CF (meaning the baby was destined to be born with the disease); neither, thus, was transferred.

The fertilized oocytes of each of the other two women produced noncarrier (clear of CF), carrier, and affected embryos. Both couples chose to have one noncarrier embryo and one carrier embryo transferred. Mrs. O'Brien became pregnant and gave birth to Chloe, who has two normal CFTR genes (both free of the three-nucleotide deletion) on her seventh chromosomes. She will never have to worry about cystic fibrosis in her line. Besides being very nice for her, that fact has implications that are almost cosmic.

Handyside has, in effect, already made the debate over germ line gene therapy moot. True, he did not alter the sperm or egg of a person to change a genetic trait specifically, but he *has been selecting for ones that are healthy.* And no matter how one views the choice, those selected embryos have resulted in a number of human beings from families whose germ line has been dramatically and purposely changed forevermore.

When she grows up, Chloe O'Brien will never specifically need to have her own babies tested prenatally for CF. She is no more likely to pass on cystic fibrosis than anyone else in the world who is not a carrier—the fates of her offspring, and all the generations to follow, have been altered.

The gene has left her branch of the O'Brien family. Handyside removed it in the name of beneficence. "The point is to do something for suffering people that doesn't involve abortion," he says. "I have patients who have been forced to have as many as *nine abortions*, rather than have another cystic fibrosis child."

But the real point is that he has done germ line therapy not once but several times, and there was not a murmur of dissent.

It wasn't very long ago that researchers were not even able to spot a colorless human ovum in the microscope, let alone tinker with it. Before PCR became available in 1985, DNA tests on the single cell would, by definition, have destroyed it. Genetic analysis normally could be performed only after conception and pregnancy. Today, as many as three hundred abnormalities may be picked up by the major prenatal screening techniques—classic amniocentesis, starting at sixteen weeks of gestation, and chorionic villi sampling (CVS), which gives answers much earlier (eight through twelve weeks) and involves going through the vagina to snip material from the developing fetal sac. Molecular diagnosis for birth defects began in earnest in 1976 when Yuet Wai Kan, a Lasker Prize winner at the University of California at San Francisco, showed that the slight biochemical differences among people can be identified by comparing the dark bands produced by their DNA when it is cut with restriction enzymes and put into an electrically charged gel. Kan's landmark test was for thalassemia, or Cooley's anemia, one of the world's most common hereditary blood disorders.

Quicker, easier tests have followed at many of the nation's research institutions, and the list seems to be growing by the month. In Germany, for example, researchers have been advancing this type of testing with an innovative use of microdissection and cloning techniques. A team at Erlangen University led by the geneticist Uwe Claussen employs a glass needle and a microscope that magnifies a chromosome 120 times to slice out the precise chromosomal band that contains the defective gene. The excised chromosomal DNA is chopped up by enzymes to reduce its size, and then implanted in bacteria, which multiply and eventually amplify the DNA stretch of interest some 20,000 times, so that it may be carefully analyzed.

DNA tests bear odd-sounding names—Southern blots, western blots, dot blots—but essentially genes may be tested for prenatally in both direct and indirect fashions. DNA probes can detect chromosomal breaks, mutations, or rearrangements. Because the genes responsible have been discovered, probes can readily home in on fetal DNA that will give rise to fragile X syndrome, alpha-1 antitrypsin deficiency, Huntington's disease, Lesch-Nyhan syndrome, neurofibromatosis, Marfan's syndrome, Tay-Sachs disease, cystic fibrosis, Gaucher's disease, hemophilia A and B, most of the muscular dystrophies, phenylketonuria, sickle-cell anemia, and the thalassemias. When mixed with a sample of fetal DNA, radioactive probes will fan out and scan for the defective gene and bond if they find it, marking its presence with dots on x-ray film. Such gene tests are designed not for broad population screening but for diagnosis in families known to be at risk because the particular disease has already struck. Southern blotting, less direct, named for its discoverer, Edward Southern of Edinburgh University, involves cutting the fetal DNA sample with restriction enzymes. The resulting fragments will be of different lengths, depending on whether the defective gene is present. Sickle-cell anemia has become the model for this type of testing because the DNA glitch in a hemoglobin gene lies in the middle of a site that would normally be cut by a restriction enzyme. The gene sequence is wrong at the site, so the enzyme will not cut it, resulting in a long fragment of 1,350 base pairs, rather than in the two short fragments that would normally occur.

When a disease-causing gene has not yet been found, as was long true in Huntington's disease, its presence still may be tested for by indirect linkage, which involves the analysis of how one or more genetic markers—restriction fragment length polymorphisms—get passed down through an individual family. By studying DNA fragments from affected and nonaffected family members, scientists are able to find patterns that differ. If fetal DNA shows the same pattern as that of affected family members, it is likely that the baby will be born with the defective gene. DNA probes also are helping doctors quickly identify viruses—including hepatitis A and B viruses, cytomegalovirus, adenoviruses, herpes simplex virus, measles virus, Epstein-Barr virus, varicella-zoster virus, human papilloma virus, leukemia-causing HTLV-1, HIV (AIDS), and rubella virus. Probes can identify many bacterial strains. And DNA tests are in the offing (they're currently used only in research settings) that can distinguish the specific genetic changes that make cells turn malignant, as well as spot the presence or absence of genes that help the body protect itself against cancer. Pathologists privately report that some assays, especially in leukemia and lymphoma research, are now sensitive enough to detect a single cancer cell in the midst of millions of normal cells, merely by a check of the bloodstream.

Genetic testing per se has been limited to a handful of rare disorders, although some form of prenatal testing is becoming increasingly common as working couples wait until their thirties and beyond to start their families, despite the higher risks of birth defects that accompany advancing maternal age. Advances in DNA testing reflect the determination of scientists to examine pregnancy as early as possible, and the two fastest growing childbearing-age groups in the United States are teenage girls and women over 35. No longer are doctors recommending that older women not have babies. For women who want to wait, the news is good. Genetic diagnosis and careful surveillance during pregnancy are allowing older women to have healthier children. Women are taking advantage of the new technology, and it works. A study of 1,328 women who were cared for at the New York Hospital–Cornell University Medical Center found that pregnancy complications were no more frequent in women aged 35 or more than in women 20 to 34 years old, although the older population tended to have more cesareans (24.5 percent, compared with 15.2 percent in younger patients). Moreover, K. L. Ales, M. L. Druzin, and D. L. Santini reported that the children of the older mothers were as healthy as those of the young mothers. "This trend," they wrote in the *Journal of Surgery, Gynecology and Obstetrics*, "was related, in part, to the choice to terminate the pregnancy by women with fetuses that had documented chromosomal anomalies.

"We conclude that advanced maternal age was not associated with an excess of adverse pregnancy outcome and suggest that . . . women aged 35 years or more can experience excellent pregnancy outcomes."

Carrier detection screening among Jews of eastern European origin for Tay-Sachs disease has generally been viewed as the first successful example of mass genetic screening. This program is designed to identify carriers of the recessive gene that can result in a child that is doomed to suffer increasing mental retardation and paralysis and usually die before the age of five. Although such testing has been voluntary, Hasidic Jews in Brooklyn, who oppose abortion and whose religious laws proscribe contraception, have virtually eradicated the disease among their group by some clever machinations.

Many Hasidic marriages are arranged by brokers. The matchmakers have compiled computerized lists of carrier status, and although the data are kept confidential—with prospective mates assigned numbers, not names—marriages between two carriers of the lethal recessive are *simply not arranged*. By the early 1980s, more than 310,000 Jews worldwide had been screened for Tay-Sachs and 286 couples identified as carriers of the lethal gene. Yet even this program has drawn fire from some critics who feel it causes needless anxiety and possibly stigmatizes people unlucky enough to carry the gene. There is no doubt, however, that Tay-Sachs screening and the attendant publicity informed millions of Jews about their vulnerability to a horrible disease that many had never heard of, and the programs have earned the support of Jewish communities in the United States, Canada, Great Britain, Israel, and South Africa.

The current issue that has been tearing geneticists apart involves the cost-benefit ratios surrounding the possibility of screening the entire Caucasian population for its most common hereditary disorder, cystic fibrosis, as a result of the discovery of the CF gene, in 1989. Several biotech companies are offering tests, even though the genetics community has come down firmly against their introduction. It had first looked as if carrier tests could pick up the simple mutation that caused most cases of the disease—an estimated 75 percent of the people who carry the gene have the mutation, while the rest have different mutations in the same gene. However, a crash program by seventy laboratories worldwide to catalog the remaining mutations itself crashed in 1990. It began to appear that the CF gene featured a plethora of possible mutations—hundreds, even—and different mutations may occur in the same person. Until all this is nailed down, CF researchers agree that widespread screening makes no sense, as it would detect only 75 percent of the carriers, and thus only half of the couples at risk of having a child with cystic fibrosis.

Yet, even if the CF tests were 100 percent accurate, screening the entire white population would be impossible, according to Neil Holtzman of Johns Hopkins, an expert in genetic screening. In fact, merely screening expectant mothers to determine whether their fetus had a chromosomal disorder or whether they themselves were carriers for just four serious disorders—cystic fibrosis, sickle-cell disease, hemophilia, and Duchenne's muscular dystrophy—would entail more than 4 million tests a year, Holtzman predicts. About 46,000 of these would be positive and require a follow-up. Yet the nation's genetics centers can serve no more than 500,000 people a year—not 4 million—and only one hundred clinical geneticists and about seventy-five genetic counselors are entering the profession annually.

Regardless, discovery of this gene may soon place the nation's physicians in an uncomfortable position. The first lawsuit stemming from a doctor's failure to tell the parents of a newborn with CF about the possibility of screening could open Pandora's box.

Nor will genetic screening rid the world of birth defects. Millions of tragedies are due to new mutations, like many of those in families with several normal children that are suddenly struck by muscular dystrophy. Although he has written extensively about the possible misuse of DNA testing, Holtzman argues that many of the proposed tests have been oversold and would not have a high predictive value. For example, DNA tests are designed to pick up defects caused by a single gene. "But the major disorders are polygenic, involving the interaction of many genes with each other and with the environment," Holtzman points out. "Such interactions are not even dimly understood."

Yet advances in genetic screening are making possible a new era of "preconception care," according to Aubrey Milunsky, director of the Center for Human Genetics at Boston University. "Typically, women now wait eight to ten weeks after conception before seeing a doctor," Milunsky says. "By that time, a fetus's fate has been basically sealed. Instead, women should be evaluated before becoming pregnant to determine their genetic risks, and what those risks may mean to their offspring. Only by knowledge can we have a chance of controlling our medical destiny."

Or perhaps it might be better to avoid those genetic risks entirely. This is the dream that keeps preimplantation pioneers like Alan Handyside, Jacques Cohen, and Yury Verlinsky poking away in the earliest hours of an embryo's life. Such scientists have created a model that has the potential to eradicate any disease caused by a single defective gene by selecting eggs that are healthy. Its major weakness hinges on its reliance on IVF. But the truth is that the research also opens the door to matters much more volatile.

Verlinsky and others may have provided a means by which a couple could select their offspring at the embryo stage for specific genetic traits unrelated to health. Babies may already be selected for gender, obviously. Perhaps someday they could be chosen for talents or certain desired physical characteristics, as the genes responsible for such things are unearthed by the Human Genome Project.

And as Verlinsky has learned to his chagrin, his preconception manipulations also raise the specter of the greatest genetic-engineering folly of all: human cloning.

The model already exists in animal science, where experimentation at the embryonic level is considered commercially essential; the lessons learned here quietly make their way into human medicine a decade or so later. The state of the art may be viewed just 125 freeway miles north of Verlinsky's Chicago lab in the laboratory of the reproductive physiologist Neal First at the University of Wisconsin at Madison. Using the same equipment as Verlinsky—the expensive microscope and micromanipulators—First's technicians routinely work with thirty-two-cell bovine embryos, each of which mingles the genes of a prize bull and a prize cow, both Holsteins, though strangers that never met. An artificially inseminated embryo is flushed from its mother's uterus at the tender age of six days; at this stage, it has not yet differentiated. Each of the thirty-two microscopic cells thus represents Total Bovine Potential; each can become an entire calf—or even be viewed as a latent herd or perhaps even as a new breed—if it later is determined that the embryo has the right genetic stuff for stupendous milk and beef production. Once the cleaving embryo is retrieved from its mother, the procedure—which took a decade to develop—becomes mechanical and rather boring. A technician merely teases away one of the embryonic cells of the fertilized ovum and carefully transfers it under the tough outer covering (the zona pellucida) of an egg that had been donated by a hapless ordinary cow, which met her maker at the slaughterhouse a few hours earlier. (This lowly animal's genes were sucked out of the egg nucleus, leaving just the shell.) A suction needle holds the scrap ovum in place, as a tiny hole is poked in the zona with a hollow glass needle containing the embryonic cell from the "superior" mating of the other two cattle. Next, the cell is ejected so that it plops out and lands on the donor zona. Then a microjolt of electricity briefly opens the pores of the plasma membrane, fusing the embryonic egg cell inside. From this point on, the slaughterhouse egg obligingly takes over the pregnancy, treating the transplanted cell as if it were its own.

By going back to the original "valuable" embryo and transplanting single undifferentiated cells into scrap eggs, the technician can ensure that each transplanted cell will be an exact genetic carbon copy—a clone—of all the others. In the meantime, the original thirty-two-cell embryo keeps dividing, replacing each departing cell with a new one. It is biology's microscopic version of a calf factory.

After transplantation, the embryonic cells in their donor eggs are encapsulated in tiny chunks of waxy agar growth medium and incubated for a few days in—of all things—the oviduct of a sheep, where growth routinely continues. When the individual embryos have themselves divided into thirty-two-cell masses, they are surgically withdrawn from the sheep and transplanted into the wombs of surrogate mother cows; 280 days later, if all goes well, a squadron of custom-built calves will be born. Cloning in this manner does not make a carbon copy of either the prize cow or the prize bull—the usual notion of a clone to laymen—but it does result in genetically identical copies of their offspring, and First has been working hard to increase the efficiency of the procedure, which succeeds less than 40 percent of the time. The first calf produced by this method was born in 1986, and since then several generations of cloned animals have been created. Cloning does not work very well yet, but it does work, and represents the latest in animal engineering. And back in Chicago, Yury Verlinsky uses the identical micromanipulations to tease away a cell from a four-cell human embryo in order to check for genetic disease before IVF.

It is, however, extremely difficult to extract a single cell from a human pre-embryo for gene analysis. Considerable skill is required to poke a glass needle into a tiny human egg that has grown to a scant four cells, and pluck out one of them. During an attempt a few years ago, as Verlinsky withdrew the cell he was after, two of the other cells stuck to his pipette and were pulled out, too. That left a single, lonely cell still in the egg. As his partner, the geneticist Strom, performed PCR analysis on the extracted material, a worried Verlinsky hovered near the incubator overnight to see what happened to that remaining cell. To his relief, it merely kept dividing and soon was back to four cells. But, in point of fact, this was truly astounding! It meant that up to four cells, at least, a human pre-embryo acts precisely like a cow pre-embryo. And, contrary to previous thinking, human eggs seem hardy enough to withstand such drastic micromanipulation.

Verlinsky has even come to believe that fertilized human eggs may lend themselves to genetic duplication almost as readily as cow eggs do. It may be possible to clone them, at least a few times. The notion understandably disturbs a lot of scientists. When Verlinsky described his preimplantation strategy at a scientific meeting in Australia a few years ago, many colleagues accused him of opening the door to cloning.

Then, as word got around, Verlinsky started getting calls from journalists demanding to know whether he was cloning people in Chicago. "Preposterous!" he hollered. "It's crazy . . . ghoulish . . . why would anybody do that?"

But the calls continue.

If anyone wanted to clone humans, this cloning strategy would not involve making a carbon copy of an individual from a body cell. That still is impossible, although it works with carrots and frogs sometimes. The problem with cells taken from someone's body is that their destinies have already been determined. They have differentiated, and there exists at present no way to make them revert back to the pre-embryonic period when their ancestors consisted of just four cells, each of which was capable of multiplying into a clone.

Moreover, Verlinsky says that his recent work with pre-embryos indicates that fertilized human eggs are turning out to be a lot fussier to work with than cow eggs, which may be cloned indefinitely, at least in theory. To make a complex biochemical story short, various factors suggest that a human egg would not permit itself to be duplicated more than a few times. "After that, the egg would die," Verlinsky predicts. "Perhaps we might consider this nature's way of protecting herself against human cloning."

But it might be possible to clone sets of human embryos. Oh, yes. That is a real possibility. The marriage of IVF and genetics has created a way to make exact genetic duplicates of embryos resulting from a man's sperm cell and a woman's egg. "A couple might be able to clone a child they had conceived," says Verlinsky, nodding vigorously. "They perhaps could store identical pre-embryos in the hospital freezer and make visits periodically to augment their family."

In this way, identical twins or triplets might result that—and the possibility strains the mind—would be separated by *several years*. Why would any couple want to do that? Well, one might ponder also the biological drives and social conditioning that forces couples to endure the humiliation, loss of privacy, pain, and monthly heartbreak of infertility therapy, not to mention the great costs and poor odds of IVF. And at least one leading ethicist, Georgetown's LeRoy Walters, can see no harm in banking genetically identical pre-embryos. "It would be as close as you can get to sort of bringing your dead child back to life," Walters says.

A less farfetched rationale was put forth, diffidently and without publicity, by scientists who presented a paper on October 13, 1993, at a meeting in Montreal of the American Fertility Society. The team, led by Jerry Hall, director of the In Vitro Fertilization and Andrology Laboratory at George Washington School of Medicine, in Washington, D.C., precipitated an ethical crisis a few weeks later when word got out that they had conjured up a workable technique to enhance the success odds of IVF by splitting embryos into twins or triplets or quadruplets before implantation, thus allowing doctors to transfer more embryos and giving women a better chance of becoming pregnant with at least one of them. The process would increase the chances of multiple births, which IVF does anyway because women are primed with drugs to produce as many eggs as possible for fertilization and implantation. Hall and colleagues had not yet employed cloned embryos for this purpose, nor would they even consider it until society set guidelines for the research, but they had accomplished the feat in tissue culture by means of embryos that were deemed unfit for implantation in IVF because they had been penetrated by multiple sperm and had extra sets of chromosomes.

The technique represented a marked improvement over animal cloning, in which individual embryonic cells must be fused with unfertilized eggs from which the nucleus has been removed because each cloned embryo must have its own intact zona pellucida, the protective jellylike outer covering that is critical for development. In earlier experiments, the George Washington researchers had found they could accomplish the same thing by coating embryonic cells with a synthetic chemical zona, thus avoiding the necessity of using other human eggs, which are considered far too valuable to have their nuclei gutted for such a purpose. The advance certainly seemed to bring human embryo cloning closer to reality.

In the experiment reported in Montreal, the researchers cloned seventeen embryos composed of from two to eight cells, separating the individual cells and coating them with the artificial zona, and placing them in a nutrient broth. Soon forty-eight new embryos (an average of three clones from each original) were growing without fuss and were left alone to see how far they would develop. Interestingly, the smallest embryos did better—those originally cloned at the two-cell stage reached the thirty-two cells a normal embryo must become before it tries to implant in the uterus. Clones from the eight-cell embryos made it to only eight cells; those from four-cell embryos stopped dividing after sixteen cells.

News of the experiment made worldwide headlines, with opinion being sharply divided, as one might expect, but after a few days the controversy fizzled away. Society, it seemed, was becoming immune to the stupefactions that were emerging from the various nooks and crannies of genetic engineering, although pioneers like Yury Verlinsky could only wonder what lies in store next. At this prospect—the first cloning of human beings—Verlinsky could do little more than raise his eyes to the heavens and shrug. "I can't deny that the possibility now exists."

23

THE ULTIMATE FRONTIER

Every day you get older.
It's a law....
—Butch Cassidy and the
Sundance Kid (1969)

When time stood still, it happened in, of all places, Waukegan, Illinois, the working-class city on Lake Michigan, where Jack Benny was born and the fantasist Ray Bradbury grew up. Bradbury immortalized his bittersweet midwestern nostalgia in *Dandelion Wine*. But to sixty-five-year-old Fred McCullough, what occurred there was more out of *The Illustrated Man*. Somewhere deep in McCullough's body, the hands of the genetic clock that ages all living things stopped . . . and began ticking backward. To his astonishment and delight, the retired auto worker found himself growing *younger*, not merely getting his body in shape but actually regaining the muscular leanness, organ fitness, and physical vigor he had lost over the preceding twenty years.

"I started feeling the changes after three months," recounts the friendly, dignified man with short, clipped white hair. "It's hard to explain. My shoes became harder to put on. Suits got small, too, as my muscles packed up, but the bathroom scale didn't show I was gaining weight." He pauses. "My wife is fifty years old—fifteen years younger— and I still could get up with her at 5 A.M. and keep up with her all day long. The main thing was I felt much stronger. . . ."

Such a burst of renewed manhood might have blown the mind of a more uptight individual, but McCullough isn't plagued by self-doubt. While undergoing a routine medical checkup at his local veterans affairs hospital in 1989, the former paint sprayer for the old American Motors Corporation plant just over the border in Kenosha, Wisconsin, was told by a male nurse of some strange but tantalizing goings-on. Curious,

McCullough made inquiries, underwent a blood test, and with few twinges of angst volunteered as a human guinea pig in what has become a medical experiment without precedent—the first no-holds-barred scientific attempt to reverse human aging.

Soon McCullough was receiving regular injections of recombinant human growth hormone (HGH), the critical muscle-maintaining substance that his body had stopped making. The idea belonged to the late Daniel Rudman, a veteran endocrinologist and professor of medicine at the Medical College of Wisconsin in Milwaukee and associate chief of staff of extended care at the Zablocki Veterans Affairs Medical Center there. Rudman believed that the brains of about a third of all middle-aged men eventually lose the ability to produce this hormone, causing them to age the fastest and end up the frailest.

For many years, growth hormone synthesized from scarce cadaver pituitary glands—the only source—had been used with success to treat a few fortunate deficient children otherwise doomed to be dwarfs. However, in the 1980s when biotech companies learned to synthesize HGH in bacteria, it suddenly became available by the truckload, although expensive. Treatments cost about $1,200 a month, which translates to roughly $10,000 an inch for youngsters who otherwise would be unable to reach their genetically preordained height.

One of the first grown-ups to receive HGH, McCullough had seemed in excellent health and kept fit by riding his bike, rowing his boat, and swimming three days a week. Still, the hormone soon revitalized him. He began to feel better than he had in years. Mind games? Self-delusion? After all, McCullough knew that every week he was getting youth shots in his buttocks from his wife, Rita Mae, who had been trained to administer them. The placebo effect?

He thinks a moment. Then he shrugs his shoulders, shakes his head. His eyes turn wistful. "I felt like a teenager again," he decides. "I mean, man, *I never felt so strong in my life!*"

The landmark experiment in progress at the Milwaukee and North Chicago Veterans Affairs medical centers has been designed to rouse the sleeping genes in aging people that once kept them young, lean, and muscular. It is a little-known aspect of gene therapy that has drawn into the fold a group of scientists who call themselves "biogerontologists" and who spent the preceding twenty years out on the fringe of biology, which is peopled by far too many kooks, crackpots, and diet faddists.

Although calcium supplements and estrogen replacement are being widely used to slow the biological clocks in postmenopausal women, the HGH venture marks the first practical application in men. In a broader scientific context, researchers in Rudman's mold are trying to retard—or even reverse—the aging process by cracking the genetic riddle that accounts for the millionfold variations in the life spans of species of higher animals and plants. Among these pioneers is the neurobiologist Caleb E. Finch, a bearded Yankee who teaches at the University of Southern California, runs a research group, directs a multi-university Alzheimer's disease center, serves on a number of national advisory panels, and plays in bluegrass bands for relaxation.

Finch has long been fascinated by the incredible diversity of life spans in nature. For 1,200 million years bacteria and protozoa have never aged, for instance, because they reproduce by splitting in two. Fruit flies, on the other hand, live for only three weeks, human red blood cells for 120 days, elephants for 70 years. Some living things, such as sponges and sea anemones, appear not to age at all. Bristlecone pines live 5,000 years, according to radiocarbon dating, but the world champs are the creosote bushes basking in California's rugged Mojave Desert that stretch back in time to the last Ice Age, some 11,000 years ago.

Finch has avidly observed his colleagues' attempts to meddle with natural life spans. Biogerontologists have, for instance, altered the reproductive saga of pacific salmon, which age quickly and die after their heroic spawning runs because of a prolonged elevation of steroids. Their fate can be prevented by castration, which removes the powerful urge to spawn and triples the length of their lives. Equally curious has been the case of the honeybee, in which a fertilized egg that would normally result in a worker bee can be turned into a long-lived (five years) queen merely by adding a hormone called the "queen substance" at an early developmental stage.

How animals respond to such manipulation is significant because humans share the same genetic material. For example, 99 percent of the genes found in men are also found in chimpanzees. Chimp genes, of course, express their chemical activity differently, which is why a chimp looks and acts like a chimp. But not one of those genes has been found to be different—not one hormone, not one enzyme.

Hormones change the activity of key genes, Finch says, by either speeding up or suppressing their mechanism. "It's not the chemistry that's changed, but merely whether the switch is turned to high, medium, or low," he explains. Growth hormone, Finch believes, "switches an off-gene on." Conversely, the human female reproductive tract atrophies after menopause because the lack of estrogen shuts off genes.

"When women are given supplementary estrogen—which can be viewed as a form of gene therapy—it turns genes back on."

In the early years of human life, genes seem to specialize in growth and development, but as the body attains maturity, many genes abruptly change their function to replacing, repairing, and maintaining cells. Such remodeling, called protein degradation and turnover, replenishes proteins damaged by metabolic by-products and other environmental hazards. New research is reaching amazing conclusions. It is suggesting that deterioration does not seem to be built into every cell and that many body parts show no ill effect at all with age. Some, like certain blood cells, are replaced every ten days, but muscle cells last for years, and blood-making stem cells show no signs of aging, nor do sperm-producing cells and those lining the intestinal tract. Still others, like heart and brain cells, last a lifetime, undergoing perpetual rebirth from within, not constant loss, as has been heretofore assumed. Work by Earl Stadtman, chief of the laboratory of biochemistry at the National Heart, Lung, and Blood Institute, indicates that each protein that goes into the membranes and inner workings of a human cell is replaced *every six days* on the average—a remarkable finding, as we shall see.

As a neurobiologist, Finch concentrates on the starring role hormones play in this drama. After decades of research, he believes that though our genes ultimately determine how we age, the process is carried out by hormones secreted mainly by the pituitary and adrenal glands and controlled by some as yet unidentified master brain clock located in the hypothalamus, a pea-sized cluster of cells situated a bit behind a spot midway between your ears.

The hypothalamus, a hormonal command and control center, is known to regulate hunger, rage, sleep, sexual desire—lots of things. A few years ago, Finch detected significant decreases in neurotransmitter chemicals in the hypothalamus tissue of old animals compared with those of young ones. Further experiments led him to hypothesize that the dwindling serves to initiate what he calls an "endocrine cascade," in which one hormone transforms the action of the next, and so on. Tinkering with this cycle of derangement, Finch has been able to trick old ovaries into working again by transplanting them into young animals or by "jump-starting" them with drugs.

But it was a more majestic effort—Finch's 1990 compendium of the mechanisms and rates of aging in thousands of organisms, *Longevity, Senescence, and the Genome*—that is likely to prove to be as momentous to the biology of aging as Victor McKusick's *Mendelian Inheritance in Man* has been to medical genetics. Finch's big book, in fact, celebrates a marriage of the two fields, for Finch has detected a "180-degree change" in the attitude of biological researchers toward the genes that underlie longevity.

"Until very recently," he says, "most people assumed that all cells in the body were destined for one type of senescence or another. In other words, it was the fate of somatic cells to degenerate and age. That was rather a hopeless view of things.

"Now, we're more optimistic. There's abundant evidence that cellular and molecular degeneration during aging is highly selective, and varies among different parts of the body. We now have found specific genes, specific hormones, specific cells. A relatively small number of changes are involved in any species, and large components of the body don't show any deterioration at all with age."

Such revelations mean that the genes that influence human life span can be identified, and Finch expects that, if current lines of research prove fruitful, "many aspects of senescence should be strongly modifiable by interventions at the level of gene expression."

Fred McCullough offers a case in point. Pathologists for more than a hundred years have noted that with advancing age the size of our internal organs tends to shrink. Brain, liver, lungs, kidneys, spleen—organs that help constitute what is known as lean body mass—all get smaller. At the same time, many physiological functions—cardiac output, kidney function, breathing capacity, and so on—decline as well. It has been speculated that the weakening of function is related to the decrease in size of the lean-body-mass organs and is due to genes that stop working or fail to receive the proper chemical messages from the aging brain. The only body component that does not decline with aging is fat; as lean body mass and its constituent organs shrink decade by decade, we get fatter and more sluggish—fat in men usually settles in the belly; in women, around the hips. Daniel Rudman was the first to try to interrupt this fact of life by giving growth hormone injections that, he believes, primarily directed the exchange of fat for muscle in his volunteers.

Growth hormone is pulse released by the pituitary in response to exercise, fasting, low blood sugar, sleep, trauma, stimulations to the brain's dopamine receptors, and other factors. Such growth, or trophic, hormones regulate body repair by streaming into cells, helping maintain both structure (muscles and bones) and functioning, including the performance of major organs. Human growth hormone promotes protein synthesis and cell division and, as it breaks down fat cells, simultaneously slows the breakdown of proteins. Healthy men and women produce this hormone all their lives, not just during growth spurts; but ac-

cording to Rudman, everyone goes through "a growth hormone menopause" in varying degrees, and some people stop producing any at all. If
you're thirty or under, you can trigger the hormone into action by exercise that reaches a brief peak of sustained muscular effort—jogging usually won't do it, for all its other benefits. Evidence from the government-
sponsored Baltimore Longitudinal Study on Aging, established in 1958
as the first research effort to examine the effects of aging in living people
instead of dead ones, shows that the growth hormone release from exercise ceases at around the age of thirty.

Rudman's first salvo against aging has worked better than anyone
dared to dream: the bodies of McCullough and twenty-seven other male
volunteers aged sixty to eighty actually began to grow more youthful
after weekly injections. McCullough's flabby skin grew taut. His soft
muscles hardened. Fat melted away. Internal organs shrunken by age
resumed their youthful size and vigor. As Rudman's team meticulously
weighed, probed, and employed skin calipers to document the progress
of their first volunteers, it compiled a spectacular 10 to 20 percent average increase in arm and leg muscles, a 14 percent decrease in fat deposits, 10 percent growth in the liver and in skin thickness, and a whopping
22 percent increase in the size of the spleen, the crucial weapons factory
of the immune defense system.

Doctors had never seen anything like it before. Rudman explained,
"We were taught in medical school that the severely shrunken bodies of
frail, elderly people were an unavoidable manifestation of the aging process. Now we know that this is not an unavoidable aspect of aging. At
least in part, it is due to the loss of human growth hormone and you can
prevent it by giving back the hormone to these people."

Rudman specialized in deficiencies that lead to weakness, poor
health, and loss of independence in older adults. His work, which became a personal crusade after he stood by helplessly and watched his
own parents grow frail and waste away in nursing homes, brought world
renown to this previously unheralded Milwaukee gerontologist. Aging
sheiks, kings, and other potentates were constantly telephoning him
and begging to cut a Faustian deal. But Rudman politely put them on
hold—it's too early yet; youth prolongation is still in its infancy.

Nevertheless, in effect, biogerontology has already managed to pull
off what only the Spanish explorer Ponce de Leon's mythic "springs of
youth" might have done: Rudman's patients lost fifteen to twenty years
of their physiological age after only six months of treatment!

Obviously, such experiments have their critics. Many traditional
gerontologists of the grow-old-with-grace school scowl at work such as

Rudman's, not to mention Caleb Finch's. They warn against trying to fool Mother Nature or tamper with Father Time and sternly quote the Greek myth of Eos, goddess of the dawn, who, after taking many lovers, including the giant Orion, married Tithonus, a mortal. Worried about her husband's frailty, Eos begged Zeus to make Tithonus immortal, but forgot to ask her boss to grant Tithonus eternal youth as well. Tithonus paid cosmically for her blunder. He grew forever withered and shriveled, thin-boned and rheumy-eyed, senile and incessantly babbling, which finally annoyed the gods so much that they changed him into a cicada.

But if the minds of researchers on aging run to fantasy at all, it is to those rejuvenated oldsters granted a second chance at life in the film *Cocoon*.

In 1991, Rudman began yanking his frisky vets back to reality: the injections were stopped after a year to see what would happen. Sure enough, their bodies began to age, suggesting that the growth hormone had indeed temporarily awakened dormant genes that had once helped keep their organs youthful. And now the genes were asleep again.

Yet the first cadre of trailblazers had been preparing for the inevitable from their first injection. For the record, at least, they seemed philosophical about growing old again. "I don't mind. It was wonderful while it lasted," says McCullough. Maybe I can try it again someday. But the experiment involved only men volunteers—we were all military vets. I think women should have a chance to experience what we did."

All the men are being followed to see whether the hormone has any aftereffects (none so far), while Rudman recruited a fresh crop of 150 debilitated male military veterans over sixty for a double-blind, placebo-controlled study that will require them to live at the hospitals for its duration. This time, doctors will also give the volunteers either the muscle-building male sex hormone testosterone, whose levels decline in many aging men, or a mysterious unisex hormone that seems right out of the pages of Ray Bradbury's books—DHEA (dehydroepiandrosterone), the most abundant steroid in the bodies of men and women alike.

DHEA is secreted by the adrenals and then converted into hormones that are used almost everywhere—the testes and ovaries, placenta, lungs, skin, brain. After the age of twenty-five or thirty, DHEA levels drop progressively, so that an eighty-year-old produces only 5 percent as much DHEA as a young person. Significantly, declining DHEA levels have been documented in families with histories of heart disease and breast cancer, and have consistently been associated with the increased risk of death from any cause—high blood pressure, serum cholesterol levels, obesity, fasting plasma glucose level (diabetes), smoking, even AIDS.

Although DHEA was discovered in urine back in 1934, it remained a mystery until fairly recently. For fifteen years, a studious, low-key Temple University microbiologist named Arthur Schwartz has been investigating the many faces of DHEA and proffering them to humanity, currently as one of the world's first cancer preventatives—it probably will be approved by the FDA for colon cancer in at-risk families. "In animals, the evidence against cancer is indisputable," Schwartz says. "If the same thing happens in humans, we really will have something!"

When veterinarians at the University of Wisconsin gave DHEA to rodents, it seemed to block development of the most common cancers—lung, colon, liver, skin. When Greg MacEwen administered it to patients at his fat-dog clinic, he discovered that the canines can eat as much as they want and still lose weight. Impressed, MacEwen occasionally takes the hormone himself, not in order to stay slim but to combat stress. Although he does not feel anything afterward, MacEwen cites research showing that DHEA is a buildup hormone that counters the tear-down, fight-or-flight hormone cortisone, which we all use for bursts of energy. Cortisone virtually pours from the thymus gland when mice are placed in wheel-like drums and spun round and round, but DHEA nullifies its effects. "The more you get into DHEA, the more it sounds like snake oil," MacEwen admits. "It's working here. It's working there. Omigod! But I think this hormone retards the aging process. It may act as a buffer that regulates or retards the development of the diseases associated with aging."

Other biogerontologists, however, believe that aging may be the price we pay to protect the body against those diseases, particularly cancer. "Every cell in your body is a stick of dynamite—if it turns neoplastic, you're a goner," points out V. K. Cristofalo, director of the Center for Gerontological Research at the Medical College of Pennsylvania, in Philadelphia. "In order to survive as a species, we had to evolve mechanisms that would allow us to control cell division long enough to reproduce."

Cristofalo studies how cells age. He believes that nature doesn't give a hoot about senescence. In fact, he says, "nature probably selected for longevity. To have a long life, and pass on your genes lots of times, evolution had to select for mechanisms that keep us from getting cancer."

The rat, for example, has the same amount of DNA per cell and the same mammalian physiology as the human, but ages thirty times as fast and gets cancer thirty times as much. "That suggests that our protective mechanisms are different."

If Cristofalo is right, a nation of Dan Rudman's lean and lusty eighty-year-olds might not be such a good idea.

Nature may have had good reasons for cutting down on such powerful chemicals as testosterone, DHEA, and growth hormone as people age. For one thing, they may need more fat to keep warm, and supplementary growth hormone, which affects specific organs only, may make them more prone to pneumonia. The possibility still appears to exist that any potent hormone that flows into aging cells screaming "Shepherds, awake!" might panic the cells and spur the development of cancer in the elderly.

From an embryo's first stirrings, genes seem to orchestrate the aging process, regulated by mysterious biological metronomes that use hormones to transmit the passage of time. Some metronomes click off metabolism in milliseconds; others *tick* our daily sleep-wake cycle; while still others *tock* our months, years, and decades—childhood, puberty, menstruation, pregnancy, midlife, menopause, retirement. The governance and periodicity of genes through time has heretofore been considered irrevocable.

Many biologists, though, are increasingly viewing the aging process not as orderly or precise but as actually quite random and capricious.

Most people in our society die from the failure of a single organ, or just a few causes, suggesting that the number of genes involved may likewise be relatively small and, if understood, controllable. The strategies being designed for such intervention are not directly aimed at eliminating all the diseases of later life. Even if those illnesses could be wiped out, recent studies from the University of Chicago predict, the average American's life expectancy would still not exceed eighty-five years, because other killer diseases would probably come to the fore. On the other hand, if one looks at the rates of molecular aging in various kinds of tissues—heart, arteries, joints—as scientists have the capacity to do now, then the potential for genetic changes becomes much greater and the campaign to forestall them more difficult.

But, thanks to gene splicing, biogerontologists are at present able to investigate whether aging hinges on any particular innate program or is merely *mechanistic*, a reflection of the relation between the destruction imposed by time and the genetic repair mechanisms of tissues, cells, and molecules. It may be that sooner or later the balance between destruction and repair simply tilts toward the former and an older person becomes less able to fix accumulated damage. Whereupon unrepaired damage elicits more damage, until the system finally succumbs to some age-related degenerative disease or just collapses.

The hormone experiments, science's initial attempts to reverse human aging, really form only the tip of the proverbial iceberg. Myriad revolutionary drugs, treatments, exercise regimens, and diagnostic tools are undergoing testing. As we have tried to show in this book, classic disorders of aging—cancer, heart disease, Alzheimer's disease—are being deciphered at the fundamental, molecular level. Likewise, scores of new pharmaceuticals either in clinical trials or under development promise to circumvent the hormonal deficiencies associated with senescence, while others aim at maladies associated with the aging brain and may in the process enhance learning and recall.

Radical concepts of nutrition and fitness—no longer the purview of faddists—are being studied by the scientific establishment and tested on volunteers to stave off diseases that fell us when our aging body's protective army falters. In the meantime, gene splicers are closing in on the genes that underlie the fantastic effects on longevity of calorie restriction in laboratory animals—the only regimen that extends the maximum life spans of species—to see whether it will work in humans.

All this could scarcely be happening at a more opportune time. The pursuit of good health and fitness has become a $40 billion industry in the United States, with about a third of the public on a diet at any given time. Sometimes this zeal can go too far—witness the nation's three to five million anorexics and bulimics—but most Americans no longer consider it vain to struggle to look good and feel young. One in five new book titles, for instance, deals with food, diets, and health, though until now there has been little authoritative information on which to base recommendations about what people should or should not do. What they're all *trying* to do, whether they recognize it or not, is to coddle, preserve, and alter their genes.

Only recently has it become apparent how much impact one's lifestyle can have, how dramatically the environment can intervene at a fundamental biochemical level. The realization has dawned on science because, without any high-tech genetic meddling, grueling workout regimens, or special diets, advances in nutrition, antibiotics, and prenatal and maternal care are letting most Americans stick around a lot longer than they used to. The average life expectancy in the United States has risen to seventy-six years, from forty-nine at the turn of the century, a 55 percent increase. The implications of this caught science by surprise, for they suggest that the genetic and environmental factors that determine longevity are subject to drastic changes. Hence, a wonderful irony seems to be emerging from the explosion in research into the genetic causes and effects of senescence: contrary to what Butch Cassidy thought, there may be no immutable biological law that decrees that human beings have to get old and sick and die.

"There is no clear reason why aging starts to occur," declares the Stanford University biochemist Elliott Crooke, who is studying the genetic "on" switch that proteins use to make cells divide and replicate. "By design, the body should go on forever."

Yet to many younger gene splicers, such modifiying factors as hormone supplements, diets, and exercise typify the thinking of a somewhat desperate old guard of scientists.

"There's a new guard, and I'm in it," says Michael West, a thirty-eight-year-old molecular biologist at the University of Texas Southwestern Medical Center, in Dallas. "In the next very few years, we will fully characterize the genes that regulate the aging of cells," predicts West, who claims to have already found two of the "ultimate genes" that cause cellular aging. "Then you'll see an aggressive application of that understanding to age-related diseases like atherosclerosis, coronary artery disease, aging of the brain, and other maladies like osteoarthritis and skin aging."

By studying skin cells, lung cells, and endothelial cells (those that line the inside of blood vessels), West has discovered a pair of "mortality genes"—M-1 and M-2—that speed aging in human cells. Many types of cells have such aging clocks built into them, he feels. "When cells grow old, a genetic program causes them not only to stop dividing but also to alter the expression of other genes that allow development of age-related disease in the tissue."

In skin, for instance, West found genes that make enzymes called collagenase and stromelysin. "Collagenase destroys the collagen in skin, and it thins out as we age; stromelysin destroys the elastin, and we get flabby." The enzymes play a role in normal cell remodeling—repair and replacement. But cellular aging seems to cause an enormous overexpression of these enzymes. "Instead of remodeling, tearing down the protein of your skin and building new protein, it starts tearing it down faster than it can build it up. So these senescence cells turn into demolishing machines, destroying the skin rather than repairing it."

By switching on and off the M-1 and M-2 genes, West says, he can make aging advance and reverse in tissue culture like a yo-yo—a mind-boggling feat, if one stops to think about it. Both genes normally seem to be switched on in aging cells. By chemically turning off the M-1 gene, West can restore the youthfulness of the cells and more than double the number of times they will divide in culture. Turning off the second gene, M-2, has an even more remarkable effect: it makes the cells switch off their alarm clocks and stay youthful forever; they go on dividing indefinitely. Then, completing the cycle, when West turns the M-1 gene back on, the cells start aging again. "We're working on technologies that can be used in people," West says, "like a pill."

If science could learn to control such genes of aging, West says, human life could be extended by 300, 400, 500 years—who knows? "By switching these genes on and off, we can cause the cells to become younger or grow older at will," he asserts. His discoveries have led him to start a small company in Hayward, California, called the Geron Corporation, the world's first biotechnology operation devoted to developing genetically engineered products to retard or reverse the aging process. And he already has a compound, GER-1428, that seems temporarily to hold cells in their youthful stage.

To West and similar investigators, aging may not be as complex as the old-car analogy. "It's more like a molecular time bomb put in key components of the car, in the engine or transmission, that causes the body to self-destruct through a genetic program," he says.

Other scientists believe, however, that no single cause will be found for the aging process. As researchers pick away at the ravages of time, dissecting them in molecular terms, the emerging picture isn't very pretty.

- A leading theory holds that cell damage accumulates as a result of voracious molecular sparks, or oxygen free radicals, promiscuous by-products of the cell's normal defense system and energy-producing activities.
- Key hormones peter out for no apparent reason, and the body slows and flounders.
- Maintenance systems that remove and replace damaged proteins billions of times during the average life of a cell falter, allowing the cell to fall into disrepair like an abandoned house. Failure to repair damaged proteins may be the end point of the biological process of aging.
- In the meantime, proteins that compose the collagen structure of skin, bones, and most other tissue lose their elasticity and become gummed up by something called cross linking, the biological equivalent of pouring sugar into a gas tank.

- Eventually, tiny cellular energy plants called mitochondria poop out, perhaps as a result of free-radical damage or mutations.
- Death genes kick in at some point, perhaps because of planned obsolesence or the accumulation of environmental insults, and cause cells to stop working.
- People unwittingly age themselves prematurely because they don't know the dietary requirements of vitamins and minerals or the need for exercise to keep a body perking along. One of the biggest accelerants of aging is overexposure to sunlight and its consequent wrinkling effect on the skin, which often can be reversed with retinoic acid, a synthetic relative of vitamin A.
- Finally, the immune system disintegrates slowly and can no longer ward off the constant threat of microbes or the danger of autoimmune disorders.

Despite so discouraging a chain of events, there exists a new understanding of the process, which is the first step in intervening. Recent biogerontological research strongly suggests that, from human conception onward, genes work in concert to protect an individual's cells until sexual maturity, when their legacy can be passed on. Somewhere around our thirtieth birthday, our genes seem to stop caring whether we age or not. A seminal thinker along these lines is the biogerontologist Richard Cutler, a research chemist with the National Institute on Aging's Gerontology Research Center, in Baltimore.

"Nature gives us thirty good years. After that all systems start to decline," says Cutler. "What happens then is really up to us."

Cutler, for much of his long and stormy career in search of what he believes to be the few "longevity determinant" genes that fix the life spans of species, has been considered something of a kook, a fringe player, an oddball. But nowadays he is being taken very seriously. "People are saying, 'Gee whiz, where have you guys been?' " he notes with pleasure, his eyes flashing behind his spectacles. People are listening to Cutler now because of an inescapable fact. "Despite medical advances, we're all living older longer, not younger longer," he sums up. "The only way to live younger longer is to directly intervene in the aging process itself."

Back in 1972, Cutler presented studies suggesting that the varying life spans of different animals—mice, guinea pigs, dogs, monkeys, men—might depend on their ability to repair damage to their DNA. Cutler has consistently shown a correlation between the life span of a dozen mammalian species and a particular enzyme, SOD, which protects DNA against damage from superoxide, one of the most toxic free-radical molecules formed during the breakdown of oxygen. Cutler argued that the total amount of SOD an animal required had to be related both to its metabolic rate and to its life span. A mouse that burns up oxygen like a house on fire, lives a short time, whereas an elephant that lumbers through life consuming far less oxygen, proportionately, lives seventy times longer.

Correlations, such as those unearthed by Cutler, are not causes, however, and mainstream science was not yet ready for him. Eminent gray-beards pooh-poohed the idea that DNA repair had anything to do with aging or that the body made special enzymes to protect its genes against metabolism, as Cutler dared to suggest although he lacked the tools then to prove it. Today the dark side of oxygen, which keeps us alive but also releases free radicals that destroy our cells much as they make metal rust or butter spoil, has come to light. Oxidative stress is under intense study for its role in the aging process, and Cutler's cherished enzyme SOD, which vigilantly quenches the vicious hydroxyl radical, has become one of the century's most intriguing chemical discoveries.

In 1991, for instance, in the latest of a series of exquisitely designed experiments that have stunned the field, the biologist Thomas Johnson, of the Institute for Behavioral Genetics at the University of Colorado at Boulder, was able to increase by 110 percent the three-week life span of one of gene researchers' favorite laboratory animals, a millimeter-long backyard roundworm called *Caenorhabditis elegans*. All Johnson did—though it took him seven years of mind-numbing toil to do it—was chemically mutate one *C. elegans* gene in ten thousand, a gene involved in oxygen metabolism.

Transparent as glass, the tiny worm known as the nematode often spends its entire life under some curious biologist's microscope. The nematode consists of precisely 959 somatic cells, no more, no fewer. The cells are rigidly outlined and defined, and their lineage has been totally deciphered—science can explain the making of a nematode step by step, from start to finish. Nematodes come in two genders, a male and a self-fertilizing hermaphrodite, which is a male for the brief part of its development and functionally female for the rest of its life. In their three-week life cycles, the animals hatch from eggs, go through several larval molts, forming either hermaphrodites or males and reaching adulthood in about three days. Each hermaphrodite can produce some three hundred offspring; in a week, a scientist who owns just one nematode can thus end up with a hundred thousand. "It's a fairly convenient system," notes Johnson. "We can get lots of biomass when that's needed."

The nematode has another stage that is highly useful to geneticists—a three-week larval stage known as a dower—which can remain for several months in a sort of suspended animation, like a spore. "When we restore it to growth conditions, it recapitulates the three-week life span," Johnson says, "so we can view the dower as basically a timeout period from aging."

After years of painstaking mapping, Johnson in 1991 announced that he had cloned the gene that he auspiciously calls age-1. It probably protects other genes against oxygen damage, he says, because by mutating age-1 with chemicals, he turned it into one that reflexively attacks dangerous oxygen molecules. The experiment was of vital interest to the biological community as a whole because it represented the first time that a mutation in a single gene had led to a longer life span in any animal.

Other Tom Johnson mutations in roundworm genes have been linked to shortened life spans, and groups of genes can collectively extend life, he has shown. But to Johnson the remarkable power of age-1 by itself proves that the life span of an animal species clearly is under genetic control. The discovery is crucial, he feels, because the elimination of any major disease is not likely to lead to an increased human life span—another illness may replace it. "But if something like age-1 exists in humans," Johnson says, "we might really be able to do something spectacular."

He certainly would like to try to juice up his own genes. "I've spent all my academic life training to perform a job I've been at now for ten years," reflects the thoughtful forty-two-year-old scientist. "In another fifteen to twenty years I'll be up for retirement. If I had it in my power, I would like to retire for ten years and then go back to school to learn a new profession and do something completely different."

Having the worm gene in hand, Johnson has set about probing the human genome to see whether a counterpart exists. If it does, he finds the doubling of the human life span "a reasonable expectation," although by drugs instead of by direct genetic intervention.

In a similar experiment with insects, James Fleming, of the Linus Pauling Institute, in Palo Alto, California, has been able to increase the life span of fruit flies substantially by merely giving them an extra copy—a double dose—of the SOD gene, which protects against oxidation. To Cutler, West, Johnson, Fleming, and other scientists tracing the complex interplay of biochemical processes that age living things, the modifying of only a few genes may permit human beings to live twice as long as they do now. "Our genetic similarity to other animals," says Cutler, "suggests that there might be just a few specific genes whose sole role is to govern how long you live. If true, we wouldn't have to go in there and rebuild the whole machine. If there really are 'longevity determinant' genes, we might intervene and change the expression of whatever they do. Then we'd really have a breakthrough in controlling aging."

Growing acceptance of the role of good nutrition, combined with exercise, in reversing and even retarding aging has reached the level of the federal government. In March of 1993, Irwin Rosenberg, director of the U.S. Department of Agriculture's Human Nutrition Research Center on Aging, at Tufts University, reported the latest findings at a nutrition meeting in Chicago. The goal of the research is to establish dietary guidelines for preventing diseases once thought to be an inevitable part of growing older. "What's exciting," said Rosenberg, "is that we are starting to get observations that say we should be able to delay or reverse many problems and symptoms associated with the aging process by our intake of nutrients that are protective."

That old well-balanced diet may suffice for most people, but many need vitamin supplements because their needs are different or their diets inadequate. A recent survey of more than two thousand elderly Bostonians, for example, found that 24 percent were at high risk for chronic malnutrition and that 38 percent were at moderate risk.

Vitamins E and B$_6$ and the mineral zinc can boost the immune systems of elderly people and reduce their risk of contracting infectious diseases, Rosenberg reported. Other studies are reinforcing earlier findings that antioxidants such as vitamins E and C and beta carotene (a precursor of vitamin A) provide protection against heart disease, cancer, and stroke. Tufts scientists were among the first to show that a combination of vitamin D and calcium is crucial for building strong bones and reducing the risk of osteoporosis in both premenopausal and postmenopausal women.

Other health benefits that can be obtained from good nutrition, Rosenberg said, include the following:

- Protection against cataracts and other eye problems from increased consumption of vitamins C and E and beta carotene.
- Reduced risk of heart disease by the consumption of foods rich in

vitamins B_6 and B_{12} and folate, as well as in soluble fiber, calcium, and potassium.

- Preservation of mental alertness with vitamins B_6 and B_{12} and folate.

Tufts researchers have pioneered studies showing that exercise is a potent weapon against many of the debilitating effects of aging. Through special exercise programs, they were able to free ninety-year-olds from their confinement to wheelchairs. The bottom line: it's never too late. Your body is always ready to give you another chance.

Homo sapiens came on the scene 100,000 to 150,000 years ago with a brain size of about 1,500 millileters, 50 percent bigger than that of his predecessor, Homo erectus. The rapid increase in brain size and tripling of life span in so short a time could have been accomplished only by a small number of favorable gene mutations, additional evidence that only a small number of genes regulate longevity. Some of the most colorful figures in modern science are devoting themselves to the search for them. None is more colorful than Roy Walford.

Even as a straight-A high school student in San Diego, Walford worried that his allotted span of life was simply too short, that there were so many wonderful things to do that he ought to live longer in order to do them. "We are cut off in the midst of our pleasures," he decided. "We are separated too soon from our loved ones, shelved at the mere beginning of our understanding."

That was in 1941, when Walford published his manifesto "The Conquest of the Future" in his school magazine. A half century later, one of the brightest stars in aging research believes he has finally found the way to defy "the laughter of the gods."

Walford has devised a diet that he says may make your life span half again as long, or longer, and prevents cancer, heart disease, diabetes, kidney failure, and even the jittery signs of aging—dry skin, gray hair, cataracts, rising cholesterol levels, and changing liver enzymes. Wait, there's even more: down at the level of your molecules, the diet defends your genes from environmental damage, tidies up your immune system, scavenges the dangerous by-products of metabolism, and keeps your body chugging smoothly in low gear. What folly is this? If you believe Walford, a pathology professor at the UCLA Medical Center, now in his late sixties and the father of the immunological theory of life extension, you can resist the ravages of time. All you need do is cut back on calories over the next five years, gradually going down to about 1,600 a day, as he has done, while supplying all the essential vitamins and nutrients. The regimen will retard aging, maintain health, and fend off doom. And, if you exercise as well, you'll retain youthful slimness and middle-aged vigor all the livelong day.

At least that's what happens to lab animals. Walford doesn't know why the diet works, exactly; nor can he prove that it will work in human beings. Nonetheless, Walford insists, to the consternation of more conservative colleagues, the first steps to lengthen one's life span can be taken right now. "They may not strike you as easy or desirable," he says. "But there's a lot of evidence they will work. It's inferential, but of a high order of probability. The choice is yours."

Walford's own Faustian pact (revealed in his book *The 120-Year Diet*) hinges on the most powerful and mysterious anti-aging mechanism known to science: dietary restriction.

As far back as 1935, the Cornell University nutrition researcher Clive MacKay showed that he could dramatically increase the life span of laboratory animals by cutting back their grub. Since then, in study after study, Walford and his younger colleagues—particularly Richard Weindruch, now at the University of Wisconsin at Madison—have shown that "*under*nutrition without *mal*nutrition" retards aging and extends life span in nearly all animal species tested so far, from protozoans to mammals.

Understand that Walford is not talking about cutting calories to increase the average life expectancy. Today, he says, everyone can act positively to ensure he or she achieves the national average of 74 years— regular exercise, low-fat–high-fiber diets, no smoking, little drinking, dietary supplements. No big deal. Walford's aim is more cosmic: to slow down biological aging and extend the maximum life span of the human species, defined as the age reached by the last few survivors, and now believed to be about 120. Walford started on his diet a decade ago, so he plans to live only to 140 or thereabouts; but if he could prolong human life for, say, 180 years, that would make 300 years a realistic goal for the next generation of scientists!

And that, he feels, would be long enough for us to avoid being shelved at the mere beginning of our understanding.

In his race against time, Walford has already combined careers as a physician, biologist and gerontologist, author, buccaneer seeker of truth, and intrepid explorer of the inner states of being and meaning. He has sailed around the world, cleaned up at gambling in Vegas, founded a Los Angeles guerrilla theater troupe, filed stories as an underground journalist, wrestled competively, learned to play classical piano, contributed poetry to the *Atlantic,* and recently completed two years as the house physician and dietician in Biosphere 2, the controversial space colony prototype–tourist attraction in the Arizona desert, the denizens of which followed his diet.

In recent years, Walford has altered his image somewhat. The gaunt build, the courtliness, the quizzical ice-blue eyes, and the rakishly shaved head remain; but gone are the Fu Manchu mustache, single earring, black leather motorcycle jacket, and biker boots that made him stand out at scientific conferences. It is his unending curiosity, his grounding in solid science represented by his mentor, the French immunologist and Nobelist Jean Dausset, and his devotion to mice and rats that have made Walford famous and his lab in Los Angeles a homeowner's nightmare.

The world of dietary restriction is a wonderfully wacky world—legions of critters are spending their days bustling about and burrowing in the straw of their little cages, even though many are incredibly ancient. *L'chaim!*—To life!—seems to be the ruling motto, as the keepers watch their captives hopefully and cheer on house favorites. Demises are carefully noted and mourned, as when a tiny female gray Walford mouse named Frieda died of natural causes at fifty-three months, falling short of the lab record by a scant two months (the average lab mouse lives about twenty-five months).

Roy Walford's mice are Methuselahs. Their body clocks should have run down long, long ago. They live nearly twice as long as other lab mice and many times longer than ordinary field mice. The only known difference is that Walford's animals have been placed on what he calls "gradually induced intermittent fasting," meaning that after two or three months of age, they are forced to fast for two days a week. Not only does this drastically slow down their aging process; it also significantly lowers their rates of all the diseases associated with senescence.

What's going on here? Why should caloric restriction have such a dynamic effect on the total animal? The mechanisms that underlie the phenomenon, obviously genetically controlled, are gradually yielding their secrets. On September 30, 1991, the National Institute on Aging, in a press release describing preliminary data on the most ambitious investigation of caloric restriction in animals ever attempted, summed up the work in a single word: "remarkable."

The experiments are being conducted in a gigantic lab that, inexplicably, is one of science's best-kept secrets—the FDA's National Center for Toxicological Research (NCTR), in Jefferson, Arkansas, which will probably go down in history for its pioneering work in gene action, cell mechanics, biological chemistry, toxicity, and diet. A sprawling, 500-

acre facility nestled in a forest between the Arkansas River and U.S. Highway 65 near Pine Bluff and south of Little Rock, the NCTR was until 1971 devoted to the production of agents for germ warfare. But for the past two decades, the laboratory has done an about-face. Scientists now study the dangers posed by synthetic chemicals in food, drugs, and the environment.

The NCTR's thirty buildings house more than 250 research projects, but the endeavor all of biogerontology is watching began in 1985 as a collaboration with the NIA to identify the biochemical and cellular signposts of aging and to determine optimal nutrition to prolong youth. NCTR scientists are studying no fewer than thirty thousand rodents. Heading the effort is the center's director, Ron Hart, fifty, a slim, magnetic, and excitable trailblazer who twenty years ago was among the first to discover that cells constantly repair their own genes and that the maximum achievable life span for animal species is directly related to their ability to perform DNA repair.

"The problem for me was that we found a correlation, but not a cause," Hart says. "Now we're narrowing down the hunt."

Hart, who eats like a sparrow and lifts weights regularly, runs the world's largest DNA repair shop. He fully expects the NCTR effort to revolutionize knowledge of human nutrition and longevity. In his mice experiments, he has found that DNA damage—aging, in his view—occurs in three principal ways: because of body temperature, as a by-product of body chemistry, or during the intermediate steps of normal metabolism, when free radicals are released as food is converted into energy.

In the NIA study announced in 1991, the animals on a limited diet ate 40 percent fewer calories than those on a typical feeding schedule. After twenty months, only 13 percent of the restricted female mice had developed cancer, while 51 percent of those without restricted diets had malignancies. By twenty-four months, mice of both sexes not on food restriction had suffered four times as many tumors. By forty-five months, when those on unrestricted diets had died, half of those on restricted diets were still alive.

Admitting that earlier studies had shown that caloric restriction could delay some tumors, Richard L. Sprott, associate director of the NIA's biology of aging program, declared, "What stunned us this time was that every single type of tumor was delayed."

Of late, Ron Hart can't wait to get to work each day. Instead of waiting around and wondering why his mice live so long, he's beginning to find out. "When you consider the changes we're seeing in the fundamental processes that not only control aging but other threats from the environment, both inside and outside the body, the implications are phenomenal," he enthuses. "We are absolutely totally flabbergasted by the results. Sometimes in science you can be just plain lucky. Everything we've looked at is turning out really significant findings."

They stem from an unusual series of collaborative agreements, initiated in 1982 by the government, that brought fourteen university laboratories into the program. That means Hart's group now boasts a total of about seventy-five M.D.'s, Ph.D.'s, and D.V.M.'s, plus technicians, animal handlers, diet preparation workers, and quality assurance specialists—a mighty contingent of mousekeepers indeed.

This army is devoted to the care and feeding of its thirty thousand rodents—five strains of mice and three of rats. All are individually housed, in sterile, temperature-controlled conditions, their food and water scrupulously screened for contaminants and constantly monitored. The scientists are looking for changes—behavioral and physiological changes, all the way down to submolecular changes, including gene expression—that result from dietary restriction. Computers juggle hormonal changes in various animals with changes in food metabolism or drug metabolism, or equate the formation of free radicals with DNA damage and repair, and deduce how these variables affect the fidelity of DNA replication.

Much of the work amplifies the findings of Roy Walford and other calorie restriction gurus. Typical is the discovery that the reduction of food intake prolongs life by lowering body temperature. Walford twenty years ago traveled to Argentina to study the world's shortest-lived fish, which complete their life cycles in a year, because their small ponds and water holes dry up during the summer. Walford extended the maximum life span of such "annual fish" by 300 percent by merely cooling their water a degree or two—an intriguing finding that he had no idea how to apply. A decade ago, armed only with the gift of gab and a rectal thermometer, he traveled through India to learn the secrets of the swamis who can lower their body temperatures at will. "They're able to get down to about ninety-four degrees," Walford recalls. "I never did discover how; maybe it's because they don't eat much." But they don't become sluggish, he says. They stay alert.

Walford's findings underscored the idea that a leading cause of gene damage, and hence of aging, is that we live internally at 98.6 degrees Fahrenheit. Nature set our temperature 50,000 to 100,000 years ago, when we all were joggers, hunting mastodons, dodging hungry saber-toothed tigers, and sprinting back to the sanctity of our caves. Our bodies are still set at a fairly fast start, and for modern genes, 98.6 may be a bit steamy, which is why our brain so fiercely controls cooling, lest pieces of our chromosomes break off in our cells, with dismal results. Walford, like the yogis (and his fish), would opt for a lower body temperature, if some way could be found to achieve it.

As Ron Hart explored the phenomena in mice, he found a significant decrease in mean body temperature in all the rodents whose diets had been restricted. "DNA is just a bunch of chemicals stuck together," he says. "Heat causes pieces of the molecule to split off randomly and must be repaired. Under calorie restriction, though, the engine runs cooler and there's less damage. Merely reducing caloric intake by 40 percent reduced this form of spontaneous DNA damage almost 25 percent!"

Normal mouse temperature is 98 to 100 degrees Fahrenheit, but at certain times during the day when the restricted animals appear to be nesting, it will plummet to 80 degrees—the ambient temperature of the room. "You'd think the animals were freezing to death," marvels Hart. "But if you touch them, they're alert. It's not torpor or hibernation; it's something else. It's a circadian rhythm—a biological clock—in reaction to stress."

Reduced body temperature, along with metabolic changes, protects the animals' genes and slows aging, Hart says. For example, energy production and the making of glucose are lower, reducing the generation of oxygen free radicals. Free radicals can induce mutations in genes, but the Hart team found that calorie restriction intervenes in the process early—each time an animal takes a breath. When oxygen is converted into energy more slowly, fewer biochemical pathways are created, which, in turn, reduces the production of these energetic radical molecules.

Another way large molecules such as DNA can be damaged is by the creation of carcinogenic compounds known as electrophiles. "In the animals, we found an increased ability to deactivate dangerous chemicals and, at the same time, an inhibitory effect on their creation," Hart says. "That means the amount of spontaneous damage is reduced, the number of free radicals, the amount of dangerous epoxides formed, fewer electrophiles—there is a uniform reduction of all those systems that can induce damage to genes."

On the other hand, Hart charted an enormous rise in antioxidants, the chemicals obtained from food that quench oxygen free radicals. He found a fourfold enhancement in catalase, a threefold enhancement in SOD, and similar enhancements in others. By any measurable standard, this is a good thing for the genes. The finding surprised Hart.

"By feeding the animals less, we thought there would be a reduction of antioxidants along with damaging oxidants. We figured that there would be less of a need for the body to produce protective chemicals. Instead, we found an increase in all these major systems. That means, during caloric restriction these molecules become more active, bolstering the entire protection system."

The benefits just seem to pile on benefits, each part of the pile affecting the chain of molecular destruction that normally would result in an aged mouse or, presumably, human being. Assessing DNA damage to the genes of restricted animals, Hart found that the genes had less need to repair themselves. "That makes sense," he says. But he was amazed to discover that the DNA repair mechanisms themselves speeded up their healing action between 50 and 200 percent. "What's fascinating," he notes, "is that reduced food intake is the only experimental paradigm ever found that enhances DNA repair."

Next, Hart's team turned to the fidelity of DNA replication. It studied various enzymes important to DNA replication in mammalian cells; they include the DNA polymerases known as alpha, beta, gamma, and epsilon. How well genes can make copies of themselves hinges on the quality of these enzymes, which serve as a template to copy material from the "maternal strand" of a gene onto a new DNA strand, known as the daughter strand. In all cases, reduced dietary intake enhanced the fidelity of replication of the polymerases between twofold and fifteenfold.

"When cells get very old, or come from an older animal, there is a natural decrease in this ability to replicate DNA," Hart says. "Dietary restriction stops this. Even in older mice, the normal age-related decline is being significantly held up."

Being a careful scientist, Hart quickly warns against taking the results of the Arkansas mice too literally. It is much too early in the game to translate these findings into one's lifestyle. "For one thing, when we cut an animal's food by 40 percent, it stops breeding." A circadian rhythm cuts in, Hart says, that may occur in nature during seasons in which food is limited. Of necessity, sex then gets repressed by the instinct for sheer survival. "Last time I checked," notes Hart, grinning, "humans liked to breed."

But he is so encouraged by the studies that in 1993 he began human trials of calorie-restricted diets, marking the first time that the government of the best-fed nation on Earth will look to see whether Roy Walford and the yogis are onto something. The ultimate ideal, Walford and

Hart agree, is not to restrict all people's calories to keep them young and healthy. Too many people enjoy food for that to become feasible. "What we want to do is to discover the mechanism that's working in these animals and then figure out how to do the same thing pharmaceutically, or by gene therapy."

To do that, of course, he has to find the genes. But that is no longer an impossible dream to biogerontologists like Hart. He says he knows where to look—in the brain. Is he looking for Caleb Finch's master brain clock? Hart does not quibble over semantics. "In fact, I suspect there's probably just one, a Master Gene of Aging," he says. "I think it regulates all the beneficial changes we see in our animals." But is this gene beyond the ken of science in the 1990s? "We have the ability to find any gene, now," sums up Hart. "I predict we'll have the aging gene in our hands very soon."

24

PROPHECY

Long is the way
And hard, that out of hell
leads up to light. . . .
—John Milton, *Paradise Lost*

But where in the vast Sahara of the human genome would a small sand hill, such as a master aging gene, lie? For that matter, where might other genes of intense interest be—not merely disease-causing genes but the ones that control our memories, or those that influence intelligence and personality? Not to mention genes that contribute to less crucial but still consequential traits or skills, such as mechanical aptitude, perfect pitch, or a killer backhand?

The future of gene therapy and its pharmaceutical adjunct, biotechnology, will hinge on an unprecedented fifteen-year international gene hunt that began in October of 1990—the Human Genome Project, with its goal of mapping, sequencing, and deciphering by the year 2006 all of the estimated 100,000 genes that compose the human blueprint. Biology's version of the Manhattan Project, the genome expedition has been crossing a chromosomal frontier so unimaginably vast that, once it is traversed, merely printing out the three billion DNA base pairs of a single person's genome would fill thirteen full sets of the *Encyclopaedia Britannica*, assuming one character per nucleotide.

At the helm of this immense effort was a man uniquely placed to guide the search—the University of Michigan's Francis Sellers Collins, chemist, physician, pioneering geneticist, collaborator par excellence in previous discoveries of critical genes, and gifted conciliator, whose gentle ways were expected to calm the troubled waters surrounding the project. It was these choppy seas that overwhelmed his predecessor, James Watson, after piloting the project through its first two, stormy years.

As originally developed by Watson and legislated by Congress, the Human Genome Project has four tiers: mapping the twenty-three human chromosomes, sequencing them in their entirety on computer databases, distributing the information freely to scientists throughout the world, and establishing an ethical component to ensure that the knowledge obtained is not used to discriminate against anyone.

Fundamental was the understanding that the venture would be of such herculean dimensions that it would have to rely on high technology not yet invented. The program would develop the tools it required as it proceeded, rather like a truck out of Dr. Seuss laying down the roadway in front of itself as it lumbered along.

Even before the project had reached its first five-year milestone, the Harvard Nobelist Walter Gilbert was already predicting imminent rewards. By 1995, declared Gilbert rather fancifully, tests would be on the market to enable people to determine their predisposition to heart disease and cancer. By the millennium, he expected to see "genetic profiles" containing 20 to 50 disease genes; by 2010, the profiles would have swollen to between 2,000 and 5,000 genes responsible for various disorders. And by 2020 or 2030, Gilbert told *Time* magazine, "you'll be able to go to a drugstore and get your own DNA sequence on a CD, which you can then analyze at home on your Macintosh."

But mass screening for disease-causing genes and possible therapy form only a small part of the anticipated bounty. The mechanisms underlying human behavior, consciousness, and intelligence should ultimately stand revealed, as should the intricate ways that genes interact with the environment, shedding light on the age-old nature-versus-nurture controversy. In fact, the story of life on Earth should become clearer as we unravel the paleontological record that is stored in our chromosomes, thereby enabling us to decipher the patterns of spontaneous mutations that seem to drive evolution.

The genome seeds were sown in the summer of 1985, when the idea, once the stuff of late-night bar chatter among gene splicers, was first floated as a serious prospect at several professional gatherings by the Nobel laureate tumor virologist Renato Dulbecco, a courtly senior statesman and former president of the Salk Institute. After a lifetime spent wrestling with the mysteries of cancer, Dulbecco saw only two approaches to defeating the late twentieth century's intractable killer: "Either to try to discover the genes involved in malignancy by a piecemeal approach, or to sequence the whole genome of a selected animal species.

"I think it will be far more useful to begin by sequencing the cellular genome," Dulbecco wrote in *Science* in 1986. "The sequence will make it possible to prepare probes for all the genes . . . [and] facilitate the identification of those involved in [cancer's] progression." Dulbecco urged the United States to throw itself into the search for the total sequence of the human genome with the same determination and spirit that had "led to the conquest of space."

By coincidence, the Department of Energy was looking for new projects to fund, and here was a high-profile, humanitarian effort that smacked of Big Science. The DOE already had some expertise, having spent billions over the years researching the effects of energy production and radiation on human heredity. Not the least of its contributions was helping establish the National Laboratory Gene Library Project, housed jointly at Lawrence Livermore and Los Alamos laboratories, a storehouse of sorted human DNA probes that are circulated without charge to researchers. The agency's David Smith observed, "We see the sequence project as a natural offshoot of the library."

Soon Dulbecco's idea and the DOE's yearnings converged in the amazing proposal to understand the human body molecule by molecule, which, depending on one's perspective, seemed either brilliant or halfbaked. After feasibility studies, Congress approved a fifteen-year project, with funding growing in steps to an eventual $200 million a year and a full budget that worked out to $3 billion, or approximately $1 per nucleotide sequenced.

Elsewhere in science were dissenting views. David Baltimore believed that the federal effort's huge expense would draw needed money away from other important, though smaller-scale, research. "I think biology has been most effective as a cottage industry and should remain so as long as it can," he declared. And Robert Weinberg, a preeminent cancer researcher at Baltimore's former home, the Whitehead Institute, opined that because 95 percent of the genome seemed to consist of socalled junk DNA—"genetic garbage," he called it—sequencing the entire thing would be a huge waste of time. "I believe it will turn out to be biologically meaningless," he said.

But at Stanford the Nobelist Paul Berg demurred. "I want to know every single nucleotide," he countered. "Only then can we decide whether it's junk or not."

Weinberg and Baltimore had hit on a crucial point of controversy: if, as is believed, only 5 percent of our DNA is truly operational, then the rest is inexplicable yet would nonetheless have to be sequenced. The genome contains molecular mile after mile of redundant sequences, chains of nucleotide pairs that are repeated endlessly for no apparent reason. Other stretches appear to be pure garble. The prolixity occurs even within genes themselves; they are interspersed by introns that seem without purpose and in fact get scissored out by cells during gene translation and transcription. Weinberg once recalled isolating a gene of 200,000 base pairs, of which only 4,000 seemed to have a function in the form of regulatory sequences and protein-encoding sequences. Consequently, genes, to Weinberg, constitute "a small archipelago of information scattered amidst a sea of drivel."

What might be the point of quiescent genetic material? There are many possibilities. During cell division, all the DNA must be duplicated precisely. Some DNA stretches probably serve as signals to make sure that chromosomes get properly separated from one another during copying and that each daughter cell gets its share. Other stretches may represent bits of viruses that sneaked into human DNA eons ago, multiplied, but could go no further and just stayed put. There are also lengths of DNA called pseudogenes that may be like bugs in a computer system. Their structure tells scientists that they are derived from genes that probably once did something. So the genome may reflect the remnants of evolution—failed genes, dead genes, old genes that have mutated into newer models. Just as the human embryo seems to climb the evolutionary ladder during gestation, making its way in anatomical stages from worm to fish to amphibian to mammal, so the entire book of human evolution appears to be writ large in our genes. The ghosts of our nightcrawler ancestors lurk therein, as may the primordial piece of DNA—or, as some paleobiologists now think, RNA—that started it all. How could anyone with an ounce of romance inveigh against learning all that?

In any case, unanimity among the scientific community was essential if genetics research was to be given a chance to pursue what Gilbert was proclaiming the "Holy Grail" of biology. Spearheading the U.S. effort would be the National Institutes of Health, the nation's keeper of the biomedical flame. To that end, the director of NIH, James Wyngaarden, created the National Center for Human Genome Research in Bethesda in October of 1988 and called upon the genetics hierarchy to close ranks. In an act of collegiality, Watson agreed to head it. By accepting the genome post, a fitting finale to a fabled career, Watson in no way gave up his position as director of the Cold Spring Harbor Laboratory, which he had held with distinction since 1968. Even though Watson and DNA were virtually synonymous, his selection to head the genome project was not purely nostalgic. He was widely regarded as the only gene scientist of sufficient luster to get so ambitious a project through Congress.

Watson, like his fellow Nobelist Dulbecco, viewed the effort as a challenge as noble as the 1961 decision by President Kennedy to send a man to the moon within the decade. "We used to think our future was in the stars," Watson said. "Now we know it's in our genes."

In the late 1980s, the only existing gene maps contained on the order of four hundred riflip signposts scattered widely among the chromosomes. These were to be used as landmarks to help in localizing genes. The map expected to be completed by the genome project in its first five years promised to have fifteen times as many signposts. "What we have now is a map like Lewis and Clark's," said MIT's Eric Lander, one of the cartographers who produced the early map. "What we will eventually get will approach the detail of something you'd get from the AAA."

In 1990, contracts were let out to various laboratories to begin to make chromosomal maps, rough maps at first with markers spanning stretches as long as a chromosomal band (about ten million base pairs), which would then become increasingly finer (two million bases) and progress to higher levels of resolution until the complete sequence of each chromosome was obtained. Within the first five years, the project was mandated to dot the human genome with either overlapping sets of cloned DNA or closely spaced markers, about thirty thousand in all. Sequencers, during this first third of the project, were to read off as many as ten million base pairs in large, continuous stretches. Project scientists were also mandated to develop effective software to support large-scale mapping and sequencing and, above all, to handle the flood of DNA sequences that would soon be filling electronic databases worldwide.

By the autumn of 1992, in a triumph set in motion during Watson's brief tenure, an international partnership known as the Collaborative Mapping Group presented the world with a unique gift—genetic linkage maps for twenty-three chromosomes (all except the Y), consisting of 1,416 loci (three times better than Lander's map), including 279 genes and their expressed sequences. The physical map, showing the relation between cloned DNA fragments, represented work by more than a hundred authors, an all-star team comprising former competitors, and demonstrated the potential for problem solving offered by Big Science. Simultaneously, Daniel Cohen at CEPH, the industrial-scale French gene-mapping center that housed nearly eighty scientists so organized that they were even using robots, weighed in with a detailed map of chromosome 21. The map consisted basically of large segments of cloned DNA that had been placed together in the right order. But it was of enormous interest because along chromosome 21 reside genes responsible for Down's syndrome, the familial form of Lou Gehrig's disease, a

type of Alzheimer's, and the free-radical quencher superoxide dismutase, to name a few. Not to be outdone, Small Science weighed in at the
same time when David Page, a geneticist at the Whitehead Institute,
contributed history's first detailed map of the male Y chromosome. As
one of the few specialists to concentrate on the male sex chromosome,
Page had spent ten years hacking his way through what he calls this
"mysterious jungle." The littlest chromosome (only sixty million base
pairs in length) holds the secret of sex in mammals—hidden on the Y is
the gene that confers maleness, the testes-determining gene (known as
SRY). Both Page and Cohen, in the interest of expediency, had skipped
sections of the two chromosomes that held monotonous repeating sequences.

For a start, this was all very impressive. But a map is not
the territory, as the semanticist Alfred Korzybski pointed out, and the
territory is what counts. Despite the momentous progress made in genetics over the past twenty-five years, science still has only the haziest
view of the big picture. There are so *many* genes in the genome. They all
work together somehow. Clumps of them interact with other clumps.
They make nucleic proteins that come back to interact with, and regulate, DNA. Some of them jump around the chromosome with unpredictable consequences. Some researchers think such "jumping genes" are
caused by environmental stress and may in part be responsible for evolutionary mutations. There are genes that switch on in fetal life and then
switch off at the moment of birth, never to turn on again. There are
master genes that control the exquisite process of differentiation, by
which a single fertilized cell becomes a full human being.

Furthermore, there is not really one human genome; there are five
billion varying genomes—a unique collection being parceled out to
every person on Earth—and most human traits rely on interactions
among many different genes and the environment. An individual talent,
to choose an example, must be carefully nurtured and then discovered
by society at large if it is to flourish. On the other hand, a genetic predisposition to the three scourges of modern industrialized society—cancer,
heart disease, and mental illness—will not result in calamity unless various mutations occur to several key genes. Even among the thousands of
diseases blamed on *single* defective genes, such as cystic fibrosis or
sickle-cell anemia, the symptoms that occur will range from mild to
severe in individual patients, illustrating the murky issue of penetrance—that is, the tendency of genes to express their traits in varying
degrees in different people. Reducing life to the molecular level, which

is what molecular biologists are trained to do, keeps resulting in landmark discoveries, but it also tends to oversimplify such complexities.

Nevertheless, Watson, when interviewed early on, did not shy away from the possibility that what he called "this ultimate tool" might be used to retrofit the human genetic program in ways that some critics might find dangerous or overweening. "Are we going to control life?" he mused one sultry afternoon in his office at picturesque Cold Spring Harbor. He ran a hand through his sparse strands of white hair, his eyebrows jumped up and down, and a puckish gleam came into his eye. "I think so," he said. "A lot of people say they're worried about changing our genetic instructions. But those are just a product of evolution designed to adapt us for certain conditions that may not exist today. We all know how imperfect we are. Why not make ourselves a little better suited to survival?"

Then he chuckled, emitting the characteristic Watson nasal snort that precedes provocative statements. "That's what we'll do," he decided: *"We'll make ourselves a little better."*

Even as the project he had set in motion slipped easily into second gear, Jim Watson felt the ground begin to tremble beneath him. In the autumn of 1991, he found himself in a public skirmish with Wyngaarden's successor, Bernadine Healy, over the work of a well-regarded NIH researcher, Craig Venter. Venter's stock in trade was making cDNA libraries, collections of unknown genes that are active in the cells from which they are derived. Because these genes are active, they are at any given time being transcribed into messenger RNA, on the way to becoming protein, and these mRNAs can be captured like fireflies. If they are thereupon exposed to the enzyme reverse transcriptase, the mRNAs will turn back into the genes whence they came—or, more accurately, into the far smaller *expressing* sequences, or "coding regions," contained in those genes, minus the intervening garble. These no-nonsense, condensed versions of genes are known as complementary DNAs, or cDNAs.

The technique had been used before, but Venter found a shortcut. He dipped randomly into his library and came up with partial sequences of these cDNAs, figuring that most of them would correspond to actual, functioning genes. He thus had what amounted to shorthand editions of all or most of the genes operating in a given kind of cell. He did this primarily because these expressed sequence tags, or ESTs, as they are called, can be used as probes to isolate entire genes and greatly speed up the process of finding tissue-specific genes.

By late 1991, Venter had collected thousands of these gene fragments, and it struck key NIH officials that it might be advantageous to seek to patent the fragments, because the American public had paid for the research and thus was entitled to the benefits. Reid Adler, NIH's director of technology transfer, therefore filed an application with the U.S. Patent Office. The only problem was that he failed, for whatever reason, to inform Watson of his intention. Watson, who erroneously assumed Healy was behind the gene-patenting decision, was greatly annoyed. He was convinced the idea would have a seriously detrimental effect on the genome project by antagonizing NIH's allies, who were, for the moment, muzzling their competitive instincts and collaborating eagerly and smoothly. Sure enough, in retaliation, the British quickly announced plans to try to patent their own fragments. Never being one to mince words, Watson told the press that NIH's notion of wholesale patenting was "sheer lunacy."

It was Healy's turn to feel the bite. Deeply nettled, she called in Watson and directed him to keep his criticisms "within the family," according to Watson. Furthermore, she designated Venter her adviser on the genome project and, Watson said, made it very clear she wanted the grand old man of DNA research to step down from his genome post.

Vulnerable despite his vast prestige and dignity, Watson now started hearing rumors that he had certain conflicts of interest, centering on his investment in a handful of biotech firms, including Du Pont–Merck Pharmaceuticals and Amgen Inc. Could Watson in good conscience keep NIH from patenting gene fragments when it might be viewed as an attempt to protect his own companies from harmful competition? Would Watson in the future recuse himself from similar thorny decisions?

As if all this weren't trouble enough, Watson also became embroiled in a spat with a Connecticut financier named Frederick Bourke, whom he criticized for trying to hire away several of the genome project's pivotal figures for a private sequencing company. According to public accounts, Bourke fired off an irate letter to Healy, accusing Watson of interfering with private business matters and implying that Watson was doing so for unethical reasons. It exacerbated the already strained relations between Healy and Watson when Healy turned the letter over to the ethics office of the Department of Health and Human Services without showing it to Watson.

Watson was completely cleared of any conflict of interest by those in charge of the ethics office. But statements emanating from Healy's office indicated she was not convinced. A proud man, Watson quit in April of

1992, registering his displeasure by faxing his resignation. He subsequently described the affair as a smear campaign aimed directly at him. "I find it sordid, awful, and very depressing," he told *Science*. "The whole thing is sickening."

It was nearly a year before NIH announced a replacement for Watson. Many insiders expressed doubt that Healy and her staff would be able to find anyone of sufficient stature to fill the great man's shoes. All the more reason why jaws dropped in the fall of 1992 when rumors began flying that NIH had found such an individual and that he was none other than Francis Collins, one of the most universally respected scientists in the field.

Why would Collins give up a dream job at Michigan for a can of worms, not to mention one that involved what he called "a significant financial hit" (his salary dropped substantially, to $135,000 per annum)? Why would he relocate his laboratory to Bethesda and take on the headaches of being director of the NCHGR? According to Collins, to do otherwise would have been unthinkable.

"I couldn't pass this up," he explained. "There is only one Human Genome Project. It will only happen once in human history, and this is that point in time. Without sounding corny, I do believe this is the most important scientific project mankind has ever mounted, this investigation into ourselves. For anybody to have a chance to direct something of that momentous import was the answer to one's dreams. I feel I've been preparing for this job my whole life."

Healy had leaned on Collins heavily to get him to take the post. Among his demands was that NIH, instead of merely being a cashbox for other institutions, have a primary research role. He was also planning to hit the ground running by drafting a new five-year plan for the genome project, noting, "Even though we haven't gotten halfway through the first five-year plan we are ahead of schedule scientifically and it is time to reevaluate."

One area that highlighted Collins's clinical background as a physician was his declared intention to harness the genome project to the furtherance of gene therapy. "We will be involved in a significant way with gene therapy here," said Collins on assuming command in April of 1993. "I'd say that at the moment, gene therapy and the genome project have been admirers of each other and haven't really gone on many dates. What I want to do with the intramural program is get them to start a serious romance so that the people who are finding disease genes and the people who are using them are working side by side."

Prizing the scriptures of nature from the living tape of DNA will have immediate and incalculable consequences. All at once, we will have answers to riddles that have tantalized the human race since the time of the ziggurats. What, for instance, makes us behave the way we do? Whence flow the currents of thought and personality that lift our otherwise inert selves from the dry creek beds of earthly form? Which genes are the ones that influence smartness? Goodness? Evil? What makes a Hitler? A Hume? A Mother Theresa?

The answers have already begun trickling in. A host of human behaviors have been shown to be rooted, to a significant extent, in the genes. In late 1992, for example, the University of Minnesota researchers Matt McGue and David Lykken reported evidence suggesting that the tendency to become divorced is in part inherited. On the basis of studies comparing the marital stability of identical twins with that of fraternal twins, the investigators calculated that divorce behavior is roughly 52 percent heritable. Environmental factors—the influence of one's own early home life—and blind chance account for the rest. The likelihood of divorce for an identical twin whose co-twin is divorced is six times greater than that of fraternal twins whose co-twins are divorced, the Minnesota study concluded. McGue and Lykken noted that while there is surely no gene for marital discord, the tendency can be expressed through a set of traits and patterns of conduct that have a biochemical basis known to be heritable. These characteristics include the capacity for happiness, a neurotic bent, impulsiveness, and sexual proclivity.

In mid-1993, investigators in the Netherlands disclosed even more startling news: they had found what seemed to be a gene for aggression. The clinical geneticist Han Brunner and associates at University Hospital in Nijmegen reported that a mutation in a gene for an enzyme known as monoamine oxidase A (MAOA) appeared to be the cause of aggressive, even violent behavior among several generations of males in a large Dutch family. Brunner had first learned of the family in 1978 when a woman came to him seeking help for her kin. Over the years, at least fourteen men in her family had shown a tendency toward unprovoked outbursts. One raped his sister and later, when sent to a mental hospital, went after the warden with a pitchfork. Another repeatedly forced his sisters to disrobe at knifepoint. A third tried to run his boss over with a car, while two others were arsonists. Each of the affected family members displayed mild mental retardation as well, with IQs of about 80.

Armed with the knowledge that the gene defect was probably on the X chromosome, because the syndrome was confined to males, Brunner's team used linkage analysis to localize the MAOA gene. Biochemical evidence supported the finding. Urine samples from affected men showed that their MAOA genes were not performing their customary task, which is to help break down certain neurotransmitters—namely, dopamine, epinephrine, norepinephrine, and serotonin. Viewed from this perspective, a mutant MAOA gene would make sense as a cause of belligerence. Each of the neurotransmitters cleared by the gene is involved in the body's reaction to stress. An overload of these same neurotransmitters would put the men's bodies into a more or less permanent "fight or flight" mode, meaning that any bit of extra stress might push them over the edge.

The Dutch team's findings must now be confirmed. A key question is whether the same mutation, or a similar one, might be found in other families. But if Brunner's assumptions hold true, they demonstrate a genetic cause for antisocial behavior, which has been identified only rarely before—notably, in the case of the HGPRT gene that underlies the Lesch-Nyhan syndrome. Since the argument for the existence of behavioral genes is still largely hypothetical, the finding would materially strengthen the case. It could also open the door to the screening of offenders for the mutation, and lead to treatments, perhaps drug or dietary therapies to block the excess neurotransmitters.

It is important to note that Brunner's "aggression" gene, even if authenticated, will not represent a true behavior gene. Its effect on human action results from its failure to perform its normal function, which is metabolic. But almost certainly there exists a wide spectrum of genes whose natural purpose is to influence and control the way we behave. By the same token, there must be other genes that have an auxiliary bearing on how we act. Testosterone, for example, produces male secondary sex characteristics, such as body hair and a deep voice. But it also produces aggression, the degree of which is probably controlled by regulatory sequences that govern expression of the gene. As the Human Genome Project proceeds, these various behavioral genes will be teased out of the mix and their functions recorded. Eventually, we will sort out the exquisitely complex way in which these genes and their proteins interact, against a background of early experience and other environmental factors, to manage our actions, temperaments, and emotional responses.

To quest for behavioral genes, though, is to invite bitter controversy, as the Salk Institute neuroscientist Simon LeVay found out in 1991 when he reported a tiny physical difference in the brains of dead heterosexual and homosexual men. Cells in the hypothalamus believed to regulate sexual activity were less than half as large in gay men as in heterosexual men (about the same size as in women), raising the possibility that sexual preference is something we are born with and that homosexuality represents biological destiny. LeVay was too good a scientist to call his finding conclusive. All nineteen homosexuals he studied, and six of sixteen heterosexuals, had died of AIDS, and the disease can devastate the brain. And, he was quick to point out, he could not say "whether the structural differences were present at birth . . . or whether they arose in adult life, perhaps as a result of the men's sexual behavior." But he did write that the findings "suggest that sexual orientation has a biological substrate."

His discovery, which followed up on earlier work by Roger Gorski, who identified two regions in the anterior hypothalamus that were more than twice as large in men as in women, ignited fiery debate, not least because LeVay is gay and began his research into sex differences in the brain after emerging from a deep depression when Richard Hersey, his lover of twenty-one years, died of AIDS. LeVay's scientific detachment became even more of an issue when he left Salk to become codirector of the West Hollywood Institute for Gay and Lesbian Education, one of the first colleges in the nation to be devoted to homosexual studies. Possible agendas aside, the study itself was open to attack—LeVay measured cell volume, instead of cell count, and researchers disagree about which is preferable; hypothalamus sizes were relatively smaller among homosexuals, but not equally small; sexual orientation was presumed, though histories were not available; there is no way to tell whether the differences observed represented cause or effect; the experiment had not yet been replicated; no animal work had preceded it—so it must be viewed merely as suggestive, nothing more.

On the other hand, LeVay does present an intriguing explanation for homosexuality. If nature's sole goal is the passing on of genes through sexual reproduction, what would be the point of "gay genes," as LeVay calls them? He leans toward a "sickle-cell" model for homosexuality. The defective hemoglobin gene responsible for the sickle-cell trait does not cause the disease in carriers, but does help protect them against malaria. Only when two carriers conceive a child can the offspring inherit two copies of the bad gene and develop sickle-cell anemia. The disease may thus be an unwanted by-product of a strategy that probably kept the gene in circulation, because risking anemia is more advantageous than risking death from malaria. By the same token, speculates LeVay, if a gay gene is recessive, it might be preserved in a population because it confers some advantage for straight people that improves their reproductive success. That means homosexuals, who inherit two copies, are actually losers in a genetic roulette game. The notion, LeVay admits, "may not appeal to many gay men or lesbian women—and it certainly doesn't appeal to me—but it nevertheless has some plausibility."

Obviously, such studies linking genes to behavior are fraught with sociological implications. Feminists, for example, might well bridle at attempts to connect certain characteristics often associated with males, such as emotional strength or decisiveness, to genes. Ethnic groups might react similarly. Imagine if it were found that the Irish have a higher than normal incidence of alcoholism genes, or if genes responsible for self-assertion or acquisitiveness were found to predominate in Jews. A taste of how such issues will be handled in this era's politically correct climate came in the fall of 1992 when NIH, under intense pressure, canceled funding for a conference at the University of Maryland entitled "Genetic Factors in Crime." The proposed conference, whose $78,000 grant NIH had initially approved on the basis of peer review, had drawn fire from a number of black opponents who argued that an attempt to link criminal behavior to genes had racist overtones. Although the university appealed the decision, the conference remained on hold as of this writing. A backlash from the scientific community can be expected. The National Research Council, no less, has recommended that biological and genetic factors be considered along with environmental in attempts to understand the roots of violence. It will be interesting to see how the need to conduct scientific investigation under the banner of free inquiry will accord with the need to avoid pejorative racial, ethnic, and gender-based distinctions in a world that has shown itself unable to handle such information.

Another area of study that may prove to be socially risky is the search for the genes behind intelligence. There is little doubt that a great many genes affect how smart we are. Estimates of the number of genes active in the brain range from 30,000 to 50,000. Of course, not all of these relate to intelligence—their tasks range from motor control to cellular housekeeping—but it is safe to say that the genes involved with intellect number in the thousands.

Some of these genes are structural, dictating how much gray or white matter a person has. Others are biochemical, affecting the ways the neurons, astrocytes, and other brain cells communicate with one another, store data, and obtain nourishment. Some genes relate to the function of the sensory organs and how their input reaches the brain cells. Gross and fine differences in many of these genes and their regulatory sequences probably account for the considerable variation in people's intelligence and also why one person is a whiz at mathematics and a failure at writing, while another excels at music but bombs at carpentry. Make no mistake, deciphering the genetics of smartness will take decades, for the issue is exquisitely complex. The genes involved are very likely acted upon by other genes and proteins in ways that cannot even be guessed at now. Moreover, influence from the environment is constantly shaping these interactions. It is probable that the human brain can truly fathom this aspect of itself only by means of computers much more powerful than those available today.

Still, it can be assumed that relatively soon, perhaps within the next ten years, we will have at least a preliminary notion of how genes control intelligence. At that time, there will begin a major effort to find a practical use for the information. And it will have repercussions affecting everyone on the planet and those yet to be born.

On balance, the campaign to demythify the brain by exposing the molecular nuts and bolts behind its operations is an enterprise of incontestable value. Simply enabling medicine to cope better with disorders of the brain would be justification enough. It has been estimated that in the United States alone, fifty million people suffer from various neurological disorders at a cost to taxpayers of upward of $300 billion a year. Thus a number of initiatives are under way to accelerate brain research at the genetic level. Among them is one sponsored by the Dana Alliance for Brain Initiatives, a consortium of more than fifty top neuroscientists who are attempting to achieve ten research objectives worked out at a Cold Spring Harbor meeting led by James Watson. Among these objectives: to develop treatments to minimize nerve cell destruction and restore healthy function following strokes or injuries to the brain or spinal cord; to find drugs and other measures to heal multiple sclerosis, Alzheimer's disease, Parkinson's disease, and epilepsy; to promote new treatments for manic depression, schizophrenia, and anxiety disorders; to devise new ways to reduce pain associated with cancer, arthritis, and migraine; to identify the genes responsible for hereditary deafness and blindness; and to elucidate the mechanisms behind learning and memory. The target date for the realization of these ambitious goals is the year 2000.

There are more puzzles to be solved, as molecular scientists return from the mountaintop with biology's equivalent of the stone tablets. What is it, for example, that drives the process of conception and human development? Obviously, genetic clues must tell the sperm to unite with the egg, and the egg to travel to the uterus for implantation. But as miracles go, these are just hors d'oeuvres. The real magic comes when the single fertilized egg cell begins flowering into the complex array of specialized cells that constitute the body: bone cells, muscle cells, nerve cells, heart cells, and so on. Something directs certain clusters of cells to specialize and then commands the irrelevant genes in those cells to shut off, so that skin cells make no thyroid hormone and kidney cells no lung surfactant. This process of differentiation is programmed down to the minute, unfolding according to a rigid timetable. While the new embryo is still but a few days old, it splits into three different layers, each destined to become one of the major venues of the body—skeleton, viscera, and epidermis. Primitive nubs, crimps, and folds then appear and gradually become diverse structures and tissues. By the end of the third week of life, the so-called neural tube starts forming, prefiguring the brain and nervous system. By the fifth week, the embryo has a head, a heart, and buds where arms and legs will be. The scenario follows a universal pattern that varies only in fine detail from individual to individual. The most deformed person is still eminently recognizable as a human being. Ears don't grow on feet. But why?

No one yet understands how a new organism can orchestrate the construction of itself. The egg cell has only a packet of instructions to go on, bequeathed from its parents in the form of DNA. But these are no ordinary, idle instructions. They are alive and dynamic. Some developmental genes, doubtless most, are concerned with specific tasks, making a nose, for example. But something else, with clipboard in hand, has to tell those genes when to start building the nose, how that particular nose is to look, and when the nose is done. Even Cyrano's nose didn't stretch into infinity. Thus there must be "executive" genes concealed within the embryonic DNA whose job is supervisory; that is, they start handing out job orders to other genes and fetal cells shortly after fertilization.

Biologists are confident that many answers to the great riddle of development will flow from the genome project. Once the executive genes are unmasked, the way they impose their will can be studied. And when the entire process is understood in its complexity, perhaps the underlying principles can be harnessed. Such an outcome would, in turn, have astonishing and, in some cases, disquieting results.

For example, if we unlock the secret by which body parts are fashioned within the womb, it may be that we can emulate the process in the laboratory. One can visualize an era in which new hearts and other organs might be grown from a patient's own cells for subsequent transplant, eliminating the agonizing and often futile wait for donor organs, not to mention the need to take antirejection drugs. Arms, legs, fingers, eyes, even spinal cord tissue, might similarly be cultivated to permit replacement after devastating injury.

By the same token, if science learns how to help cells recapture the plasticity they enjoyed prenatally, there is theoretically nothing to prevent the correction of many kinds of birth defects. Such flaws often occur because, on the crucial days during gestation when the schedule calls for making certain body structures, something—a drug or an environmental toxin perhaps—gums up the process. Nature never goes back to repair the loss. It just picks up at the next point in the blueprint. That is what happened with the infamous drug thalidomide during the early 1960s. The tranquilizer, taken for morning sickness in early pregnancy, jumbled key biochemical instructions during the critical time period when arms and legs were being fashioned, leaving the resulting children with grotesquely deformed limbs. But it is now possible to imagine intervening after birth, or even before, to compensate for the omitted developmental step, perhaps by carefully turning on the requisite growth factors in a painstaking procedure not unlike creating a sculpture. Such a gene-based procedure might be superior to surgery for conditions such as cleft palate or certain heart deformities.

But why stop there? Why not create an entire human being? Today's test-tube babies are not really such. They are conceived in a glass dish, it is true, but their continued viability and maturation depend on their being placed in a welcoming uterus for nine months. And what happens then—after the fertilized ovum is placed via catheter into the uterus—remains a total mystery, which accounts for the lack of success that plagues the procedure. Mastery over the genetics of development might permit artificial gestation, where a fetus is literally grown and nourished in some kind of high-tech, simulated womb. The benefits of such a capability are difficult to envision at present, and it does have its nightmarish side, as the chilling portrayal of baby farms in Huxley's *Brave New World* amply demonstrates. Yet there may come a time when the social and economic needs of the species makes it necessary, and culturally acceptable, for pregnancy as we now know it to become obsolete.

Of all the uses for genetic knowledge, surely one of the most compelling would be the healing of AIDS. Yet, the predatory immune disease has stubbornly resisted efforts at gene manipulation.

For a time in the early 1990s, there was great excitement over a proposed molecular cure for AIDS that involved tricking the human immunodeficiency virus (HIV) that causes the fatal illness. The major target of HIV in the human body are the CD4 T cells, the so-called helper cells that alert the rest of the immune system to the presence of an invader. The virus gains entry to these crucial cells because it has a surface protein that bonds to a receptor on the CD4 cell membrane. Once the virus is inside the cell, of course, it eventually incapacitates it. Researchers came up with a cunning proposal. Having isolated the gene for the CD4 receptor, they proposed to inject patients with sham constructs outfitted with CD4 receptors. The virus would be decoyed by the bogus receptors, bond to them, and drift off to oblivion. Or so the theory went. Unfortunately, technical problems proved insurmountable. The decoys didn't last long in the body, requiring injections every few hours, and various schemes to increase their life span proved fruitless. The strategy had to be abandoned.

The application of straightforward gene therapy to AIDS is hampered by a number of problems. Ordinary retroviral vectors will not work, because CD4 cells do not, as a rule, divide. Moreover, the gene to be inserted would most likely be not a standard human gene but some antiviral concoction that would deactivate HIV. Since such a gene would not be native to the human body, it might raise safety issues. In addition, HIV mutates so quickly that it may well circumvent the genetic obstacles with which investigators keep trying to foil it. Finally, there are the headaches that face gene therapy in general, such as the difficulty one has in getting enough cells to take up new genes to make a difference against disease. For such reasons, gene therapy has lagged behind other modalities, notably drugs and vaccines, in the effort to defeat AIDS. Of nearly sixty human gene therapy experiments that had been approved by federal regulators by the end of 1993, only four related to AIDS.

Nevertheless, several of these showed promise. In San Diego, the biotechnology firm Viagene, Inc., was proceeding with a protocol that called for putting the gene for an HIV surface protein into HIV-infected patients in hopes that it would stimulate an immune response against the virus that would slow progression of the disease. An approach by Gary Nabel's group at the University of Michigan called for collecting T cells from AIDS patients and genetically engineering some of the cells so that they produce a mutated form of an HIV protein called rev. Rev puts a damper on viral replication and theoretically would give the cells some protection from HIV when they are placed back in the patient.

Other treatments were also pending. At the Dana-Farber Cancer Institute in Boston, the virologist Joseph Sodrowski has announced plans to use HIV itself—rendered harmless, of course—as a vector to insert anti-HIV genes into cells. Despite the irony, HIV would make a good vector because it would be eminently capable of getting genes into the nondividing CD4 cells. It is almost alone among retroviruses in this capability.

While none of these specific proposals promises to vanquish AIDS outright, science stands a very good chance one day soon of learning enough about viruses—and about ourselves through the genome project—to bring the terrible illness to its knees.

Another area where we are certain to gain immeasurably from the Human Genome project is in our knowledge of evolution. Indeed, one bonus of molecular biology has been the development of the DNA equivalent of radiocarbon dating. By comparing the genes of living species, paleobiologists are now able to pinpoint the moment in prehistory when the common ancestors of various plants and animals split off and went their separate ways. But these comparisons have yielded an even bigger surprise, and that is how closely related all living things are. Each human being has genes that can be found in every plant or animal alive today, and some of these genes date back to the beginning of life. The more highly developed the organism, the more genes we have in common. For example, we share perhaps 80 to 90 percent of our DNA with chickens and cows but an estimated 99 percent with the chimpanzee. An astonishing thought, that fewer than one thousand genes distinguish us from our hairy, jug-eared, banana-eating sibling species!

Nature is frugal and hates throwing anything away. When a gene proves itself, she distributes it to various evolving organisms and makes gradual modifications as necessary, according to the evolutionary biochemist Russell Doolittle, of the University of California at San Diego. "Evolution is primarily chance and opportunity," says Doolittle. "If you have a gene that works, you can make it do many different things through natural selection. It's sort of like duct tape. You can use it to wrap ducts, if you want. But you can also use it to fix your running shoes, patch up walls, all sorts of things."

This tendency of nature to recycle and build upon genes is known as conservation. An example of a highly conserved gene is that for fibrinogen, an essential blood-clotting factor whose history Doolittle is tracing. Blood clotting marked a crucial evolutionary milestone that permitted one-celled organisms to evolve into multicellular creatures between 1.4 and 1.8 billion years ago. It accompanied the development of the circulatory systems necessary to feed and maintain the new partnerships being formed by cells. Fibrinogen's purpose was to plug leaks in the circulation that would otherwise have caused the cell colonies to suffocate or starve. On the surface, humans would seem to have little in common with lampreys, eel-like fish whose ancestors split off from ours nearly half a billion years ago, and even less with the sea cucumber, which was left behind by man's forebears 550 million years ago. But 50 percent of our fibrinogen gene is identical to the lamprey's and 40 percent the same as the sea cucumber's. These humble creatures are quite literally our blood brothers. We may use our duct tape differently than the lamprey— but then consider our different lifestyles. When was the last time you needed to clot blood at fifty fathoms?

The ultimate aim of tracing our genetic origins is, of course, to answer the eternal question of how we all got here. One of the things evolutionary biology is trying to tell us is something that every child of the 1960s already knows—we are stardust.

As stars die, they give off a brilliant flash known as a supernova, which spews an enormous dust cloud into the cosmos. This cloud contains all the known chemical elements, which formed as the dying star's core collapsed in a burst of neutrons. Eventually, a number of dust clouds coalesce to form new star systems. It was one of these dust clouds that some 4.6 billion years ago came together and formed the sun and all of its planets. From the rich spectrum of chemicals borne by the cloud, life evolved on Earth. All it took was a little water and some catalytic help from an energy source, to wit, lightning and the sun's ultraviolet rays.

In a famous 1953 experiment, Stanley Miller and Harold Urey of the University of Chicago mixed together methane, ammonia, hydrogen, and water—the fabled primordial soup—and shot sparks through it to simulate lightning. To their amazement, they discovered in the sizzling ooze traces of amino acids, the building blocks of proteins. A decade later, Cyril Ponnamberuma, of NASA, while tinkering with a similar soup, discovered that he had created adenine, one of the five nucleotides, including uracil, that make up DNA and RNA. Then, in 1983, Ponnamberuma, by now at the University of Maryland, learned he could synthesize all five nucleotides in the laboratory under conditions believed to have existed on the infant Earth. In a spectacular encore that same year, he identified each of these nucleotide building blocks in a piece of the Murchison meteorite that landed in Australia in 1969. The message was plain enough: the components of life are so easy to make that they could have been created here on Earth or else in outer space and brought here by comets or meteors. Take your pick.

Nature doubtless experimented over numberless eons until she finally hit upon a chemical combination that had the power to make copies of itself. This ability to reproduce is what we now think of as the first life. But what was this mother chemical of us all? For many years, it was assumed that it must have been DNA, but recent findings indicate otherwise. The fact is that RNA, long disparaged as the slavish messenger boy for DNA, may actually have been created first.

Thomas Cech, of the University of Colorado, and Sidney Altman, of Yale, shared a Nobel Prize for shedding light on RNA's new role. Their work appears to solve the conundrum of which came first, DNA or proteins. The riddle is not so simple as it sounds, for while DNA is undeniably required for making proteins, so are proteins necessary for DNA to carry out its job, helping with replication, self-repair, and many other tasks. Enter Cech and Altman. RNA, they showed, is capable of acting in either capacity. It can function as a protein, catalyzing chemical reactions the way enzymes do. It can operate upon itself, making changes in its own genetic material. And it can make proteins on its own. Evolutionists like Cech therefore theorize that RNA, being more versatile, was the aboriginal molecule of life. But being error prone, it created DNA as a way to store genetic information safely.

The thesis that our biological odyssey began with an "RNA World" is not universally accepted. In any case, the forerunner molecules of genes and proteins, whatever they were made of, eventually came together for the mutual good, taking on the qualities of an organized system that we associate with living organisms. "At some point," says Cech, "there was a breakthrough organism that survived and became the predecessor of all life on Earth." But what was that breakthrough organism? Most likely, we will never know. Certainly, it consisted of a single cell and dwelled in the oceans. It might have been bacteria, or perhaps blue-green algae. Fossilized examples of these earliest life forms have been found in Australian rocks that date back more than three billions years.

For more than a billion of these years, the single-cell organisms luxuriated in their good fortune. Random mutations in their genetic material—occurring either during cell division or as a result of environmental exposure—created new genes, which, acting in concert with natural selection, produced a host of variant species. Then, somewhere around 1.8 billion years ago, a simple but fateful event occurred. In what scientists have come to call, whimsically, the "big gulp," a large single-cell organism swallowed up a smaller one, which promptly set up housekeeping inside, a process that we can still observe in microscopic creatures today. In one of these scenarios, the smaller cell had already evolved genes for transforming certain chemicals into energy. It was a vital trait that the small cell brought to its host, making multicell life possible.

Modern humans still carry those ancient genes in each of their cells, because they provide the principal source of energy for metabolism. These genes exist outside the nucleus in the cell's cytoplasm in what are essentially extra chromosomes, known as mitochondria.

In another scenario, smaller cells conferred on larger hosts other genes for converting sunlight into energy, facilitating development of plant life. This process, as every schoolchild knows, is called photosynthesis, and from carbon dioxide and water it creates oxygen. Thus these early precursors of plant life transformed the atmosphere, making it breathable for future animal life.

Just as primordial genes and proteins, most likely mediated by RNA, joined forces to create one-cell organisms, so may bacteria have been the mothers of us all. At least that is the theory of Lynn Margulis, of the University of Massachusetts, who says that the tendency of bacteria to unite in a form of nonsexual mating hastened the evolution of multicellular creatures. The complex functions of human cells, she argues, were formerly the properties of individual bacteria that linked up to form collaborative units. Her theory, once considered heterodoxy, is becoming widely accepted. "We are made, if you will, of conglomerated bacteria," Margulis insists. If we behave with the selfish rapacity of microorganisms, it may be because we are walking colonies of one hundred trillion tiny beastics.

Like the single-cell creatures before them, the new multicell organisms coalesced into yet larger colonies, their genes merging, expanding, mutating, until new species began to be generated in extraordinary numbers. Ancient rocks from the Canadian Rockies attest silently to the explosion of new life forms that erupted some 570 million years ago. Fossils etched into the rocks that were once on the ocean bottom demonstrate that all the animal body types were created then, including a number that, for unknown reasons, did not survive.

Nature has periodically undone her creations in fits of mindless rage. About 200 million years ago, some 96 percent of all life was wiped out in a mass extinction. The creatures that by chance survived gave rise to the age of dinosaurs. Dinosaurs then proceeded to dominate the world until their own luck ran out 65 million years ago. Somehow, small mammals escaped annihilation that time, and the way was clear for the proliferation of mammalian life that ultimately led to man.

This disturbing tendency of nature to erase her handiwork has led to a new twist on an old theme. Evolution is no longer considered solely in terms of the survival of the fittest, nor do many paleobiologists still think that the ascent of man was inevitable. The genetic mutations that drive evolutionary change occur at random and move in all directions, backward as well as forward. The species that seem to survive happen to be in the right place at the right time. The arbitrary mutations that produced them suited them to survival in their particular environment. Thus the central hypothesis of evolution has been restated as "the survival of the luckiest," by Motoo Kimura, of the National Institute of Genetics in Mishima, Japan, a seminal thinker whom many consider to be the new Darwin. It is Kimura's belief that the emergence of man was essentially a fluke, made possible by the demise of the giant reptiles.

Nevertheless, mankind did evolve and with surprising rapidity. It has been determined that man's direct ancestors and the great African apes diverged only five million years ago, instead of twenty million years ago as originally thought. As many as five different prehuman hominids may have emerged from that point on because of random mutations, but only one of them survived to become *Homo sapiens.* The modern human body was essentially completed about one million years ago, but the brain itself has doubled in size since then, an unprecedented rate of growth for any body structure in the fossil record. To many authorities, that is evidence that evolution selects strongly for intelligence.

"I would conclude from that that intelligence is an inexorable consequence once life starts anywhere," says the astronomer Robert Jastrow, who argues that intelligent beings almost certainly exist on other worlds. Think of it! The dinosaurs, had they endured, might have developed the same capacity for self-awareness and thought that we now possess.

What is the new story of evolution telling us? "The message is that there is no message of intrinsic human superiority," observes the Harvard paleontologist Stephen Jay Gould. "We're just a lucky, contingent result of evolution, and we ought to be quite pleased with that. We shouldn't seek the source of morality in nature, because it isn't there."

Where is it then? "It's in ourselves!" asserts Gould. "We have to define the meaning of life for ourselves." It may behoove us to do this as speedily as possible, it seems germane to mention, for with our growing mastery over genes, we are approaching a time when we may control forevermore the process of evolution itself. The genome project will very probably advance greatly our knowledge of how genes change to reformulate life forms. Barring a suicidal act, nuclear or otherwise, or else an extraterrestrial event such as that which may have wiped out the dinosaurs, it is probable that we will at length be equipped to make evolution our servant, producing new species at will, or making alterations in our own DNA, accomplishing in a matter of days or weeks what used to take thousands of years. Will we do this with wisdom or wickedness? We had best consider such matters as we proceed with the genome project, for it may ultimately place a godlike power in our hands.

If control over evolution, once unthinkable, is possible in the not too distant future, what can we say about other capabilities that may drop into our lap as we come to understand the genome? It is not too hard to imagine that the putative aging genes will be isolated and subject to laboratory manipulation. Will that portend the indefinite extension of the human life span?

May we subsequently come to understand the means by which the lifeless chemicals within DNA and RNA can be reanimated? In other words, are we destined to divine someday the very secret of initiating life? Will it become possible to raise the dead from their remains? Will feats such as the resurrection of the dinosaurs as fancifully portrayed in Michael Crichton's thriller *Jurassic Park* become reality?

Lest we become giddy with possibility, the state of the art for targeting genes and altering them bears examination. For several years, a particularly promising research theme has kept resonating through molecular genetics like a leitmotif in a Wagnerian opera. Homologous recombination, as it is known, is nature's own gene-healing mechanism, and mastering its secrets occupies scientists intent on inserting foreign DNA into *specific* places in *specific* chromosomes. Pioneers include the Italian-born, Harvard-trained Mario Capecchi, a Howard Hughes investigator at the University of Utah, who has spent fourteen years learning how to place genes in animal bodies with precision—sans a retrovirus or other vector that dumped them willy-nilly—in order to kick out the bad copy of a gene while leaving only the good. If gene therapists could mimic nature, no longer would they have to smash into unsuspecting chromosomes; they could quietly go in and surgically alter a gene. Such visionary researchers as Capecchi, Oliver Smithies, and the Nobel laureate Paul Berg are obsessed by a fact of life: homologous recombination occurs in simple organisms all the time. "When a piece of foreign DNA is introduced into yeast, or *E. coli,*" explains Berg, "the foreign segment finds its homologue, its matching partner, wherever that segment is situated on the chromosome, and recombines. The DNA doesn't ever get incorporated into an irrelevant, or nonhomologous, site."

The trouble, says Berg, is that mammalian cells, with their boundless genome, do not respond in this manner. Recombination runs rampant.

Copies of DNA end up pasted everywhere, without rhyme or reason. But DNA is DNA; in all living organisms it is the same and perpetually under repair by special enzymes that cut and splice it flawlessly. With all the licentious recombining going on, how does the body know how to do that? "Our cells are constantly irradiated by ultraviolet rays from sunlight," Capecchi points out. "We eat chemicals that are mutagenic, inhale pollutants, smoke cigarettes. If nature didn't have a way of repairing the damage, we wouldn't have evolved beyond a few cells."

In 1985, Smithies and Raju Kucherlapati published in *Nature* a landmark experiment showing that the genetic machinery of a mammalian cell is indeed sometimes willing to accept a transplanted gene *spontaneously*, and then even go so far as to use the new gene as a spare part to replace or repair a resident gene. Smithies and Kucherlapati plunked a hemoglobin gene precisely where they wanted it to go, in the hemoglobin gene site on chromosome 11. They did this by making their gene "sticky," by exaggerating the natural magnetism that causes genes of similar makeup to attract one another. What the researchers had created in a petri dish was, in effect, the cure for sickle-cell anemia and other lethal blood disorders, but their system was inefficient, to say the least. That was the hitch. Only one gene in a million would home to the right spot, though Smithies was quick to point out that because there are three billion base pairs of DNA in every cell, the experiment resulted in targeting that was three million times better than chance.

The penchant of DNA to recombine homologously might explain a riddle that has long perplexed geneticists—namely, the double-stranded nature of the genetic material itself. Why are there two complementary strands? Only one is involved in protein synthesis. Viruses, in fact, possess but a single strand and are pros at survival. Perhaps in higher organisms, the double strand evolved into nature's way of providing emergency road service for gene breakdowns, with the second strand furnishing a template of information to repair the first?

For scientists working in this arduous field, merely developing the ability to recognize the rare homologous recombination events that occur in mammalian cells has been no small matter. Elaborate selection chemistry had to be invented so that the few repaired genes that exist in what Capecchi calls "a vast arena of scattered, nonhomologous recombination activity" could be singled out and grown up in culture. Nevertheless, in November 1988, Capecchi edged the technology from basic gene targeting to the next plateau, gene repair. He transplanted new genes into cultured mouse cells at the precise locations of defective genes, whereupon the healthy genes set about fixing or exchanging the old, using themselves as the model. As a target, Capecchi singled out a mouse oncogene called *int-2* and tried to replace it with a cloned version. It worked like a charm. To introduce the genes into cells, Capecchi jolted them with a shock of electricity, a technique called electroporation, which has become standard in this research. "The electricity opens up little holes in the cell membrane and allows the DNA to go inside," Capecchi says. "Then, it closes up back again. Remarkably, the new DNA scans all the chromosomes until it finds the counterpart gene and stops next to it. Then an amazing natural phenomenon occurs: the two genes exchange chemical information and combine."

Again, such a wondrous event did not occur very often—only about one in two thousand cells that had accepted the inserted genetic material had situated it in the right place, although that was a stellar improvement over the one-in-a-million opening shot from Smithies and Kucherlapati in 1985. Capecchi then submitted all the cells to the tough chemistry exam that the gene-repaired cells must pass if they are to survive, finding to his delight that nineteen cells had the correct gene mutation at the correct address on the chromosome. He had done it! "We altered the genes the way we wanted to," says Capecchi. What was happening inside the cell? "We activated the repair mechanism. Sometimes, it can fix things by using the opposite strand on the DNA helix. If only one strand is broken, it can use the other strand to correct it. But if both strands are broken, the repair enzymes must go to the complement, the counterpart gene on the partner chromosome. Remember, we have two copies of every gene."

The next step for scientists was to insert mutant genes into three-day-old mouse embryos, a stage when all the embryonic mouse cells are still "pleuripotent," meaning they had not yet differentiated but would soon strike off to make a whole mouse. On those occasions when the technique worked, it resulted in an animal that carried in every cell of its body a specific, preplanned genetic alteration. The selective breeding of such animals resulted in a new class of expensive germ line chimeras known as knockout mice, and by the early 1990s they were revolutionizing medical research.

Knockout technology can be applied to any single-gene defect, in theory, and animals soon were created that carried targeted disruptions in a host of genes (HGPRT, *ab1*, *en-2*, *n-myc*, B-2 microglobin, *IGF-2*, *int-1*.) The study of immune disorders, in particular, lends itself to homologous recombination experiments because lack of a single gene is rarely lethal, and, in any case, such mice can be kept in sterile conditions.

But should you want to study the CFTR gene that causes cystic fibrosis, for example, you merely need to knock it out of the mouse's genome at the embryonic stage and see what happens—a medical team at the University of North Carolina, in Chapel Hill, working with Smithies, created the first litter of CF mice in 1992. Other scientists have been specializing in the brain, studying such infinitely complex matters as the biochemistry of memory. Teams led by the MIT Nobelist Susumu Tonegawa and Eric Kandel of Columbia University knocked out brain protein genes for what are known as kinase enzymes, resulting in mice that seemed perfectly normal but were learning disabled, no longer able to remember an important task, such as how to swim out of a water maze. Such experiments have produced the strongest linkage yet between the action of certain kinases in mammals and a brain process called long-term potentiation that stores spatial information and enables it to be recalled.

By 1993, knockout technicians were busily eliminating genes during various stages of mammalian embryonic development and determining how genes control it, particularly the tightly conserved "homeobox" genes (intensively studied in fruit flies), which help specify where and when specific parts of the body will develop.

It is the fervent wish of pacesetters like Mario Capecchi that mice—even fancy designer mutant mice—lead to men. But Capecchi cautions that it will be years before homologous recombination is used for gene therapy. "We must target the stem cells of embryonic animals in order to alter their genome," he explains. "Human stem cells for the various body organs have not yet been found. So, even if society permitted me to attempt germ line gene therapy on a human embryo, I wouldn't be able to do it. But when stem cells are located for the skin, heart, brain, liver, and other systems—ah, then the story will be different."

For the time being, Capecchi believes that the retroviral and other vector systems being employed for gene therapy will be "fine." But they won't compare to his system, once the bugs are worked out. "Eventually," he states, firmly, "homologous recombination for human gene therapy is the only way to go."

There are many things science can do to help people even before the technical gymnastics of direct gene replacement have been mastered. Since the 1970s, the biotech revolution has occurred because science has learned how to tap into the body's own drugstore and synthesize the proteins that the body makes, often in quantities too small even to be measured. Not long ago, a researcher would spend an entire career trying to purify a few micrograms, hardly enough to be tested.

Now, once the genes are cloned, the proteins they carry the recipes for can be made in vats. So content are human genes in new habitats that they will even mass produce their proteins when spliced into plants and animals. Already the first genetically engineered goats, sheep, and cows are producing important human proteins in their milk; crops are being devised that will be harvested for the drugs they yield rather than the food; and plants soon will be marketed that have developed natural resistance to pests because various insecticidal and antiviral genes have been implanted in them. Nowadays scientists can even be found sitting at keyboards like modern-day Mendels, punching out computer-generated models for designer proteins that, someday, will improve on their natural counterparts. By artfully adding or eliminating an amino acid or two—or, for that matter, by constructing new amino acids that nature never thought of—science may produce lifesaving proteins that are smoother, stickier, or otherwise more efficient.

The list of new gene-based medicines already making an impact on clinical practice is long. Monoclonal antibodies, for instance, are true magic bullets with the unique ability to home in on any specific cell in the body. Scientists in 1975 invented a way to make endless copies of antibodies by fusing the cells that produce them with continuously multiplying cancer cells. Monoclonals have proven to be extremely useful in medicine because as they find their way to diseased cells, they can be tracked or used for the diagnosis and treatment of many disorders. Spotting the AIDS virus in patients and donated blood was made possible because of monoclonal antibodies, for example, as were home pregnancy tests. Mini-antibodies promise to be even better magic bullets. In 1990, British and American scientists discovered how to make key parts of antibodies. Instead of mating cells to produce monoclonals, they splice the key antibody genes into bacteria, thus producing with greater speed and accuracy a wider assortment of disease fighters.

Another strategy involves gene foolers, in which genetic messages are blocked from causing diseases, ranging from cancer and AIDS to the common cold. The first step in genetic message transmission occurs when a gene translates its DNA into the mirror image molecule, RNA. If the message is wrong, or garbled, and the misprint occurs in sufficient numbers of cells, the result may be disease. But researchers now are blocking the disease message by creating synthetic RNA copies of the gene and putting them into cells. There the artificial genes recognize the reverse-image RNA copies and stick to them like Velcro, thereby obliterating the destructive message and preventing it from being read. The technique is called "antisense" medicine because it blocks the "sense" or message of a gene. As of this writing, Jack Roth at M. D. Anderson, in Houston, is poised to begin the first human experiments on lung cancer by means of the antisense technique. In practice, blocking genes this way proved problematical because the antisense segments did not bind tightly enough to messenger RNA. However, in June of 1993, researchers at Gilead Sciences Inc., of Foster City, California, reported in *Science* that they could strengthen the antisense molecule so that it bound from ten to one hundred times more firmly. A better Velcro. At the same time, another private company, Isis Pharmaceuticals, Inc., of Carlsbad, California, had begun clinical trails of an antisense drug to treat the papilloma virus that causes genital warts.

Other tactics under way include gene decoys, which trick viruses and other invaders into not attacking key cells, and bogus proteins that look, to a virus, like proteins that it needs to construct new viruses but that are built to be faulty. It may be possible to treat some diseases by putting missing genes into skin cells and implanting them. Tests of transplantable gene drugs in animals have shown remarkable promise against hemophilia, thalassemia, and other diseases. Vaccines, in the past, were made by killing or weakening a virus to provoke an immune response in someone before a virus invaded. Now recombinant DNA techniques are being used to make a new breed of vaccines that should be safer, more effective, and more versatile against a wide variety of disease-causing microbes. By disassembling the genetic codes of the germs, scientists can isolate and clone those genes that are harmless but that can still stimulate immune protection. The new vaccines are being designed to rid the world of many ancient diseases such as hepatitis, malaria, gonorrhea, and schistosomiasis.

These gene drugs will revolutionize medicine to a far greater extent than the discovery of antibiotics did some five decades ago. Antibiotics typify the old hit-or-miss method of finding drugs. Almost all compounds in use today are chemicals extracted from nature—molds, yeast, plants, and microbes—and, being foreign to the body, they often cause side effects. The new drugs, by contrast, are synthesized copies of the body's own potent chemicals; because they are natural, they heal cleanly, allowing researchers to replace or correct proteins that are missing, overproduced, or functioning improperly. This permits treatment of the causes of diseases, rather than just the symptoms.

Finding genes and using them effectively will merely be the preliminary issue raised by the Human Genome Project. Many legal and ethical issues are sure to arise. As genes are isolated, for example, who will own them? In the early days of gene therapy, attempts were already being made to cash in. Will the benefits accrued from the knowledge be open to everyone or just to the rich? So far, genetically engineered drugs have not been cheap or widely available.

In broader terms, what will the explosion of information spawned by the genome project mean to society? What impact will it have on the soul of science?

Back in 1983, Health and Human Services Secretary Margaret Heckler restated the classic mission of the National Institutes of Health. "NIH," proclaimed Heckler on a visit to the campus, "is an island of objective and pristine scientific research excellence untainted by commercialization influences."

Objective. Pristine. Excellence untainted. What a difference a decade makes! By 1994, even though gene therapy had yet to cure a single human being, it was the hottest ticket in medicine. Most of the stars of molecular biology had affiliations with companies—French Anderson, Richard Mulligan, David Baltimore, Ronald Crystal, Leroy Hood, Walter Gilbert, Craig Venter, and Mark Skolnick, to name a few. Fame and fortune were there for the taking, with the nation's handful of original gene therapists constantly on the move to hastily built institutes, taking their entire labs and workers' families with them. As new players jumped into the game, floods of new proposals—many of them me-toos—were sailing through the once superfinicky RAC every three months, on the basis of what many observers deemed the flimsiest of scientific preparation. Moreover, charges were being made that the hype regarding the new therapy had sparked a gold rush replete with divided loyalties, potential conflicts of interest, and hidden agendas.

In the long run, it may be that simple greed will damage the genome project and human gene therapy far more than scaremongering and eugenics scenarios ever could. Scientists traditionally resist any limits to their freedom. Society concurs, feeds them well, funds their academic research, grants them tenure, and rarely even asks for an accounting. Scientists, unlike artists, take public funding for granted; the only issue is how much it will be. Moreover, the public has shown great tolerance for the lust for fame that drives scientists, but there is little tolerance for those who have been caught profiteering. The pursuit of pecuniary gain interferes with the process of seeking truth, most people believe. According to David Blake, senior associate dean of the Johns Hopkins School of Medicine, finding appropriate rules for stock ownership by academic researchers "is one of the most ethically charged areas of medicine." Each institution has its own rules; each case is judged on its own merits; but no matter how the ethics are sliced, a university researcher affiliated with a private company is serving two masters. Graduate students often are unsure whom they are working for. Is the direction of the research being driven by the desire for monetary gain or by the Socratic pursuit of knowledge? Is it asking too much for money not to be part of the equation?

Researchers in private industry have never earned the respect that "pure" academic researchers command, because secrecy and proprietary rights preclude the openness and publication that are at the heart of the scientific enterprise. The inherent ethical conflict has dogged the biotechnology industry since the 1970s, but the enormous possibilities offered by the Human Genome Project brought it to a boil in the early 1990s.

Wrestling with so many competing interests, the regulating bodies of biomedicine were displaying deep ambivalence. They had a problem: simply put, biologists, always the underpaid sisters of science, suddenly had the chance to make a bundle with discoveries that resulted directly from publicly funded academic research. And they were shamelessly doing so, often shattering the boundaries of good taste in their zeal to form private companies, join advisory boards of start-ups, welcome venture capitalists into their labs, and accept equity positions with the promise of lucrative stock deals down the road.

What was wrong with this? Nothing, according to the scientists, the universities that backed them (which stood to cash in on licensing and royalty income), and advocates within the federal government who since the 1980s had been seeking ways to promote technology transfer. Smart people, the thinking went, were going to be able to cash in on their discoveries, and in the process the public would ultimately benefit.

Yet, behind closed doors, many senior scientists were worried that the public would perceive things differently. For one thing, congressional overseers, led by Representative John Dingell, a Michigan Democrat, had put the fear of God into NIH with their hunts for irregularities (government researchers in 1992 were even afraid to accept free doughnuts). For another, the mere appearance of conflict of interest could make officials vulnerable, as James Watson learned to his dismay. And he had not even joined a company, but merely owned stock in a couple of blue chips.

It is ironic that Watson was skewered for a supposed ethical lapse. For he had taken a bold step to insulate the Human Genome Project from any criticism that it was insensitive to issues of propriety. When the project was being organized, it seemed imperative to Watson and others that there be some sort of watchdog unit to address major ethical and social policy questions as they arose. Given that the highly charged information expected to flow from the project was likely to have a profound effect on society, the planners sought to do the social and philosophical equivalent of environmental impact studies well ahead of time. The idea was to avoid the kind of moral swamp that followed the development of atomic energy in the 1940s. Thus was the Ethical, Legal, and Social Implications Branch (ELSI) of the genome office established and placed under the leadership of Nancy Wexler, of the Hereditary Disease Foundation. It was given a multimillion-dollar budget, which for ethicists seemed like the moon.

Early on, the ELSI working group identified four areas it considered to be of the highest priority for study. The first of these was fairness: the need for guarantees to ensure that genetic information would not be used to discriminate against people on the basis of genetic endowment. The second was the allied issue of privacy, by which was meant the right of individuals to control disclosure of genetic information about themselves. The third area involved the delivery of health care and embraced concerns that genetic services, including counseling, be available to all. The final area was education, that is, the raising of public consciousness about the new biology and the opportunities and problems it creates.

The launching of ELSI was a first in the annals of science. Heretofore, scientific inquiry had operated more or less in a vacuum. Practitioners distanced themselves from responsibility for their creations with the notion that their work was essentially value neutral and that it was the public's job to appraise and use the results as it saw fit. Now scientists were joining with experts in the law, health care, the social sciences, and the humanities to weigh the potential fallout of their research before, not after, the genie was out of the bottle.

For Wexler, science *was* value neutral. The danger lay in ill-advised or malign applications. "Genetic information itself is not going to hurt the public," Wexler said. "What could hurt the public is existing social structures, policies, and prejudices against which information can ricochet. We need genetic information right now in order to make better choices so we can live better lives. We need the improved treatments that will eventually be developed using genetic information. So I think the answer is certainly not to slow down the advancing science, but to try, somehow, to make the social system more accommodating to the new knowledge."

Indeed, these were not idle anxieties. There were disquieting signs that the twin fields of genome research and gene therapy were drifting into ethical anarchy, where existing rules did not apply and new ones had yet to be worked out. In late 1992, the well-established, if sometimes boisterous, process by which proposals to perform gene therapy were approved was threatened by the exercise of naked clout. Senator Tom Harkin, a Democrat, wrote Bernadine Healy in October of that year on behalf of a San Diego woman who was dying of a brain tumor. He asked that the woman's doctor, Ivor Royston, of the San Diego Regional Cancer Center, be allowed to perform a kind of gene therapy on the woman by rigging her cancer cells with genes for interleukin-2. The goal, typical of many such experiments, was to immunize her against her own tumor. The problem was that Royston had already sought RAC approval for a similar experiment and the RAC had rebuffed him, arguing that he had not shown its safety or efficacy. Now Harkin was trying to go around the RAC, asking for a special dispensation known as a compassionate exemption. Why should Harkin take such a stance? It turned out that the patient and her family, originally from Harkin's home state of Iowa, had a long association with Democratic politics in general and with Harkin in particular. The woman's husband was a politically astute attorney who had roots in the White House dating back to the Kennedy administration, and her sister-in-law had worked in Harkin's campaigns.

In and of itself, the idea that a U.S. senator might try to help an old friend was not terribly sinister. It happens all the time. Moreover, Harkin had cloaked his appeal, making it part of a blanket request that NIH adopt a policy expediting review and approval of gene therapy proposals for terminally ill patients who might otherwise die before the normal RAC review process could be completed. What disturbed many observers was that Harkin had more leverage than most arm-twisters. He was chairman of the Senate appropriations subcommittee that controlled the NIH budget. And, sure enough, after first flatly rejecting Harkin's plea, Healy did a dramatic about-face, approving it on December 28, 1992. At the same time, she called for an emergency meeting of the RAC to draft new guidelines for future compassionate exemptions of unapproved gene therapy protocols. Many gene therapists, afraid that the RAC was being subverted by politics, and that the new technology was about to spill prematurely into doctors' offices in every hamlet and byway in America, began speaking out. "This is monstrous," protested the RAC member Robert Haselkorn, of the University of Chicago. "The worst thing I have seen," fumed Dusty Miller, also of the RAC. "It throws the RAC review to the wind. . . . If somebody wants to put recombinant split pea soup into people, that's fine."

As for conflict-of-interest rules, in 1989, sensing the changing winds, NIH proposed a tough policy for grant recipients that disallowed "personal equity holdings or options in any company that would be affected by the outcome of the research or that produces a product or equipment being evaluated in the research projects." There was such a howl of protest—more than eight hundred angry letters from scientists and research institutions—that the government retracted the guidelines and consigned the idea to limbo. Such a ban on stock ownership, warned the Association of American Medical Colleges, "would discourage attraction of the best minds into biomedical research, stifle creativity and diminish technology transfer at a time when advances in research offer unprecedented opportunities for the commercialization of research results for the public benefit."

Chastened but still worried, the NIH in mid-1991 came up with another set of guidelines, but these never made it past Director Bernadine Healy's desk, because she deemed them too weak. Again the rules of the game remained unclear, although researchers considered no guidelines good guidelines.

But then Irv Weissman got drummed out by the Howard Hughes Medical Institute, and things got very complicated. Weissman, a Stanford pathologist, was everyone's best bet to pull off the crowning feat of immunology—identification of the elusive bone marrow stem cells that reconstitute the human blood system. Such a discovery would hold enormous implications for the treatment of cancer and immune system disorders. And nearly everyone knew that Weissman had formed a company, SyStemix, Inc., and that it was pursuing parallel research. This was fine with Stanford. But in late 1991, when Weissman announced to the world that human stem cells were at last in hand, the great news emerged not from the university but in a SyStemix press release crowing that the firm had been granted a patent on this vitally important technology.

Weissman's claim, in fact, has yet to be replicated, but what mattered was that the story appeared not in the news pages of the nation's newspapers but in the business pages. Wall Street tagged SyStemix as a comer. Within months, the pharmaceutical giant Sandoz had bought 60 percent of SyStemix's stock for a walloping $392 million, boosting Weissman's net worth overnight by an estimated $16 million—the dream of every scientist who got rich by the sweat of his brow.

But in October 1992 Weissman was forced to resign his coveted post as Hughes investigator after the institute decided that his link to SyStemix constituted a possible conflict of interest. Weissman contended that the Hughes people had known about SyStemix when they began funding his thirteen-member lab at Stanford in 1989; Hughes officials, in turn, said they hadn't known until they read about it in the *Wall Street Journal* in December 1991.

As the most prestigious private philanthropist in biomedical research and the major funder of the top two hundred labs in the world, Hughes does not allow "significant" equity holdings by its investigators—5 percent being about the maximum, although the greater the value of the company, the smaller the piece the institute permits its superstars to own. By 1993, it had become apparent that what Hughes actually had done in the Weissman case was to blow the whistle on the capricious and inconsistent financial ties between academe and the private biotechnology industry. It also called attention to the bewildering, federally mandated notion of technology transfer, which by then had resulted in two hundred CRADAs between NIH and the private sector, leading to dark mutterings about government researchers fattening their portfolios while feeding at the public trough.

There is not a scientist in the world who does not know in his or her heart what constitutes conflict of interest. But was conscience enough? "Our primary interest is that a scientist's work is science driven," and not motivated by personal gain, stated a Hughes official, explaining the institute's position. But this position might have to be reversed in the years ahead. Most Hughes investigators are of Irv Weissman's caliber. If sufficient numbers of the top scientists in the world become minimoguls, what will Hughes do? Fire them all?

Individually upsetting as these instances of ethical drift are, their larger implications are far more troubling. For at the very moment when it ought to be proving itself qualified by dint of wisdom and character to handle the promethean knowledge that biology will soon deliver into its hands, the human race has been acting like . . . , well, the human race. Greed, hypocrisy, political muscle, double standards, the nod and the wink—all the familiar shortcomings of our species are on full and garish display, before the new technology is even out of its swaddling clothes. What's more, the high priests themselves, the scientists and physicians who always claimed to put principle first, now give every sign of being co-opted by the system. What it portends seems frightening. Is the ultimate power over life and death to be treated like just another side of pork, to be divvied up equally among the priests and the denizens of the Beltway? Is control over the human blueprint to be ceded to a collection of financiers, finaglers, bureaucrats, and Potomac insiders who will steer the benefits to a chosen few and determine future policy on the basis of personal and parochial agendas.

The problem might seem trifling when the issues are relatively small. But the patterns being established today seem to bode ill for the era when policymakers will have to deal with questions of much greater magnitude. For example, when and if the genes and regulatory sequences that dictate native intelligence are understood, and a safe, workable, and painless system for delivering genes to the brain is perfected, how will the knowledge be used? There will be strong pressure to try and boost the intellect of children and others by means of the techniques of gene therapy. Who will make the determination that such manipulations are desirable? And if they are so deemed, on what basis will we decide who benefits from the new technique? Let us say a safe procedure is worked out for adding five or ten crucial genes that would boost IQ by, say, fifty points. Certainly, there would be a hefty price for such a procedure, probably too high to make "smart-lifts" universal. Will the technique then go only to the sons and daughters of the well-to-do, who, being smarter, will probably succeed in life even more than they might have otherwise, further widening the gap between the economic classes? Is such a possibility acceptable in a democratic society? Will the ensuing debate threaten the social fabric?

The same dilemma will arise with other human qualities. Suppose it becomes possible to guarantee greater height, athletic prowess, musical skill, or literary talent. How will these be apportioned? And might it not create a synthetic and somehow degraded culture if anyone could become a Mozart or a Michael Jordan on demand? If there were hundreds of Shakespeares, would the excess supply of bards diminish the demand for their output? Would a *Hamlet* seem as admirable if any high school sophomore could crank one out?

Assuming that the aging genes are eventually isolated, and found to be subject to manipulation, can we say for sure that bringing longevity-enhancing procedures on-line will be advantageous? Obviously, the prospect of longer life will be very alluring. Again, though, questions of equity arise. If the procedure cannot be made available to all, should it be available to some? And if it can be made universal, will that not have an overwhelmingly negative impact on society? The planet cannot support the population it hosts now. How would it handle the resulting glut if death were virtually eradicated? The supply of food, jobs, goods, and living space itself would quickly become depleted.

And when at last the secrets of evolution are revealed, and it becomes possible to alter the nature of the species itself, how will we deal with that? The power to tamper with our humanness, of course, presupposes that the current dispute over germ line versus somatic cell gene therapy is resolved in favor of germ line. One cannot hope to influence the course of evolution without altering the genes not only of a large and diverse segment of humanity but of their descendants as well. Otherwise, the intervention will not "take." But assuming that germ line techniques become permissible, not to mention successful, then a startling array of possibilities is opened up.

It might, for example, become possible to create specialized subspecies of humans—"metahumans"—who are better adapted to certain tasks. Astronauts on interstellar voyages would benefit if they were able to subsist on a plentiful, nonperishable food supply. Thus we might want to outfit them with termite digestive genes so that they could live on a diet of cellulose. People who work in desert climes might weather the dryness better if their genome were judiciously enhanced with certain key genes from the camel or the prairie dog. The list could go on, of course, but the point is that it is almost assuredly going to be possible to produce human hybrids with capacities far beyond the norm. Clearly, such a technology will involve ethical questions that dwarf virtually anything we have had to deal with before. Begin with the question of consent. Who would be subject to such procedures—only those who volunteered? If their germ lines were altered, would their children be considered "volunteers" as well? What about status? Would underwater farmers with webbed feet and gills be considered as fully human as the rest of us?

Beyond such novel questions lie the possibilities for misuse. One can easily imagine armies of genetically engineered soldiers with hides impervious to shellfire, or with aquiline eyesight and gorillalike strength. How can we ensure that the ability to hybridize people does not get into the hands of an unscrupulous chief of state?

On a broader level, suppose as a species we deemed it desirable to hasten the wholesale evolution of *Homo sapiens* into a still more sapient form by seeding today's population with the genes for greater intelligence. Presumably, and without the tedious process of natural selection, the entire species would become vastly brainier in several generations. Or we might imagine similar programs to make us taller, more agile, more resistant to disease. For that matter, a host of chimeric possibilities come to mind. Why shouldn't humans realize the dream of flight? The addition of a set of bird genes—or the stimulation of latent avian genes already within our genomes—might provide us with wings, feathers, lightweight bone structure, a more aerodynamic shape—indeed, the full array of organic paraphernalia necessary to fly. Fanciful, but certainly possible.

It is to be hoped, however, that those who may someday control the managed evolution of humankind will put the technology to worthier use than to turn us into a race of ospreys. We might, for example, want to banish from our essential natures the "reptilian brain," a hangover from our distant past that is thought responsible for much of our aggressive and bestial behavior. Or we might well consider it useful to eliminate on a macro scale the potential for disease. We could in a few generations do away with certain mental illnesses, perhaps, or diabetes, or high blood pressure, or almost any affliction we selected.

The important thing to keep in mind is that the quality of decision making dictates whether the choices to be made are going to be wise and just. True, most nightmarish scenarios ignore the innate good sense of people; common sense and an instinct for fair play, backed by the power of federal laws, it is assumed, will enable us to do good and protect us from doing ill. Nevertheless, the body politic still works its will through its representatives in government and the judiciary. These representatives in turn are influenced by powerful forces in the professions, business, and academe. The rather inglorious way that the scientific and administrative elite are handling the earliest fruits of gene therapy is ominous. Can our leaders be trusted in the years ahead to steer the spirited steed of molecular biology in the right direction? Will they handle the wisdom about to be entrusted to them with the high-mindedness and noble idealism that it deserves? Or will they be unable to clamber out of the low, crass bayous of the spirit so familiar in leadership circles nowadays.

It is a problem that, interestingly enough, is genetic. We humans have evolved intellectually to the point that, relatively soon, we will be able to understand the composition, function, and dynamics of the genome in much of its intimidating complexity. Emotionally, however, we are still apes, with all the behavioral baggage that brings to the issue. Perhaps the ultimate form of gene therapy would be for our species to rise above its baser heritage and learn to apply its new knowledge wisely and benignly. It calls to mind once again Jim Watson's unabashed desire to wield gene therapy as a tool to improve on the evolution of human nature. As he said, only half kiddingly, when asked what the future holds, "We'll make ourselves a little better. That's what we'll do. We'll make ourselves a little better."

ACKNOWLEDGMENTS

We would like to thank the hundreds of researchers who welcomed us into their laboratories and gave so generously of their time during the seven years it took us to complete this work. Mary Cunnane, our editor at W. W. Norton, besides shaping our manuscript, refused to lose faith in two journalists who were trying to write a book about a major story that was breaking before their eyes. She saw the possibilities early on and never wavered in the course of various drafts, updates, and new directions imposed on us—constantly, it seemed—by events. Our agent, Peter Shepherd, of Harold Ober & Assoc., gave us enormous support and encouragement at every step along the way. The same must be said for the *Chicago Tribune*, particularly our editors Jim Squires, Colleen Dishon, and Tom Stites, without whom there would have been no "Altered Fates" project in the first place, and the publisher, Jack Fuller, for suggesting that there might be a book in it somewhere. A friend and colleague, the *Tribune* science writer Ron Kotulak, offered indispensable help at many critical stages. Dr. Arthur Kohrman, of the University of Chicago and La Rabida Children's Hospital and Research Center, gave of his many years of medical experience in reading this manuscript and provided invaluable advice and insight. However, any errors contained herein are solely the responsibility of the authors.

The boundless patience and understanding and unflagging enthusiasm of our families kept reminding us how fortunate we are.

NOTES

CHAPTER 1: A GATHERING OF GEESE

pp. 18–21 *As a bookish:* Biographical material on William French Anderson here and elsewhere in this work is drawn from personal interviews with him over a period from July 1985 to June 1993, the authors' own observations, and interviews with various of Anderson's colleagues and acquaintances past and present.

p. 24 *"The floodgates":* Interview with W. French Anderson, March 8, 1992.

p. 24 *"This was before anyone":* Interview with Martin Eglitis, Sept. 4, 1987.

p. 28 *"I start with the premise":* Interview with Paul Berg, Aug. 22, 1988.

p. 31 *"It's basically":* Interview with Stuart Orkin, Sept. 12, 1985.

p. 34 *"I'll admit I":* Telephone interview with Richard C. Mulligan, Jan. 14, 1992.

p. 35 *"something of a stunt":* Orkin interview.

p. 35 *"Over the next":* Interview with Leroy Hood, Oct. 11, 1989.

p. 36 *A reporter once:* B. Siegel, "Gene Therapy Race: Egos, Prizes on the Line for Scientists" and "More Labs Than Patients: Desire to Be First Colors Gene Studies," *Los Angeles Times*, Dec. 13, 1987.

CHAPTER 2: THE LONG AND WINDING ROAD

p. 38 *Hemophilia was well known: The Babylonian Talmud*, trans. E. Epstein (London: Concino Press, 1961), Seder Nashim VI, Yebamoth 64B, p. 431.

p. 38 *It was this instrument:* F. Miescher, "Ueber die chemische Zusammensetzung der Eiterzellen," in *Medicinisch-chemische Untersuchungen*, ed. F. Hoppe-Seyler, vol. 4 (Berlin, 1871), 441–60.

p. 39 *Even as young Miescher:* The classic biography of Mendel is H. Iltis, *Life of Mendel* (New York: W. W. Norton, 1932). See also Mendel's landmark paper, *Experiments in Plant-Hybridisation*, translated by the Royal Horticultural Society of London, corrected and annotated by W. Bateson (Cambridge: Harvard Univ. Press, 1946).

p. 40 *"Biochemists weren't interested":* G. Beadle and M. Beadle, *The Language of Life* (New York: Doubleday, 1966).

p. 40 *Meanwhile, across the Atlantic:* A. Garrod, "The Incidence of Alkaptonuria: A Study in Chemical Individuality," *Lancet*, Dec. 13, 1902, 1616–20; idem, *Inborn Errors of Metabolism*, 2d ed. (London: Oxford Univ. Press, 1923; 1st ed., 1909).

p. 41 *More than thirty years:* G. W. Beadle and E. L. Tatum, "Genetic Control of Biochemical Reactions in *Neurospora*," *Proceedings of the National Academy of Sciences* 27 (1941): 499–506. See also G. Beadle, "Genes and the Chemistry of the Organism," *American Scientist* 34 (1946): 31–53, 76.

p. 41 *These longstanding mysteries:* O. T. Avery, C. M. MacLeod, and M. McCarty, "Studies on the Chemical Nature of the Substance Inducing Transformation of Pneumococcal Types," *Journal of Experimental Medicine* 79 (1944): 137–58.

p. 42 *Now that science knew:* J. D. Watson and F. H. C. Crick, "A Structure for Deoxyribose Nucleic Acid," *Nature* 171 (1953): 737–38.

p. 44 *Ultimately, Watson:* E. Chargaff, "Chemical Specificity of Nucleic Acids and Mechanism of Their Enzymatic Degradation," *Experientia* 6 (1950): 201–9; S. Zamenhof, G. Brawerman, and E. Chargaff, "On the Desoxypentose Nucleic Acids from Several Microorganisms," *Biochimica et Biophysica Acta* 9 (1952): 402–5.

p. 46 *Tatum and Lederberg made:* J. Lederberg and E. L. Tatum, "Novel Genotypes in Mixed Cultures of Biochemical Mutants of Bacteria," *Cold Spring Harbor Symposia on Quantitative Biology* 11 (1946): 113–14; idem, "Gene Recombination in *Escherichia coli*," *Nature* 158 (1946): 558.

p. 46 *The investigators had:* N. Zinder and J. Lederberg, "Genetic Exchange in Salmonella," *Journal of Bacteriology* 64 (1952): 679–99.

p. 48 *What the researchers had discovered:* M. L. Morse, E. M. Lederberg, and J. Lederberg, "Transductional Heterogenotes in *Escherichia coli*," *Genetics* 41 (1956): 758–59.

p. 48 *Lederberg recognized:* J. Lederberg, "Participatory Evolution: What Controls for Genetic Engineering," *Current* 121 (Sept. 1970): 49.

p. 48 *Other farsighted individuals:* R. D. Hotchkiss, "Portents for a Genetic Engineering," *Journal of Heredity* 56 (1965): 199.

p. 49 *A visionary to the bone:* "Repair of Genetic Defects Predicted," *Pediatric News*, June 1968.

p. 50 *As sometimes happens:* "Good Genes for Bad," *Newsweek,* June 19, 1967, 92.

p. 51 *Rogers later wrote:* S. Rogers, "Reflections on Issues Posed by Recombinant DNA Molecule Technology—II," *Annals of the New York Academy of Sciences* 265 (1976): 66–70.

p. 51 *"I'm pathologically":* Rogers, cited in *Newsweek,* June 19, 1967, 92.

p. 51 *"with much trepidation":* Rogers, "Genetic Engineering," *Human Genetics: Proceedings of the 4th International Congress of Human Genetics, Paris, Sept. 6–11, 1971* (Amsterdam: Excerpta Medica, 1971), 36–40.

p. 52 *"In our view":* T. Friedmann and R. Roblin, "Gene Therapy for Human Genetic Disease?" *Science* 175 (1972): 949–55.

p. 52 *Even as the Rogers debacle:* C. R. Merrill, M. R. Geier, and J. C. Petricciani, "Bacterial Virus Gene Expression in Human Cells," *Nature* 233 (1971): 398–400.

p. 52 *Molecular biology has become:* Telephone interview with Victor McKusick, June 17, 1993.

p. 54 *"Salad is virtually":* T. M. Powledge, *The AAAS Observer, Supplement to Science,* Nov. 3, 1989, 5–7. Lander reportedly made the remark during Human Genome I, the first annual meeting designed to acquaint biologists with the progress of the Human Genome Project, held in San Diego in Oct. 1989.

p. 55 *In 1972, Stanford's:* D. A. Jackson, R. H. Symons, and P. Berg, "Biochemical Method for Inserting New Genetic Information into DNA of Simian Virus 40: Circular SV40 DNA Molecules Containing Lambda Phage Genes and the Galactose Operon of *Escherichia coli,*" *Proceedings of the National Academy of Sciences* 69 (1972): 2904–9.

p. 55 *In 1973 came:* S. Cohen et al., "Construction of Biologically Functional Bacterial Plasmids in Vitro," *Proceedings of the National Academy of Sciences* 70 (1973): 3240–44.

p. 56 *Not surprisingly, 1975:* T. Maniatis et al., "Amplification and Characterization of a B-Globin Gene Synthesized in Vitro," *Cell* 8 (1976): 163–82; T. Rabbits, "Bacterial Cloning of Plasmids Carrying Copies of Rabbit Globin Messenger RNA," *Nature* 260 (1976): 221–25.

p. 56 *There was the occasional:* W. F. Anderson, "Arithmetic in Maya Numerals," *American Antiquity* 36 (1971): 54–63; A. W. Nienhuis, D. G. Laycock, and W. F. Anderson, "Translation of Rabbit Haemoglobin Messenger RNA by Thalassaemic and Non-thalassaemic Ribosomes," *Nature New Biology* 231 (1971): 205–8; A. W. Nienhuis and W. F. Anderson, "Hemoglobin Switching in Sheep and Goats: Change in Functional Globin Messenger RNA in Reticulocytes and Bone Marrow Cells," *Proceedings of the National Academy of Sciences* 69 (1972): 2184–88.

p. 56 *The effort culminated:* A. Diesseroth et al., "Localization of the Human Alpha-Globin Structural Gene to Chromosome 16 in Somatic Cell Hybrids by Molecular Hybridization Assay," *Cell* 12 (1977): 205–18.

p. 57 *"When she showed me":* Anderson interviews.

p. 58 *By the end of 1979:* Ibid.

CHAPTER 3: A SIMPLE LITTLE SYSTEM

This chapter is based primarily on personal and telephone interviews with Richard C. Mulligan, beginning on Sept. 10, 1985, in Cambridge, Mass., and continuing over the years in sites as diverse as Menlo Park, Cal., Toronto, Boston, Chicago, and Bethesda.

p. 59 *"the most exciting discipline":* Mulligan interview, Sept. 10, 1985.

p. 60 *Mulligan's chimeric feat:* R. C. Mulligan, B. H. Howard, and P. Berg, "Synthesis of Rabbit Beta-Globin in Cultured Monkey Kidney Cells following Infection with an SV40 Beta-Globin Recombinant Genome," *Nature* 277 (1979): 108–14.

p. 60 *Five years earlier:* The story of the International Conference on Recombinant DNA Molecules, held Feb. 24–27, 1975, at the Asilomar Conference Center, Pacific Grove, Cal., and its aftermath has been told many times. Two excellent resources are J. D. Watson and J. Tooze, *The DNA Story* (San Francisco: W. H. Freeman, 1981), and M. Rogers, *Biohazard* (New York: Alfred A. Knopf, 1977). For a critical analysis, see S. Krimsky, *Genetic Alchemy: The Social History of the Recombinant DNA Controversy* (Cambridge: MIT Press, 1982).

p. 61 *The scientists came away:* P. Berg et al., "Summary Statement of the Asilomar Conference on Recombinant DNA Molecules," *Proceedings of the National Academy of Sciences* 72 (1975): 1981–84.

p. 62 *The worries have never:* U.S. Department of Health, Education, and Welfare, *Recombinant DNA Research,* vol. 3, *Documents Relating to NIH Guidelines for Research involving Recombinant DNA Molecules, February 1976–1976* (Washington, D.C.: Government Printing Office, 1976).

p. 64 *Enter a peculiar:* The distinctive characteristics and evolution of retroviruses were elucidated for the authors by Howard Temin, in interviews on Aug. 21, 1988, in Toronto, and on Oct. 2, 1989, in Madison, Wis.

pp. 68–75 *Cline was a brilliant:* The story of Cline's experiment appears in several sources and is ably reported in Y. Baskin, *The Gene Doctors: Medical Ethics at the Frontier* (New York: William Morrow, 1984), 168–76; summarized in the federal report *Human Gene Therapy: A Background Paper* (Washington, D.C.: Office of Technology Assessment, 1984); and evaluated in J. C. Fletcher, "Moral Problems and Ethical Issues in Prospective Human Gene Therapy," *Virginia Law Review* 69 (1983): 515–46.

p. 69 *moving into live mice:* M. J. Cline et al., "Gene Transfer into Intact Animals," *Nature* 284 (1980): 422–25.

p. 75 *As the furor raged:* N. Wade, "Gene Therapy Caught in More Entanglements," *Science* 212 (1981): 24–25. See also idem, "Gene Therapy Pioneer

Draws Mikadoesque Rap," *Science* 212 (1981): 1253, and M. Sun, "Cline Loses Two NIH Grants," *Science* 214 (1981): 1219–20.

p. 76 *"The experiment was destined"*: Maniatis quoted in Baskin, *Gene Doctors*, 173.

p. 77 *Observes Paul Berg*: Interview with Paul Berg, Aug. 22, 1988.

p. 78 *"It's a matter"*: Interview with Inder Verma, Aug. 9, 1985.

p. 78 *"We were glad"*: Interview with David Baltimore, Sept. 10, 1985.

pp. 78–81 *The payoff for science*: R. Mann, R. Mulligan, and D. Baltimore, "Construction of a Retrovirus Packaging Mutant and Its Use to Produce Helper-Free Defective Retroviruses," *Cell* 33 (1983): 153–59.

p. 84 *Yet in an interview*: Mulligan interview, June 8, 1989.

CHAPTER 4: THE HORROR

p. 86 *"All these children"*: The historic paper is M. Lesch and W. L. Nyhan, "A Familial Disorder of Uric Acid Metabolism and Central Nervous System Function," *American Journal of Medicine* 36 (1964): 561–70. Other important papers include W. L. Nyhan, "A Disorder of Uric Acid Metabolism and Cerebral Function in Childhood," *Arthritis and Rheumatism* 8 (1965): 659–64; W. L. Nyhan, W. J. Oliver, and M. Lesch, "A Familial Disorder of Uric Acid Metabolism and Central Nervous System Function," *Journal of Pediatrics* 67 (1965): 257–63; William L. Nyhan, "Clinical Features of the Lesch-Nyhan Syndrome," *Archives of Internal Medicine* 130 (1972): 186–92; idem, "The Lesch-Nyhan Syndrome," in *The Handbook of Clinical Neurology*, ed. P. J. Vinken and G. W. Bruyn (New York: North Holland, 1977), 263–78.

p. 87 *"The kids I treat"*: The history and the personal material throughout the chapter come from interviews with William Nyhan, starting on Aug. 7, 1985, in San Diego, followed by several telephone interviews, most recently on April 11, 1993.

p. 87 *enzyme deficiency that causes it*: J. E. Seegmiller, F. M. Rosenbloom, and W. N. Kelley, "Enzyme Defect Associated with a Sex-Linked Human Neurological Disorder and Excessive Purine Synthesis," *Science* 155 (1967): 1682–83.

p. 87 *Born in Vienna*: The biographical and subsequent materials in the chapter are based on interviews with Theodore Friedmann, from Aug. 7, 1985. See also his major review articles, including T. Friedmann and R. Roblin, "Gene Therapy for Human Genetic Disease?" *Science* 175 (1972): 949–55, and T. Friedmann, "Progress toward Human Gene Therapy," *Science* 244 (1989): 1275–81.

pp. 88–93 *"the special plague"*: Nyhan 1985 interview.

p. 94 *Nyhan and Lesch published*: Nyhan, "Familial Disorder."

pp. 95–97 *Weiner knew something*: Interview with Felice Weiner, Aug. 8, 1985, and telephone interview, May 13, 1993.

p. 96 *"Can't control it"*: Interview with Craig Weiner, Aug. 8, 1985.

pp. 97–103 *battle to free people:* The cloning of the HGPRT gene was recounted by the team member Douglas Jolly, interviewed on Aug. 8, 1985. The paper is D. J. Jolly et al., "Isolation of a Genomic Clone Partially Encoding Human Hypoxanthine Guanine Phosphoribosyl Transferase," *Proceedings of the National Academy of Sciences* 79 (1982): 5038–41. Other relevant papers are J. E. Seegmiller, "Contribution of Lesch-Nyhan Syndrome to the Understanding of Purine Metabolism," *Journal of Inherited Metabolic Diseases* 12 (1989): 184–96; C. T. Caskey and J. T. Stout, "Molecular Genetics of HPRT Deficiency," *Seminars in Nephrology* 9 (1989): 162–67 (a review of the molecular biology of the Lesch-Nyhan syndrome).

p. 104 *"I want to die":* Craig Weiner interview.

CHAPTER 5: THE UNDERSTUDY STEPS IN

p. 105 *With little charity:* Interview with Stuart Orkin, Sept. 12, 1985.

pp. 106–7 *Upon meeting Alison Ashcraft:* The material on Alison Ashcraft is drawn from a series of interviews with her and her family, conducted over a period from Jan. 1986 to the present.

p. 107 *Life with ADA deficiency:* Interviews with the DeSilva family from March 1991 through June 1993.

p. 108 *Cindy began displaying:* Interviews with the Cutshall family, July 1991.

p. 110 *Soon there would be:* Orkin interview.

p. 110 *Ever the cheerleader:* W. F. Anderson and J. C. Fletcher, *New England Journal of Medicine* 303 (1980): 1293. The updated article appeared as W. F. Anderson, "Prospects for Human Gene Therapy," *Science* 226 (1984): 401–9.

p. 111 *Anderson was ready:* Interview with W. French Anderson, Sept. 4, 1987.

p. 111 *The Orkin/Mulligan version:* Interview with Richard C. Mulligan, June 8, 1989.

pp. 112–13 *Much of what we today know:* A thorough account is provided by R. A. Good, in *Birth Defects: Original Articles Series* (1975): 377–79.

p. 114 *Great excitement followed:* G. J. Spangrude, S. Heimfeld, and I. L. Weissman, "Purification and Characterization of Mouse Hematopoietic Stem Cells," *Science* 241 (1988): 58–62.

p. 116 *"We were leery":* Interview with Susan Cutshall, July 25, 1991.

p. 116 *Early returns:* M. S. Hershfield et al., "Treatment of Adenosine Deaminase Deficiency with Polyethylene Glycol-Modified Adenosine Deaminase," *New England Journal of Medicine* 316 (1987): 589–96.

p. 116 *Several youngsters placed:* Telephone interview with Kenneth Culver, Sept. 5, 1991.

p. 116 *"PEG-ADA seems to bathe":* Telephone interview with Michael Hershfield, March 15, 1990.

p. 118 *"Cindy hasn't done":* Susan Cutshall interview.

p. 119 *"They always make me cry":* Ashanthi DeSilva interview.

p. 119 *PEG-ADA helped:* Alison Ashcraft interview, March 18, 1991.

p. 119 *"I'd barely even":* Interview with W. French Anderson, July 26, 1991.

CHAPTER 6: THE ROAD FORKS

p. 120 *"Talk to Mike Blaese":* Interview with W. French Anderson, July 26, 1991.

p. 120 *disastrous hereditary immune disorder:* R. M. Blaese et al., "Hypercatabolism of IgG, IgA, IgM, and Albumin in the Wiskott-Aldrich Syndrome: A Unique Disorder of Serum Protein Metabolism," in *Immunologic Deficiency Syndromes,* ed. P.J. Edelson (New York: Mss. Corporation, 1973), 61–77.

pp. 120–122 *"I had my lab switch":* Telephone interviews with R. Michael Blaese, Dec. 5–6, 1990.

p. 121 *strain of mice:* E. Gilboa et al., "Transfer and Expression of Cloned Genes Using Retroviral Vectors," *Biotechniques* 4 (1986): 504–12.

p. 123 *admirable retroviral vector:* P. Kantoff et al., "Correction of Adenosine Deaminase Deficiency in Cultured Human T and B Cells by Retrovirus-Mediated Gene Transfer," *Proceedings of the National Academy of Sciences* 83 (1986): 6563–67. The "parlor trick" quotation comes from the telephone interview with Michael Hershfield, March 15, 1990.

p. 123 *Up in Boston:* D. A. Williams et al., "Introduction of New Genetic Material into Pluripotent Haematopoietic Stem Cells of the Mouse," *Nature* 310 (1984): 476–80.

p. 124 *Despite its failure:* Interview with W. French Anderson, Feb. 13, 1988.

p. 125 *By this time:* Larry Thompson, "Body Builders: The Breakthrough Research at Genetic Therapy," *Warfield's: The Baltimore Business Monthly,* Aug. 1990. See also B. J. Culliton, "NIH, Inc.: The CRADA Boom," *Science* 245 (1989): 1034–36. Other material about his affiliation with Genetic Therapy Inc. was supplied to the authors by W. French Anderson.

p. 126 *Whenever Anderson has a question:* Anderson interview, July 26, 1991.

pp. 129–30 *"It's a huge deal":* The saga of Monkey Robert and his fellows is recounted from interviews with W. French Anderson, Oct. 3, 1985 and May 5 and Aug. 3, 1988. It is also summarized in K. Cornetta, R. Wieder, and W. F. Anderson, "Gene Transfer into Primates and Prospects for Gene Therapy in Humans," *Progress in Nucleic Acid Research and Molecular Biology* 36 (1989): 311–22.

p. 130 *Ironically, he later decided:* F. Anderson, "Musings on the Struggle—Part I: The 'Phonebook,' " *Human Gene Therapy* 3 (1992): 251–52.

pp. 130–33 *the impatient French Anderson:* Anderson interview, May 5, 1988. See also W. F. Anderson et al., "Human Gene Therapy Preclinical Data Document," submitted to the Recombinant DNA Advisory Committee of NIH, April 24, 1987.

p. 133 *"It was a depressing time":* Blaese interview, April 1, 1991.

p. 134 *Bearded, blow-dried, energetic:* The remainder of the chapter is compiled from numerous interviews with Kenneth Culver (particularly those of April 3, 1990, Jan. 24, May 16, and Sept. 5, 1991, June 3 and Dec. 29, 1992, and March 15, 1993); from published papers, including "Lymphocytes as Vehicles for Gene Therapy," in *Cellular Immunity and the Immunotherapy of Cancer,* ed. M. T. Lotze and O. J. Finn (New York: Wiley-Liss, 1990), 29–137; and from unpublished reminiscences graciously supplied by him.

CHAPTER 7: THE TROIKA FORMS

pp. 138–41 *Blaese wanted to work:* Telephone interviews with R. Michael Blaese, Dec. 5–6, 1990.

pp. 141–51 *Steven Rosenberg's T-cell monomania:* The struggle of Steven A. Rosenberg to develop a new major tool in the battle against cancer is compiled from many sources. Pertinent material comes from interviews with Rosenberg, primarily that of July 26, 1991, but also those of May 22, 1989, April 17, 1990, and Jan. 29, 1991. An excellent overview article of his quest utilizing gene-altered cells may be found in S. Rosenberg, "Adoptive Immunotherapy for Cancer," *Scientific American,* May 1990, 62–69. See also B. J. Culliton, "Fighting Cancer with Designer Cells," *Science* 244 (1989): 1430–33. For Rosenberg's triumphs and tribulations with TILs, see S. L. Topalian and S. A. Rosenberg, "Tumor Infiltrating Lymphocytes: Evidence for Specific Immune Reactions against Growing Cancers in Mice and Humans," in *Important Advances in Oncology, 1990,* ed. V. T. DeVita, S. A. Hellman, and S. A. Rosenberg (Philadelphia: J. B. Lippincott, 1990), 19–41. See also A. S. Kasid et al., "Human Gene Transfer: Characterization of Human Tumor-Infiltrating Lymphocytes as Vehicles for Retroviral-Mediated Gene Transfer in Man," *Proceedings of the National Academy of Sciences* 87 (1990): 473–77; S. A. Rosenberg et al., Treatment of 283 Consecutive Patients with Metastatic Melanoma or Renal Cell Cancer Using High-Dose Bolus Interleukin 2," *Journal of the American Medical Association* 271 (1994): 907–13.

p. 152 *As David Baltimore once:* Interview with Baltimore, Feb. 6, 1989.

pp. 152–54 *Since 1974, the government:* The following is the chronology of the review process for the "N2-TIL Human Gene Transfer" clinical protocol, whose primary investigators were Steven A. Rosenberg, R. Michael Blaese, and W. French Anderson:

   6/10/88: protocol submitted to clinical research subpanel, National Cancer Institute (NCI)

   6/20/88: NCI IRB grants conditional approval

   6/21/88: National Heart, Lung, and Blood Institute (NHLBI) IRB grants conditional approval.

7/13/88: NIH IBC grants conditional approval

7/29/88: Human Gene Therapy Subcommittee defers decision

10/3/88: RAC grants approval, 16 to 5

10/6/88: IND is submitted to Food and Drug Administration (FDA)

11/4/88: NIH IBC grants final approval, unanimously

11/21/88: NCI IRB grants final approval, unanimously, pending final informed-consent document

11/28/88: NCI IRB subcommittee finalizes informed-consent document

12/8/88: Anderson et al. meet with FDA IND representatives

12/9/88: Human Gene Therapy Subcommittee grants final approval, 12 to 0

12/13/88: NHLBI IRB defers decision

12/17/88: NHLBI IRB subcommittee hears testimony

12/19/88: FDA Advisory Committee–Vaccines and Related Biologic Products grants approval, unanimously

12/20/88: NHLBI IRB grants final approval, unanimously

1/89: RAC mail ballot results give final approval, 21 to 0, with three abstentions

1/19/89: NIH director and the FDA grant approval

1/30/89: NIH director reports approval; Jeremy Rifkin files lawsuit

5/15/89: Rifkin's lawsuit is settled out of court

5/22/89: first patient, Maurice W. Kuntz, receives TIL therapy with gene transfer

pp. 153–54 *first gene transfer proposal:* "Minutes of the Human Gene Therapy Subcommittee of the Recombinant DNA Advisory Committee, 29 July 1988," *Recombinant DNA Research* 13 (1988): 180–84. See also "N2-TIL Human Gene Transfer Clinical Protocol," *Human Gene Therapy* 1 (1990): 73–92.

p. 155 *The RAC subcommittee expressed:* B. J. Culliton, "NIH Delays Gene Transfer Experiment" and "Journals and Data Disclosure," *Science* 242 (1988): 856–57.

p. 156 *Rosenberg's* New England Journal *paper:* S. A. Rosenberg et al., "Use of Tumor-Infiltrating Lymphocytes and Interleukin-2 in the Immunotherapy of Patients with Metastatic Melanoma: A Preliminary Report," *New England Journal of Medicine* 319 (1988): 1676–80.

p. 156 *"We didn't have everything":* Interview with W. French Anderson, Jan. 23, 1989.

p. 157 *It wasn't until 1993:* W. F. Anderson, "Musings on the Struggle—Part III: The October 3 RAC Meeting," *Human Gene Therapy* 4 (1993): 401–2.

p. 157 *meeting of the parent RAC:* For final RAC approval, see "Minutes of the Recombinant DNA Advisory Committee, Oct. 3, 1988," *Recombinant DNA Research* 13 (1988): 227–54.

p. 161 *Anderson and Rosenberg insist:* Anderson interview, Jan. 23, 1989.

p. 161 *"We had our patient":* P. Gorner, "Human Testing May Open Door to Gene Therapy," *Chicago Tribune,* May 21, 1989.

CHAPTER 8: GENE TRANSFER IS BORN

The material in this chapter is derived from interviews with Sharon Kuntz, W. French Anderson, and Steven Rosenberg. Kuntz kindly shared her experiences, memorabilia, and the videotape (of her late husband after the gene transfer) with the authors on Sept. 29, 1991, at her home in Franklin, Ind.

p. 168 *"Maurice seemed serene":* Interview with W. French Anderson, May 22, 1989.

p. 169 *Recalling his mood:* Ibid.

p. 169 *Then they added:* K. B. Mullis and F. A. Faloona, "Specific Synthesis of DNA *in Vitro* via a Polymerase-Catalyzed Chain Reaction," *Methods in Enzymology,* vol. 155, *Recombinant DNA,* pt. F (San Diego: Academic Press, 1987, 335–50.

p. 170 *"The vials":* Anderson interview.

p. 173 *All of them:* Documented reports of the first five TIL patients may be found in S. A. Rosenberg et al., "Gene Transfer into Humans: Immunotherapy of Patients with Advanced Melanoma, Using Tumor-Infiltrating Lymphocytes Modified by Retroviral Gene Transduction," *New England Journal of Medicine* 323 (1990): 570–78; D. Cournoyer and C. T. Caskey, "Gene Transfer into Humans—A First Step" (editorial), ibid., 601–2.

p. 174 *"In general, about half":* Telephone interview with Steven Rosenberg, Sept. 18, 1989.

CHAPTER 9: GENE THERAPY MEETS THE PUBLIC

This chapter consists primarily of the authors' coverage of the meeting of the Human Gene Therapy Subcommittee, March 30, 1990, National Institutes of Health, Bethesda, Md.

p. 177 *A major stumbling block:* M. S. Hershfield et al., "Treatment of Adenosine Deaminase Deficiency with Polyethylene Glycol-Modified Adenosine Deaminase," *New England Journal of Medicine* 316 (1987): 589–96.

pp. 177–78 *First, the gene therapists:* For a summary of the first human gene therapy protocol, see R. M. Blaese, K. W. Culver, and W. F. Anderson, "Treatment of Severe Combined Immunodeficiency Disease (SCID) Due to Adenosine Deaminase (ADA) Deficiency with Autologous Lymphocytes Transduced with a Human ADA Gene," *Human Gene Therapy* 1 (1990): 331–62.

p. 179 *"To try to cure":* Telephone interview with W. French Anderson, March 30, 1990.

p. 179 *Despite Anderson's elucidations:* B. J. Culliton, "ADA Deficiency: A Prime Candidate," *Science* 246 (1989): 751. A rebuttal by Michael Hershfield appears in a letter to the editor, ibid., p. 1373. The other citation is from N.

Angier, "Gene Implant Therapy Is Backed for a Rare Disease," *New York Times*, March 8, 1990.

p. 180 *Seated in the audience:* Michael Hershfield's presence was recounted by Richard C. Mulligan, in an interview, April 7, 1990.

pp. 180–295 *Hershfield claimed:* Interview with Michael Hershfield, March 15, 1990.

p. 189 *RAC's trademark federal guidelines:* National Institutes of Health, "Recombinant DNA Research: Request for Public Comment on 'Points to Consider in the Design and Submission of Human Somatic-Cell Gene Therapy Protocols,' " *Federal Register* 50 (1985): 2940–45. Revised version: "Points to Consider in the Design and Submission of Protocols for the Transfer of Recombinant DNA into the Genome of Human Subjects," ibid., 55 (1990): 7443–47.

p. 189 *in bone marrow:* P. W. Kantoff et al., "Expression of Human Adenosine Deaminase in Nonhuman Primates after Retrovirus-Mediated Gene Transfer," *Journal of Experimental Medicine* 166 (1987): 219–34.

p. 193 *the lab data indicated:* K. W. Culver et al., "Retroviral-Mediated Gene Transfer into Cultured Lymphoid Cells as a Vehicle for Gene Therapy," *Journal of Cell Biochemistry* 12B, suppl. (1988): 171.

p. 194 *in five rhesus monkeys:* K. W. Culver et al., "In Vivo Expression and Survival of Gene-Modified Rhesus T Lymphocytes," *Human Gene Therapy* 1 (1991): 399–410.

p. 194 *lived substantially longer:* K. W. Culver et al., "Correction of ADA Deficiency in Human T-Lymphocytes Using Retroviral-Mediated Gene Transfer," *Transplantation Proceedings* 23 (1991): 170–71.

p. 197 *as optimal a selection:* The ADA Human Gene Therapy Clinical Protocol, first version, submitted to the Recombinant DNA Advisory Committee of NIH, March 30, 1990. The graphic "Phenotype of ADA(-) D1010 T-lymphocytes," Fig. 5, 13.2, is on p. 58.

p. 198 *"This was amazing":* Interview with Richard C. Mulligan, July 30, 1990.

p. 199 *"nice white lab coats":* Mulligan interview, April 7, 1990.

p. 200 *"We never intended":* Interview, W. French Anderson, Aug. 2, 1990.

p. 201 *The official reason:* Interview with Kenneth Culver, Sept. 4, 1990.

p. 201 *he had run into trouble:* Interview with Michael Hershfield, April 16, 1991.

CHAPTER 10: VICTORY

pp. 202–4 *Perhaps it was jet lag:* The authors' coverage of the meeting of the Human Gene Therapy Subcommittee July 30, 1990, National Institutes of Health, Bethesda, Md.

p. 205 *"We show that":* G. Ferrari et al., "An in Vivo Model of Somatic Cell Gene Therapy for Human Severe Combined Immunodeficiency," *Science* 251 (1991): 1363–66.

p. 205 *Typical was the reaction:* B. J. Culliton, "Conflict at the RAC," *Science* 248 (1990): 159.

p. 208 *Among the first patients:* Interview with Alison Ashcraft, May 13, 1988.

p. 209 *"I have nothing":* Interview with Melvin Berger, Sept. 13, 1990.

p. 210 *In an interview:* Interview with Steven Rosenberg, April 20, 1990.

pp. 212–13 *In a letter:* Richard C. Mulligan to Nelson A. Wivel, Office of Recombinant DNA Activities, NIH, May 31, 1990 (letter provided by Wivel).

p. 212 *The meeting opened:* NIH, "Minutes of the June 1, 1990 Meeting of the Human Gene Therapy Subcommittee," *Recombinant DNA Technical Bulletin* 13 (1990): 245–73.

p. 214 *"Also from the data presented":* Mulligan letter to Wivel.

pp. 214–16 *It confused Mulligan:* Interview with Richard C. Mulligan, July 30, 1990.

p. 219 *"I feel great":* Telephone interview with W. French Anderson, moments after subcommittee approval, June 1, 1990.

p. 219 *Looking back:* Telephone interview with Charles Epstein, June 26, 1991.

p. 222 *Rosenberg's latest proposal:* "TNF/TIL Human Gene Therapy Clinical Protocol," *Human Gene Therapy* (1990): 443–80.

p. 224 *The next day:* Authors' coverage of the July 31, 1990, meeting of the Recombinant DNA Advisory Committee, also available in "Minutes of Recombinant DNA Advisory Committee Meeting, July 31, 1990," *Human Gene Therapy* 2 (1991): 177–85.

pp. 224–26 *A few hours:* Interview with Richard C. Mulligan, July 30, 1990.

CHAPTER 11: SEPTEMBER 14, 1990

pp. 228–29 *In a matter of hours:* Press release from Rainbow Hospital, Aug. 1990.

p. 228 *Raj DeSilva later:* Interview with Raj DeSilva, June 15, 1991.

pp. 229–40 *Thus it was:* This account is drawn from the recollections of W. French Anderson, R. Michael Blaese, Kenneth Culver, Melvin Berger, the DeSilva family, and the Cutshall family, as well as personal observation by the authors and videotape footage shot by NIH staff.

CHAPTER 12: ONE ERRANT CELL

p. 241 *In his own account:* S. A. Rosenberg and J. M. Barry, *The Transformed Cell: Unlocking the Mysteries of Cancer* (New York: G. P. Putnam's Sons, 1992).

p. 242 *Much of a physician's life:* Interview with W. French Anderson, Sept. 17, 1990.

p. 242 *In the aftermath:* L. Thompson, "French Anderson's Genetic Destiny,"

Washington Post Magazine, Jan. 20, 1991; R. M. Henig, "Dr. Anderson's Gene Machine," *New York Times Magazine,* March 31, 1991.

p. 243 *Blaese gulped:* Interview with R. Michael Blaese, June 19, 1991, Chicago.

p. 243 *Coincidentally with the infusion:* P. Gorner and R. Kotulak, "Scientists Criticize Human Gene Therapy," *Chicago Tribune,* Sept. 20, 1990. Authors' coverage of the First International Symposium on Preimplantation Genetics, Chicago, Sept. 14–19, 1990. Interviews with W. French Anderson, Arthur Bank, Stuart Orkin, and Richard C. Mulligan.

p. 245 *Interviewed that morning:* Interview with R. Michael Blaese, April 1, 1991.

p. 246 *"Philosophically, I would hate":* Blaese interview, June 19, 1991.

p. 246 *Ken Culver, for his part:* Interview with Kenneth Culver, Nov. 30, 1990.

p. 249 *"We've had major problems":* R. Michael Blaese quoted in *The Blue Sheet,* F-D-C Reports, Inc., Feb. 5, 1992, p. 17.

p. 249 *Despite everything, however:* P. Gorner, "For 1st Time, Cancer Patients Get Gene Therapy," *Chicago Tribune,* Jan. 30, 1991.

p. 249 *He refused to identify:* Interview with Steven Rosenberg, Jan. 29, 1991.

pp. 250–54 *From then on:* Rosenberg and Barry, *Transformed Cell,* 317–30.

pp. 252–53 *But was the remission:* C. Anderson, "Gene Therapy Researcher under Fire over Controversial Cancer Trials," *Nature* 360 (1992): 399–400.

p. 255 *Another strategy:* P. T. Golumbek et al., "Treatment of Established Renal Cancer by Tumor Cells Engineered to Secrete Interleukin-4," *Science* 254 (1991): 713–16.

p. 255 *"This is the first time":* The remarks of Golumbek and Pardoll are from a special communication, Johns Hopkins University, Nov. 5, 1991.

p. 256 *On October 8:* Rosenberg and Barry, *Transformed Cell,* 334.

pp. 257–58 *In the meantime:* R. Kotulak, "Vaccine to Arrest Cancer Shows High Promise in Tests," *Chicago Tribune,* April 7, 1991.

CHAPTER 13: BELT AND SUSPENDERS

p. 260 *Several months into:* Unpublished data from R. Michael Blaese, W. French Anderson, and Kenneth Culver.

p. 260 *From a purely:* Ibid.

p. 261 *"Our purpose":* Interview with R. Michael Blaese, July 24, 1991.

p. 262 *"Before we started":* Ibid.

p. 262 *Prior to therapy:* Unpublished data from R. Michael Blaese. W. French Anderson, and Kenneth Culver.

p. 262 *"Saying we only":* Telephone interview with Kenneth Culver, Sept. 5, 1991.

p. 263 *"If we can":* Ibid.

p. 263 *"We're at least":* Blaese interview.

p. 263 *"We couldn't be":* Ibid.

pp. 263–64 *"There are not":* Interview with W. French Anderson, June 3, 1991.

pp. 266–70 *It is later:* This section is from the authors' personal observations, July 26, 1991.

pp. 271–73 *Right from the start:* Unpublished data from R. Michael Blaese, W. French Anderson, and Kenneth Culver.

p. 271 *"It wasn't until":* Telephone interview with Kenneth Culver, Sept. 5, 1991.

p. 271 *He was seconded:* Telephone interview with Raj DeSilva, June 1, 1993.

p. 273 *"Doctors have been waiting":* Gerard McGarrity, on the occasion of final RAC approval of the first gene therapy experiment, July 31, 1990.

p. 273 *Whether the protocol:* From Blaese et al., stem cell therapy protocol approved Feb. 11, 1992.

p. 274 *"We're going for":* Telephone interview with Kenneth Culver, Dec. 29, 1992.

pp. 275–76 *On Thursday:* News accounts and W. French Anderson telephone interview, May 17, 1993.

p. 276 *"It's kind of":* Raj DeSilva interview.

CHAPTER 14: AFTERMATH

p. 277 *"So I am starting":* Interview with W. French Anderson, July 26, 1991.

p. 278 *By mid-1992:* Running total courtesy of NIH Office of Recombinant DNA Activities and W. French Anderson, who was keeping an independent count.

p. 278 *Its star virologist:* J. Crewdson, *Chicago Tribune* series on Robert Gallo and the HIV virus, Sept. 1989.

p. 279 *"For twenty-five years":* Telephone interview with W. French Anderson, March 18, 1992.

pp. 279–80 *For example, at St. Jude:* Gene transfer protocols submitted to the Recombinant DNA Advisory Committee in 1991–92, including Malcolm K. Brenner et al., St. Jude Children's Research Hospital, "Autologous Bone Marrow Transplant for Children with Acute Myelogenous Leukemia in First Complete Remission: Use of Marker Genes to Investigate the Biology of Marrow Reconstitution and the Mechanism of Relapse," gene transfer protocol approved by the RAC on Feb. 4, 1991; Malcolm K. Brenner et al., "A Phase I/II Trial of High Dose Carboplatin and Etoposide with Autologous Marrow Support for Treatment of Stage D Neuroblastoma in First Remission: Use of Marker Genes to Investigate the Biology of Marrow Reconstitution and the Mechanism of Relapse," gene transfer protocol approved by the RAC on May 31, 1991; Malcolm K. Brenner et al., "A Phase II Trial of High-Dose Carboplatin and Etoposide with Autologous Marrow Support for Treatment of Relapse/Refactory Neuroblastoma without Apparent Bone Marrow Involvement," gene transfer protocol approved by the RAC on May 31, 1991; Albert B. Diesseroth et al., M. D. Anderson Cancer Center, "Autologous Bone Marrow

Transplantation for Chronic Myelogenous Leukemia in which Retroviral Markers Are Used to Discriminate between Relapse Which Arises from Systemic Disease Remaining after Preventive Therapy Versus Relapse Due to Residual Leukemic Cells in Autologous Marrow: A Pilot Trial," gene transfer protocol approved by the RAC on May 31, 1991; Fred D. Ledley and Savio L. C. Woo, Baylor College of Medicine, "Hepatocellular Transplantation in Acute Hepatic Failure and Targeting Genetic Markers to Hepatic Cells," a gene transfer protocol approved by the RAC on May 31, 1991; Michael T. Lotze et al., University of Pittsburgh School of Medicine, "The Administration of Interleukin-2, Interleukin-4, and Tumor Infiltrating Lymphocytes to Patients with Melanoma," gene transfer protocol approved by the RAC on May 31, 1991.

pp. 280–81 *"I love it here":* Telephone interview with W. French Anderson, Jan. 20, 1993.

p. 282 *Ann Arbor experiments:* This section is compiled from interviews with James Wilson on Sept. 20, 1989, Sept. 18, 1990, and Feb. 18 and June 19, 1992. See also J. M. Wilson et al., "Correction of the Genetic Defect in Hepatocytes from the Watanabe Heritable Hyperlipidemic Rabbit," *Proceedings of the National Academy of Sciences* 85 (1988): 4421–25; and J. M. Wilson et al., "Ex Vivo Gene Therapy of Familial Hypercholesterolemia," gene therapy protocol approved by the RAC on Oct. 8, 1991.

p. 283 *The second experiment:* This section is derived from interviews with Gary Nabel, April 13 and June 11, 1992, and March 18, 1993. See also G. J. Nabel et al., "Immunotherapy of Malignancy by *in Vivo* Gene Transfer into Tumors," gene therapy protocol approved by the RAC on Feb. 10, 1992.

pp. 285–87 *But it was Culver's:* From interviews with Kenneth Culver, Sept. 5, 1991, and June 3 and Dec. 29, 1992, and unpublished data supplied by Culver. See also K. W. Culver et al., "In Vivo Gene Transfer with Retroviral Vector Producer Cells for Testing of Experimental Brain Tumors," *Science* 256 (1992): 1550 52; and Edward Oldfield et al., "Gene Therapy for the Treatment of Brain Tumors Using Intra-Tumoral Transduction with the Thymidine Kinase Gene and Intravenous Ganciclovir," gene therapy protocol approved by the RAC on June 1, 1992.

pp. 288–94 *As others were concentrating:* This section is based on interviews with W. French Anderson between July 1991 and May 1993.

CHAPTER 15: A MATTER OF THE HEART

pp. 299–302 *"Pigs and ourselves":* Interviews with Jan Rapacz, Oct. 2–3, 1989.

p. 300 *The role of genes:* K. Berg, "Genetics of Coronary Heart Disease," in *Progress in Medical Genetics,* vol. 5 (Philadelphia: W. B. Saunders, 1983), 35.

p. 300 *A series of rather wacky:* S. W. Lesko, L. Rosenberg, and S. Shapiro, "A Case-Controlled Study of Baldness in Relation to Myocardial Infarction in

Men," *Journal of the American Medical Association* 269 (1993): 998–1003; W. J. Elliott and T. Karrison, "Increased All-Cause and Cardiac Morbidity and Mortality Associated with the Diagonal Earlobe Crease: A Prospective Cohort Study," *American Journal of Medicine* 3 (1989): 247–54.

p. 300 *"Every disease is":* Interview with Michael Brown, May 4, 1990.

p. 301 *"and probably five":* Telephone interview with Peter Kwiterovich, Nov. 28, 1989.

p. 301 *Jan Rapacz is:* N. Maeda et al., "Molecular Genetics of the Apolipoprotein B Gene in Pigs in Relation to Atherosclerosis," *Gene* 70 (1988): 213–29; D. L. Ebert et al., "Primary Structure Comparison of the Proposed Low Density Lipoprotein (LDL) Receptor Binding Domain of Human and Pig Apolipoprotein B: Implications for LDL-Receptor Interactions," *Journal of Lipid Research* 29 (1988): 1501–9.

p. 301 *"My teachers told":* Rapacz interview.

p. 302 *most highly decorated:* M. S. Brown and J. L. Goldstein, "A Receptor-Mediated Pathway for Cholesterol Homeostasis," *Science* 232 (1986): 34–47.

p. 304 *The earliest suspicions:* A. Keys, "Diet and the Epidemiology of Coronary Heart Disease," *Journal of the American Medical Association* 164 (1957): 1912–19.

p. 304 *Soon, long-range studies:* W. B. Kannel et al., "Serum Cholesterol Lipoproteins and the Risk of Coronary Heart Disease: 'The Framingham Study,' " *Annals of Internal Medicine* 74 (1971): 1–12; R. D. Abbott et al., "High Density Lipoprotein Cholesterol, Total Cholesterol, and Myocardial Infarction: The Framingham Study," *Arteriosclerosis* 8 (1988): 207–11; Ancel Keys, *Seven Countries: A Multivariate Analysis of Death and Coronary Heart Disease* (Cambridge: Harvard Univ. Press, 1980): 1–381.

p. 304 *Two later studies:* M. H. Frick et al., "Helsinki Heart Study: Primary Prevention Trial with Gemfibrozil in Middle-Aged Men with Dislipidemia," *New England Journal of Medicine* 317 (1987): 1237–45. A five-year study of 4,081 healthy middle-aged men with high cholesterol levels found that a drug that lowers "bad" cholesterol reduced the risk of heart attack by 34 percent. See also Report of the Committee of Principal Investigators, "WHO Cooperative Trial in Primary Prevention of Ischaemic Heart Disease with Clofibrate to Lower Serum Cholesterol: Final Mortality Follow-up," *Lancet* ii (1984): 600–604; Lipid Research Clinics Coronary Primary Prevention Trial, "Reduction in Incidence of Coronary Heart Disease" and "The Relationship of Reduction in Incidence of Coronary Heart Disease to Cholesterol Lowering," *Journal of the American Medical Association* 251 (1984): 351–74; S. Dayton et al., "A Controlled Trial of a Diet High in Unsaturated Fat in Preventing Complications of Atherosclerosis," *Circulation* 39–40, suppl. II (1969): 1–63.

p. 306 *Joseph Goldstein and: Current Biography Yearbook, 1987,* 208–11; Brown and Goldstein, "Receptor-Mediated Pathway."

p. 308 *In 1974:* M. S. Brown and J. L. Goldstein, "Familial Hypercholesterolemia: Defective Binding of Lipoproteins to Cultured Fibroblasts Associated with Impaired Regulation of 3-Hydroxy-3-Methylglutanyl Coenzyme A Reductase Activity," *Proceedings of the National Academy of Sciences* 71 (1974): 788–92; M. S. Brown, S. E. Dana, and J. L. Goldstein, "Regulation of 3-Hydroxy-3-Methylglutanyl Coenzyme A Reductase Activity in Cultured Human Fibroblasts," *Journal of Biological Chemistry* 249 (1974): 789–96.

p. 309 *"Think of it":* Brown interview.

p. 310 *A unique case:* R. A. Norum et al., "Familial Deficiency of Apolipoproteins A-I and C-III and Precocious Coronary Artery Disease," *New England Journal of Medicine* 306 (1982): 1513–19; J. L. Breslow, "Apolipoprotein Genetic Variation and Human Disease," *Physiological Review* 68 (1988): 85–132.

p. 310 *Still another example:* J. W. McLean et al., "cDNA Sequence of Human Apolipoprotein(a) Is Homologous to Plasminogen," *Nature* 330 (1987): 132–37; K. A. Hajjar et al., "Lipoprotein(a) Modulation of Endothclial Cell Surface Fibrinolysis and Its Potential Role in Atherosclerosis," *Nature* 339 (1989): 303–5.

p. 311 *"That observation":* Telephone interview with Jan Rapacz, June 24, 1990.

p. 311 *Throughout the 1970s:* J. Rapacz et al., "Lipoprotein Mutations in Pigs Are Associated with Elevated Plasma Cholesterol and Atherosclerosis," *Science* 234 (1986): 1573–77; S. W. Lowe et al., "Defective Receptor Binding of Low Density Lipoprotein from Pigs Possessing Mutant Apolipoprotein B Alleles," *Journal of Biological Chemistry* 263 (1988): 15467–73; W. J. Checovich et al., "Defective Catabolism and Abnormal Composition of Low Density Lipoproteins from Mutant Pigs with Hypercholesterolemia," *Biochemistry* 27 (1988):1934–41.

p. 312 *"Nobody accepted it":* Rapacz interview.

p. 312 *"These findings suggest":* A. D. Attie and Margaret Prescott, "The Spontaneously Hypercholesterocmic Pig as an Animal Model for Human Atherosclerosis," *ILAR News* 30 (Fall 1988): 5–12.

p. 313 *Yet a third possibility:* T. Kodama et al., "Type I Macrophage Scavenger Receptor Contains α-helical and Collagen-like Coiled Coils," *Nature* 343 (1990): 532–35; L. Rohrer et al., "Coiled-Coil Fibrous Domains Mediate Ligand Binding by Macrophage Scavenger Receptor Type II," ibid., 570–72; M. S. Brown and J. L. Goldstein, "Scavenging for Receptors," ibid., 508–9.

p. 313 *"We have a long":* Telephone interview with Mahley.

p. 313 *"Once we know":* Rapacz interview.

p. 313 *By 1989:* Interview with James Wilson, Sept. 20, 1989. See also Peter Gorner, "Gene Therapy Enters New Field: 1st Transfer into Liver Cells; Early Heart Attacks Targeted," *Chicago Tribune,* June 15, 1988; J. M. Wilson et al., "Correction of the Genetic Defect in Hepatocytes from the Watanabe Heritable Hyperlipidemic Rabbit," *Proceedings of the National Academy of Sciences* 85 (1988): 4421–25.

p. 314 *It was a combination:* Telephone interview with James Wilson, Feb. 18, 1992.

p. 314 *The patient not only survived:* Telephone interview with James Wilson, July 20, 1994.

p. 315 *At about the same time:* Interview with W. French Anderson, Feb. 21, 1989.

p. 315 *"Our idea":* Interviews with James Wilson and Richard C. Mulligan, May 1990. J. A. Zweibel et al., "High-Level Recombinant Gene Expression in Rabbit Endothelial Cells Transduced by Retroviral Vectors," *Science* 24 (1989): 220–22; D. A. Dichek et al., "Seeding of Intravascular Stents with Genetically Engineered Endothelial Cells," *Circulation* 80 (1989): 1347–53. See also J. M. Wilson et al., "Implantation of Vascular Grafts Lined with Genetically Modified Endothelial Cells," *Science* 244 (1989): 1344–46; B. J. Culliton, "Designing Cells to Deliver Drugs," *Science* 246 (1989): 746; J. L. Swain, "Gene Therapy: A New Approach to the Treatment of Cardiovascular Disease," *Circulation* 80 (1989): 1495–96.

p. 316 *At the same time:* E. G. Nabel et al., "Recombinant Gene Expression *in Vivo* within Endothelial Cells of the Arterial Wall," *Science* 244 (1989): 1342–44; interview with Gary Nabel, March 18, 1993.

p. 317 *"That's how I":* Interview with Elizabeth Nabel, March 25, 1993.

p. 319 *In 1989 and 1990:* E. G. Nabel et al., "Recombinant Gene Expression"; E. G. Nabel, G. Plautz, and G. J. Nabel, "Site-Specific Gene Expression *in Vivo* by Direct Gene Transfer into the Arterial Wall," *Science* 249 (1990): 1285–88. Also see the following: G. Plautz, E. G. Nabel, and G. J. Nabel, "Introduction of Vascular Smooth Muscle Cells Expressing Recombinant Genes *in Vivo*," *Circulation* 83 (1991): 578–83; G. Plautz, E. G. Nabel, and G. J. Nabel, "Selective Elimination of Recombinant Genes *in Vivo* with a Suicide Retroviral Vector," *New Biology* 3 (1991): 709–15; M. J. Stewart, et al., "Gene Transfer *in Vivo* with DNA Liposome Complexes: Safety and Toxicity in Mice," *Human Gene Therapy* 3 (1992): 267–75; E. G. Nabel, G. E. Plautz, and G. J. Nabel, "Transduction of a Foreign Histocompatability Gene in the Arterial Wall Induces Vasculitis," *Proceedings of the National Academy of Sciences* 89 (1992): 5157–61.

pp. 319–20 *"These studies told us":* Nabel interview.

p. 320 *Michael Brown is:* Brown interview.

p. 321 *"Genetics has the potential":* Kwiterovich interview.

p. 321 *"As it is":* Brown interview.

p. 321 *"I have two":* Rapacz interview.

p. 321 *"My cholesterol":* Brown interview.

CHAPTER 16: JEKYLL AND HYDE

pp. 322–23 *If all goes:* Telephone interviews with Jack Roth, June 15, 1992, and April 13, 1993. See also Jack A. Roth, "New Approaches to Treating Early Lung Cancer," *Cancer Research (Suppl.)* 52 (1992): 2652s–57s; J. A. Roth et al., "A Molecular Approach to Lung Cancer Therapy," *Journal of Cell Biochemistry* 15F, suppl. (1991): 4.

p. 324 *In writing up:* P. Rous, "Transmission of a Malignant New Growth by Means of a Cell-Free Filtrate," *Journal of the American Medical Association* 56 (1911): 198.

p. 325 *The Biochemist Bruce Ames:* B. Ames, J. McCann, and E. Yamasaki, "Methods for Detecting Carcinogens and Mutagens with the *Salmonella*/Mammalian-Microsome Mutagenicity Test," *Mutation Research* 31 (1975): 347–64; J. McCann and B. N. Ames, "Detection of Carcinogens as Mutagens in the *Salmonella*/Microsome Test: Assay of 300 Chemicals, Discussion," *Proceedings of the National Academy of Sciences* 73 (1976): 950–54.

p. 325 *In 1969:* R. J. Huebner and G. J. Todaro, "Oncogenes of RNA Tumor Viruses as Determinants of Cancer," *Proceedings of the National Academy of Sciences* 64 (1969): 1087–93.

p. 325 *"It wasn't clear":* Interview with J. Michael Bishop, May 5, 1989.

p. 326 *Sure enough:* D. Syehlin et al., "DNA Related to the Transforming Gene(s) of Avian Sarcoma Viruses Is Present in Normal Avian DNA," *Nature* 260 (1976): 170–73.

p. 326 *But the field did not:* M. Goldfarb et al., "Isolation and Preliminary Characterization of a Human Transforming Gene from T24 Bladder Carcinoma Cells," *Nature* 296 (1982): 404–9.

p. 326 *To everyone's surprise:* L. F. Parada et al., "Human EJ Bladder Carcinoma Oncogene Is Homologue of Harvey Sarcoma Virus *ras* Gene," *Nature* 297 (1982): 474–79.

p. 326 *When scientists:* C. J. Tabin et al., "Mechanism of Activation of a Human Oncogene," *Nature* 300 (1982): 143–49; E. P. Reddy et al., "A Point Mutation Is Responsible for the Acquisition of Transforming Properties by the T24 Human Bladder Carcinoma Oncogene," *Nature* 300 (1982): 149–52.

p. 327–28 *At present:* L. C. Cantley et al., "Oncogenes and Signal Transduction," *Cell* 64 (1991): 281–302.

p. 329 *Consider a proto-oncogene:* B. Lewin, "Oncogenic Conversion by Regulatory Changes in Transcription Factors," *Cell* 64 (1991): 303–12.

pp. 329–30 *The finding was so:* Telephone interview with Bert Vogelstein, Sept. 24, 1990; S. Baker et al., "Suppression of Human Colorectal Carcinoma Cell Growth by Wild-Type p53," *Science* 249 (1990): 912–15.

p. 331 *For the better part:* A. G. Knudson, "Mutation and Cancer: Statistical Study of Retinoblastoma," *Proceedings of the National Academy of Sciences* 68 (1971): 820–23.

p. 331 *The first clue:* J. Yunis, "The Chromosomal Basis of Human Neoplasia," *Science* 221 (1983): 227–36.

p. 331 *Soon other researchers:* W. K. Cavenee et al., "Expression of Recessive Alleles by Chromosomal Mechanisms in Retinoblastoma," *Nature* 305 (1983): 779–84.

p. 332 *The discovery in 1986:* T. P. Dryja et al., "Molecular Detection of Deletions Involving Band q14 of Chromosome 13 in Retinoblastoma," *Proceedings of the National Academy of Sciences* 83 (1986): 7391–94.

p. 332 *In pinpointing:* S. H. Friend et al., "A Human DNA Segment with Properties of the Gene That Predisposes to Retinoblastoma and Osteosarcoma," *Nature* 323 (1986): 643–46; telephone interview with Thaddeus P. Dryja, Nov. 30, 1987.

pp. 333–34 *In 1988, a paradigm:* Vogelstein interview; interview with Ray White, Aug. 6, 1991; J. Marx, "Many Gene Changes Found in Cancer," *Science* 246 (1989): 1386–88; E. J. Stanbridge, "Identifying Tumor Suppressor Genes in Human Colorectal Cancer," *Science* 247 (1990): 12–13; S. F. Baker et al., "Chromosome 17 Deletions and p53 Gene Mutations in Colorectal Carcinomas," *Science* 244 (1989): 217–21; E. Fearon et al., "Identification of a Chromosome 18q Gene That Is Altered in Colorectal Cancers," *Science* 247 (1990): 49–56; E. Solomon, "Colorectal Cancer Genes," *Nature* 343 (1990): 412–15; S. E. Kern et al., "Allelic Loss in Colorectal Carcinoma," *Journal of the American Medical Association* 261 (1989): 3099–103; K. Kinzler et al., "Identification of FAP Locus Genes from Chromosome 5q21," *Science* 253 (1991): 661–63; M. Leppert et al., "Genetic Analysis of an Inherited Predisposition to Colon Cancer in a Family with a Variable Number of Adenomatous Polyps," *New England Journal of Medicine* 322 (1990): 904–8.

p. 334 *But even as the team:* J. D. Minna et al., "Recessive Oncogenes and Chromosomal Deletions in Human Lung Cancer," *Current Communications in Molecular Biology: Recessive Oncogenes and Tumor Suppressors*, ed. W. Cavenee, N. Hastie, and E. Stanbridge (Cold Spring Harbor, N.Y.: Cold Spring Harbor Laboratory Press, 1989), 57–65; "Clues to Lung Cancer Found in Gene Defect," *Chicago Tribune*, Sept. 27, 1989.

p. 334 *Discovery of p53's footprints:* J. M. Nigro et al., "Mutations in the p53 Gene Occur in Diverse Tumour Types," *Nature* 342 (1989): 705–8; A. von Deimling et al., "p53 Mutations Are Associated with 17p Allelic Loss in Grade II and Grade III Astrocytoma," *Cancer Research* 52 (1992): 2987–90.

p. 334 *What seems increasingly likely:* M. Hollstein et al., "p53 Mutations in Human Cancers," *Science* 253 (1991): 49–53.

p. 335 *The current thinking:* Tony Hunter, "Cooperation between Oncogenes," *Cell* 64 (1991): 249–70.

p. 335 *Among the most dreadful:* D. Malkin et al., "Germ Line p53 Mutations in a Familial Syndrome of Breast Cancer, Sarcomas, and Other Neoplasms," *Science* 250 (1990): 1233–38.

p. 335 *"The suspicion":* J. Marx, "Genetic Defect Identified in Rare Cancer Syndrome," *Science* 250 (1990): 1209.

p. 336 *The Cancer researcher:* J. M. Hall et al., "Linkage of Early-Onset Familial Breast Cancer to Chromosome 17q21," *Science* 250 (1990): 1684–89; D. Sidranski et al., "Inherited p53 Gene Mutations in Breast Cancer," *Cancer Research* 52 (1992): 2984–86.

p. 337 *A clue to how:* Telephone interview with Francis Collins, June 15, 1992; L. B. Andersen et al., "Mutations in the Neurofibromatosis 1 Gene in Sporadic Malignant Melanoma Cell Lines," *Nature Genetics* 3 (1993): 118–21.

p. 337 *But there appear:* Lewin, "Oncogenic Conversion."

p. 339 *Subsequently, it was learned:* S. S. Clark et al., "Expression of a Distinctive BCR-ABL Oncogene in Ph1-Positive Acute Lymphocytic Leukemia (ALL)," *Science* 239 (1988): 775–77.

p. 339 *In 1982, a translocation:* P. Leder et al., "Human Immunoglobulin Heavy Chain Genes Map to a Region of Translocations in Malignant B Lymphocytes," *Science* 216 (1982): 301–3; R. Dalia-Favera et al., "Human C-Myc Onc [*sic*] Gene Is Located on the Region of the Chromosome 8 That Is Translocated in Burkitt's Lymphoma Cells," *Proceedings of the National Academy of Sciences* 79 (1982): 7824–27. (The latter is a paper by Carlo Croce.) Also see J. A. Goldstein and R. L. Bernstein, "Burkitt's Lymphoma and the Role of Epstein-Barr Virus," *Journal of Tropical Pediatrics* 36 (1990): 114–20.

p. 339 *"It's as if":* Interviews with Philip Leder, Sept. 11, 1985, and Aug. 25, 1988.

p. 340 *"Myc is one":* Interview with Janet Rowley, Sept. 18, 1985.

p. 343 *"People very likely":* R. Kotulak, "Some Good News about Cancer," *Chicago Tribune*, April 24, 1988.

p. 343 *In April 1992:* D. Sidransky et al., "Identification of *ras* Oncogene Mutations in the Stool of Patients with Curable Colorectal Tumors," *Science* 256 (1992): 102–4; J. Marx, "Test Could Yield Improved Colon Cancer Detection," *Science* 256 (1992): 32.

pp. 344–47 *Among the most:* Roth interview.

p. 347 *Later, when the existence:* P. J. Saxon et al., "Introduction of Normal Chromosome 11 into Wilms' Tumor Cell Line Controls Its Tumorigenic Expression," *Science* 236 (1987): 175; J. M. Trent et al., "Tumorigenicity in Human Melanoma Cell Lines Controlled by Introduction of Human Chromosome 6," *Science* 247 (1990): 568–71; R. Bookstein et al., "Suppression of Tumorigenicity of Human Prostate Carcinoma Cells by Replacing a Mutated RB Gene," ibid., 712–15; J. Cheng et al., "Suppression of Acute Lymphoblastic Leukemia by the Human Wild-Type p53 Gene," *Cancer Research* 52 (1992): 222–26.

p. 347 *As Roth labored:* Telephone interviews with Kenneth Culver, Dec. 29, 1992, and March 15, 1993.

CHAPTER 17: TRIUMPH OR TRAGEDY?

p. 349 *Then, suddenly:* Material provided by Cell Therapy Research Foundation, "Breakthrough Therapy for Muscular Dystrophy," March 5, 1992; P. K. Law et al., "Feasibility, Safety, and Efficacy of Myoblast Transfer Therapy on Duchenne Muscular Dystrophy Boys," *Cell Transplantation* 1 (1992): 235–44. Peter Law's reminiscences and reflections in this chapter come from several interviews with the authors.

p. 350 *Despite the Peter Law:* P. Gorner and J. Lyon, "Scientists Find Muscular

Dystrophy Gene," *Chicago Tribune*, Oct. 16, 1986; Anthony P. Monaco et al., "Isolation of Candidate cDNAs for Portions of the Duchenne Muscular Dystrophy Gene," *Nature* 323 (1986): 646–50. See also M. Koenig et al., "Complete Cloning of the Duchenne Muscular Dystrophy (DMD) cDNA and Preliminary Genomic Organization of the DMD Gene in Normal and Afflicted Individuals," *Cell* 50 (1987): 509–17. The definition of the protein is from E. P. Hoffman, R. H. Brown, Jr., and L. M. Kunkel, "Dystrophin: The Protein Product of the Duchenne Muscular Dystrophy Locus," *Cell* 51 (1987): 919–28.

pp. 350–51 *Both research milestones:* The Louis Kunkel biographical and methodological material throughout the chapter is derived from interviews by the authors with the researcher and his team, beginning in Boston, on Sept. 23, 1985, and continuing with several telephone interviews through 1993.

p. 351 *Another late starter:* Interview with Susumu Ohno, Aug. 8, 1985.

p. 352 *That Duchenne's is caused:* G. B. A. Duchenne, "Recherches sur la paralysie musculaire pseudo-hypertrophique ou paralysie myosclerosique," *Archives Générales de Médecine* 11 (1868): 5, 178, 305, 421, 552. For the history of the disease and its discoverer, see L. C. McHenry, Jr., *Garrison's History of Neurology* (Springfield, Ill.: Charles C. Thomas, 1969), 278–82, and Cathy Carlson, "Biography of a Maverick," *MDA Newsmagazine*, Nov. 1985.

p. 353 *The first clue:* A. E. H. Emergy and S. Holloway, "Use of Normal Daughters' and Sisters' Creatine Kinase Levels in Estimating Heterozygosity in Duchenne Muscular Dystrophy," *Human Heredity* 27 (1977): 118–26.

pp. 354–55 *The MDA, founded:* The history of the Muscular Dystrophy Association and the hunt for the gene comes from several interviews with Donald S. Wood, director of science technology for the Muscular Dystrophy Association (Oct. 17 and 22, 1986, April 1 and 6, and Dec. 21, 1987, May 2 and Oct. 25, 1988, etc.), and several articles, including "Finding the Duchenne Gene: Success Is Close at Hand," *MDA Newsmagazine*, Nov. 1985, 8–10; "Duchenne Muscular Dystrophy—Fatal Muscle Flaw Found," ibid., Spring 1988, 8–12; L. P. Rowland, "Duchenne Dystrophy: Behind the Discoveries," ibid., Spring 1989, 16–20.

p. 356 *Early in 1985:* U. Francke et al., "Minor Xp21 Chromosome Deletion in a Male Associated with Expression of Duchenne Muscular Dystrophy, Chronic Granulomatous Disease, Retinitis Pigmentosa, and McLeod Syndrome," *American Journal of Human Genetics* 37 (1985): 250–67.

p. 361 *In the spring of 1989:* D. R. Love et al., "An Autosomal Transcript in Skeletal Muscle with Homology to Dystrophin," *Nature* 339 (1989): 55–57.

p. 363 *The mdx mouse:* T. A. Partridge et al., "Conversion of mdx Myofibres from Dystrophin-Negative to -Positive by Injection of Normal Myoblasts," *Nature* 337 (1989): 176–79.

p. 363 *In early 1990:* P. Gorner, "Gene Injections to Fight Muscular Dystrophy," *Chicago Tribune*, April 26, 1991.

p. 364 *In Montreal, meanwhile:* G. Karpati et al., "Myoblast Transfer in Duchenne Muscular Dystrophy," *Annals of Neurology* 34, no. 1 (1993): 8–17; telephone interview with George Karpati, Nov. 22, 1993.

p. 366 *"We have already"*: Interview with C. Thomas Caskey, July 27, 1990.

p. 367 *Other delivery systems:* Interview with Jon Wolff, March 19, 1990; J. A. Wolff et al., "Direct Gene Transfer into Mouse Muscle in Vivo," *Science* 247 (1991): 1465–68.

p. 367 *Early in 1991:* Gorner, "Gene Injections."

p. 368 *At any rate:* Ibid.

p. 368 *Accordingly, on March 5:* Law et al., "Feasibility."

p. 369 *viewed Law's study:* Paul R. Sanberg interview in Gorner, "Gene Injections."

p. 369 *It is not generally known:* See M. H. Brooke et al., "Clinical Trial in Duchenne Dystrophy: 1. The Design of the Protocol," *Muscle Nerve* 4 (1981): 186–97, and M. H. Brooke et al., "Clinical Investigation in Duchenne Dystrophy: 2. Determination of the 'Power' of Therapeutic Trials Based on Natural History," *Muscle Nerve* 6 (1983): 91–103.

p. 370 *Over the years:* Interview with Donald Wood, Sept. 4, 1992.

p. 370 *As recently as 1989:* J. R. Mendell et al., "Randomized, Double-Blind Six-Month Trial of Prednisone in Duchenne Muscular Dystrophy," *New England Journal of Medicine* 320 (1989): 1592–97. See also R. M. Sklar and R. H. Brown, "Methylprednisone Increases Dystrophin Levels by Inhibiting Myotube Death during Myogenesis of Normal Human Muscle in Vitro," *Journal of Neurological Science* 101 (1991): 73–81.

p. 371 *At any rate:* This material comes from the FDA report of its on-site investigation of the Cell Therapy Research Foundation, in Memphis, obtained by the authors. "Inspectional Observations," by David C. Benstein, Department of Health and Human Services, Public Health Service, Food and Drug Administration, May 21, 1991.

p. 372 *However, without controls:* Interview with Robert Brown, March 2, 1993.

p. 372 *"The MDA loves":* Interview with Peter Law, April 2, 1992.

p. 373 *Stunned, the MDA's:* Interview with Leon Charash, April 2, 1992.

p. 373 *On March 31:* Material from Stanford University Medical Center: "Cell Therapy in Muscular Dystrophy Patients Yields Missing Gene Product" (April 1, 1992); R. G. Miller et al., "Myoblast Implantation in Duchenne Muscular Dystrophy: The San Francisco Study," *Neurology* 42, suppl. 3 (1992): 189.

p. 374 *Law termed:* Law interview.

p. 374 *a neutral party:* Interview with Terence Partridge, April 3, 1992.

p. 375 *On July 14:* The announcement of the patent was provided by the Cell Therapy Research Foundation, July 14, 1992, and by a letter from Hill, Van Santen, Steadman & Simpson, Attorneys and Counselors at Law, Chicago. The Peter Law material (including the words about Jonas Salk) comes from a Law telephone interview, July 14, 1992.

p. 375 *In the March 1993 issue:* K. R. Sharma, M. A. Mynhier, and R. G. Miller,

"Cyclosporine Increases Muscular Force Generation in Duchenne Muscular Dystrophy," *Neurology* 43 (1993): 527–31.

p. 376 *In November of 1993:* P. K. Law et al., "Cell Transplantation as an Experimental Treatment for Duchenne Muscular Dystrophy," *Cell Transplantation* 2 (1993): 485–505; telephone interviews with Peter Law, Nov. 8, 1993; with George Karpati, Nov. 22, 1993; with Michael Brooke, Nov. 23, 1993; and with Jerry Mendell, Nov. 29, 1993.

p. 380 *Two months later:* Brown study announced in material from the MDA, Dec. 20, 1992.

p. 381 *No one was talking:* T. Ragot et al., "Efficient Adenovirus-Mediated Transfer of a Human Minidystrophin Gene to Skeletal Muscle of *mdx* Mice," *Nature* 361 (1993): 647–50.

p. 382 *In August of 1993:* C. C. Lee et al., "Expression of Recombinant Dystrophin and Its Localization to the Cell Membrane," *Nature* 349 (1991): 334–36.

CHAPTER 18: THE THIEF OF BREATH

p. 384 *"It looks too promising":* Telephone interview with Ronald Crystal, June 30, 1992

p. 384 *"Gene therapy is":* Telephone interview with Francis Collins, Sept. 21, 1990.

pp. 385–86 *Cystic fibrosis has caused:* Interviews with Bottorff family, Aug. 5, 1987.

pp. 386–95 *Events began unfolding:* This section is based on interviews over a number of years with all of the major participants, including Lap-Chee Tsui, Aug. 24, 1988; Ray White, Nov. 22, 1985, and, by telephone, Aug. 6, 1991; Helen Donis-Keller, Nov. 3, 1985, by telephone; Robert Williamson, Feb. 13, 1988, and Aug. 24, 1988; and Francis Collins, Sept. 20, 1989. We also acknowledge a debt to the reporting of Leslie Roberts, of *Science*, whose two-part article "The Race for the Cystic Fibrosis Gene," *Science* 240 (1988): 141–44, 282–85, gives an excellent overview of both the early efforts to locate the CF gene and the intense, sometimes unseemly competition that drives scientific investigation.

p. 388 *"It was wonderful":* Donis-Keller interview, Nov. 3, 1985.

p. 388 *"I think it is very clear":* Roberts, "Race."

p. 389 *According to Donis-Keller:* Telephone interviews with Helen Donis-Keller, Nov. 3, 1985, and Dec. 3, 1993.

p. 390 *"We ran the probe":* Telephone interview with Ray White, Nov. 22, 1985.

p. 391 *Donis-Keller is philosophical:* Donis-Keller interview, Dec. 3, 1993.

p. 392 *"I can't emphasize":* Interview with Robert Williamson, Feb. 13, 1988.

p. 392 *"The Cystic Fibrosis Trust":* Ibid.

p. 393 *Williamson was soaring:* X. Estivill et al., "A Candidate for the Cystic Fibrosis Locus Isolated by Selection for Methylation-Free Islands," *Nature* 326 (1987): 840–45.

p. 394 *"We never issued":* Interview with Robert Williamson, Aug. 24, 1988.

p. 394 *"We all did terrible":* Interview with Francis Collins, Sept. 20, 1989.

p. 394 *"It was an awful":* Williamson interview, Feb. 18, 1988.

p. 394 *"But what really"* Ibid.

p. 395 *In October of:* Ibid.

pp. 395–98 *Walking along:* This section is based on interviews with Francis Collins, Sept. 20, 1989, and with Lap-Chee Tsui, Aug. 23, 1988. Also see J. M. Rommens et al., "Identification of the Cystic Fibrosis Gene: Chromosome Walking and Jumping," *Science* 245 (1989): 1059–65; J. R. Riordan et al., "Identification of the Cystic Fibrosis Gene: Cloning and Characterization of Complementary DNA," ibid., 1066–73; B. Kerem et al., "Identification of the Cystic Fibrosis Gene: Genetic Analysis," ibid., 1073–79; and an excellent summing-up by J. Marx, "The Cystic Fibrosis Gene Is Found," ibid., 923–25.

p. 396 *But in 1983:* F. S. Collins et al., "Construction of a General Human Chromosome Jumping Library, with Application to Cystic Fibrosis," *Science* 235 (1987): 1046–49.

p. 400 *"Gene therapy has been":* Telephone interview with Robert Beall, Sept. 4, 1989.

p. 401 *In early 1992:* M. Rosenfeld et al., "*In Vivo* Transfer of the Human Cystic Fibrosis Transmembrane Conductance Regulator Gene to the Airway Epithelium," *Cell* 68 (1992): 143–55.

p. 402 *"We can only get":* Crystal 1992 interview.

p. 402 *"I'm a little less":* Telephone interview with Francis Collins, June 15, 1992. See also F. S. Collins, "Cystic Fibrosis: Molecular Biology and Therapeutic Implications," *Science* 256 (1992): 774–79.

p. 402 *"But when Jim":* J. Engelhardt et al., "Submucosal Glands Are the Predominant Site of CFTR Expression in the Human Bronchus," *Nature Genetics* 4 (May 1993): 27–34.

p. 402 *But as time passed:* Telephone interview with James Wilson, April 9, 1993, and unpublished data. See also telephone interview with Francis Collins, March 15, 1993.

p. 403 *Crystal put:* Telephone interview with Ronald Crystal, March 23, 1993, and unpublished data.

p. 403 *"So while":* Crystal interview.

p. 404 *"Few studies have":* Wilson interview.

p. 404 *"I think we can":* Crystal interview.

p. 404 *"We expect we":* Ibid.

p. 405 *"We don't want":* Wilson interview. See also J. Whitsett et al., "Human Cystic Fibrosis Transmembrane Conductance Regulator Directed to Respiratory Epithelial Cells of Transgenic Mice," *Nature Genetics* 2 (1992): 919–23.

p. 405 *Just as mindful:* M. Welsh, "Cystic Fibrosis Gene Therapy: *In Vivo* Safety and Efficacy in Nasal Epithelium," a protocol submitted to the RAC and reviewed Sept. 10, 1992.

p. 405 *"A huge amount":* Wilson interview.

p. 405 *By the middle of 1994:* Telephone interviews with Ronald Crystal, July 19, 1994, and with James Wilson, July 20, 1994.

p. 405 *"I can't answer":* Crystal interview.

p. 406 *Bob Williamson, who:* Williamson interview.

CHAPTER 19: IT TAKES A WORRIED MAN

This chapter is based on interviews since 1985 with Nancy Wexler, Herbert Pardes, James Gusella, David Housman, Francis Collins, and other members of the team that hunted for the gene. Several Huntington's families also contributed, though their request for anonymity is being honored. Insights into problems surrounding history's first presymptomatic test for a fatal genetic disease were provided by Jason Brandt (July 14, 1988), Kimberly Quaid (April 16, 1988), Nancy Wexler, and patients in the United States and Canada—particularly "Catherine" of Boston and Karen Sweeney of Fort Lee, Va.—who have undergone the screening.

p. 408 *In 1983:* J. F. Gusella et al., "A Polymorphic DNA Marker Genetically Linked to Huntington's Disease," *Nature* 306 (1983): 234–38.

p. 408 *In plain English:* James Gusella, interview, Sept. 13, 1985.

p. 410 *The disease, its namesake:* G. Huntington, "On Chorea," *Medical and Surgical Reporter* 26 (1872): 317–21.

p. 410 *"What we did":* P. Gorner, "Out of the Shadow: A New Genetic Test Can Foretell Agonizing Death," *Chicago Tribune,* Aug. 4, 1988. The controversial genetic marker test for carriers of the Huntington's disease gene has amassed its own body of literature. Especially recommended are M. Steinmann, "In the Shadow of Huntington's Disease," *Columbia, the Magazine of Columbia University* 13, no. 2 (1987): 15–19; A. Newman, "The Legacy on Chromosome 4," *Johns Hopkins Magazine* 20 (1988): 30–39. A comprehensive review paper is J. Brandt et al., "Presymptomatic Diagnosis of Delayed-Onset Disease with Linked DNA Markers: The Experience with Huntington's Disease," *Journal of the American Medical Association* 261 (1989): 3108–14.

pp. 411–12 *From the outset:* Interview with Nancy Wexler, July 31, 1985.

pp. 414–16 *The suicide rate:* Ibid.

p. 416 *"Basically, the idea":* Interview with David Bostein, May 9, 1989, South San Francisco. For the power of the technique, see D. Housman and J. Gusella, "Molecular Genetic Approaches to Neural Degenerative Disorders," in *Molecular Genetic Neuroscience,* ed. F. O. Schmitt, S. J. Bird, and F. E. Bloom (New York: Raven Press, 1982), chap. 34.

pp. 420–21 *Finally, in January:* L. Roberts, "Huntington's Gene: So Near, Yet So

Far: Three Steps Forward, Two Steps Back," *Science* 247 (1990): 4943. See also idem, "The Huntington's Gene Quest Goes On," *Science* 258 (1992): 740–41.

pp. 422–24 *The Huntington's gel:* The discovery of the gene was recounted by James Gusella, interview, April 30, 1993.

p. 424 *Michigan's Francis Collins:* The reactions from Francis Collins and Nancy Wexler come from interviews, March 21, 1993.

p. 425 *"Interestingly, no first author:* The Huntington's Disease Collaborative Research Group, "A Novel Gene Containing a Trinucleotide Repeat That Is Expanded and Unstable on Huntington's Disease Chromosomes," *Cell* 72 (1993): 971–83.

CHAPTER 20: DELIVERING THE GOODS

p. 429 *"Lesch-Nyhan patients":* Telephone interview with Beverly Davidson and Blake Roessler, March 17, 1993.

p. 430 *"It turns out":* Ibid.

p. 430 *"There is no":* Ibid.

p. 430 *"We had very high":* Ibid.

p. 431 *Finally there was:* B. L. Davidson et al., "A Model System for *in Vivo* Gene Transfer into the Central Nervous System Using an Adenoviral Vector," *Nature Genetics* 3 (1993): 219–223.

p. 431 *Within weeks:* S. Akli et al., "Transfer of a Foreign Gene into the Brain Using Adenovirus Vectors," *Nature Genetics* 3 (1993): 224–28; G. Le Gal La Salle et al., "An Adenovirus Vector for Gene Transfer into Neurons and Glia in the Brain," *Science* 259 (1993): 988–90.

p. 432 *Almost at the same time:* G. Bajocchi et al., "Direct *in Vivo* Gene Transfer to Ependymal Cells in the Central Nervous System Using Recombinant Adenovirus Vectors," *Nature Genetics* 3 (1993): 229–34.

p. 432 *Adenovirus vectors:* S. Akli et al., "Transfer."

p. 434 *Recent experiments:* S. Jiao, V. Gurevich, and J. A. Wolff, "Long-Term Correction of Rat Model of Parkinson's Disease by Gene Therapy," *Nature* 362 (1993): 450–53.

p. 434 *"Assuming the animal":* Davidson and Roessler interview.

p. 436 *"ALS is so lethal":* Telephone interview with Robert Brown, March 15, 1993.

p. 437 *On May 16, 1991:* T. Siddique et al., "Linkage of a Gene Causing Familial Amyotrophic Lateral Sclerosis to Chromosome 21 and Evidence of Genetic-Locus Heterogeneity," *New England Journal of Medicine* 324 (1991): 1381–84; P. M. Conneally, "A First Step toward a Molecular Genetic Analysis of Amyotrophic Lateral Sclerosis" (editorial), ibid., 1430–32.

p. 438 *"But so far as anyone knew":* Interview with Teepu Siddique, Feb. 25, 1993.

p. 439 *This was big news:* The narrative is based on interviews with Siddique

(Feb. 25, 1993) and Robert Brown (March 15, 1993). See also D. R. Rosen et al., "Mutations in Cu/Zn Superoxide Dismutase Gene Are Associated with Familial Amyotrophic Lateral Sclerosis," *Nature* 362 (1993): 59.

p. 439 *Massachusetts General announced:* Press release distributed by Massachusetts General Hospital, March 2, 1993.

p. 439 *The discovery means:* P. Gorner, "Scientists Hail Discovery of Gene That Triggers ALS," *Chicago Tribune,* March 4, 1993; N. Angier, "Scientists Find Long-Sought Gene That Causes Lou Gehrig's Disease," *New York Times,* March 4, 1993; "Happy Birthday, Double Helix," *Time,* March 15, 1993, 58.

p. 440 *In fact, such a mouse:* Z. S. Xu et al., "Increased Expression of Neurofilament Subunit (NF-L) Produces Morphological Alterations That Resemble the Pathology of Human Motor Neuron Disease," *Cell* 73 (1993): 23–33.

p. 440 *Besides causing:* Gene Cohen statement issued by National Institute on Aging, March 3, 1993.

p. 442 *"Who knows":* Telephone interview with James Gusella, April 3, 1987.

p. 443 *The initial findings:* P. St. George-Hyslop et al., "The Genetic Defect Causing Familial Alzheimer's Disease Maps on Chromosome 21," *Science* 235 (1987): 885–89.

p. 443 *That very week:* R. Tanzi et al., "Amyloide B Protein Gene: cDNA, mRNA Distribution and Genetic Linkage Near the Alzheimer Locus," *Science* 235 (1987): 880–84; J. Kang et al., *Nature* 325 (1987) 733–35; D. Goldgaber et al., *Science* 235 (1987): 877–80; K. Robakis et al., "Chromosome 21q21 Sublocalization of Gene Encoding Beta Amyloid Peptide in Cerebral Vessels and Neuritic (Senile) Plaques of People with Alzheimer Disease and Down Syndrome," *Lancet,* Feb. 14, 1987, 384–85.

p. 443 *Moreover, studies by:* C. Van Broeckhoven et al., "Failure of Familial Alzheimer's Disease to Segregate with the A4-Amyloid Gene in Several European Families," *Nature* 329 (1987): 153–55; R. Tanzi et al., "The Genetic Defect in Familial Alzheimer's Disease Is Not Tightly Linked to the Amyloid B-Protein Gene," *Nature* 329 (1987): 156–57.

p. 443–44 *"It would have been":* Telephone interview with Peter St. George-Hyslop, Feb. 4, 1990.

p. 444 *"For some reason":* Gusella interview.

pp. 444–45 *"If there is":* Ibid.

p. 445 *Before long:* E. H. Corder, "Gene Dose of Apo E Type 4 Allele and the Risk of Alzheimer's Disease in Late Onset Families," *Science* 261 (1993): 828–29.

p. 446 *"One possibility":* Telephone interview with Rudolph Tanzi, May 6, 1993.

p. 447 *"Any disruption":* Telephone interview with James Gusella, April 30, 1993.

p. 447 *"We are on the verge":* Tanzi interview.

p. 447 *One study:* W. Fischer, "Amelioration of Cholinergic Neutron Atrophy and Spatial Memory Impairment in Aged Rats by Nerve Growth Factor," *Nature* 329 (1987): 65–68.

p. 447 *Franz Hefti:* R. Kotulak, "Alzheimer's Studies Break Ground," *Chicago Tribune,* Feb. 11, 1990.

p. 447 *sure enough:* Telephone interview with Lars Olson, March 1991.

p. 448 *Enthusiasm for the:* J. Marx, "NGF and Alzheimer's: Hopes and Fears," *Science* 247 (1990): 408–10; J. S. Whitson, D. J. Selkoe, and C. W. Cotman, "Amyloid Beta Protein Enhances the Survival of Hippocampal Neurons in Vitro," *Science* 243 (1989): 1488–90.

p. 448 *On the other hand:* M. B. Rosenberg et al., "Grafting Genetically Modified Cells to the Damaged Brain: Restorative Effects of Nerve Growth Factor Expression," *Science* 242 (1988): 1575–78.

p. 448 *Recently, another:* L. Joachim, H. Mori, and D. J. Selkoe, "Amyloid β-Protein Deposition in Tissues Other Than Brain in Alzheimer's Disease," *Nature* 341 (1989): 226–30.

p. 449 *The ability to look:* H. M. Schmeck, Jr., "Research Suggests Virus Link to Alzheimer's Disease," *New York Times,* July 23, 1988.

p. 450 *"We know you get":* Interview with Charles Epstein, May 8, 1989.

p. 451 *Recently, in collaboration:* C. J. Epstein et al., "Transgenic Mice with Increased Cu/Zn-Superoxide Dismutase Activity: Animal Model of Dosage Effects in Down Syndrome," *Proceedings of the National Academy of Sciences* 84 (1987): 8044–48.

p. 451 *"Then you're in trouble":* Epstein interview.

pp. 451–52 *Working on his own:* K. B. Avraham et al., "Down's Syndrome: Abnormal Neuromuscular Junction in Tongue of Transgenic Mice with Elevated Levels of Human Cu/Zn-Superoxide Dismutase," *Cell* 54 (1988): 823–29.

p. 452 *In 1989, Groner:* M. Schickler et al., "Diminished Serotonin Uptake in Platelets of Transgenic Mice with Increased Cu/Zn-Superoxide Dismutase Activity," *EMBO Journal* 8 (1989): 1385–92.

p. 452 *"I would use PKU":* Telephone interview with David Cox, Jan. 14, 1986; interview with David Cox, May 9, 1989. See also R. Allore et al., "Gene Encoding the B Subunit of S100 Protein Is on Chromosome 21: Implications for Down Syndrome," *Science* 239 (1988): 1311–13.

pp. 453 *"Whether the function":* Epstein interview.

p. 453 *It has been known:* H.-P. Hartung et al., "Familial Multiple Sclerosis," *Journal of the Neurological Sciences* 83 (1988): 259–68; E. Kinnunen et al., "Multiple Sclerosis in a Nationwide Series of Twins," *Neurology* 37 (1987): 1627–29; A. Sadovnick, P. Baird, and R. Ward, "Multiple Sclerosis: Updated Risks for Relatives," *American Journal of Medical Genetics* 29 (1988): 533–41; A. Sadovnick and P. Baird, "The Familial Nature of Multiple Sclerosis: Age-Corrected Empiric Recurrence Risks for Children and Siblings of Patients," *Neurology* 38 (1988): 990–91.

p. 454 *Over the past few:* N. Odom et al., "HLA-DP Antigens Are Involved in the Susceptibility to Multiple Sclerosis," *Tissue Antigens* 31 (1988): 235–37.

p. 454 *In early 1989:* S. Beall et al., "The Germline Repertoire of T Cell Receptor

B-Chain Genes in Patients with Chronic Progressive Multiple Sclerosis," *Journal of Neuroimmunology* 21 (1989): 59–66.

p. 454 *"What this suggests":* Paul Recer, of the Associated Press, from a telephone interview with Dale McFarlin, in a wire story moved on Jan. 19, 1989.

p. 455 *Sequence similarities:* E. Reddy et al., "Amplification and Molecular Cloning of HTLV-I Sequences from DNA of Multiple Sclerosis Patients," *Science* 243 (1989): 529–33. This paper finds evidence of HTLV-1 infection in MS patients. But see L. Grimaldi et al., "HTLV-I-Associated Myelopathy: Oligoclonal Immunoglobulin G. Bands Contain Anti-HTLV-I p24 Antibody," *Annals of Neurology* 24 (1988): 727–31. This paper finds evidence of the virus in a central nervous system disorder called HTLV-1-Associated Myelopathy (HAM) but not in MS.

p. 455 *Much of this work:* C. Readhead et al., "Expression of a Myelin Basic Protein in Transgenic Shiverer Mice: Correction of the Dysmyelinating Phenotype," *Cell* 48 (1987): 703–12.

pp. 455–56 *Another shock came:* A. Vogt, "Panel Tells FDA: Speed Approval of MS Drug," *Chicago Tribune,* March 20, 1993.

p. 456 *"If you look":* Telephone interview with Steven Rheingold, Jan. 1990.

CHAPTER 21: SOMETHING IN THE BLOOD

p. 457 *A research group:* J. Egeland et al., "Bipolar Affective Disorders Linked to DNA Markers on Chromosome 11," *Nature* 325 (1987): 783–87.

p. 457 *Within three weeks:* M. Baron et al., "Genetic Linkage between X-Chromosome Markers and Bipolar Affective Illness," *Nature* 326 (1987): 289–92. See also J. Mendlewicz et al., "Polymorphic DNA Marker on X Chromosome and Manic Depression," *Lancet,* May 30, 1987, 1230–31.

p. 457 *The dust had scarcely:* A. Bassett et al., "Partial Trisomy Chromosome 5 Cosegregating with Schizophrenia," *Lancet,* April 9, 1988, 799–801.

pp. 458–61 *The human laboratory:* Telephone interview with Janice Egeland, June 14, 1990; "New Findings on the Heritability of Bipolar Disorder: An Interview with Janice A. Egeland, Ph.D.," *Currents in Affective Illness* 6 (1987): 5–12; J. Egeland and A. Hostetter, "Amish Study I: Affective Disorders among the Amish, 1976–1980," *American Journal of Psychiatry* 140 (1983): 56–61; J. Egeland and J. Sussex, "Suicide and Family Loading for Affective Disorders," *Journal of the American Medical Association* 254 (1985): 915–18.

p. 461 *In a 1987 interview:* "New Findings on the Heritability."

pp. 461–62 *Unfortunately, the assumption:* S. D. Detera-Wadleigh et al., "Close Linkage of C-Harvey *Ras*-1 and the Insulin Gene to Affective Disorder Is Ruled Out in Three North American Pedigrees," *Nature* 325 (1987): 806–8; J. Kelsoe et al., "Reevaluation of the Linkage Relationship between Chromosome 11p Loci and the Gene for Bipolar Affective Disorder in the Old Order Amish,"

Nature 342 (1989): 238–43; M. Barinaga, "Manic Depression Gene Put in Limbo," *Science* 246 (1989): 886–87; Egeland telephone interview; H. M. Schmeck, Jr., "Scientists Now Doubt They Found Faulty Gene Linked to Mental Illness," *New York Times*, Nov. 7, 1989; J. Egeland et al., "The Impact of Diagnoses on Genetic Linkage Study for Bipolar Affective Disorders among the Amish," *Psychiatric Genetics* 1 (1990): 5–18; E. Ginns et al., "Update on the Search for DNA Markers Linked to Manic-Depressive Illness in the Old Order Amish," *Psychiatric Research* 26 (1992): 305–8; J. Kelsoe et al., "Studies Search for a Gene for Bipolar Affective Disorder in the Old Order Amish," *Psychiatric Times*, June 1990; telephone interview with Janice Egeland, March 14, 1993.

p. 463 *"The highest probability"*: Egeland 1990 interview.

pp. 464–65 *"That's the most sobering"*: Egeland 1993 interview.

p. 465 *"The X chromosome"*: Telephone interview with Miron Baron, June 3, 1990.

p. 465 *The two conditions:* T. Reich, P. J. Clayton, and G. Winokur, "Family History Studies: V. The Genetics of Mania," *American Journal of Psychiatry* 125 (1969): 1358–69.

p. 466 *An extraordinary degree:* Baron et al., "Genetic Linkage."

p. 466 *"We don't want to say"*: Baron interview.

p. 466 *It was a good thing:* M. Baron et al., "Diminished Support for Linkage between Manic Depressive Illness and X-Chromosome Markers in Three Israeli Pedigrees," *Nature Genetics* 3 (1993): 49–55.

p. 467 *As Egeland later:* David George, press release from the American Mental Health Fund, sponsors of the Victor M. Cannon Award, Nov. 17, 1988.

p. 467 *Schizophrenia has been:* "Where Next with Psychiatric Illness?" (editorial), *Nature* 336 (1988): 95–96.

p. 468 *In November 1988:* R. Sherrington et al., "Localization of a Susceptibility Locus for Schizophrenia on Chromosome 5," *Nature* 336 (1988): 164–67.

p. 469 *But, alas.* J. L. Kennedy et al., "Evidence against Linkage of Schizophrenia to Markers on Chromosome 5 in a Northern Swedish Pedigree," *Nature* 336 (1988): 167–70. Also see S. Detera-Wadleigh et al., "Exclusion of Linkage to 5q11–13 in Families with Schizophrenia and Other Psychiatric Disorders," *Nature* 340 (1989): 391–93.

pp. 469–70 *Comings's views set off:* P. Thomas, "Tourette's Debate Boils Over," *Medical World News*, Dec. 26, 1988; Paul Raeburn, Associated Press report, Nov. 28, 1988; telephone interview with David Comings, Dec. 6, 1988; "Tourette's Syndrome and Attention Deficit Disorder with Hyperactivity," an exchange of views between the Comingses and Pauls et al., *Archives of General Psychiatry* 44 (1987): 1023–26. See also P. Gorner, "Tourette Theory in Hot Dispute," *Chicago Tribune*, Jan. 15, 1989.

pp. 471–72 *The evidence for their:* D. E. Comings and B. G. Comings, "A Controlled Study of Tourette Syndrome," *American Journal of Human Genetics*

41 (1987): 701–41. See also idem, "Hereditary Agoraphobia and Obsessive-Compulsive Behavior in Relatives of Patients with Gilles de la Tourette's Syndrome," *British Journal of Psychiatry* 151 (1987): 195–99; idem, "Tourette's Syndrome and Attention Deficit Disorder with Hyperactivity: Are They Genetically Related?" *Journal of the American Academy of Child Psychiatry* 23 (1984): 138–46.

p. 471 *Typically, says David Comings:* Interview with David Comings, Aug. 5, 1985; telephone interviews with David Comings, Dec. 6, 1988, and May 4, 1993.

p. 472 *"Our data do not":* Thomas, "Tourette's Debate."

pp. 472–73 *Comings thinks he may:* Comings 1988 and 1993 interviews.

p. 473 *Among the first:* E. J. Devor and C. R. Cloninger, "Genetics of Alcoholism," *Annual Review of Genetics* 23 (1989): 19–36.

pp. 473–74 *In the 1960s:* H. Begleiter and B. Porjesz, "Potential Biological Markers in Individuals at High Risk for Developing Alcoholism," *Alcoholism* 12 (1986): 488–93.

pp. 473–74 *This correlated with:* R. C. Cloninger et al., "Psychopathology in Adopted-out Children of Alcoholics, the Stockholm Adoption Study," *Recent Developments in Alcoholism*, vol. 3, ed. M. Galenter (New York: Plenum, 1985), 37–51. See also G. Kolata, "Alcoholism: Genetic Links Grow Clearer," *New York Times*, Nov. 10, 1987.

p. 474 *At the University of California:* M. A. Schuckit, "A Longitudinal Study of Children of Alcoholics," *Recent Developments in Alcoholism* 9 (1991): 5–19.

pp. 474–75 *But no one had:* Kenneth Blum et al., "Allelic Association of Human Dopamine D2 Receptor Gene in Alcoholism," *Journal of the American Medical Association* 263 (1990): 2055–60. See also "Finding the Gene(s) for Alcoholism" (editorial), ibid., 2094–95; Peter Gorner, "Scientists Find Gene Involved in Alcoholism," *Chicago Tribune*, April 18, 1990; Lawrence K. Altman, "Researchers Identify Gene Linked to Alcoholism," *New York Times*, April 18, 1990; Shauna Roberts, "A Marker Gene for Alcoholism," *Journal of NIH Research* 2 (1990): 24–25.

p. 475 *Meanwhile, early 1993:* Peter Hauser et al., "Attention Deficit–Hyperactivity Disorder in People with Generalized Resistance to Thyroid Hormone," *New England Journal of Medicine* 328 (1993): 997–1001.

p. 475 *"Thyroid hormone is":* Mark Sampson, press release from National Institute of Diabetes and Digestive and Kidney Disease, April 7, 1993.

p. 476 *"It's not bad parenting":* Ibid.

p. 477 *"Molecular biology will":* Steven E. Hyman, "The Role of Molecular Biology in Psychiatry," *Psychosomatics* 29 (1988): 328–32.

p. 477 *"Looking down the":* Telephone interview with Herbert Pardes, April 26, 1990. See also H. Pardes et al., "Genetics and Psychiatry: Past Discoveries, Current Dilemma, and Future Directions," *American Journal of Psychiatry* 146 (1989): 435–43.

CHAPTER 22 GYPSY FORTUNE-TELLERS OF TECHNOLOGY

p. 480 *"It has been very consuming":* Interview with Francis Collins, April 9, 1993.

pp. 480–81 *Collins's pioneering venture:* Interviews with Barbara Weber, June 18, 1992, and with members of the Michigan family that led researchers to the first genetic marker for breast cancer, including "Susan" (Aug. 18, 1993) and her sister "Janet" (Aug. 19, 1993).

p. 483 *In light of:* Interview with Henry Lynch Sept. 17, 1992. See also H. T. Lynch, *Genetics and Breast Cancer* (New York: V. N. Reinhold, 1981); H. T. Lynch and A. J. Krush, "Carcinoma of the Breast and Ovary in Three Families," *Journal of Surgery, Gynecology and Obstetrics* 133 (1971): 644.

p. 485 *Already, though:* D. Goodman, "Discrimination Linked to Genetic Screening," *American Medical News*, June 9, 1989 (interview with Paul Billings).

p. 485 *In the meantime:* C. T. Caskey, "Disease Diagnosis by Recombinant DNA Methods," *Science* 236 (1987): 1223–29; S. H. Orkin, "Genetic Diagnosis by DNA Analysis: Progress through Amplification," *New England Journal of Medicine* 317 (1987): 1023–25; S. E. Antonarakis, "Diagnosis of Genetic Disorders at the DNA Level," ibid., 320 (1989): 153–63. The landmark paper for mass genetic screening of newborns for metabolic disease is R. Guthrie and A. Susi, "A Simple Phenylalanine Method for the Detection of Phenylketonuria in Large Populations of Newborn Infants," *Pediatrics* 32 (1963): 338–43. See also W. I. Nyhan, "Neonatal Screening for Inherited Disease" (editorial), *New England Journal of Medicine* 313 (1985): 43–44.

p. 486 *"We all know":* Interview with Kimberly A. Quaid, April 16, 1988. See also K. Quaid, "The Decision to be Tested for Huntington's Disease," *Journal of the American Medical Association* 257 (1987): 3362; P. Gorner, "Out of the Shadow: A New Genetic Test Can Foretell Agonizing Death," *Chicago Tribune*, Aug. 4, 1988; D. Grady, "The Ticking of a Time Bomb in the Genes," *Discover*, June 1987, 26. For review papers, see G. Meissen et al., "Predictions Testing for Huntington's Disease with Use of a United Marker," *New England Journal of Medicine* 318 (1988): 535; J. Brandt et al., "Presymptomatic Diagnosis of Delayed-Onset Disease with Linked DNA Markers: The Experience with Huntington's Disease," *Journal of the American Medical Association* 261 (1989): 3108–14.

p. 487 *From its inception:* Interview with Nancy Wexler, July 31, 1985. See also N. S. Wexler, "Huntington's Disease and Other Late Onset Genetic Disorders," in *Psychological Aspects of Genetic Counseling*, ed. A. H. Emery and I. M. Pullen (New York: Academic Press, 1984), 125–46.

p. 487 *The feat, which also:* Interviews with Norman Arnheim, Sandra Carson, and John Buster, Sept. 14–19, 1990, Chicago. See Norman Arnheim et al., "PCR Analysis of DNA Sequences in Single Cells: Single Sperm Gene Mapping and Genetic Disease Diagnosis," and S. A. Carson et al., "Recovery of Blastocysts by Uterine Lavage following Superovulatory Drugs" (Papers presented at the First International Symposium on Preimplantation Genetics, Chicago, Sept. 14–19, 1990). See also J. E. Buster and S. A. Carson, "Genetic Diagnosis of the Preimplantation Embryo," *American Journal of Medical Genetics* 34 (1989): 211–16.

p. 488 *Since 1985:* P. Gorner, "Guilty or Innocent? Genetic Fingerprints Leave No Doubt," *Chicago Tribune,* March 6, 1988. DNA fingerprinting was explained to the authors by two pioneers, the British researcher Alec Jeffreys, April 25, 1989, in Chicago, and the MIT professor David Housman, in a telephone interview, Feb. 25, 1988.

pp. 488–89 *The brainstorm struck:* The Yury Verlinsky material in this chapter is derived from several interviews, beginning Oct. 20, 1989. See Y. Verlinsky and A. Kuliev, eds., *Preimplantation Genetics* (New York: Plenum, 1991).

p. 490 *IVF boils down to:* Interview with Robert G. Edwards, Sept. 17, 1990. An update of French IVF procedures was provided the same day by Michelle Plachot, of the Hospital Necker, Paris. See interview with Jacques Cohen, Sept. 18, 1990.

pp. 490–91 *No sooner had Verlinsky:* Y. Verlinsky et al., "Genetic Analysis of Polar Body DNA: A New Approach to Preimplantation Genetic Diagnosis" (abstract), *American Journal of Human Genetics* 45, suppl. (1989): 1072.

pp. 494–96 *The initial triumph:* Interview with Alan Handyside, Sept. 17, 1990.

p. 496 *"Basically, we're sexing embryos":* A. H. Handyside, "Preimplantation Diagnosis," *Current Obstetrics and Gynecology* 2 (1992): 85–90; K. Hardy et al., "Human Preimplantation Development in Vitro Is Not Adversely Affected by Biopsy at the 8-Cell Stage," *Human Reproduction* 5 (1990): 708–14; K. Hardy and A. H. Handyside, "Biopsy of Cleavage Stage Human Embryos and Diagnosis of Single Gene Defects by DNA Amplication," *Archives of Pathology and Laboratory Medicine* 116 (1992): 388–92. For the clearest explanation of the discovery of PCR, see K. B. Mullis, "The Unusual Origin of the Polymerase Chain Reaction," *Scientific American,* April 1990, 56–65.

p. 496 *Handyside's live births:* A. H. Handyside et al., "Pregnancies from Biopsied Human Preimplantation Embryos Sexed by Y-Specific DNA Amplification," *Nature* 344 (1990): 768–70; C. Coutelle et al., "Genetic Analysis of DNA from Single Human Oocytes: A Model for Preimplantation Diagnosis of Cystic Fibrosis," *British Medical Journal* 299 (1989): 22–24.

p. 496 *In March of 1992:* A. H. Handyside et al., "Birth of a Normal Girl after in Vitro Fertilization and Preimplantation Diagnostic Testing for Cystic Fibrosis," *New England Journal of Medicine* 327 (1992): 905–9; J. L. Simpson and S. A. Carson, "Preimplantation Genetic Diagnosis" (editorial), ibid., 951–53.

p. 498 *It wasn't very long:* M. Upadhyaya et al., "DNA Polymorphisms and Fetal

Sexing for X Linked Disorders with Chorion," in *First Trimester Fetal Diagnosis,* ed. M. Fraccaro, G. Simoni, and B. Brambati (Heidelberg: Springer-Verlag, 1985), 286–94.

p. 498 *Molecular diagnosis:* Y. W. Kan and A. M. Dozy, "Polymorphism of DNA Sequence Adjacent to Human B-Globin Structural Gene: Relationship to Sickle Mutation," *Proceedings of the National Academy of Sciences* 75 (1978): 5631–35.

p. 499 *When a disease-causing gene:* W. A. Check, "DNA Probes May Mean New Roles for Internists," *American College of Physicians Observer,* July–Aug. 1990, 1, 22. The landmark study that led to the revolution in gene screening is D. Botstein et al., "Construction of a Genetic Linkage Map in Man Using Restriction Fragment Length Polymorphisms," *American Journal of Human Genetics* 32 (1980): 314–31.

p. 500 *Women are taking advantage:* K. L. Ales, M. L. Druzin, and D. L. Santini, "Impact of Advanced Maternal Age on the Outcome of Pregnancy," *Journal of Surgery, Gynecology and Obstetrics* (1990): 209–16.

p. 501 *Yet, even if:* Interview with Neil A. Holtzman, Sept. 1989. See also N. A. Holtzman, *Proceed with Caution* (Baltimore: Johns Hopkins University Press, 1989), a standard work on the practice and ramifications of genetic screening, as is D. Nelkin and L. Tancredi, *Dangerous Diagnostics: The Social Power of Biological Information* (New York: Basic Books, 1989).

pp. 501–2 *"But the major orders":* Holtzman interview.

pp. 502–3 *Yet advances in:* Aubrey Milunsky interview, Dec. 1, 1989. *The model already exists:* The story of cattle cloning is derived from an interview with the animal physiologist Neal First, Oct. 2, 1989, Madison, Wis. See also P. Gorner and R. Kotulak, "Life by Design," *Chicago Tribune,* April 10, 1990 (part 2 of a series).

p. 504 *It is, however:* Verlinsky interview. See also Y. Verlinsky, E. Pergament, and C. Strom, "The Preimplantation Diagnosis of Genetic Diseases," *Journal of in Vitro Fertilization Embryo Transfer* 7 (1990): 1–5.

p. 505 *And at least one:* Interview with LeRoy Walters, Feb. 10, 1990.

CHAPTER 23: THE ULTIMATE FRONTIER

The story of the modern scientific pursuit of longevity was compiled from more than two hundred interviews with geneticists and biogerontologists for "Aging on Hold," a *Chicago Tribune* series by Peter Gorner and Ronald Kotulak, Dec. 8–15, 1991. See *Aging on Hold: Secrets of Living Younger Longer* (Orlando: Tribune Publishing, 1992). The authors acknowledge their debts to several pioneers, although their words do not appear in the chapter. These include Bruce Ames, Jeffrey B. Blumberg, Robert Butler, William J. Evans, Peter Davies, Bess Dawson-Hughes, Robert A. Floyd, James Fozard, Judith Hallfrisch, Denham Harman, Leonard Hayflick, Zaven S. Khachaturian, Linus Pauling, Walter Pierpoli, Eugene Roberts, Michael Rose, Edward L. Schneider, Richard Sprott, and Huber Warner, all of whom graciously shared their insights and experience.

pp. 507–8 *"I started feeling":* Interview with Frederick McCullough, March 15, 1991.

p. 508 *The landmark experiment:* Interviews with Daniel Rudman, Nov. 17, 1990, and March 15 and June 6, 1991. See D. Rudman et al., "Impaired Growth Hormone Secretion in the Adult Population: Relation to Age and Adiposity," *Journal of Clinical Investigation* 67 (1981): 1361–69.

p. 509 *Among these pioneers:* Telephone interview with Caleb E. Finch, June 10, 1991.

p. 510 *Work by Earl Stadtman:* Telephone interview with Earl Stadtman, June 3, 1991. See E. R. Stadtman, "Minireview: Protein Modification in Aging," *Journal of Gerontology* 43 (1988): B112–B120.

p. 510 *The hypothalamus, a hormonal:* C. E. Finch, "Neural and Endocrine Determinants of Senescence: Investigation of Causality and Reversibility by Laboratory and Clinical Interventions," in *Aging,* vol. 31, *Modern Biological Theories of Aging,* ed. H. R. Warner et al. (New York: Raven Press, 1987), 261–306; C. E. Finch et al., "Ovarian and Steroidal Influences on Neuroendocrine Aging Processes in Female Rodents," *Endocrine Review* 5 (1984): 467–97; J. A. Severson et al., "Age-Correlated Loss of Dopamine Binding Sites in Human Basal Ganglia," *Journal of Neurochemistry,* 39 (1982): 1623–31.

p. 510 *But it was a more:* C. E. Finch, *Longevity, Senescence, and the Genome* (Chicago: University of Chicago Press, 1990).

p. 511 *Growth hormone is pulse released:* Rudman interviews.

p. 512 *Evidence from the:* N. W. Shock et al., *Normal Human Aging: The Baltimore Longitudinal Study* (Washington, D.C.: Department of Health and Human Services, 1984).

p. 513 *They warn against:* B. Evans, *Dictionary of Mythology* (New York: Dell, 1970), 101.

p. 513 *DHEA is secreted:* N. Orentreich et al., "Age Changes and Sex Differences in Serum Dehydroepiandrosterone Sulfate Concentrations throughout Adulthood," *Journal of Clinical Endocrinology and Metabolism* 59 (1984): 551–55.

p. 514 *Although DHEA was discovered:* Telephone interview with Arthur Schwartz, June 6, 1991. See also A. G. Schwartz et al., "Novel Dehydroepiandrosterone Analogues with Enhanced Biological Activity and Reduced Side Effects in Mice and Rats," *Cancer Research* 48 (1988): 4817–22; idem, "Dehydroepiandrosterone and Structural Analogs: A New Class of Cancer Chemopreventive Agents," *Advances in Cancer Research* 51 (1988): 391–423.

p. 514 *When veterinarians at:* Interview with Greg MacEwen, March 20, 1991.

p. 514 *Other biogerontologists:* Telephone interview with V. K. Cristofalo, Aug. 26, 1991. See V. J. Cristofalo, "The Destiny of Cells: Mechanisms and Implications of Senescence" (the 1983 Robert W. Kleemeier Award Lecture), *Gerontologist* 25 (1985): 577–83; V. J. Cristofalo and B. M. Stanulis-Praeger, "Cellular

Senescence *in Vitro,"* in *Advances in Tissue Culture,* vol. 2, ed. K. Maramsch (New York: Academic Press, 1982), 1–68.

p. 515 *Even if those illnesses:* Interview with S. Jay Olshansky, Oct. 30, 1990. See S. J. Olshansky, B. A. Carnes, and C. Cassel, "In Search of Methuselah: Estimating the Upper Limits to Human Longevity," *Science* 250 (1990): 634–40; idem, "The Aging of the Human Species," *Scientific American,* April 1993, 46–52.

p. 517 *"There is no clear reason":* Telephone interview, with Elliott Crooke, Oct. 16, 1990.

p. 517 *"There's a new guard":* Telephone interview with Michael West, Oct. 15, 1991. See M. D. West, O. M. Pereira-Smith, and J. R. Smith, "Replicative Senescence of Human Skin Fibroblasts Correlates with a Loss of Regulation and Overexpression of Collagenase Activity," *Experimental Cell Research* 184 (1989): 138–47.

p. 519 *A seminal thinker:* Interview with Richard Cutler, Nov. 14, 1990.

p. 519 *Back in 1972:* R. G. Cutler, "Transcription of Unique and Reiterated DNA Sequences in Mouse Liver and Brain Tissues as a Function of Age," *Experimental Gerontology* 10 (1975): 37–60; idem, "Evolution of Longevity in Primates," *Journal of Human Evolution* 5 (1976). 169–202; idem, "Nature of Aging and Life Maintenance Processes," in *Interdisciplinary Topics in Geron tology,* vol. 9, ed. R. G. Cutler (Basel: Karger, 1976), 83–133; idem, "Evolution of Longevity in Ungulates and Carnivores," *Gerontology* 25 (1979): 69–86; idem, "The Dysdifferentiative Hypothesis of Mammalian Aging and Longevity," in *The Aging Brain: Cellular and Molecular Mechanisms of Aging in the Nervous System,* ed. E. Giacobini et al. (New York: Raven Press, 1982), 1–19; idem, "Superoxide Dismutase, Longevity, and Specific Metabolic Rate," *Gerontology* 29 (1983): 113–20; idem, "Antioxidants and Longevity of Mammalian Species," *Basic Life Sciences* 35 (1985): 15–73.

pp. 520–21 *In 1991, for instance:* Interview with Thomas Johnson, June 18, 1991. See T. E. Johnson, "Molecular and Genetic Analyses of a Multivariate System Specifying Behavior and Life Span," *Behavioral Genetics* 16 (1986): 221–35; T. E. Johnson, "Aging Can Be Genetically Dissected into Component Processes Using Long-Lived Lines of *Caenorhabditis elegans,"* *Proceedings of the National Academy of Sciences* 84 (1987): 3777–81; T. E. Johnson et al., "Arresting Development Arrests Aging in the Nematode *Caenorhabditis elegans,"* *Mechanisms of Aging and Development* 28 (1984): 23–40; T. E. Johnson and W. B. Wood, "Genetic Analysis of Lifespan in *Caenorhabditis elegans,"* *Proceedings of the National Academy of Sciences* 79 (1982): 6603–7; T. E. Johnson, "Age-1 Mutants of *Caenorhabditis elegans* Prolong Life by Modifying the Gompertz Rate of Aging," *Science* 249 (1990): 908–12.

pp. 521–22 *In a similar experiment:* Telephone interview with James Fleming, June 6, 1991. See I. Reveillaud et al., "Expression of Bovine Superoxide Dismutase in *Drosophila melanogaster* Augments Resistance to Oxidative Stress," *Molecular and Cell Biology* 11 (1991): 632–40; J. E. Fleming et al., "Aging Results in an Unusual Expression of *Drosophila* Heat Shock Proteins," *Proceedings of the National Academy of Sciences* 85 (1988): 4099–103.

pp. 522–23 *Growing acceptance of:* Interviews with Irwin H. Rosenberg, Nov. 13, 1990, and March 25, 1993. The guidelines cited were developed by Rosenberg, as director of the USDA Human Nutrition Research Center on Aging, at Tufts University.

p. 523 *Even as a straight-A:* The Roy Walford material in the chapter is based on several sources, including a telephone interview on June 24, 1991, and an interview with his colleague Richard Weindruch, March 21, 1991. For background material, the authors gratefully acknowledge their debt to Carol Kahn, whose *Beyond the Helix: DNA and the Quest for Longevity* (New York: Times Books, 1985) predicted the revolution in biogerontology. See R. L. Walford, *The Immunologic Theory of Aging* (Copenhagen: Munksgaard, 1969); idem, *The 120-Year Diet: How to Double Your Vital Years* (New York: Simon & Schuster, 1986); R. Weindruch and R. L. Walford, "Dietary Restriction in Mice Beginning at 1 Year of Age: Effect on Life-Span and Spontaneous Cancer Incidence," *Science* 215 (1992): 1415–18.

pp. 525–30 *What's going on here:* Telephone interviews with Ronald Hart, Aug. 12–13, 1991. See the landmark paper on DNA repair, R. W. Hart and R. B. Setlow, "Correlation between Deoxyribonucleic Acid Excision-Repair and Life-span in a Number of Mammalian Species," *Proceedings of the National Academy of Sciences* 71 (1974): 2169–74; P. H. Duffy et al., "Effect of Chronic Caloric Restriction on Physiological Variables Related to Energy Metabolism in the Male Fischer 344 Rat," *Mechanisms of Aging and Development* 48 (1989): 117–33; J. M. Lipman, A. Turturro, and R. W. Hart, "The Influence of Dietary Restriction on DNA Repair in Rodents: A Preliminary Study," ibid., 135–43; J. E. A. Leakey et al., "Effects of Aging and Caloric Restriction on Hepatic Drug Metabolizing Enzymes in the Fischer 344 Rat—I: The Cytochrome P-450 Dependent Monooxygenase System," ibid., 145–55; idem, "Effects of Aging and Caloric Restriction on Hepatic Drug Metabolizing Enzymes in the Fischer 344 Rat—II: Effects on Conjugating Enzymes," ibid., 157–66; R. J. Feuers et al., "Effect of Chronic Caloric Restriction on Hepatic Enzymes of Intermediary Metabolism in the Male Fischer 344 Rat," ibid., 179–89; R. Hullahalli, M. H. Lu, and R. W. Hart, "Status of Dehydroepiandrosterone and Hepatic Metabolism of Aflatoxin B-1 in Food Restriction Rats," *Biochemistry Archives* 6 (1990): 419–27; R. W. Hart and F. B. Daniel, "Genetic Stability in Vitro and in Vivo," *Advances in Pathobiology* 7 (1980): 123–41.

CHAPTER 24: PROPHECY

p. 532 *As originally developed: Understanding Our Genetic Inheritance: The U.S. Human Genome Project: The First Five Years, FY 1991–1995,* NIH Publication No. 90-1590 (April 1990).

p. 532 *Even before the project:* Walter Gilbert quoted in L. Jaroff, "Happy Birthday, Double Helix," *Time,* March 15, 1993. Gilbert explained his hopes for the genome project in an interview of Aug. 26, 1988.

p. 532 *The genome seeds:* R. Dulbecco, "A Turning Point in Cancer Research: Sequencing the Human Genome," *Science* 231 (1986): 1055–56.

p. 533 *By coincidence the Department:* For the scientific and political history of the genome project, see C. Wills, *Exons, Introns and Talking Genes: The Science behind the Human Genome Project* (New York: Basic Books, 1991); U.S. Congress, Office of Technology Assessment, *Mapping Our Genes—The Genome Projects: How Big, How Fast?* (Washington, D.C: Government Printing Office, April 1988); National Academy of Sciences, *Mapping and Sequencing the Human Genome* (Washington, D.C.: National Academy Press, 1988). Various protagonists also have written accounts, including C. DeLisi, "Overview of Human Genome Research," *Basic Life Sciences* 46 (1988): 5–10; J. D. Watson, "The Human Genome Project: Past, Present and Future," *Science* 248 (1990): 44–49; C. R. Cantor, "Orchestrating the Human Genome Project," ibid., 49–51. The DOE's participation was further explained to the authors by the agency's Benjamin Barnhart, Jan. 18, 1989, in Boston; by Anthony Carrano of Lawrence Livermore Laboratory, Aug. 24, 1988, in Toronto; and by Charles Cantor that same day in Toronto and on May 10, 1989, in Berkeley, Cal.

p. 533 *Elsewhere in science:* David Baltimore's well-known opposition to sequencing the human genome is quoted in L. Jaroff, "The Gene Hunt," *Time,* March 20, 1989. See also R. A. Weinberg, "The Case against Gene Sequencing," *Scientist,* Nov. 16, 1987; idem, "The Human Genome Sequence: What Will It Do for Us?" *BioEssays* 9 (Aug.–Sept. 1988): 91–92.

p. 533 *But at Stanford:* Interview with Paul Berg, May 10, 1989, Stanford.

p. 535 *"We used to think":* Cited in Jaroff, "Gene Hunt" and "Happy Birthday."

p. 535 *"What we have now":* Telephone interview with Eric Lander, Oct. 1987.

p. 535 *By the autumn of 1992:* L. Roberts, "Two Chromosomes Down, 22 to Go," *Science* 258 (1992): 28–30. See also NIH/CEPH Collaborative Mapping Group, "A Comprehensive Genetic Linkage Map of the Human Genome," ibid., 67–86; D. Vollrath et al., "The Human Y Chromosome: A 43-Interval Map Based on Naturally Occurring Deletions," ibid., 52–59.

p. 536 *As one of the few:* S. Foote et al., "The Human Y Chromosome: Overlapping DNA Clones Spanning the Euchromatic Region," *Science* 258 (1992): 60–66.

p. 537 *Nevertheless, Watson:* Interview with James Watson, Aug. 1, 1988.

pp. 537–39 *Even as the project:* For the work of Craig Venter, see C. Wills *Exons,* 144–46; M. D. Adams et al., "3,400 New Expressed Sequence Tags Identify Diversity of Transcripts in Human Brain," *Nature Genetics* 4 (1993): 256. James Watson's travails and resignation on April 10, 1992, were reported by L. Roberts, "Why Watson Quit as Project Head," *Science* 456 (1992): 301–2; M. Waldholz and H. Stout, "Rights to Life: A New Debate Rages over the Patenting of Gene Discoveries," *Wall Street Journal,* May 11, 1992.

p. 539 *Why would Collins:* Telephone interview with Francis Collins, March 21, 1993.

p. 540 *The answers have:* C. Holden, "Why Divorce Runs in Families," *Science* 258 (1992): 1734.

p. 540 *In mid-1993:* V. Morell, "Evidence Found for a Possible 'Aggression Gene,' " *Science* 260 (1993): 1722–23.

pp. 541–43 *To quest for:* S. LeVay, "A Difference in Hypothalamic Structure between Heterosexual and Homosexual Men," *Science* 253 (1991): 1036. See also L. S. Allen et al., "Two Sexually Dimorphic Cell Groups in the Human Brain," *Journal of Neuroscience* 9 (1989): 497–506; M. Bailey and R. C. Pillard, "A Genetic Study of Male Sexual Orientation," *Archives of General Psychiatry* 48 (1991): 1093; S. LeVay, *The Sexual Brain* (Cambridge: MIT Press, 1993).

p. 543 *A taste of how:* R. Stone, "NIH Wrestles with Furor over Conference," *Science* 257 (1992): 739; C. Holden, "Back to the Drawing Board, Says NIH," ibid., 1474.

p. 543 *The National Research:* "Consider Genetic Factors, Violence Study Says," *Chicago Tribune*, Nov. 13, 1992.

p. 543 *Another area where:* R. Kotulak and P. Gorner, "Life by Design," *Chicago Tribune*, April 8–13, 1990. In addition to the named sources, valuable insights for this section were provided by the astrophysicist Stanford E. Woolsey; the geoscientists Stanley M. Awramik, Karl W. Flessa, and Preston Cloud; the botanist Peter H. Raven; and the zoologist David W. Deamer—interviewed Jan. 17–19, 1989, in Boston.

p. 548 *Nature is frugal:* Telephone interview with Russell F. Doolittle, Jan. 17, 1990.

p. 550 *The thesis that:* Telephone interview with Thomas Cech, Jan. 3, 1990.

p. 551 *Just as primordial:* Telephone interview with Lynn Margulis, Jan. 1990.

p. 552 *"The survival of the luckiest":* Interview with Motoo Kimura, Aug. 26, 1988.

p. 552 *"I would conclude":* Interview with Robert Jastrow, Oct. 19, 1989.

p. 552 *What is the new:* Interview with Stephen Jay Gould, Oct. 26, 1989.

p. 553 *Lest we become:* The section on homologous recombination is based primarily on several interviews with Mario Capecchi, 1985–93, and Oliver Smithies, 1988–93. Capecchi's work is featured in J. Travis, "Scoring a Technical Knockout in Mice," *Science* 256 (1992): 1392–94. See also M. R. Capecchi, "Altering the Genome by Homologous Recombination," *Science* 244 (1989): 1288–92; idem, "Tapping the Cellular Telephone," *Nature* 344 (1990): 105.

p. 553 *"When a piece":* Berg interview.

p. 554 *In 1985, Smithies:* O. Smithies et al., "Insertion of DNA Sequences into the Human Chromosomal Beta-Globin Locus by Homologous Recombination," *Nature* 317 (1985): 230–34.

p. 555 *Knockout technology can:* M. Barinaga, "Knockouts Shed Light on Learning," *Science* 257 (1992): 182–83. For the cystic fibrosis mouse, see J. N.

Snouwaert et al., "An Animal Model for Cystic Fibrosis Made by Gene Targeting," ibid., 1083–88; 1992), L. L. Clarke et al., "Defective Epithelial Chloride Transport in a Gene-Targeted Mouse Model of Cystic Fibrosis," ibid., 1125–28. For the Tonegawa learning-disabled mouse, see A. J. Silva et al., "Deficient Hippocampal Long-term Potentiation in alpha-Calcium-Calmodulin Kinase II Mutant Mice," ibid., 204–8; A. J. Silva, et al., "Impaired Spatial Learning in alpha-Calcium-Calmodulin Kinase II Mutant Mice," ibid., 209–13.

pp. 556–58 *There are many things:* Kotulak and Gorner, "Life." See also A. Gibbons, "Biotech's Second Generation," *Science* 256 (1992): 766–68.

p. 557 *Already the first:* I. Amato, ed., "Random Samples: A Biotech Bonanza on the Hoof," *Science* 259 (1993): 1698. See also U.S. Congress, Office of Technology Assessment, *New Developments in Biotechnology: Patenting Life—Special Report* (Washington, D.C.: Government Printing Office, April 1989).

p. 557 *Another strategy involves:* H. M. Weintraub, "Antisense RNA and DNA," *Scientific American*, Jan. 1990, 40–46; L. M. Fisher, "Search Advances for 'Antisense' Drugs," *New York Times*, June 8, 1993.

p. 559 *In the long run:* "Stock Ownership: Sullying Research Purity?" *Medicine and Health Perspectives*, Feb. 8, 1993.

p. 561 *Early on, the:* G. Friedman and R. Reichelt, "ELSI: Ethical, Legal, and Social Implications," *Los Alamos Science* 20 (1992): 304.

p. 561 *For Wexler, science:* Ibid.

p. 561–62 *Indeed, these were:* The plight of Clemma Hewitt, the San Diego woman suffering from glioblastoma, and Harkin's pressuring of Healy for compassionate-use provisions in RAC gene therapy protocols was recounted by L. Thompson, "Harkin Seeks Compassionate Use of Unproven Treatments," *Science* 258 (1992): 1728. See also "U.S. Clears Gene Therapy for Woman's Brain Tumor," *Chicago Tribune*, Dec. 30, 1992.

pp. 563–64 *As for conflict:* J. Mervis, "NIH Rebutted, Rethinks New Ethics Regulations," and D. E. Chubin and M. C. LaFollette, "Science Community Needs Its Conduct Rules to Be Explicit," *Scientist*, Feb. 5, 1990; Christopher Anderson, "NIH Scientists Chafe under Ethics Rules on Industry Ties," *Nature* 357 (1992): 180.

p. 563 *Such a ban:* "Stock Ownership."

p. 563 *But then Irv Weissman:* C. Anderson, "Hughes' Tough Stand on Industry Ties" and "Federal Conflict Rules Nearing Completion," *Science* 259 (1993): 884.

p. 563 *Weissman's claim:* For the announcement of the SyStemix patent, see M. Chase, "SyStemix Wins a Controversial Patent on a Human Cell; Shares Rise Sharply," *Wall Street Journal*, Nov. 1, 1991. Business writers received a twenty-three page press release announcing the patent from Kekst and Company on Nov. 1 (release courtesy of Michael Millenson, who wrote the story for the *Chicago Tribune*). Patent 5,061,620 was listed in the *New York Times* the following day.

p. 564 *There is not:* Anderson, "Stand." See also B. J. Culliton, "NIH, Inc.: The Crada Boom" and "CRADAs Raise Conflict Issues," *Science* 245 (1989): 1034–36.

p. 564 *Most Hughes investigators:* Irving Weissman earlier stated his position about conflict of interest in an interview on May 10, 1989. The following is from the taped transcript of that interview: ". . . the way things are going in universities right now, and especially at Stanford Medical School, the over-whelming issue for the dean is conflict of interest: 'Let's make sure we don't have even the appearance of conflict of interest.' And the second issue is, 'let's get the most rapid transfer of technology to the patient.' The transfer of all these technologies occurs outside of the medical school. It has to. There's no way you can commit the resources or justify committing the space and gradu-ate students, or postdoctoral fellows, to doing things that are the 'D' part of 'R and D.' We do the research here, and what the medical school and the univer-sity are careful to do is to make sure we don't turn our laboratory into the 'R' part of the 'R and D' for a particular company with which we're associated. They bend over backward—sometimes I think too far backward—because I still think the most important thing is the most efficient rapid transfer of the technology to the patient."

INDEX

AAT deficiency, 494, 498
Abl oncogene, 339
acetylcholine, 412, 441, 448, 452
acute monocytic leukemia (AML), 339
"ADA Levels in LASN Transduced T-Lymphocytes," 220
Addolorata, Maria, 75–76
adenine, 43, 44–45, 549
adenosine deaminase (ADA) deficiency, 106–10, 198, 478
  Anderson-Mulligan competition and, 110, 111
  bone marrow transplants and, 112–13, 114
  Bordignon's research and, 202–5
  cause of, 108
  chromosomes and, 108
  gene replacement proposal and, 177–79, 191–93
  human research issue and, 123–25
  immune system and, 107–9, 112
  as model for gene therapy, 112–13, 114
  PEG-ADA therapy and, 115–16, 177–79
  point mutation and, 423
  primate research and, 128–32
  protein replacement therapy and, 115–16
  stem cell therapy and, 273–74
  T cells and, 177–78
  temporary gene therapy and, 139–40
adenosine triphosphate (ATP), 400
adenoviruses, 289, 314, 477, 499
  immune response and, 403

  as vector for brain research, 425–32, 434
  as vector for cystic fibrosis, 401–4
adenylosuccinic acid, 370
Adler, Reid, 538
adoptive immunotherapy, 142
aggression, 540–41
aging, 304, 488, 507–30, 565
  bacteria and, 509
  body temperature and, 527–28
  cancer and, 514–15
  DHEA and, 513–15
  diet and, 516, 522–24
  dietary restrictions and, 524–25, 528–29
  DNA replication and, 529
  environment and, 516–17
  gene for, 530
  gene repair and, 515–16, 517, 519, 520, 526, 529
  hormones and activity of genes in, 509–13
  human growth hormone and, 508–10, 511, 512, 513
  immune system and, 519
  lifestyle and, 516–17, 524
  "mortality genes" and, 517–18
  NCTR and, 525–26
  nematode experiment and, 520–22
  nutrition and, 516, 522–24
  oxygen free radicals and, 437–38, 520
  process of, 518–19
  SOD and, 440, 519–20, 528
Agriculture Department, U.S., 522
AIDS, 64, 109, 122, 145, 224, 426, 514, 557
  gene therapy and, 546–48
  Kaposi's sarcoma and, 341

alcoholism, 473–75
Ales, K. L., 500
Algeny (Rifkin), 161
alkaptonuria, 40–41
allopurinol, 92, 100
Alloway, Cynthia, 134
alpha globins, 56, 82–83
alpha interferon, 256, 339
alpha-1 antitrypsin, 432
alpha thalassemia, 82
Altman, Lawrence K., 155
Alzheimer, Alois, 441
Alzheimer's Association, 449
Alzheimer's disease, 27, 29, 407, 433,
    437–38, 440–49, 535–36, 544
  APLP and, 446
  apo E and, 445–46
  APP and, 444–45
  brain abnormalities and, 441
  chromosome 11 and, 446
  chromosome 21 and, 443
  diagnostic tests for, 448
  Down's syndrome and, 443
  early-onset form of, 442
  environment and, 445, 446
  familial form of, 443–44
  genetic basis of, 442, 445–46
  late-onset form of, 442
  nerve growth factor and, 447–48
  neurotransmitter shortage and,
    441–42
  plaque and, 441, 446
  prevalence of, 441
  traits of, 441
  viruses and, 442, 449
American Cancer Society, 152, 343
American Fertility Society, 505
American Heart Association, 321
American Journal of Medicine, 94
American Society of Human Genetics,
    388, 395, 469, 485, 491
Ames, Bruce, 325
Amish community, 458–62
amplification, 338
amyloid precursor-like proteins (APLPs),
    446
amyloid precursor protein (APP), 443,
    444–45
amyotrophic lateral sclerosis (ALS), 25,
    27, 54, 407, 433–40, 535–36
  characteristic traits of, 434–35
  chromosome 21 and, 437, 438
  familial form of, 436–39

free radicals and, 439, 440
future and, 439–40
gene for, 436–37
motor nerves and, 435
point mutation and, 438
rate of progression of, 436
SOD and, 437, 439–40
vitamin therapy and, 439
Anderson, Daniel French, 18
Anderson, Kathryn, 36, 165, 233, 275,
    279, 280, 316
Anderson, LaVere, 18
Anderson, William French, 18, 32–33,
    45, 60, 62, 78, 85, 88, 119, 124, 134,
    143, 176, 179, 203, 204, 206, 241,
    258, 374, 489, 559
  ADA research and, 110–11, 120,
    128–33
  Ashanthi DeSilva met by, 229–30
  athletic interests of, 19, 36–37
  background and personality of, 18–19
  Bank's criticism of, 243–44, 245
  Blaese's T-cell idea and, 139–41
  criticism of, 24, 26, 34–36, 243–45
  described, 23
  Dumbo incident and, 235
  early social problems of, 181–82
  endothelial cells and, 315, 316
  gene replacement debate and, 180,
    184, 187, 188–90, 198, 200, 216–19
  gene transfer proposal of, 151–53,
    155–56, 158–59, 161
  human gene experiment and, 232–35,
    238–40, 262, 263–64
  human research proposal of, 131–33
  injectable vector sought by, 277–78,
    288, 289–92
  Kant admired by, 293–94
  Kuntz's therapy and, 165, 166–73
  marker experiments and, 280–81
  media and, 242–43, 245, 246
  microinjection and, 57–58
  Mulligan contrasted with, 34–35
  new gene therapy procedures and, 280
  NIH lab of, 21–24
  Orkin's criticism of, 244–45
  performance charts of, 36
  private industry and, 126–28
  publications by, 56
  religion and, 293
  research by, 49–50, 56–58
  Rosenberg and, 150–51, 212, 242
  SAX retrovirus and, 123

self-assessment of, 292–93
on TIL therapy results, 174
Wilson and, 283
xenografts and, 281
Angier, Natalie, 236
angiogenesis factors, 317–18
Anitschkow, Nikolai, 304
Annals of Neurology, 364, 368
antibodies, 113, 209, 262, 455
in immune system, 260
monoclonal, 114, 273, 557
nebulin and, 359
antigens, 254–55, 256, 261, 283
hidden, 257
anti-oncogenes, 330
cell differentiation and, 336–37
function of, 332
"antisense" strategy, 345, 346–47, 426,
557–58
Antrim, Robert, 250–51
A-1 allele, 474
apheresis, 230, 236
apo (a), 310–11
apoprotein A (apo A), 310
apoprotein B (apo B), 301, 309, 311–12,
313, 321
apoprotein E (apo E), 309–10, 445–46
Areen, Judith, 195
arginase, 50, 51
"Arithmetic in Maya Numerals"
(Anderson), 56
Ashcraft, Aaron, 116–17
Ashcraft, Alison, 106–7, 108, 112, 115,
119
PEG-ADA therapy and, 116–18, 195,
208, 274
as potential gene therapy patient, 208
Ashcraft, Corinne, 116
Ashcraft, David, 117
Asilomar Conference (1975), 61, 62, 71
Association of American Medical
Colleges, 563
atherosclerosis, 301, 303–4, 307, 313,
314, 315
LDL and, 308–9
attention deficit disorder, 475–76
Attie, Alan, 312
A-2 allele, 474
Audrey (cancer patient), 481
autism, 407, 473
Avery, Oswald T., 41
Avila Giron, Ramón, 418
Axel, Richard, 76

bacteria, 46
aging and, 509
conjugation and, 46
evolution and, 551
phages and, 47–48
restriction enzymes and, 54–55
viral invasion of, 47–48
balance, 179
Baldwin, Ernest, 406
balloon angioplasty, 318, 320
Baltimore, David, 33, 78, 79–80, 84, 152,
278, 284, 533, 559
Baltimore Longitudinal Study on Aging,
512
Bank, Arthur, 243–44, 245
Bannister, Roger, 19
Baron, Miron, 465–66
Barrett, Jim, 127
basal cell carcinoma, 342
Bassett, Ann, 468
Bates, Gillian, 420
B-cells, 113, 121, 193, 261, 262
Beadle, George, 40, 41, 46
Beall, Robert, 383, 400
Beaudet, Arthur, 25
Becker muscular dystrophy, 360
Beckman Center for Molecular and
Genetic Medicine, 28
Begleiter, Henri, 473
Belmont, John, 25
Benstein, David, C., 371–72
Benzer, Seymour, 415
Berg, Paul, 28, 55, 59–60, 61, 67, 77,
553–54
Berger, Melvin, 209, 228, 234, 235, 236,
243, 268, 269
Berlin, Nathaniel, 142
Bernstein, Alan, 26
beta amyloid, 441, 443, 444–45, 446,
448
beta galactosidase enzyme, 315, 316,
317, 367, 430, 431–32
beta globins, 56, 69, 82–83
beta interferon, 339, 455–56
beta thalassemia (Cooley's anemia),
56–57, 58, 68–69, 72, 73–74, 82–83,
114, 423, 498
"big gulp," 550
Billings, Paul, 485
Bioblaster gene gun, 366
biological therapy, 144
birth defects, 501–2, 546
Bishop, J. Michael, 84, 325–26, 327

Blaese, R. Michael, 23, 26, 123, 129,
    133–34, 137, 158, 167, 174, 176,
    203, 205, 208, 209, 227–28, 231,
    232, 241, 243, 249, 258, 260,
    262–63, 265, 269, 270, 272, 275,
    292, 374, 433
  Anderson-Rosenberg meeting and,
    150–51
  Ashanthi DeSilva met by, 229–30
  background of, 121–23
  Bordignon and, 204
  bystander effect and, 286–87
  credit in media and, 245–46
  on disappearing data, 221
  gene replacement debate and, 180,
    184, 191–93, 195–198, 213, 215,
    216–17, 219, 221
  gene transfer proposal and, 151–52
  human gene experiment and, 233–35,
    238–40
  T-cell concept of, 138–41, 190–91
  virus vector experiment and, 285,
    286–87
Blair, Deeda McCormick, 126
Blake, David, 559
Blau, Helen, 318, 373
blood-brain barrier, 101, 164, 285, 428,
    430, 448
Blum, Kenneth, 474–75
bone marrow transplant, 99–100, 102,
    124, 133–35, 212–13, 281, 430
  ADA deficiency and, 112–13, 114
  first, 112–13, 112
  ordeal of, 112
  primate research and, 128–29
  retrovirus and, 114–15
  stem cells and, 112–14, 273
Boon, Thierry, 255
Bordignon, Claudio, 202–5, 206, 207,
    208, 213, 217, 218, 219, 222, 240,
    246, 262
  stem cells and, 274–75
Botstein, David, 416
Bottorff, Eric, 385–86
Bottorff, John, 385–86
Bottorff, Mary Kaye, 385–86
Bourke, Frederick, 538
Boyer, Herbert, 55, 352
brain, 101, 285–86, 543–44
  of alcoholics, 473
  Alzheimer's disease and, 441
  cells of, 431
  genes and function of, 432–34

HGPRT and, 430
Huntington's disease and, 407
Lesch-Nyhan syndrome and, 103
mind and, 476
mutation and, 407
NGF in, 447
serotonin and, 472
as target for gene therapy, 428–29
ventricles in, 432–33
viral disease and, 449
Brandt, Jason, 487
Brave New World (Huxley), 546
BRCA1 gene, 7–8, 480, 482
BRCA2 gene, 7–8
breast cancer, 25, 29, 333, 344
  gene for, 7–8, 480
  genetic screening for, 480–84
  inherited form of, 336, 480–81
  medical insurance and, 484–85
Brenner, Malcolm, 25–26, 279
Breslow, Jan, 310
Brooke, Michael, 369, 378–79
Brown, Jeff, 364
Brown, Jeremy, 364
Brown, Judy, 364
Brown, Louise Joy, 490
Brown, Michael, 300–301, 306, 307–9,
    313, 320–21
  on cholesterol, 302–3
Brown, Robert, 24–25, 372, 380–81, 382,
    436–39
Brunner, Hans, 540–41
Bryer, Bruce, 356
Buchwald, Manuel, 386–87, 388
Bulfield, Graham, 355, 363
Burkitt's lymphoma, 339, 340, 341
Bush, George, 265
bypass grafts, 315, 320
  restenosis and, 317
bystander effect, 286–87, 346, 347, 433

Calabro, Susan, 167, 168
calcium precipitation, 68
California Pacific Medical Center, 373,
    375
Camp, David, 112–13
Campbell, Kevin, 361
cancer, 25, 28, 29, 32, 283, 285, 322–48,
    437, 544
  aging and, 514–15
  anti-oncogenes and, 330
  "antisense" strategy and, 345, 347
  body's defense system and, 323

of breast, *see* breast cancer
carcinogens and, 340–41
cell differentiation and, 336–37
cell division and, 324–25, 328, 329
"cooperation" and, 335
DHEA and, 514
DNA repair and, 342–43
early detection of, 343–46
forms of, 340
gene therapy for, 151–52, 222–23
growth factor and, 328–29, 340
heredity and, 341–42
HIV and, 255
immune system and, 144, 254–55
LAKs therapy and, 146–48
of lung, *see* lung cancer
oncogenes and, 324–28
p53 gene and, 329–30, 333–34
polyps and, 333–34
proto-oncogenes and, 327–29, 341
retinoblastoma studies and, 330–32
retroviruses and, 64, 65
Rous and, 324–25
smoking and, 342, 344–45
spontaneous remission and, 141–42
TIL therapy and, 148–50, 151
translocation phenomenon and,
    338–40
tumor suppressor genes and, 332–38
"vaccines" for, 254–57
viruses and, 324, 328, 341
Candida, 260–61
Capecchi, Mario, 363, 553–55, 556
carcinomas, 340, 342
cardiovascular disease, 25, 28, 29,
    299–321, 437
apo B mutation and, 311–12
apoproteins and, 305–7, 310–12
bypass grafts and, 315–16
causes of, 300–301
cholesterol and, 301–7
diet and, 304, 312
FH and, 306–8
Framingham study on, 304
gene therapy and, 316–320
genetic basis for, 300, 306, 307–12
LDL and, 312–14
mutations and, 309–10
prevention and, 321
smoking and, 313
Cardon-Cardo, Carlos, 257
Carner, Donald C., 159–60
Carter, Charlie, 263

Carter, Jimmy, 341–42
Carter, Lillian, 342
Caskey, Thomas, 25, 355, 366–67
Cavenee, Webster, 331
CBS, 232
CD34 marker, 273–75
Cech, Thomas, 550
cell(s), 60
    aging of, 514–15, 517–18
    of brain, 431
    chloride ions and, 399
    cholesterol and, 303, 304–5, 307–8
    "daughter," 192
    endothelial, *see* endothelial cells
    foam, 313
    fusion of, 98
    "ghost," 257
    human, 44
    of lung, 404
    nerve, 412–13, 435, 448
    nucleus of, 45
    "packaging," 81, 156
    protein in, 510
    receptors on, 254, 311
    sensory, 435
    smooth-muscle, 318–19
    src gene and, 326
    stem, *see* stem cells
    tumor suppressor genes and, 330
    viral invasion of, 63–64
Cell, 360, 425, 440
cell differentiation, 328, 335, 536
    anti-oncogenes and, 336–37
    cancer and, 336–37
    cloning and, 504
    Human Genome Project and, 544–46
    oncogenes and, 338
    p53 gene and, 337–38
    tumors and, 336–37
cell division, 261–62, 335, 338
    cancer and, 324–25, 328, 329
    chromosomes and, 534
    DNA repair and, 342–43
Cell Pro Inc., 273
cell therapy, 368–69, 376
Cell Therapy Research Foundation, 367
Cell Transplantation, 368, 376
Cell Transplant Society, 368
Center for Cancer Research, 78
c-erbA genes, 337
c-ets-1 proto-oncogene, 339–40
Cetus Corporation, 146, 170, 210–12,
    247, 248–49, 250

Chamberlain, Jeffrey, 382
Charash, Leon, 373, 374
Charcot, Jean Martin, 435
Charcot-Marie-Tooth syndrome, 54
Chargaff, Erwin, 44
chelators, 46–57
Chelsea (patient), 135–37
Chicago Tribune, 12, 343, 447
Children's Inn, 266
Childress, James F., 207
chloride channel, 399–400
cholesterol, 282, 299, 366
  ailments and, 303–4
  apoproteins and, 301, 309–10, 311–13
  in blood stream, 304–6
  Brown on, 302–3
  cardiovascular disease and, 301–7
  cells and, 303, 304–5, 307–8
  forms of, 305
  genetically high levels of, 306
  lipoproteins and, 305–6
  liver and, 307
  receptors and, 308
  research on, 302–4
  "slag," 303
chorea, 413
chorionic villi sampling (CVS), 498
chromosome 1, 468
chromosome 2, 339
chromosome 3, 334
chromosome 4, 408–9, 410, 420, 424,
    425, 461
chromosome 5, 331, 468–69
chromosome 6, 361
chromosome 7, 389–90
chromosome 8, 339
chromosome 9, 338, 339
chromosome 11:
  Alzheimer's disease and, 446
  dopamine receptor gene on, 474
  hemoglobin gene on, 554
  leukemia and, 339
  manic depression and, 457, 461–62,
    463
  Wilms' tumor and, 347
chromosome 12, 333
chromosome 13, 7, 331, 334
chromosome 14, 339, 446
chromosome 16, 450, 454
chromosome 17, 7, 333, 336, 480
chromosome 18, 333
chromosome 19, 445, 446
chromosome 20, 108

chromosome 21:
  ALS and, 437, 438
  Alzheimer's disease and, 443
  Down's syndrome and, 443, 450,
    451–52
  map of, 535
  muscular dystrophy and, 353
chromosome 22, 338, 339
chromosomes, 31–32, 53, 65, 115, 217,
    332
  ADA deficiency and, 108
  cell division and, 534
  DNA and, 44
  gene mapping and, 40, 396
  genetic linkage and, 535
  globin produced by, 83
  host, 79–80
  "jumping" technique and, 396–97, 482
  mapping of, 393, 395, 535
  Philadelphia, 338
  telomere of, 420
  "walking" along, 395–96
chronic granulomatous disease, 356, 358
chronic myelogenous leukemia (CML),
    338–39
c-jun proto-oncogene, 329
Claussen, Uwe, 498
Cleveland, Don W., 440
Cline, Martin, 52, 68–77, 79, 85, 203
  background of, 68
  censure of, 75
cloning, 243, 310, 387, 447
  cell differentiation and, 504
  chromosome walking and, 395–96
  of genes, 53–54, 55
  genetic screening and, 502–6
  of humans, 504–6
  IT-15 and, 421
clots, 314–15, 316
c-myc gene, 339–40
codons, 7
Cohen, Daniel, 535, 536
Cohen, Gene, 440
Cohen, Jacques, 493, 502
Cohen, Stanley, 55
Cold Spring Harbor Laboratory, 17–18,
    35, 57, 534
Collaborative Mapping Group, 535
Collaborative Research, Inc., 387–92, 464
Collins, Francis, 25, 486, 489
  breast cancer screening and, 480–84
  cystic fibrosis gene and, 383, 384, 394,
    395, 396–97, 398, 402

Human Genome Project and, 531, 539
Huntington's disease gene and, 410,
    411, 424, 425
ras proto-oncogene and, 337
Collins, Margaret, 398
colon cancer, 54, 329, 333, 334, 343, 344
Comings, Brenda, 470–72
Comings, David, 469–73
complement, blood, 290
Concepción, María, 412
Congress, U.S., 278, 484, 533, 534
Conneally, P. Michael, 419
"Conquest of the Future, The"
    (Walford), 523
Cook, David, 117
Cook-Deegan, Robert, 195–96
Cooley's anemia, see beta thalassemia
Cooley's Anemia Blood and Research
    Foundation for Children, 57, 245
cooperative research and development
    agreements (CRADAs), 125, 216,
    279, 564
"cord" blood, 275
Cotman, Carl, 448
Cox, David, 25, 411, 452, 485
CpG islands, 397
creatine kinase, 353
c-rel genes, 337
Creutzfeldt-Jakob disease, 449
Crick, Francis, 19, 21, 37, 42, 43–44, 49
Cristofalo, V. K., 514–15
Croce, Carlo, 339
Crooke, Elliott, 517
crossovers, 394, 396–97
Crystal, Ronald, 24, 384, 401–6, 429, 559
Culliton, Barbara J., 155, 236
Culver, Kenneth W., 23, 26, 138, 139,
    176, 191, 201, 203, 205, 208, 209,
    215, 258, 260, 262–63, 269, 271,
    272, 274, 276, 282, 322, 347–48,
    431, 433
  ADA research and, 193–94
  background and personality of, 134
  bystander effect and, 286–87
  Chelsea case and, 134–35
  human gene experiment and, 232,
    234–38, 240
  party and, 264–65
  patent incident and, 246–47
  Rosenberg and, 246
  vector virus experiment of, 285–88
"Current Potential for Modification of
    Genetic Defects" (Anderson), 50

Cutler, Richard, 519–22
Cutshall, Cynthia, 108, 112, 115, 116,
    118, 119, 216, 228–29, 243, 264–65,
    272, 274, 280
  at Children's Inn, 266–67, 268, 270
  post-experiment testing of, 260, 261,
    271
  as potential gene therapy patient, 209
  stem cell therapy and, 275, 276
  tonsils of, 262
Cutshall, Laurie, 266
Cutshall, Susan, 108, 116, 118, 266, 268,
    269, 270
Cutshall, William, 267, 268
cyclophosphamide, 364
cyclosporine, 367, 370, 374, 375–76, 377
cystic fibrosis, 25, 26, 54, 228, 281, 289,
    383–406, 478, 498
  adenovirus and, 401–4
  ancestral roots of, 384–85
  ATP and, 400
  CFTR and, 397–98, 401–3
  chloride channel and, 399–400
  chromosome "walking" and, 395–96
  crossovers and, 394, 396–97
  gene for, 393–95, 398–99, 482, 555–56
  gene therapy and, 400–406
  genetic screening and, 485, 496–97,
    501
  germ line therapy and, 496–98
  int-related protein and, 393
  J. 311 probe and, 390–93
  "jumping" technique and, 396
  met oncogene and, 389–90
  Mormons and, 25, 387–88
  mutation and, 398–99, 423, 497, 501
  riflip analysis and, 386–87
  sweat test and, 385
  zoo blot and, 397
Cystic Fibrosis Foundation, 392, 400
cystic fibrosis transmembrane
    conductance regulator (CFTR),
    397–398, 401–3
cystinuria, 89, 90–91
cytokines, 145, 183, 186, 248, 255, 371,
    433
cytomegalovirus (CMV), 261, 499
cytoplasm, 45
"cytoplasmic tail," 291
cytosine, 43, 44–45, 397

"damper" genes, 337
Dana Alliance for Brain Initiatives, 544

"daughter cells," 192
Dausset, Jean, 525
David (bubble boy), 109–10
Davidson, Beverly, 428–32, 433, 434
Davies, Kay, 353–54, 355, 361, 366
Davis, Bernard, 159
DCC gene, 333
Dean, Michael, 389–90
Deford, Frank, 384
deletion, 338, 355–57
DeltaF508 mutation, 497
dementia, 448
deoxyadenosine, 108–9, 116, 193, 262
Desferal (desferrioxamine), 57
DeSilva, Anoushka, 107
DeSilva, Ashanthi, 107–8, 112, 115,
    118–19, 216, 227–40, 241, 242, 243,
    264, 265, 272, 274, 280, 294
  Anderson met by, 229–30
  at Children's Inn, 266–69, 270
  experiment results and, 267–68, 271
  post-experiment testing of, 260–62
  as potential gene therapy patient,
    209–10
  stem cell therapy and, 275–76
  tonsils of, 262
DeSilva, Dilani, 107, 267
DeSilva, Raja, 107, 228–31, 234–36,
    238–39, 265, 266–69, 271, 275–76
DeSilva, Van, 107, 228, 231, 233, 234,
    236, 238, 265, 266–67, 270
DHEA (dehydroepiandrosterone), 513–15
diabetes, 29, 300, 380
Diacumakos, Elaine, 57–58
Dingel, John, 278, 560
disease:
  of blood system, 81–82
  of brain, 449
  candidate, 81–82
  genetic basis of, 26–28, 40–41
  gene transfer technology and, 279
  inherited, 28–29
  oxygen free radicals and, 437–38
  retroviruses and, 64–65
disequilibrium test, 393, 394
DNA (deoxyribonucleic acid), 12
  bacterial conjugation and, 46
  chromosomes and, 44
  as component of genes, 41–42
  double helix form of, 20–21, 43–44
  double strandedness of, 554
  enzymes and, 91
  flexibility of, 42

forensics and, 488
genetic testing and, 498–99
homologous recombination and,
    553–54
immortal nature of, 42–43
in situ hybridization and 409
interchangeability of, 43
"junk," 98–99, 496, 533–34
libraries of, 537–38
liposomes and, 284
marker, 69, 331
"naked," 317
nucleotide bases of, 43
plasmid, 317
provirus form of, 79
pseudogenes and, 534
repair of, 342–43
reflips and, 386
RNA and, 45
self-reproduction by, 44–45
spacer, 401
splicing techniques and, 55
unclonable, 396
vaccines and, 558
viruses and, 47–48, 50–52, 64–65
zoo blot and, 397
DNase, 47, 400
dominance, 39
Donis-Keller, Helen, 351, 387, 388–91
Donohue, Kathy, 147
donor-specific tolerance, 281
Doolittle, Russell, 548
dopamine, 289, 413, 430, 434, 472,
    474–75, 541
double helix, 20, 42, 43–44, 425
Down's syndrome, 25, 217, 407, 437,
    535–36
  Alzheimer's disease and, 443
  characteristic traits of, 450
  chromosome 21 and, 443, 450, 451–52
  genetics and, 449–51
  genetic screening for, 485
  serotonin and, 452
  SOD and, 451–52
  treatment of, 452–53
"Dr. Anderson's Gene Machine," 243
Dreyer, William, 415
DR2 antigen, 454, 455
Druzin, M. L., 500
Dryja, Thaddeus, 332
D2 receptor, 474–75
Duchenne, Guillaume Benjamin
    Amand, 352–53

Duchenne's muscular dystrophy, 54,
    288, 349–82, 501
  cause of, 352–53
  chromosome 21 and, 353
  course of, 350
  cyclosporine and, 375–76, 377
  deletion and, 356–57
  dystrophin and, 359–63
  gene for, 357–59, 366, 380–81
  gene therapy and, 365–67
  Law's myoblast therapy and, 367–72,
    376–78
  MDA's crash program and, 354–55
  mutation and, 350
  myoblast transfer and, 363–66
  prednisone and, 370–71
  reverse genetics and, 353–54
  X chromosome and, 352–54, 355,
    356–57
Dulbecco, Renato, 532–33, 535
dystrophin, 359–63, 373
  Bioblaster and, 366–67
  Kunkel's research and, 359–60
  role of, 360–61
  scaled down version of, 381–82

Easton, Doug, 7
Eco RI enzyme, 55
Edgar (patient), 88, 92, 93–94, 96
Edsall, John T., 20
Edwards, Robert, 490
Efstratiadis, Argiris, 56
Egeland, Janice, 26, 457, 458–64, 465,
    467
Eglitis, Martin, 22, 24
Einstein, Albert, 44, 285, 375
Eisenreich, Jim, 470
electrophiles, 528
electroporation, 68, 555
Eli Lilly and Company, 7
"endocrine cascade," 510
endothelial cells, 314, 319, 517
  bypass grafts and, 315–16
  growth factor and, 317–18
Energy Department, U.S., 533
"enhancer" genes, 328, 329
enhancer LTR, 79–80
env gene, 80
Enzon, Inc., 115, 180–81, 191, 196, 228,
    270
enzymes, 41, 309, 367, 529
  for ADA, 108–9
  for blood clots, 315

DNA and, 91
  for DNA repair, 342–43
  Eco RI, 55
  function of, 91, 108
  genes and, 41
  growth factors and, 328–29
  HGPRT, see HGPRT enzyme
  HMG-CoA, 308
  protease, 444
  proto-oncogenes and, 327
  restriction, 54–55, 71, 386, 409, 498,
    499
episome, 317
Epstein, Charles, 25, 196–97, 214,
    217–18, 222, 223, 224, 450–51, 453,
    478
Epstein-Barr virus, 341, 499
erb oncogene, 326
Escherichia coli, 46, 48, 52, 55, 61, 63,
    80, 128, 315, 358
  beta galactosidase and, 430
  gene mapping of, 54
estrogen, 483, 509–10
Ethical, Legal, and Social Implications
    Branch (ELSI), 561
eugenics, 26
evoked potentials (EPs), 473
evolution, 39
  bacteria and, 551
  Human Genome Project and, 548–53,
    565–67
  mutation and, 550–52
  RNA and, 550
expressed sequence tags (ESTs), 537

factor VIII, 38, 380
factor IX, 380
familial ALS (FALS), 436, 438, 439
familial Alzheimer's disease (FAD), 442,
    443–44
Familial Amyotrophic Lateral Sclerosis
    Collaborative, 436
familial hypercholesterolemia (FH),
    282–83, 306–9, 314
familial retinoblastoma, 331
Fauci, Anthony S., 126
Fay (cancer patient), 481
Federal Bureau of Investigation, 488
Feldman, Michael, 257
fes oncogene, 326
FGF-5, 318
fibrinogen, 549–50
fibroblast growth factor (FGF), 318, 319

Finch, Caleb E., 509, 510–11, 513, 530
First, Neal, 502–3
First International Symposium on
    Preimplantation Genetics, 243
5-prime (5') end, 421
Fleming, James, 521–22
Fletcher, John, 110
foam cells, 313
Food and Drug Administration (FDA),
    153, 188, 219, 224, 318, 380
    cell therapy and, 376
    DHEA and, 514
    human gene therapy experiment
        approved by, 230–31, 234–35
    myoblast transfer therapy and,
        371–72, 376
    PEG-ADA approved by, 179, 195
    TIL/TNF research and, 247–49,
        251–53
forensic medicine, 488
Foundation on Economic Trends, 161
founder effect, 385
Fox, Jimmy, 123
fragile X syndrome, 54, 422, 424, 498
Framingham study, 304
Francke, Uta, 356, 357
Fraumeni, Joseph, 335
Fred Hutchinson Cancer Research
    Center, 78, 273
free radicals, see oxygen free radicals
"French Anderson's Genetic Destiny,"
    242
Freud, Sigmund, 244, 458
Fridovich, Irwin, 437
Friedman, Orrie, 387, 391–92
Friedmann, Theodore, 25, 52, 78, 87–88,
    98, 99, 101–2, 132, 448
Friend, Stephen, 332, 335
Frischauf, Anna Marie, 410
Fuks, Zvi, 257
Furlong, Chris, 380

Gage, Fred, 25, 447, 448
gag gene, 80
galactose, 48
galactosemia, 52
Gallo, Robert, 122, 145, 278, 292
gamma aminobutyric acid (GABA), 412
ganciclovir, 286, 287, 347
Garini, Luigi, 49
Garrod, Archibald, 40–41
Gartland, William, 132, 133
Gaucher's disease, 498

G. B. (ADA patient), 203, 204, 275
Gehrig, Eleanor, 435–36
Gehrig, Lou, 27, 435–36
G8 probe, 409
gene(s):
    for aggression, 540–41
    for aging, 530
    aging and repair of, 515–16, 517, 519,
        520, 526, 529, 554–55
    alpha globin, 56
    for ALS, 436–37
    for Alzheimer's disease, 442, 445–46
    amplification and, 338
    bacterial, 47–48
    for beta globins, 56, 69
    BRCA1, 480, 482
    for breast cancer, 480
    cardiovascular disease and, 300, 306,
        307–12
    catalog of, 53–54
    c-erbA, 337
    chemical code of, 32
    chromosomes and, 44
    cloning of, 53–54, 55
    c-myc, 339–40
    conservation of, 548
    c-rel, 337
    for cystic fibrosis, 393–95, 398–99,
        482, 555–56
    "damper," 337
    DCC, 333
    deletion and, 338
    disequilibrium test and, 393
    dominant, 39
    double helix form of, 19–20, 23, 42,
        43–44
    as drugs, 315, 403
    D21S58, 437, 438
    "enhancer," 328, 329
    env, 80
    enzymes and, 41
    first attempt to transfer, 50–51
    function of, 41, 44
    gag, 80
    "gun" for, 366–67
    for hemophilia, 358
    HLA-B7, 283–84
    "homeobox," 556
    hormones and activity of, 509–13
    housekeeping, 92
    H2K produced by, 257
    human, 31, 53
    human behavior and, 540

for Huntington's disease, 482, 486
for IL-2, 148
for intelligence, 543–44
"junk" DNA and, 98–99
knockout technology and, 555–56
K-ras, 323, 344–45, 346
lacZ, 315
linkage analysis of, 356
mammalian, 56, 60
for manic depression, 457, 461, 463, 464
MAOA, 540–41
mapping of, see gene mapping
M-1, 517–18
"mortality," 517–18
M-2, 517–18
Mulligan's recombinant experiment and, 60–61
for muscular dystrophy, 357–59, 366, 380–81
mutation of, 32, 108
naming of, 39
neomycin resistance, 80, 128
penetrance and, 536
p53, see p53 gene
pol, 80
as probes, 344
"promoter," 328, 340
psychiatric diseases and, 408
ras, see ras oncogene
recessive, 39
relics of, 43
"reporter," 315, 367
for retinoblastoma, 332
role of, 31–32
spacing of, 98–99
splicing of, 60–61
src, 325–26, 329
SRY, 536
"suicide," 286
TGF beta, 319
translocation and, 338–39
tumor suppressor, 330, 332–38
gene mapping, 40, 53–54, 388–89
chromosomes and, 40, 396
CpG islands and, 397
in situ hybridization and, 409
"jumping" technique and, 396–97, 482
riflip analysis and, 386, 409, 411, 416–17
saturation, 393
zoo blot test and, 397

gene therapy:
blood disease and, 68–69
controversy and, 60–62, 71, 75–77
current research in, 24–26
ethical concerns and, 30–31
goal of, 32–33
growth of, 21–24, 278, 281
human disease and, 26–28
infant mortality and, 30
nonhealth issues and, 30–31
popular press and, 50
potential beneficiaries of, 29–30
potential disaster and, 30
progress in, 12–13
public reaction to, 13, 60–62, 71
TNF and, 183
tools of, 54–55, 62–63
two-part system for, 81
"genetic imprinting," 413
genetic screening:
birth defects and, 501–2
for breast cancer, 480–84
cost-benefit ratios and, 501
critics of, 485
for cystic fibrosis, 485, 496–97, 501
for Down's syndrome, 485
forensic medicine and, 488
medical insurance and, 484–85
prenatal, 498–500
sperm and, 487
uterine lavage and, 487–88
virus identification and, 499
Genetic Therapy, Inc., 127–28, 212, 215–16, 247, 248–49, 250, 279, 315
gene transfer experiment, 163–73
attempts to halt, 159–61
PCR and, 169–71
proposal for, 154–59
safety issue and, 156–58
germ line, 32, 49–50, 496–98
Geron Corporation, 518
"ghost" cells, 257
Gilbert, Walter, 532, 534, 559
Gilboa, Eli, 120–21, 257, 292
Gilead Sciences, Inc., 558
Ginns, Edward, 434, 464
glioblastoma multiforme, 285
globin, 56, 58, 82
glutamine, 425
glycine, 326
glycoproteins, 79
Gobea, Andrew, 275
Goldgar, David, 7

Goldstein, Joseph, 302, 306, 307–9, 313
Golumbek, Paul T., 256
Gómez, Fidela, 412
Good, Robert, 112, 121, 122, 128
"Good Genes for Bad," 50
Goodman, Howard, 352
Gorski, Roger, 542
Gould, Stephen Jay, 552
gout, 91, 92–94
GPT enzyme, 79
graft-versus-host disease, 112
Gramm-Rudman-Hollings deficit
    reduction act, 125
Greenberg, Philip, 253
Greenblatt, Jay, 234–35
Groner, Yoram, 451–52
Grossman, Mary Ann, 314
growth factors, 317–19, 337
    fetus and, 328
    receptors and, 327, 328
    signaling enzymes and, 328–29
G6PD deficiency, 465
GTPase, 328–29
guanine, 43, 44–45, 397
Gurling, Hugh, 468
Gusella, James, 24–25, 408–10, 417, 419,
    420, 421, 422–25, 442, 443, 444,
    445, 446, 447
Guthrie, Woody, 412, 415

Haijar, Katherine, 311
Hall, Jerry, 505
Hamburger, Rahel, 465
Hammer, Armand, 142
Handyside, Alan, 494–98, 502
Hardy, J. A., 443–44, 445
Harkin, Tom, 562
Harlet, Janet, 292
Harper, Peter, 410, 420–21
Hart, Ron, 526–30
Harvey, William, 491
Haselkorn, Robert, 562
HAT culture medium, 98
Hauser, Peter, 475
Hawking, Steven, 436
Hayden, Michael, 411, 420–21, 425
Hayes, William, 48
Healthcare Ventures, 126–27
Healy, Bernadine, 537–38, 539, 562, 563
heart disease, see cardiovascular disease
Heckler, Margaret, 559
Hefti, Franz, 447
Heisenberg, Werner, 285

helper (CD-4) T cells, 179, 197, 199,
    214–15, 255, 454, 456, 546–48
    imbalance of, 226
    ratio and, 220–22
Helsinki committee, 73
hemoglobin, 56, 82–83
"Hemoglobin Switching in Sheep and
    Goats: Change in Functional
    Globin Messenger RNA in
    Reticulocytes in Bone Marrow
    Cells" (Anderson), 56
hemophilia, 25, 26, 38, 380, 496, 498,
    501
    gene for, 358
Hemophilus influenzae B, 261
hepatitis B, 290, 341, 499
hepatocytes, 314
hereditary breast-ovarian cancer
    syndrome, 483
heredity:
    alcoholism and, 473–75
    cancer and, 341–42
    machinery of, 38–39
    mental illness and, 476–77
    multiple sclerosis and, 453–54
Hereditary Disease Foundation, 411,
    415, 477
herpes simplex virus, 285–86, 287, 341,
    347, 428, 431, 499
Hersey, Richard, 542
Hershfield, Michael, 115, 116, 118, 132,
    180, 192, 207, 208, 213, 217, 272
    gene replacement debate and, 195–97,
    199, 200–201
HER-2/neu oncogene, 344
HGPRT enzyme, 98, 105, 154, 541, 555
    human bone marrow and, 99–100,
    102–3
    Lesch-Nyhan syndrome and, 92–93,
    99–102, 429–30
high-density lipoprotein (HDL), 305–6,
    310
HindIII enzyme, 408, 409
HIV retrovirus, 64, 122, 145, 222, 278,
    426, 499
    cancer and, 255
    gene therapy and, 546–48
    immune system and, 290
HLA (human leukocyte antigen), 454
HLA-B₇ gene, 283–84
HMG-CoA reductase, 308
Hodgkin's disease, 25
Hoffman, Eric, 358, 359, 362, 374

Holtzman, Neil, 501–2
"homeobox" genes, 556
homologous recombination, 553–56
homosexuality, 541–43
Hood, Leroy, 35, 455, 559
Horvitz, Oscar, 439
Horvitz, Robert, 439
host range, 289–90
Hotchkiss, Rollin, 48–49
housekeeping genes, 92
Housman, David, 410, 416, 417
Howard Hughes Medical Institute, 25,
    281–82, 283
    conflict-of-interest issue and, 563–64
H₂K, 257
HTLV-I retrovirus, 123, 290, 499
    cancer and, 341
    multiple sclerosis and, 455
Huebner, Robert, 325
human experimentation, 99–100, 102–3
    Anderson's "selective advantage" and,
        123–24
    animal research and, 157–58
    Cline's work and, 67–77
    Rogers's work and, 50–52
human gene experiment:
    Anderson and, 232–35, 238–40, 262,
        263–64
    animal model for, 207, 219
    assessment of, 272–73
    Blaese and, 233–35, 238–40
    Culver and, 232, 234–38, 240
    disappearing data and, 220–21
    FDA approval of, 230–31, 234–35
    first anniversary party and, 264–66
    infusion site of, 237
    Milan data and, 202–5, 206, 207, 217
    Mulligan and, 214–16, 219, 220–21
    Orkin's assessment of, 244–45
    PEG-ADA and, 207, 262–64
    performance of, 227–40
    potential patients for, 208–9
    press and, 228, 232, 236–37, 240
    results of, 260, 267–72
    Rosenberg and, 241–42
Human Gene Therapy, 133, 157
Human Gene Therapy Subcommittee,
    see Recombinant DNA Advisory
    Committee
Human Genome Project, 54, 320, 407,
    412, 502, 531–67
    aggression and, 540–41
    AIDS research and, 546–48

anticipated bounty of, 532–33, 536–37
antisense medicine and, 557–58
artificial gestation and, 546
birth defects and, 546
cell differentiation and, 544–46
chromosome maps and, 535–36
Collaborative Mapping Group and,
    535
Collins and, 531, 539
conflict-of-interest rules and, 563–64
ELSI and, 561
Energy Department and, 533
ethical issues and, 558–67
evolution research and, 548–53,
    565–67
fetal development and, 546
future of, 564–67
gene repair and, 554
Harkin incident and, 562
homologous recombination and,
    553–56
homosexuality and, 541–43
human behavior and, 540–43
intelligence and, 543–44
junk DNA and, 533–34
monoclonal antibodies and, 557
NIH and, 534, 539, 543, 559, 562
opposition to, 533
political correctness and, 543
private industry and, 559–60
proteins and, 556–57
selective mutation and, 555
tiers of, 532
vaccines and, 558
Watson controversy and, 537–39
human growth hormone (HGH), 318,
    508–12
    aging and, 508–10, 511, 512, 513
    exercise and, 512
    synthetic, 508
Human Nutrition Research Center on
    Aging, 522
human papilloma virus, 341, 499
Huntington, George Sumner, 410
Huntington's disease, 54, 407–27, 434,
    435, 498
    brain and, 407
    chromosome 4 and, 408–9, 410, 420,
        424, 425, 461
    drugs for, 414
    gene for, 482, 486
    gene therapy and, 426–27
    genetic imprinting and, 413

Huntington's disease (continued)
   genetic markers for, 499
   IT-15 and, 421
   linkage disequilibrium and, 420–21
   nerve cells and, 412–13
   onset and symptoms of, 410, 413–14,
      423–24
   predictive testing and, 486–87
   riflip analysis and, 409, 411, 416–17,
      419–20
   stutter defect and, 421–25
   suicide and, 414
   Venezuela study and, 412, 414,
      417–19, 461, 477
   Wexler and, 411–12
Huntington's Disease Collaborative
      Research Group, 411, 425
Hutchinson Cancer Research Center, 25
Hutton, John, 110–11, 119, 120
hydrogen peroxide, 437, 438
hydroxyl, 437, 438, 451, 520
Hyman, Steven, 477
hypercholesterolemia, 306, 312
hypospadias, 54

immune system:
   ADA deficiency and, 107–9, 112
   adenoviruses and, 403
   aging and, 519
   antibodies in, 260
   antigens and, 254
   cancer and, 144, 254–55
   cystic fibrosis gene therapy and,
      403–4
   cytokines and, 145–46
   HIV and, 290
   multiple sclerosis and, 454–55, 456
   T cells in, 178–79
   TNF and, 183
   tonsils and, 262
immunity, 138–39
   balance and, 179
   repertoire and, 178–79
immunoglobulin, 340
immunotherapy, 144–45
Imuran, 371
infant mortality, 30
Ingelfinger, Franz, 50, 155
Institute of Cancer Research, 7
institutional biosafety committees
      (IBCs), 153, 200, 211
institutional review board (IRB), 71, 73,
      153

insulin, 115
intelligence, 543–44, 565
interesting transcript-15 (IT-15), 421,
      425
interferon, 256, 339–40, 455–56
interleukin-1 (IL-1), 248
interleukin-2 (IL-2), 122, 144, 193, 207,
      211, 214, 229, 248, 249, 250, 251,
      252, 254, 256, 257
   cancer therapy and, 147–49
   discovery of, 145
   gene for, 148
   genetically engineered, 146
   T cells and, 122, 145, 193
   TILS and, 165
interleukin-4 (IL-4), 255, 256
intermediate-density lipoprotein (IDL),
      305, 309
International AIDS Conference (1991),
      278
int-related protein (IRP), 393–95
in situ hybridization, 409
int-2 oncogene, 555
in vitro fertilization (IVF), 243, 489–94,
      497, 502, 505
   cost of, 490
   ethical controversy and, 489–90
   polar body analysis and, 490–92
ionizing radiation, 325
Isis Pharmaceuticals, Inc., 558
isoleucine, 444

Jackson, David, 55
Jaenisch, Rudolf, 33, 363
Janet (cancer patient), 481–82
Jastrow, Robert, 552
Javits, Jacob, 27
Johannsen, Wilhelm, 39
Johnson, Thomas, 520–21
Jolly, Douglas, 25
Journal of Experimental Medicine, 133
Journal of Surgery, Gynecology and
      Obstetrics, 500
Journal of the American Medical
      Association, 474
J3.11 probe, 390, 392–93, 395, 397
"jumping" technique, 396–97, 482
"junk" DNA, 98–99, 496, 533–34

Kafatos, Fotis, 56
Kan, Yuet Wai, 498
Kandel, Eric, 556
Kant, Immanuel, 293–94

Kantoff, Phil, 123
Kaposi's sarcoma, 341
Karpati, George, 26, 363, 377, 378, 381
Kathy (patient), 481
Kelley, William N., 154–55, 157,
    186–87, 197, 205, 214, 264, 429
Keys, Ancel, 304
Kidd, Kenneth, 469
killer (CD-8) T cells, 179, 194, 197,
    214–15, 255, 455
    imbalance of, 226
    ratio of, 220–22
Kimura, Motoo, 552
kinases, 400, 556
King, Mary-Claire, 25, 336, 480, 484
Knapp, Lois, 236
Knudson, Alfred, Jr., 331, 333
Kohn, Donald, 123, 134, 137, 275
Konopka, Ronald, 415
Kornberg, Arthur, 59
Korzybski, Alfred, 536
Koshland, Daniel E., Jr., 155
Kotulak, Ron, 343
K-ras cancer genes, 323, 344–45, 346
Kreigler, Michael, 210
Kucherlapati, Raju, 554, 555
Kunkel, Louis, 24–25, 362, 374
    background of, 351–52
    deletion research of, 355–57
    dystrophin and, 359–60
    myoblast transfer and, 365–66
    subtraction hybridization technique
        and, 357
Kuntz, Maurice, 162–73, 236, 258
Kuntz, Sharon, 162–69, 171–73, 236
kuru (brain disease), 449
Kwiterovich, Peter, 301, 321
Kyle (monkey), 130

lacZ genes, 315
lambda phage, 55, 61
Lander, Eric, 33, 54, 535
Lap-Chee Tsui, 26
La Rochefoucauld, François de, 241
Lasker Foundation, 126
Latt, Samuel, 355
Laura (patient), 134–35
Law, Peter, 349–50, 363–64
    criticism of, 378–80, 381
    FDA and, 371–72
    MDA and, 368–69, 372–73, 375
    myoblast therapy of, 367–72, 376–78
    patent of, 375

Lawson, Becky, 216
L-dopa, 434
"leader" region, 291
lean body mass, 511
Leder, Philip, 49, 339
Lederberg, Esther, 46–47
Lederberg, Joshua, 45–47, 48, 51, 63
Lederhoff, Fred, 433–34
Ledley, Fred, 25
Lehrach, Hans, 410, 420
Leiden, Jeffrey, 25, 318
Lenfant, Claude, 245
Lesch, Michael, 87, 90–91, 93–94
Lesch-Nyhan syndrome, 86–104, 154,
    198, 431, 432, 198, 541
    brain defects and, 103
    cell fusion and, 98–99
    dopamine and, 430
    gene deletion and, 357
    HGPRT enzyme and, 92–93, 99–102,
        429–30
    as model for gene therapy, 101–2, 105
    point mutation and, 423
    symptoms of, 88–89
Les Turner ALS Foundation, 436
leukemia, 64, 68, 279, 281, 326, 455,
    499
    chromosome 7 and, 389–90
    chromosome 11 and, 339
    CML form of, 338–39
    T-cell, 341
Levay, Simon, 541–43
Leventhal, Brigid, 224
Levy, Ronald, 253
Lewis, Jerry, 349
Lewis, Richard, 302
Li, Frederick, 335–36
Liebert, Mary Ann, 133
Life Technologies, 127
Li-Fraumeni syndrome, 335–36, 337
ligase enzymes, 55
Lincoln, Abraham, 42–43
Linda (cancer patient), 147
linkage analysis, 356, 387–88
linkage disequilibrium, 420–21
lipids, 319
lipoproteins, 305, 309
liposomes, 284, 308
lod scores, 462, 466, 468
Longevity, Senescence, and the Genome
    (Finch), 510–11
long terminal repeats (LTRs), 79–80
Lotze, Michael, 24

low-density lipoprotein (LDL), 282–83,
    305, 307, 308–9, 320
    apo B and, 312–13
    smoking and, 313
    Wilson's research and, 313–15
luciferase, 367
lung cancer, 25, 322–23, 333, 334, 426,
    558
    mutation and, 344–45
lupus, 281
Lykken, David, 540
lymphocyte alteration, 272
lymphokine-activated killer cells
    (LAKs), 146–49, 254
lymphoma, 25, 64, 326, 341, 455
Lynch, Henry, 483

McCullough, Fred, 507–8, 511, 512
McCullough, Rita Mae, 508
McCune, Michael, 363
MacDonald, Marcy, 421–24, 426
MacEwen, Greg, 514
McFarlin, Dale, 454–55
McGarrity, Gerard J., 128, 154, 273
McGill University, 7
McGue, Matt, 540
McIvor, Scott, 154, 155, 157, 158, 187,
    214, 222–23
MacKay, Clive, 524
McKusick, Victor, 53–54, 460, 510
macrophages, 254, 313, 371, 455
Maeterlinck, Maurice, 211
Mahley, Robert, 309–10, 313
major histocompatability complex
    (MHC), 454
Mandel, Batsheva, 465
Maniatis, Tom, 56, 69, 77, 84
manic depression, 457–64, 544
    Amish study and, 458–62, 466, 467
    chromosome 11 and, 457, 461–62, 463
    environment and, 466–67
    gene for, 457, 461, 463, 464
    Jerusalem study on, 465–67
    riflips and, 459, 461, 462, 465
    X chromosome and, 457, 462, 465–66
Mann, Richard, 78
March of Dimes, 354, 373
Marfan's syndrome, 42–43, 498
Margulis, Lynn, 551
marker DNA, 69, 331
marker gene, 79, 80, 173, 315
marker proteins, 261
marking experiments, 279–80

Marotto, Suzanne, 250–51
Martin, Joseph, 416, 417
Mary (Alzheimer's patient), 447–48
Maxam, Allan, 56
mdx mice, 355, 359, 361, 363, 366, 371,
    382
Medawar, Peter, 79
Medicaid, 270
medical insurance, 484–85
melanoma, 25, 164, 251, 255, 283–84
    diagnosis of, 163
Mendel, Gregor, 39–40
Mendelian Inheritance in Man
    (McKusick), 53–54, 510
Mendell, Jerry R., 379
mental illness, 457–61
    environment and, 469
    future treatment of, 477–79
    gene therapy and, 477
    genetic link to, 464–65, 469, 472
    heredity and, 476–77
    nature vs. nurture debate and, 458, 476
    see also manic depression;
        schizophrenia
Merril, Carl, 52
messenger RNA, 345
Methotrexate, 69
met oncogene, 389–90, 392–93, 395, 397
Meyers, Abbey, 219, 224
Michael (patient), 88–93, 96
microinjection, 58, 68, 110, 455
Miescher, Friedrich, 38–40
Milan data, 202–5, 206, 207, 217
Miller, A. Dusty, 25, 78, 156–57, 160,
    165, 292, 562
Miller, Henry I., 188
Miller, Robert, 375–76
Miller, Stanley, 549
Milunsky, Aubrey, 502
Minna, John, 334
Mintz, Beatrice, 113–14
Miró, Joan, 490–91
mitochondria, 551
mock infection, 204
molecular biology, 28–29
molecular research, 22
Moloney murine (mouse) leukemia
    virus, 80, 290
Monaco, Tony, 358
M-1 gene, 517–18
monoamine oxidase A (MAOA), 540–41
monoclonal antibodies, 114, 273, 557
monocytes, 339–40

Montagnier, Luc, 278
Montreal Neurological Institute, 363
Morduch, Ora, 73–75
Morgan, Thomas Hunt, 40, 331
Mormons, 25, 387–88
"mortality genes," 517–18
motor neurons, 435
Motulsky, Arno, 470
M-2 gene, 517–18
Müller, Carl, 306
Mulligan, Jim, 66
Mulligan, Richard, 24, 63, 64, 88, 110,
    121, 123, 132, 140, 180, 181, 212, 244,
    264, 271, 313, 314, 315, 316, 489, 559
  academic career of, 59–60, 62, 66–67
  Anderson contrasted with, 34–35
  background of, 33–34, 65–67
  cancer research by, 255
  disappearing data and, 220–22
  doctoral thesis of, 62
  gene replacement debate and, 188–89,
    198–200
  gene splicing experiment of, 60–61
  gene transfer proposal opposed by,
    154–57, 158, 160
  human gene transfer experiment
    approval and, 214–16, 219, 220–21
  liver gene therapy and, 282
  packaging cell and, 81
  Parker and, 84–85, 279
  post-doctoral research by, 77–81
  on TNF proposal, 224–26
  Wilson and, 282
Mullis, Kary, 169–70
multiple sclerosis, 407, 435, 544
  beta interferon and, 455–56
  environment and, 454
  heredity and, 453–54
  HLA markers and, 454
  immune system and, 454–55, 456
  T cells and, 454–55
  traits of, 453
  treatment of, 455–56
  viruses and, 454, 455
Multiple Sclerosis Society, 456
Murray, Robert F., 207
Muscle & Nerve, 378
muscular dystrophy, 25, 26, 281, 288,
    289, 496, 498
  see also Duchenne's muscular
    dystrophy
Muscular Dystrophy Association, 349,
    352, 355, 360

crash program of, 354–55
  Law and, 368, 369, 372–73, 375
"Musings on the Struggle" (Anderson),
    157
mutation, 32, 40, 108
  aggression gene and, 540–41
  amplification and, 338
  of apo B, 311–12
  of APP genes, 444, 445, 446
  birth defects and, 501
  brain and, 407
  carcinogenesis and, 325
  carcinogens and, 340–41
  cardiovascular disease and, 309–12
  cystic fibrosis and, 398–99, 423, 497,
    501
  deletion and, 338
  DeltaF508, 497
  early detection of, 343–44
  evolution and, 330–32
  founder effect and, 385
  of gene for LDL receptor, 308–9
  genetic testing and, 498
  "jumping" genes and, 536
  k-ras and, 344–45, 346
  lung cancer and, 344–45
  muscular dystrophy and, 350
  of p53 gene, 346
  point, see point mutation
  ras oncogene and, 343–45
  retinoblastoma and, 331–32
  selective, 555
  spot, 358
  translocations and, 338
myc oncogene, 326, 334, 340
mycoplasma, 234
Myers, Richard, 25, 411
myoblasts, 318, 349, 362, 363–66
myoblast transfer therapy:
  arterial injection and, 380–81
  FDA and, 371–72, 376
  Kunkel and, 365–66
  Law and, 367–74, 376–78
  safety and, 371–72
  Stanford study and, 373–74
myocytes, 317
myotonic muscular dystrophy, 54, 422,
    424
Myriad Genetics, Inc., 7

Nabel, Elizabeth, 25, 284, 316–19
Nabel, Gary, 25, 280, 283–85, 316, 319,
    344, 547

"naked" DNA, 317
National Academy of Sciences, 61, 488
National Cancer Institute, 122, 142,
    200, 224, 286, 287, 325, 389
  Division of Cancer Treatment of, 253
  Rosenberg's TIL/TNF funding and,
    252–53
National Center for Human Genome
    Research, 482, 534, 539
National Center for Toxicological
    Research (NCTR), 525–26
National Cystic Fibrosis Foundation,
    383
National Heart, Lung, and Blood
    Institute, 22, 200, 216, 245
National Institute of Allergy and
    Infectious Diseases, 126
National Institute on Aging, 440, 525,
    526
National Institutes of Health (NIH),
    21–22, 36, 45, 56, 61, 122, 156, 161,
    227, 242, 272, 306, 354, 373, 404,
    429, 432, 434, 485
  ADD gene identified by, 475
  Anderson's departure from, 278–79
  Cline censured by, 75
  CRADA agreements and, 127
  gene therapy controversy and, 76–77
  gene transfer proposal and, 152–53
  Human Genome Project and, 534,
    539, 543, 559, 562
  Institutional Biosafety Committee of,
    211
  national deficit and, 125–27
  private industry and, 126–28
  repressive climate at, 278–79, 281–82
  Tsui's funding and, 394, 395
National Laboratory Gene Library
    Project, 533
National Organization for Rare
    Disorders, 219
National Research Council, 543
Nature, 70, 133, 247, 253, 254, 301, 360,
    361, 368, 372, 373, 390–91, 393,
    394, 408, 411, 425, 439, 457, 467,
    468–69, 554
Nature Genetics, 382, 432
nebulin, 358–59
Negrette, Americo, 418
Neiman, Paul, 207, 215
nematodes, 54, 520–22
Neo marker, 130, 153, 157, 187, 193, 194
neomycin resistance gene, 80, 128

nerve growth factor (NGF), 447–48
neuroblastoma, 279
neurofibromatosis, 54, 337, 482, 498
neurofilaments, 440
Neurology, 375, 378, 380
neurotransmitters, 452, 541
  Alzheimer's disease and, 441–42
New England Journal of Medicine, 50,
    147, 177, 180, 437
  Ingelfinger Rule and, 155, 156
Newsweek, 50, 51, 147
Newton, Isaac, 20, 91
New York Times, 155, 179–80, 236, 245,
    439
New York Times Magazine, 242–43
NF-1 protein, 337
Nirenberg, Marshall, 45, 49, 125
Niven, David, 27
Noble, Ernest, 474–75
noradrenaline, 441, 452
Norum, Robert, 310
nucleotides, 43–44, 338, 356, 386
  "junk," 98–99
  in ras gene, 326
  synthesis of, 549
Nyhan, William, 25, 86–87, 88, 89–91,
    93–95, 96, 100, 101–2

O'Brien, Chloe, 496–97
O'Brien, Martin, 497
O'Brien, Michelle, 496, 497
O'Brien, Paul, 496
obsessive-compulsive disorder, 472
Oedipus Rex (Sophocles), 480
Office of Recombinant DNA Activities,
    13
Office of Technology Assessment, U.S.,
    484
Ohno, Susumu, 351
Old, Lloyd J., 183
Oldfield, Edward, 287
oligodendrocytes, 455
Olson, Lars, 447–48
oncogenes, 291, 330
  "antisense" strategy and, 345, 346–47
  cancer and, 324–28
  cell differentiation and, 338
  classes of, 327–28
  first human, 326
  function of, 327
  tumor cells and, 335
  types of, 326–27
  in viruses, 325–26

120-Year Diet, The (Walford), 524
O'Reilly, Richard J., 128, 203–4
organ transplantation, 281
Orkin, Stuart, 31, 35, 105, 110, 111, 121, 123, 132
    human gene experiment assessed by, 244–45
Orloff, Jack, 125, 126
osteosarcomas, 332–33
ovarian cancer, 7, 336
oxygen free radicals, 451, 528
    aging and, 437–38, 520
    ALS, 439, 440
    disease and, 437–38

"packaging" cell, 81, 156
Page, David, 536
pancreatic cancer, 341
Paradise Lost (Milton), 531
paraquat, 451
Pardes, Herbert, 416, 477–79
Pardoll, Drew, 255–56
Parker, Richard, 84–85, 279
Parkinson's disease, 25, 26, 27, 289, 413, 431, 437, 544
Parkman, Robertson, 102, 105, 198, 207, 212–213, 218, 220
partial zona dissection, 493–96
Participatory Evolution, 48
Partridge, Terence, 374–75
Pasteur, Louis, 244
Pauls, David, 464, 472
Pedigree 110, 461, 462
PEG-ADA therapy, 132, 133, 203, 204–5, 213, 215, 228, 229, 271, 272
    ADA deficiency and, 115–16, 177–79
    Ashcraft and, 116–18, 195, 208, 274
    cost of, 118, 191, 269–70
    efficacy of, 118–19
    in Europe, 208
    FDA approval of, 179, 195
    gene replacement therapy and, 192–93, 195, 197, 199–201
    Hershfield and, 180–81
    human gene experiment and, 207, 262–64
    in media, 179–80
    revised gene replacement proposal and, 180–81
    success of, 177
Pergament, Eugene, 490
Perricaudet, Michel, 382
pERT87, 357

Peschanski, Marc, 432
Peschle, Cesare, 72–73, 75
p53 gene, 329–30, 333–38, 345, 347
    cell differentiation and, 337–38
    cell growth and, 338
    Li-Fraumeni syndrome and, 335–36
    mutation of, 346
phages, 47–48, 52–53, 55, 61
Phelps, Creighton, 449
phenylalanine, 398, 400
phenylketonuria (PKU), 90, 101, 103, 452, 498
Philadelphia chromosome, 338
Planck, Max, 285
plaque, 319–20
    Alzheimer's disease and, 441, 446
plasmid DNA, 317
plasminogen, 311
platelet-derived growth factor (PDGF), 319, 327
point mutation, 326, 338, 385
    ALS and, 438
    cystic fibrosis and, 398, 423
    Lesch-Nyhan syndrome and, 423
polar body, 490–94
pol gene, 80
polio, 261, 289, 353
polymerase chain reaction analysis (PCR), 169–71, 174, 194, 343–44, 346, 373, 422, 491, 496, 498
"Polymorphic DNA Marker Genetically Linked to Huntington's Disease, A" (Gusella), 408
"polymorphic trinucleotide repeat," 422
polymorphism, 311, 385, 386
"polypropylene hinge," 291
polyps, 333–34
Ponnamberuma, Cyril, 549
Potter, Huntington, 444
Potts, John, 78
Poulletier de la Salle, 303
prednisone, 370–71
preimplantation genetics, 488, 494–96
prenatal diagnosis, 28–29, 31
Prescott, Margaret, 312
presenile dementia, 441
President's Commission for the Study of Ethical Problems in Medicine, 76–77
Price, Donald, 447
progenitor cells, *see* stem cells
"promoter" genes, 328, 340
promoter LTR, 79–80

propionic acidemia, 87, 90
proteases, 444
protein, 54
    in cells, 510
    Human Genome Project and, 556–57
protein replacement therapy, 115–16,
    362
proto-oncogenes, 65, 318, 330, 332
    cancer and, 327–29, 341
    classes of, 327–28
    function of, 332, 336–7
provirus, 79, 80
pseudogenes, 534
psi sequence, 79, 81
P22 virus, 47–48
Public Health Service, U.S., 71
purines, 43–44, 91, 92, 100
pyrimidines, 43–44

Quaid, Kimberly, 486–87
"queen substance," 509

rabies virus, 289
Rachmilewitz, Eliezer, 72, 73, 74
Ragot, Thierry, 381
Rainbow Babies and Children's
    Hospital, 208, 228, 232
Rapacz, Jan, 299–300, 313, 321
    apo B mutation and, 311–12
    background of, 301–2
    polymorphism and, 311–12
ras oncogene, 326, 329, 333, 334, 337,
    461
    mutation and, 343–45
Raucher, Frank, Jr., 343
Reagan, Ronald, 142
recessive genes, 39
Recombinant DNA Advisory
    Committee, 61, 77, 160, 211, 272,
    323, 405, 559
    additional gene therapy proposals and,
        278, 280
    Anderson's human research proposal
        and, 131–33
    brain tumor experiments and, 287
    compassionate exemptions and, 562
    federal guidelines and, 189–90
    gene transfer debate and, 153–59,
        174–77, 182–83, 186–90, 200–201,
        202, 205–6
    Harkin incident and, 562
    June meeting of, 212–19
    marking experiment approved by, 279
    meetings of, 152–54
    Roth's lung cancer study and, 345
    stem cell proposal approved by, 273
    TIL/TNF protocol approved by, 247
    TNF proposal and, 183, 222–24
Relman, Arnold, 155
Renee (cancer patient), 481
repertoire, 178–79, 199
"reporter" gene, 315, 367
Reproductive Genetics Institute, 489
restenosis, 317, 318–19
restriction enzymes, 54–55, 71, 386,
    498, 499
restriction fragment length
    polymorphism (riflip), 388, 535
    cystic fibrosis and, 386–87
    DNA and, 386
    gene mapping and, 386, 409, 411,
        416–17
    Huntington's disease and, 409, 411,
        416–17, 419–20
    manic depression and, 459, 461, 462,
        465
retinal degeneration, 281
retinitis pigmentosa, 356, 358
retinoblastoma, 330–33, 335
retroviruses, 77, 79, 101, 341, 366, 402,
    428, 402, 428
    bone marrow transplant and, 114–15
    cancer and, 64, 65
    disease and, 64–65
    function of, 64–65
    genetic package of, 79–80
    see also HIV retrovirus; HTLV-1
        retrovirus
reverse genetics, 353–54
reverse transcriptase, 537
rev protein, 547
Rheingold, Steven, 456
rhinoviruses, 289
ribosomes, 45, 56
Rich, Alexander, 66–67
Rifkin, Jeremy, 161
Riordan, John, 397–98
Risch, Neil, 466
RNA (ribonucleic acid), 56
    DNA and, 45
    evolution and, 550
    messenger, 345
    in retroviruses, 64–65
Robert (monkey), 129–30, 131, 194
Roberts, Brian, 67
Roblin, Richard, 52

Roessler, Blake, 428–32, 433, 434
Rogers, Stanfield, 50–52, 67
Roman, Mark, 380
Rosen, Daniel R., 438, 439
Rosenberg, Irwin, 522
Rosenberg, Rachel, 242
Rosenberg, Steven, 23, 26, 85, 123, 176,
    200, 219, 229, 232, 240, 258, 265,
    265, 283, 292, 344, 346
  Anderson and, 150–51, 212, 242
  autobiography of, 241
  cancer gene therapy proposal of,
    210–12
  cancer vaccine and, 256
  career of, 141–42, 145
  Culver and, 246
  first human gene experiment and,
    241 42
  gene replacement debate and, 180,
    182–86, 188, 189
  gene therapy as seen by, 145
  gene transfer proposal and, 151–52,
    153, 155, 156, 158–59, 161
  Kuntz' therapy and, 165, 166, 168,
    170, 171
  LAKs therapy and, 146–48
  personality and ambition of, 142–44
  TILs therapy and, 148–50, 173–74
  TIL/TNF experiment of, 247–54
  TNF protocol of, 222–25, 247–49
Rosenthal, Alan, 122
Roses, Allen, 436, 445
Roth, Jack, 25, 280, 322–23, 344–46,
    426, 558
Rous, Francis Peyton, 324–25
Rous sarcoma virus, 325, 326, 328
Rowley, Janet, 338–39, 340, 389–90
Royston, Ivor, 562
Rubin, A. Harry, 324–25
Rudman, Daniel, 508, 509, 511–13, 514
Rutter, Bill, 352
Ryan, Alan, 292

St. George-Hyslop, Peter, 443–44
Saint Vitus' dance, see Huntington's
    disease
Salk, Jonas, 375
Salk Institute for Biological Studies, 25,
    78, 87, 380
Salmonella, 46–47, 325
Salser, Winston, 68, 69, 70–71
Sanberg, Paul R., 368–69
Sandoz Pharma, Ltd., 128, 373, 563

Santini, D. L., 500
Santos, George, 85
sarcomas, 325, 340, 393
saturated fats, 304
saturation mapping, 393, 395
SAX retroviral vector, 123, 128
Scanu, Angelo, 310
schizophrenia, 26, 27, 437, 457, 458,
    476, 477, 478, 479, 544
  genetics and, 467–69
  gene translocation and, 468
  traits of, 467
Schuckit, Marc, 474
Schuening, Friedrich, 25
Schwartz, Arthur, 514
Science, 70, 110, 126, 205, 255, 334,
    335–36, 388–89, 391, 533, 539,
    558
  Anderson as covered by, 243
  Ingelfinger Rule and, 155
  PEG-ADA overlooked in, 179
Seegmiller, J. Edwin, 87, 154
"selective advantage," 124
Selkoe, Dennis, 448
Senate, U.S., 435, 562
serotonin, 441, 452, 472, 541
serum proteins, 318
severe combined immunodeficiency
    (SCID), 109
Sharp, Phil, 78
Sherrington, Robin, 468
Shevach, Ethan, 122
Shope papilloma virus, 50–53
short tandem repeats, 464
sickle-cell anemia, 26, 56, 68–69, 72,
    82–83, 228, 423, 485, 498, 499, 501,
    542, 554
Siddique, Teepu, 25, 436–40
Simpson, O. J., 13, 488
sis oncogene, 326
"site specific integration," 290–91
Skolnick, Ed, 292
Skolnick, Mark, 7–8, 25, 333, 559
"slag" cholesterol, 303
Slamon, Dennis, 344
Smith, David, 533
Smithies, Oliver, 553–56
SmithKline Beecham, 373
smoking, 304
  cancer and, 342, 344–45
  LDL and, 313
Snyder, Jimmy "the Greek," 384
Sodrowski, Joseph, 547

somatostatin, 441
Sophocles, 480
Southern, Edward, 499
Southern blots, 498, 499
Southern California, University of, 278,
    279, 281, 487
spacer DNA, 401
Spengler, Barbara, 252, 256
spina bifida, 485
spinal muscular atrophy, 54, 353
spinobulbar muscular atrophy, 422
spontaneous remission, 141–42
spot mutation, 358
Sprott, Richard L., 526
src gene, 325–26, 329
SRY gene, 536
Stadtman, Earl, 510
Stanford University, 373–74
Starzl, Thomas, 375
Steinberg, Wallace, 126–27
stem cells, 70–71, 83–84, 109, 124,
    128–29, 151, 261, 281
  bone marrow transplant and, 112–14,
    273
  Bordignon and, 274–75
  function of, 112
  immunological balance and, 179
stem cell therapy, 273
  ADA deficiency and, 273–74
  CD34 cells and, 273–74
  Cutshall and, 275, 276
  DeSilva and, 275–76
  Kohn's research on, 275
  newborns and, 275
Stern, Larry, 368
steroids, 211, 303
Stratton, Michael, 7
Strehler, Bernard, 295
stroke, 28, 300, 315, 437
Strom, Charles, 491, 494, 504
stutter defect, 421–25
subtraction hybridization, 357
"suicide gene," 286
superoxide dismutase (SOD), 438,
    536
  aging and, 440, 519–20, 528
  ALS and, 437, 439–40
  Down's syndrome and, 451–52
Susan (cancer patient), 481, 482
SV 40 virus, 55, 61, 67, 77, 78, 79,
    430
sweat test, 385
Symons, Robert, 55

Syntex Corporation, 447
SyStemix, Inc., 563–64

Taniguchi, Tada, 146
Tanzi, Rudolph, 443, 446–47
"targeting," 290–91
Tatum, Edward, 41, 46
Tay-Sachs disease, 485, 498, 500
T cells, 109, 112, 114, 116, 118, 121,
    123, 124, 130, 131, 146, 151, 191,
    196–98, 204, 229, 231, 239, 255, 256
  ADA therapy and, 177–78
  animal research and, 193–94
  balance and, 179
  CD-4 (helper), *see* helper T cells
  CD-8 (killer), *see* killer T cells
  "daughter" cells of, 192
  function of, 113, 261
  IL-2 and, 122, 145, 193
  imbalance of, 214–15
  immunological repertoire and, 178–79
  leukemia and, 341
  long-term memory of, 138–39, 140, 192
  mature, 138–39, 192
  multiple sclerosis and, 454–55
Technology Transfer Act (1986), 125–26
telomere, 420
Temin, Howard, 132
temporary gene therapy, 139–40
testosterone, 513, 514, 540
test tube babies, *see* in vitro fertilization
TGF-beta growth factor gene, 319
thalidomide, 546
Thomas, Lewis, 62, 106, 113
thymidine kinase (TK), 58, 69, 70, 76,
    285, 286, 287
thymine, 43, 44–45
thymus gland, 109, 118
TIL therapy:
  cancer and, 148–50, 151
  IL-2 and, 165
  Kuntz and, 164–168, 172, 173
  Rosenberg and, 148–50, 173–74
TIL/TNF experiment, 249–54
  criticism of, 253–54
  FDA and, 247–49, 251–53
  NCI and, 252–53
  protocol for, 247–49
TIL/TNF therapy, 222–25
Time, 439, 532
tissue plasminogen activator (TPA), 315
Tobin, Allan, 415, 416
Todaro, George, 325

Tonegawa, Susumu, 556
Tourette's syndrome, 470–73
TPA, 316
transcription, 79–80
transcriptional factors, 328, 329, 337
"Transduction of Genes into Animal
    Cells" (Mulligan), 62
Transformed Cell, The: Unlocking the
    Mysteries of Cancer (Rosenberg),
    241
"Translation of Rabbit Haemoglobin
    Messenger RNA by Thalassaemic
    and Non-Thalassaemic Ribosomes"
    (Anderson), 56
translocation, 338–39, 468
"Treatment of Adenosine Deaminase
    Deficiency with Polyethylene
    Glycol-Modified Adenosine
    Deaminase," 177
"Treatment of Severe Combined
    Immunodeficiency Disease (SCID)
    Due to Adenosine Deaminase
    (ADA) Deficiency with Autologous
    Lymphocytes Transduced with a
    Human ADA Gene," 176
triglycerides, 305, 309, 310
Trisha (patient), 481
tropical spastic paresis, 455
tropic virus, 288–89
tryptophan, 472
Tsien Sen Li, 433
Tsui, Lap-Chee, 383, 386–91, 393, 394,
    395, 396–97
Tulsa World, 18
tumor antigens, 254–55
tumor infiltrating lymphocytes (TILs),
    148–50, 151, 229, 279
  cancer gene therapy and, 210–11
  gene replacement debate and, 186–88
tumor necrosis factor (TNF), 183, 187
  Rosenberg's protocol on, 222–25
  sensitivity to, 210–11
tumor suppressor genes, 330, 332–37
Tuskegee study, 71, 72
tyrosine hydroxylase, 434
tyrosine kinase, 328, 339

University of Utah Medical Center, 7
uracil, 549
Urey, Harold, 549
uric acid, 91, 92, 100
uterine lavage, 487–88
utrophin, 361

vaccine:
  cancer, 254–57
  DNA and, 558
Valerio, Dinko, 110
valine, 83, 326, 444
Vande Woude, George, 389
Varmus, Harold, 326
vector virus experiments, 285–91
  site specific integration and, 290–91
Venter, Craig, 537, 559
Verlinsky, Yury, 488–94, 502, 503–5,
    506
Verma, Inder, 78, 87, 99, 380
very-low-density lipoprotein (VLDL),
    305, 309
Viagene, Inc., 25, 547
virion, 81
viruses, 27
  Alzheimer's disease and, 442, 449
  antigens of, 254
  bacteria invaded by, 47–48
  cancer and, 324, 328, 341
  cell infiltration by, 63–64
  DNA and, 47–48, 50–52, 64–65
  envelope of, 79–81
  gene decoys and, 558
  gene splicing and, 60–61
  in genetic engineering, 63–64
  genetic testing and, 499
  for hepatitis B, 290
  host range of, 289–90
  multiple sclerosis and, 454, 455
  oncogenes in, 325–26
  Shope, 50–53
  src gene and, 325–26
  SV40, 55
  tropic, 288–89
  types of, 289
Vogelstein, Bert, 24, 329–30, 333–34,
    338, 343–44, 347
Vogelzang, Nicholas, 25

Waldmann, Thomas, 121, 122
Walford, Roy, 523–25, 527, 528, 529–30
Wall Street Journal, 564
Walter, LeRoy, 154, 182, 189, 213–14,
    216, 218–19, 222, 505
Washington Post, 242, 246, 368
Wasmuth, John, 410
Watanabe, Yoshio, 313
Watson, James, 19, 21, 42, 43–44, 57,
    354, 355, 367, 412, 482, 531–32,
    534–35, 536, 544, 560–61, 567

Watson, James (*continued*)
    Human Genome Project controversy
        and, 537–39
Weber, Barbara, 481
Weichselbaum, Ralph, 254
Weinberg, Robert, 33, 326, 332, 336,
    533–34
Weindruch, Richard, 524
Weiner, Brett, 95
Weiner, Brock, 95
Weiner, Craig, 95–98, 103–4
Weiner, Felice, 95–96, 97, 104
Weiner, Herman, 97
Weiner, Scott, 95
Weintraub, Bruce, 475
Weissman, Irving, 114, 273, 563–64
Weest, Michael, 517–18, 521
Welsh, Michael, 25, 405
Wexler, Alice, 411, 415, 416
Wexler, Milton, 415
Wexler, Nancy, 411, 413, 414–15, 416,
    417, 418, 419, 424, 426–27, 477–78,
    487, 561
White, Ray, 25, 331, 387, 390, 391, 394,
    416
Whitehead Institute for Biomedical
    Research, 33–34, 62
Whitman, Walt, 322
William Keck Foundation, 478
Williamson, Robert, 387, 390–95, 405,
    406
Wilms' tumor, 347
Wilson, James, 24, 282–84, 315–16, 366,
    384, 401–5, 430

LDL receptor research of, 313–15
    Mulligan and, 282
Wiskott-Aldrich syndrome, 120, 121,
    191
Wivel, Nelson A., 13, 212
Wolff, Jon, 25, 367, 434
Woo, Savio, 25
Wood, Donald, 354–55, 360, 361, 370
Working Group on Human Gene
    Therapy, 77
Worton, Ron, 26, 355, 360
Wu, George, 282
Wyngaarden, James, 155, 160–61, 169,
    408, 534, 537

X chromosome, 40, 87, 496
    gene mapping of, 53
    MADA gene and, 540–41
    manic depression and, 457, 462,
        465–66
    muscular dystrophy and, 352–54, 355,
        356–57
    Wiskott-Aldrich syndrome and, 121
xenografts, 281
x-ray diffraction, 19–20

Yaron, Rena, 452
Y chromosome, 40, 356, 465, 496, 535
    map of, 536
Yunis, Jorge, 331, 340

Zametkin, Alan, 476
Zinder, Norton, 46–47
zoo blot test, 397